Lecture Notes in Computer Sci

Edited by G. Goos, J. Hartmanis, and J. va

T0238426

Springer

Berlin
Heidelberg
New York
Hong Kong
London
Milan
Paris
Tokyo

Sanjeev Arora   Klaus Jansen
José D.P. Rolim   Amit Sahai (Eds.)

Approximation, Randomization, and Combinatorial Optimization

Algorithms and Techniques

6th International Workshop on Approximation Algorithms
for Combinatorial Optimization Problems, APPROX 2003
and 7th International Workshop on Randomization
and Approximation Techniques in Computer Science, RANDOM 2003
Princeton, NJ, USA, August 24-26, 2003
Proceedings

Springer

Series Editors

Gerhard Goos, Karlsruhe University, Germany
Juris Hartmanis, Cornell University, NY, USA
Jan van Leeuwen, Utrecht University, The Netherlands

Volume Editors

Sanjeev Arora
Amit Sahai
Princeton University, Department of Computer Science
35 Olden Street, Princeton, NJ 08544-2087, USA
E-mail: {arora,sahai}@cs.princeton.edu

Klaus Jansen
Christian-Albrechts-Universität zu Kiel
Institut für Informatik und Praktische Mathematik
Olshausenstraße 40, 24098 Kiel, Germany
E-mail: kj@informatik.uni-kiel.de

José D.P. Rolim
Université de Genève, Centre Universitaire d'Informatique
24, rue du Général-Dufour, 1211 Genève 4, Switzerland
E-mail: Jose.Rolim@cui.unige.ch

Cataloging-in-Publication Data applied for

Bibliographic information published by Die Deutsche Bibliothek
Die Deutsche Bibliothek lists this publication in the Deutsche Nationalbibliografie;
detailed bibliographic data is available in the Internet at <http://dnb.ddb.de>.

CR Subject Classification (1998): F.2, G.2, G.1

ISSN 0302-9743
ISBN 3-540-40770-7 Springer-Verlag Berlin Heidelberg New York

Springer-Verlag Berlin Heidelberg New York
a member of BertelsmannSpringer Science+Business Media GmbH

http://www.springer.de

© Springer-Verlag Berlin Heidelberg 2003
Printed in Germany

Typesetting: Camera-ready by author, data conversion by Boller Mediendesign
Printed on acid-free paper   SPIN: 10930939   06/3142   5 4 3 2 1 0

Preface

This volume contains the papers presented at the 6th International Workshop on Approximation Algorithms for Combinatorial Optimization Problems (APPROX 2003) and the 7th International Workshop on Randomization and Approximation Techniques in Computer Science (RANDOM 2003), which took place concurrently at Princeton University during August 24–26, 2003. APPROX focuses on algorithmic and complexity issues surrounding the development of efficient approximate solutions to computationally hard problems, and this was the sixth in the series, after Aalborg (1998), Berkeley (1999), Saarbrücken (2000), Berkeley (2001), and Rome (2002). RANDOM is concerned with applications of randomness to computational and combinatorial problems, and this was the seventh workshop in the series, following Bologna (1997), Barcelona (1998), Berkeley (1999), Geneva (2000), Berkeley (2001), and Harvard (2002).

Topics of interest for APPROX and RANDOM are: design and analysis of randomized algorithms, randomized complexity theory, design and analysis of approximation and online algorithms, complexity of approximation problems, random combinatorial structures, error-correcting codes, pseudorandomness and derandomization, network models and algorithms, average-case analysis, property testing, expander graphs and randomness extractors, random walks, Markov chains, probabilistic proof systems, random projections and embeddings, computational learning, randomness in cryptography, and various applications.

The volume contains 16+17 (APPROX + RANDOM) contributed papers, selected by the two program committees from 40+34 submissions received in response to the call for papers.

We would like to thank all of the authors who submitted papers, the members of the program committees

APPROX 2003
Sanjeev Arora, Princeton, Chair
Yossi Azar, Tel Aviv
Gruia Calinescu, Illinois IT
Chandra Chekuri, Bell Labs
Anupam Gupta, CMU
George Karakostas, McMaster
Philip Klein, Brown
Robert Krauthgamer, Berkeley
Kamal Jain, Microsoft
Stefano Leonardi, Rome
Yuri Rabinovich, Haifa
David Williamson, IBM

RANDOM 2003
Amit Sahai, Princeton, Chair
Paul Beame, Seattle
Bernard Chazelle, Princeton
Jennifer Chayes, Microsoft
Alan Frieze, CMU
Joe Kilian, NEC
Eyal Kushilevitz, Technion
Dana Randall, Georgia Tech
Ran Raz, Weizmann and Princeton
Dana Ron, Tel Aviv
Michael Saks, Rutgers
Alistair Sinclair, Berkeley
Chris Umans, Cal Tech

and the external subreferees: Dimitris Achlioptas, Andris Ambainis, Matthew Andrews, Aaron Archer, Nikhil Bansal, Luca Becchetti, Christian Borgs, Moses Charikar, Shuchi Chawla, Bernard Chazelle, Joseph Cheriyan, Don Coppersmith, Artur Czumaj, Bhaskar Dasgupta, Nikhil Devanur, Adrian Dumitrescu, Martin Dyer, Leah Epstein, Eldar Fischer, Rosario Gennaro, Catherine Greenhill, Sudipto Guha, Shirley Halevy, Shlomo Hoory, Sandy Irani, Yuval Ishai, Mark Jerrum, Ryan Johnston, Ravi Kannan, Anna Karlin, Howard Karloff, Michal Karonski, Claire Kenyon, Sanjeev Khanna, Subhash Khot, Alexei Kitaev, Michael Krivelevich, Amit Kumar, Vijay Kumar, Xiang-Yang Li, Vincenzo Liberatore, Laci Lovasz, Avner Magen, Mohammad Mahdian, Adam Meyerson, Micheal Mitzenmacher, Kousha Moaveni-Nejad, Michael Molloy, Cris Moore, Elchanan Mossel, Moni Naor, Ashwin Nayak, Gaia Nicosia, Andrew Odlyzko, Alessandro Panconesi, Christos Papadimitriou, Rene Peralta, Yuval Rabani, R. Ravi, Oded Regev, Yossi Richter, Adi Rosen, Alex Russell, Amin Saberi, Mohammad R. Salavatipour, Guido Schaefer, Rene Sitters, Angelika Steger, Kunal Talwar, Prasad Tetali, Luca Trevisan, Kasturi Varadarajan, Umesh Vazirani, Santosh Vempala, Jacques Verstraete, Anastasios Viglas, Eric Vigoda, Berthold Voecking, Peng-Jun Wan, Peter Winkler, and David Zuckerman.

We gratefully acknowledge support from the Computer Science Department of Princeton University, the Institute of Computer Science of the Christian-Albrechts-Universität zu Kiel, and the Department of Computer Science of the University of Geneva. We also thank Ute Iaquinto, Marian Margraf and Parvaneh Karimi Massouleh for their help, and Mitra Kelly for the local arrangements.

August 2003   Sanjeev Arora and Amit Sahai, Program Chairs
                       Klaus Jansen and José D.P. Rolim, Workshop Chairs

Table of Contents

Contributed Talks of APPROX

Contributed Talks of RANDOM

Correlation Clustering with Partial Information

Erik D. Demaine and Nicole Immorlica*

Laboratory for Computer Science, MIT, Cambridge, MA 02139, USA.
edemaine, nickle@theory.lcs.mit.edu.

Abstract. We consider the following general *correlation-clustering problem* [1]: given a graph with real edge weights (both positive and negative), partition the vertices into clusters to minimize the total absolute weight of cut positive edges and uncut negative edges. Thus, large positive weights (representing strong correlations between endpoints) encourage those endpoints to belong to a common cluster; large negative weights encourage the endpoints to belong to different clusters; and weights with small absolute value represent little information. In contrast to most clustering problems, correlation clustering specifies neither the desired number of clusters nor a distance threshold for clustering; both of these parameters are effectively chosen to be the best possible by the problem definition.

Correlation clustering was introduced by Bansal, Blum, and Chawla [1], motivated by both document clustering and agnostic learning. They proved NP-hardness and gave constant-factor approximation algorithms for the special case in which the graph is complete (full information) and every edge has weight $+1$ or -1. We give an $O(\log n)$-approximation algorithm for the general case based on a linear-programming rounding and the "region-growing" technique. We also prove that this linear program has a gap of $\Omega(\log n)$, and therefore our approximation is tight under this approach. We also give an $O(r^3)$-approximation algorithm for $K_{r,r}$-minor-free graphs. On the other hand, we show that the problem is APX-hard, and any $o(\log n)$-approximation would require improving the best approximation algorithms known for minimum multicut.

1 Introduction

Clustering objects into groups is a common task that arises in many applications such as data mining, web analysis, computational biology, facility location, data compression, marketing, machine learning, pattern recognition, and computer vision. Clustering algorithms for these and other objectives have been heavily investigated in the literature. For partial surveys, see e.g. [6, 11, 14, 15, 16, 18].

In a theoretical setting, the objects are usually viewed as points in either a metric space (typically finite) or a general distance matrix, or as vertices in a graph. Typical objectives include minimizing the maximum diameter of a cluster (k-clustering) [8], minimizing the average distance between pairs of clustered

---

* Research was supported in part by an NSF GRF.

points (k-clustering sum) [17], minimizing the maximum distance to a "centroid object" chosen for each cluster (k-center) [8], minimizing the average distance to such a centroid object (k-median) [10], minimizing the average squared distance to an arbitrary centroid point (k-means) [11], and maximizing the sum of distances between pairs of objects in different clusters (maximum k-cut) [13]. These objectives interpret the distance between points as a measure of their dissimilarity: the larger the distance, the more dissimilar the objects. Another line of clustering algorithms interprets the distance between points as a measure of their similarity: the larger the distance, the more similar the objects. In this case, the typical objective is to find a k-clustering that minimizes the sum of distances between pairs of objects in different clusters (minimum k-cut) [13]. All of these clustering problems are parameterized by the desired number k of clusters. Without such a restriction, these clustering objective functions would be optimized when $k = n$ (every object is in a separate cluster) or when $k = 1$ (all objects belong to a single cluster).

In the *correlation-clustering problem*, introduced by Bansal et al. [1], the underlying model is that objects can be truly categorized, and we are given probabilities about pairs of objects belonging to common categories. For example, the multiset of objects might consist of all authors of English literature, and two authors belong to the same category if they correspond to the same real person. This task would be easy if authors published papers consistently under the same name. However, some authors might publish under several different names such as William Shakespeare, W. Shakespeare, Bill Shakespeare, Sir Francis Bacon, Edward de Vere, and Queen Elizabeth I. Given some confidence about the similarity and dissimilarity of the names, our goal is to cluster the objects to maximize the probability of correctness. As we consider both similarity and dissimilarity measures, our objective is in a sense a generalization of the typical clustering objectives mentioned above. In fact, an appropriate interpretation of our problem instance suggests that our objective is a combination of the minimum k-clustering sum and minimum k-cut clustering objectives.

An interesting property of our problem is that the number k of clusters is no longer a parameter of the input; there is some "ideal" k which the algorithm must find. Another clustering problem with this property is *location area design*, a problem arising in cell phone network design. As formulated by Bejerano et al. [2], this problem attempts to minimize the sum of the sizes squared of the clusters plus the weight of the cut induced by the clustering. The authors provide an $O(\log n)$ approximation for this problem in general graphs using region-growing techniques and an $O(r^3)$ approximation in $K_{r,r}$-minor-free graphs using a lemma of Klein et al. [12]. The similarities between these two problems allow us to apply many of the same techniques.

For our lower bounds, we exploit the similarities between the correlation-clustering problem and the minimum-multicut problem, introduced by Hu [9]. In the minimum-multicut problem, we are given an edge-weighted graph and a list of k pairs of vertices, and the goal is to remove edges of minimum total weight such that the resulting graph disconnects all k input pairs. For $k = 1$,

this is simply the standard minimum-cut problem. For $k = 2$, Yannakakis et al. [21] gave a polynomial time algorithm. For $k \geq 3$, this problem was shown to be APX-hard by Dahlhaus et al. [4]. Currently, the best known approximation for this problem in general graphs is an $O(\log k)$ approximation algorithm by Garg et al. [7]. For graphs excluding any $K_{r,r}$ minor, Tardos and Vazirani [19] use a lemma of Klein et al. [12] to provide an $O(r^3)$ approximation (and thus constant approximation for planar graphs).

The only previous work on the correlation-clustering problem is that of Bansal et al. [1]. Their paper considers correlation clustering in a complete graph with all edges assigned weights from $\{+1, -1\}$, representing that every pair of objects has an estimate of either "similar" or "dissimilar". They address two main objective functions, minimizing disagreements and maximizing agreements between the input estimates and the output clustering. The decision versions of these two optimization problems are identical and shown to be NP-complete. For minimizing disagreements, they give a constant-factor approximation via a combinatorial algorithm. For maximizing agreements, they give a PTAS. Both algorithms assume the input graph is complete. However, in many applications, estimate information is incomplete.

In this paper, we consider minimizing disagreements in general graphs and with arbitrary weights. We give an $O(\log n)$-approximation algorithm for general graphs and an $O(r^3)$-approximation algorithm for graphs excluding the complete bipartite graph $K_{r,r}$ as a minor (e.g., graphs embeddable in surfaces of genus $\Theta(r^2)$). Our $O(\log n)$ approximation is based on linear programming, rounding, and the "region growing" technique [13, 7]. Using ideas developed in Bejerano et al. [2], we are able to prove that this rounding technique yields a good approximation. We then use a lemma of Klein et al. [12] to extend our results to an $O(r^3)$ approximation for $K_{r,r}$-minor-free graphs [19, 2]. We further prove that the gap in the linear program can be $\Omega(\log n)$, and therefore our bounds are tight for any algorithm based on rounding this linear program. We also prove that our problem is as hard as the APX-hard problem minimum multicut [4], for which the current best approximation is $O(\log k)$ for a certain parameter k [7]. Any $o(\log n)$-approximation algorithm for our problem would require improving the state-of-the-art for approximating minimum multicut.

Almost simultaneously, two groups of researchers independently obtained results similar to this paper. Charikar et al. [3] and Emanuel and Fiat [5] both give $O(\log n)$ approximations for the minimization version and approximation-preserving reductions from minimum multicut, as we do. In addition, Charikar et al. [3] improve the Bansal et al. [1] result for complete graphs and give a constant factor approximation for the maximization version in general graphs. Emanuel and Fiat [5] also prove the equivalence of this problem with the minimum-multicut problem.

The rest of this paper is organized as follows. Section 2 formalizes the correlation-clustering problem, the objective of minimizing disagreements, and presents the linear-programming formulation. Section 3 demonstrates a rounding technique that yields an $O(\log n)$ approximation for this linear program in general

graphs. Section 4 considers the special case of $K_{r,r}$-minor-free graphs and uses an alternate rounding technique to get an $O(r^3)$ approximation in these graphs. In Section 5, we prove lower bounds, establishing APX-hardness and a logarithmic gap in the linear program. We conclude with open problems in Section 6.

2   Problem Definition and Linear-Programming Formulation

An instance of the correlation-clustering problem is an undirected graph $G = (V, E)$ with edge weights $c_e \in (-\infty, +\infty)$ for each $e \in E$. Each edge weight can be interpreted as a confidence measure of the similarity or dissimilarity of the edge's endpoints. For example, if there is a function $f(u, v)$ that outputs the probability of u and v being similar, then a natural assignment of weight to edge $e = (u, v)$ is $c_e = \log \frac{f(u,v)}{1-f(u,v)}$ [1]. Hence, an edge $e = (u, v)$ of weight $c_e > 0$ corresponds to a belief that nodes u and v are similar. Larger c_e indicate higher confidence in this belief. Similarly, an edge weight of $c_e < 0$ suggests that u and v are dissimilar. An edge weight of $c_e = 0$ (or, equivalently, the lack of an edge between u and v), indicates no belief about the similarity of u and v.

In this paper, our goal is to output a partition or *clustering* $\mathbb{S} = \{S_1, \ldots, S_k\}$ of the vertices that minimizes disagreements. The disagreements or cost of a partition is the total weight of the "mistakes", that is, the weight of positive edges between clusters and the absolute value of the weight of negative edges within clusters. In the case $c_e \in \{-1, 0, +1\}$, the cost of a partition is simply the number of cut positive edges plus uncut negative edges. Intuitively, this objective penalizes the clustering whenever presumed similar objects are in different clusters and presumed dissimilar objects are in the same cluster. For the purposes of approximation algorithms, minimizing disagreements is different from maximizing agreements (the weight of cut negative edges plus uncut positive edges).

We introduce the following notation for the cost of a clustering:

$$\text{cost}(\mathbb{S}) = \text{cost}_p(\mathbb{S}) + \text{cost}_m(\mathbb{S}),$$
$$\text{cost}_p(\mathbb{S}) = \sum \left\{ |c_e| \; : \; e = (u, v) \in E; \; c_e > 0; \text{ and } \forall i, |\{u, v\} \cap S_i| \leq 1 \right\},$$
$$\text{cost}_m(\mathbb{S}) = \sum \left\{ |c_e| \; : \; e = (u, v) \in E; \; c_e < 0; \text{ and } \exists i, |\{u, v\} \cap S_i| = 2 \right\}.$$

We will refer to the optimal clustering as OPT and its cost as cost(OPT).

Previous approximation algorithms. Bansal et al. [1] give a constant factor approximation for this problem in the special case of complete graphs with edge weights in $\{-1, +1\}$. Their algorithm is combinatorial. It iteratively "cleans" clusters until every cluster C is δ-*clean* (i.e. for every vertex $v \in C$, v has at least $(1 - \delta)|C|$ plus neighbors in C and at most $\delta|C|$ plus neighbors outside C). They bound the approximation factor of their algorithm by counting the number of "bad" triangles (triangles with two +1 edges and one −1 edge) in a δ-clean clustering and use the existence of these bad triangles to lower bound OPT.

   Complete graphs have many triangles, and the counting arguments for counting bad triangles rely heavily on this fact. When we generalize the problem to
graphs that are not necessarily complete, bad triangles no longer form a good
lower bound on OPT. It may be possible to find a combinatorial algorithm for
this problem that bounds the approximation factor by counting bad cycles (cycles with exactly one minus edge). However, in this paper, we formulate the
problem as a linear program, round it, and use its optimal solution to bound our
approximation factor.

Linear-programming formulation. Consider assigning a zero-one variable x_{uv} to
each pair of vertices (hence $x_{uv} = x_{vu}$). When $(u, v) \in E$, we will sometimes
write x_{uv} as x_e where it is understood that $e = (u, v)$. Given a clustering, set
$x_{uv} = 0$ if u and v are in a common cluster, and $x_{uv} = 1$ otherwise. To express
cost($\mathbb{S}$) in this notation, notice that $1 - x_e$ is 1 if edge e is within a cluster and
0 if edge e is between clusters. Define constants

$$m_e = \begin{cases} |c_e| & \text{if } c_e < 0, \\ 0 & \text{if } c_e \geq 0, \end{cases}$$

and

$$p_e = \begin{cases} |c_e| & \text{if } c_e > 0, \\ 0 & \text{if } c_e \leq 0. \end{cases}$$

Then

$$\text{cost}(\mathbb{S}) = \sum_{e \in E} m_e(1 - x_e) + \sum_{e \in E} p_e x_e.$$

Our goal is to find a valid assignment of x_{uv}'s to minimize this cost. An assignment of x_{uv}'s is valid (corresponds to a clustering) if $x_{uv} \in \{0, 1\}$ and the x_{uv}'s
satisfy the triangle inequality.
   We relax this integer program to the following linear program:

$$\text{minimize } \sum_{e \in E} m_e(1 - x_e) + \sum_{e \in E} p_e x_e$$

$$\text{subject to } x_{uv} \in [0, 1]$$

$$x_{uv} + x_{vw} \geq x_{uw}$$

$$x_{uv} = x_{vu}$$

Because the solution set to this linear program contains the solution set to the
integer program, the optimal solution to the linear program is a lower bound on
the cost of the optimal clustering.

3   Approximation in General Graphs

We use the linear-programming formulation of this problem to design an approximation algorithm. The algorithm first solves the linear program. The resulting

fractional values are interpreted as distances between vertices; close vertices are most likely similar, far vertices are most likely different. The algorithm then uses region-growing techniques to group close vertices and thus round the fractional variables. Using ideas from Bejerano et al. [2], we are able to show that this approach yields an $O(\log n)$ approximation. A modification to this approach, outlined in Section 4, will yield an $O(r^3)$ approximation for $K_{r,r}$-minor-free graphs.

Region growing. We iteratively grow balls of at most some fixed radius (computed according to the fractional x_{uv} values) around nodes of the graph until all nodes are included in some ball. These balls define the clusters in the final approximate solution. As high x_{uv} values hint that u and v should be in separate clusters, this approach seems plausible. The fixed radius guarantees an approximation ratio on disagreements within clusters while the region-growing technique itself guarantees an approximation ratio on disagreements between clusters.

First we present some notation that we need to define the algorithm. A *ball* $B(u, r)$ of radius r around node u consists of all nodes v such that $x_{uv} \le r$, the subgraph induced by these nodes, and the fraction $(r - x_{uv})/x_{vw}$ of edges (v, w) with only endpoint $v \in B(u, r)$. The *cut* cut(S) of a set S of nodes is the cost of the positive edges with exactly one endpoint in S, i.e.,

$$\text{cut}(S) = \sum_{|\{v,w\} \cap S| = 1, \ (v,w) \in E} p_{vw}.$$

The *cut* of a ball is the cut induced by the set of vertices included in the ball. The *volume* vol(S) of a set S of nodes is the weighted distance of the edges with both endpoints in S, i.e.,

$$\text{vol}(S) = \sum_{\{v,w\} \subset S, \ (v,w) \in E} p_{vw} x_{vw}.$$

Finally, the *volume* of a ball is the volume of $B(u, r)$ including the fractional weighted distance of edges leaving $B(u, r)$. In other words, if $(v, w) \in E$ is a cut edge of ball $B(u, r)$ with $v \in B(u, r)$ and $w \notin B(u, r)$, then (v, w) contributes $p_{vw} \cdot x_{vw} \cdot (r - x_{uv})$ weight to the volume of ball $B(u, r)$. For technical reasons, we also include an initial volume I to the volume of every ball (i.e., ball $B(u, 0)$ has volume I).

Algorithm. We can now present the algorithm for rounding a fractional solution FRAC to an integral solution SOL. Suppose the volume of the entire graph is F, and thus cost$_p$(FRAC) $= F$. Let the initial volume I of the balls defined in the algorithm be F/n.

Algorithm ROUND

1. Pick any node u in G.
2. Initialize r to 0.

3. Grow r by $\min\{(d_{uv} - r) > 0 : v \notin B(u,r)\}$ so that $B(u,r)$ includes another entire edge, and repeat until $\text{cut}(B(u,r)) \leq c\ln(n+1) \times \text{vol}(B(u,r))$.
4. Output the vertices in $B(u,r)$ as one of the clusters in $\mathbb{S}$.
5. Remove vertices in $B(u,r)$ (and incident edges) from G.
6. Repeat Steps 1–5 until G is empty.

In this algorithm, c is some constant which we will determine later. This algorithm is clearly polynomial and terminates with a solution that satisfies the constraints. We must show that the resulting cost is not much more than the original fractional cost. Throughout the analysis section, we will refer to the optimal fractional solution as FRAC, the solution our algorithm returns as SOL, and the optimal integral solution as OPT. We also use $\text{FRAC}(x_{uv})$ and $\text{SOL}(x_{uv})$ to denote the fractional and rounded solution to the linear program.

Positive edges. The termination condition on the region-growing procedure guarantees an $O(\log n)$ approximation to the cost of positive edges (between clusters):

$$
\begin{aligned}
\text{cost}_p(\text{SOL}) &= \sum_{(u,v)\in E} p_{uv}\,\text{SOL}(x_{uv}) \\
&= \tfrac{1}{2}\sum_{\text{ball } B} \text{cut}(B) \\
&\leq \tfrac{c}{2}\ln(n+1) \times \sum_{\text{ball } B} \text{vol}(B) \\
&\leq \tfrac{c}{2}\ln(n+1) \times \left(\sum_{(u,v)\in E} p_{uv}\,\text{FRAC}(x_{uv}) + \sum_{\text{ball } B} \frac{F}{n} \right) \\
&\leq \tfrac{c}{2}\ln(n+1) \times \left(\text{cost}_p(\text{FRAC}) + F\right) \\
&\leq c\ln(n+1) \times \text{cost}_p(\text{OPT})
\end{aligned}
$$

where the fourth line follows from the fact that the balls found by the algorithm are disjoint.

The rest of our analysis hinges on the fact that the balls returned by this algorithm have radius at most $1/c$. This fact follows from the following known lemma [20]:

Lemma 1. *For any vertex u and family of balls $B(u,r)$, the condition $\text{cut}(B(u,r)) \leq c\ln(n+1) \times \text{vol}(B(u,r))$ is achieved for some $r \leq 1/c$.*

Negative edges. As in Bejerano et al. [2], we can use this radius guarantee to bound the remaining component of our objective function. We see that our solution gives a $\frac{c}{c-2}$-approximation to the cost of negative edges (within clusters):

$$
\begin{aligned}
\text{cost}_m(\text{FRAC}) &= \sum_{(u,v)\in E} m_{uv}(1 - \text{FRAC}(x_{uv})) \\
&\geq \sum_{\text{balls } B} \sum_{(u,v)\in B\cap E} m_{uv}(1 - \text{FRAC}(x_{uv}))
\end{aligned}
$$

$$\geq \sum_{\text{balls } B} \sum_{(u,v)\in B\cap E} m_{uv}(1 - 2/c)$$

$$\geq (1 - 2/c) \sum_{\text{balls } B} \sum_{(u,v)\in B\cap E} m_{uv}$$

$$= \frac{c-2}{c} \text{cost}_m(\text{SOL})$$

where the third line follows from the triangle inequality and the $1/c$ bound on the radius of the balls. The approximation ratio $\frac{c}{c-2}$ is $O(1)$ provided $c > 2$.

Overall approximation. Combining these results, we pay a total of

$$\text{cost}(\text{SOL}) = \text{cost}_p(\text{SOL}) + \text{cost}_m(\text{SOL})$$

$$\leq \frac{c}{2} \ln(n+1) \times \text{cost}_p(\text{OPT}) + \frac{c-2}{c} \text{cost}_m(OPT)$$

$$\leq \max\left\{\frac{c}{2}\ln(n+1), \frac{c-2}{c}\right\} \text{cost}(\text{OPT})$$

and thus we have an $O(\ln n)$ approximation, where the lead constant, $c/2$, is just slightly larger than 1.

4   Approximation in $K_{r,r}$-Minor-Free Graphs

In $K_{r,r}$-minor-free graphs, we can use a theorem of Klein et al. [12] to round our linear program in a way that guarantees an $O(r^3)$ approximation to the cost of disagreements between clusters. The clusters produced by this rounding have radius at most $1/c$, and thus the rest of the results from the previous section follow trivially. The theorem states that, in graphs with unit-length edges, there is an algorithm to find a "small" cut such that the remaining graph has "small" connected components:

Theorem 1. [12] *In a graph G with weight u on the edges which satisfy the triangle inequality, one can find in polynomial time either a $K_{r,r}$ minor or an edge cut of weight $O(rU/\delta)$ whose removal yields connected components of weak diameter[1] $O(r^2\delta)$ where U is the total weight of all edges in G.*

As in the case of the region-growing technique, this theorem allows us to cluster the graph cheaply subject to some radius guarantee. As this clustering cost is independent of n, this technique is typically applied in place of the region-growing technique to get better approximations for $K_{r,r}$-minor-free graphs (see, for example, Tardos and Vazirani [19] or Bejerano et al. [2]). In particular, this implies constant factor approximations for planar graphs.

---

[1] The *weak diameter* of a connected component in a modified graph is the maximum distance between two vertices in that connected component as measured in the original graph. For our purposes, distances are computed according to the $x_{u,v}$ which satisfy the triangle inequality and are defined on all pairs of vertices, so the weak diameter equals the diameter.

The idea is as follows. Given a $K_{r,r}$-minor-free graph G with weights p_e and edge lengths x_e as defined by the linear program, we subdivide each edge e into a chain of $\lceil kx_e \rceil$ edges of the same weight p_e for some appropriate k, yielding a new graph G'. We apply Theorem 1 to G', getting an edge cut F' which maps to an edge cut F in G of at most the same weight. This creates the natural correspondence between the resulting components of G' and G. Note two nodes at distance d in G are at distance kd in G'. Hence, if we take δ such that $O(r^2\delta) < 2k/c$, the components in G will have diameter at most $2/c$. It is sufficient to take $\delta = O(k/r^2)$. To bound the weight of the cut F, we just need to bound the total weight U' of the graph G'. Let $U = \sum_{e \in G} p_e$ be the total weight of edges in G and recall $\mathrm{vol}(G) = \sum_{e \in G} p_e x_e$. Then

$$
\begin{aligned}
U' &= \sum_{e \in G'} p_e \\
&= \sum_{e \in G} \lceil kx_e \rceil p_e \\
&\leq \sum_{e \in G} (kx_e + 1)p_e \\
&= k\,\mathrm{vol}(G) + U.
\end{aligned}
$$

By Theorem 1, the weight of F is $O(rU'/\delta) = O(r^3(k\,\mathrm{vol}(G) + U)/k)$. Taking $k = U/\mathrm{vol}(G)$, this becomes $O(r^3\,\mathrm{vol}(G))$ and is thus an $O(r^3)$ approximation of the cost of disagreements between clusters, as desired. The size of G' may be pseudopolynomial in the size of G. However, the algorithm of Klein et al. [12] consists of r breath-first searches of G', and these can be implemented without explicitly subdividing G. Thus, the algorithm is polynomial.

5   Lower Bounds

We prove that it is APX-hard to minimize disagreements in correlation clustering. We use a reduction from the APX-hard problem minimum multicut [4]: given a graph G and k pairs of nodes $P = \{(u_1, v_1), \ldots, (u_k, v_k)\}$, find a set of edges of minimum weight that, when removed, separate each pair of nodes $p \in P$.

Theorem 2. *An r-approximation for minimizing disagreements in correlation clustering implies an r-approximation for the minimum-multicut problem.*

Proof. Given a multicut instance G', construct graph G as follows. For every edge in G' of weight c'_e, add an edge to G of weight $c_e = c'_e$. Note all these c_e are positive. Let M be the maximum c_e. For each pair $p = (u_i, v_i)$, add an edge e between u_i and v_i of weight $c_e = -(M + 1)n^2$. Note we have added at most n^2 edges and increased the maximum weight by factor at most n^2 so G is polynomial in the size of G'.

We claim that the cost of the optimal multicut in G' equals the cost of the optimal clustering in G. A correlation clustering of G that puts every vertex in its own component costs at most Mn^2. However, any solution that does not separate all pairs costs at least $(M + 1)n^2$, and so the optimum solution must separate all pairs. As the only negative edges in G are those between these pairs, the optimum solution only makes mistakes on positive edges (disagreements between clusters). Therefore the optimum clustering in G induces a multicut of the same cost in G'. In fact, any clustering which only makes positive mistakes induces a multicut of the same cost in G'. Furthermore, any multicut in G' cuts all negative edges in G and thus induces a clustering in G of the same cost. In particular, the optimum multicut in G' induces a clustering in G of the same cost, and the claim follows.

Now suppose we have an r-approximation algorithm for the correlation-clustering problem. Consider the output of this algorithm on graph G. If the outputted clustering only makes mistakes on positive edges (and so separates all pairs), then the above arguments show that this clustering induces a multicut which is an r-approximation to the optimal multicut in G'. If the output clustering does not cut some negative edge, then the cost is at least $(M + 1)n^2$. In this case, the clustering which places every node in a separate cluster costs at most Mn^2 and is an r-approximation. Therefore, cutting every edge in G' is an r-approximation to the optimal multicut in G'. Thus, given an r-approximation algorithm for the correlation-clustering problem, we can design an r-approximation algorithm for the minimum-multicut problem. □

Because the minimum-multicut problem is APX-hard, this theorem shows that there is no PTAS for minimizing disagreements in correlation clustering unless $P = NP$. Furthermore, it shows that this problem is as hard as minimum multicut. The current best approximation for minimum multicut is $O(\log k)$ [7]. Because k can be $\Omega(n^2)$ in the worst case, an $o(\log n)$ approximation for our problem would require improving the $O(\log k)$ approximation for minimum multicut, which a long-standing open problem.

The above reduction is also useful in leading us to find difficult instances for the correlation-clustering problem. Garg, Vazirani, and Yannakakis [7] construct an example that shows that the ratio between the value of the minimum multicut and maximum multicommodity flow (i.e., optimal multicut linear-program value) can be as large as $\Omega(\log k)$. The example uses a *bounded-degree expander*.

Definition 1. *A graph G is a* bounded-degree expander *if there is some constant d such that all nodes have degree at most d and for any set S of vertices, $|S| < n/2$, the number of edges that cross S is at least $c|S|$ for some constant c.*

We can use the same example to prove that the gap of our linear program (the ratio between OPT and FRAC) can be $\Omega(\log n)$, suggesting that it is probably not possible to obtain a $o(\log n)$ approximation by rounding this linear program.

Theorem 3. *The gap of the linear program presented in Section 2 is $\Omega(\log n)$ in the worst case.*

Proof. Consider a bounded-degree expander G'. Note since the degree of each node is at most d, there are at least $n - \sqrt{n}$ vertices at a distance of at least

$\log_d n/2$ from any vertex v. Construct $O(n^2)$ pairs of vertices as follows: for each vertex v, add the $O(n)$ pairs (v, u) where u is a vertex of distance at least $(\log_d n)/2$ from v. Assign all edges in the graph weight $c_e = 1$. Perform the above reduction to get graph G. As discussed, the optimal integral solution separates all the $O(n^2)$ pairs of vertices. Hence, the diameters of the resulting clusters must be $o(\log_d n)$. Because the vertices have bounded degree, the size of the clusters is bounded by $n/2$. By the expansion property of G', we must cut at least $c \sum_{S \in \mathbb{S}} |S| = cn$ positive edges, and so cost(OPT) $= \Omega(n)$.

On the other hand, assigning $x_e = 2/\log_d n$ for positive edges and $x_e = 1$ for negative edges is a feasible fractional solution of value at most $(dn/2) \times (2/\log_d n)$, and so cost(FRAC) $= O(n/\log n)$. The theorem follows. $\square$

6   Conclusion

In this paper, we have investigated the problem of minimizing disagreements in the correlation-clustering problem. We gave an $O(\log n)$ approximation for general graphs, and an $O(r^3)$ approximation for $K_{r,r}$-minor-free graphs. We also showed that this problem is as hard as minimum multicut, and that the natural linear-programming formulation has a gap of $\Omega(\log n)$.

A natural extension of this work would be to improve the approximation factor for minimizing disagreements. Given our hardness result and the history of the minimum-multicut problem, this goal is probably very difficult. Another option is to improve the lower bound, but for the same reason, this goal is probably very difficult. On the other hand, one might try to design an alternate $O(\log n)$-approximation algorithm that is combinatorial, perhaps by counting "bad" cycles in a cycle cover of the graph.

Another interesting direction is to explore other objective functions of the correlation-clustering problem. Bansal et al. [1] give a PTAS for maximizing agreements in complete graphs with edge weights in $\{-1, +1\}$. In *maximizing agreements*, the cost of a solution is the weight of positive agreements (uncut positive edges) plus negative agreements (cut negative edges). They also mention the objective of maximizing agreements minus disagreements. This objective is of particular practical interest. However, there are no known approximation algorithms for this objective, even for complete graphs.

Finally, it would be interesting to apply the techniques presented here to other problems. The region-growing technique and Klein et al. [12] rounding technique both provide a radius guarantee on the outputted clusters. Many papers have used this radius guarantee to demonstrate that the solution is *feasible*, i.e. satisfies the constraints. Inspired by Bejerano et al. [2], we also use the radius guarantee to *bound the approximation factor*. This idea might be applicable to other problems.

Acknowledgements. Many thanks go to Shuchi Chawla, Avrim Blum, Mohammad Mahdian, David Liben-Nowell, and Grant Wang. Many results in this paper were inspired by conversations with Seffi Naor.

References

[1] Nikhil Bansal, Avrim Blum, and Shuchi Chawla. Correlation clustering. *IEEE Symp. on Foundations of Computer Science*, 2002.

[2] Y. Bejerano, N. Immorlica, S. Naor, and M. Smith. Location area design in cellular networks. *International Conference on Mobile Computing and Networking*, 2003.

[3] Moses Charikar, Venkatesan Guruswami, and Anthony Wirth. Clustering with qualitative information. Unpublished Manuscript.

[4] E. Dahlhaus, D. S. Johnson, C. H. Papadimitriou, P. D. Seymour, and M. Yannakakis. The complexity of multiway cuts. *ACM Symp. on Theory of Comp.*, 1992.

[5] Dotan Emanuel and Amos Fiat. Correlation clustering — minimizing disagreements on arbitrary weighted graphs. *European Symp. on Algorithms*, 2003.

[6] Martin Ester, Hans-Peter Kriegel, Jörg Sander, and Xiaowei Xu. Clustering for mining in large spatial databases. *KI-Journal*, 1, 1998. Special Issue on Data Mining. ScienTec Publishing.

[7] N. Garg, V. V. Vazirani, and M. Yannakakis. Approximate max-flow min-(multi)cut theorems and their applications. *SIAM J. Comp.*, 25, 1996.

[8] D. S. Hochbaum and D. B. Shmoys. A unified approach to approximation algorithms for bottleneck problems. *Journal of the ACM*, 33, 1986.

[9] T. C. Hu. Multicommodity network flows. *Operations Research*, 1963.

[10] Kamal Jain and Vijay V. Vazirani. Primal-dual approximation algorithms for metric facility location and k-median problems. *IEEE Symp. on Foundations of Computer Science*, 1999.

[11] Tapas Kanungo, David M. Mount, Nathan S. Netanyahu, Christine D. Piatko, Ruth Silverman, and Angela Y. Wu. An efficient k-means clustering algorithm: Analysis and implementation. *IEEE Transactions on Pattern Analysis and Machine Intelligence*, 24(7), 2002.

[12] Philip N. Klein, Serge A. Plotkin, and Satish Rao. Excluded minors, network decomposition, and multicommodity flow. *ACM Symp. on Theory of Comp.*, 1993.

[13] Tom Leighton and Satish Rao. Multicommodity max-flow min-cut theorems and their use in designing approximation algorithms. *Journal of the ACM*, 46(6), 1999.

[14] Marina Meila and David Heckerman. An experimental comparison of several clustering and initialization methods. *Conference on Uncertainty in Artificial Intelligence*, 1998.

[15] F. Murtagh. A survey of recent advances in hierarchical clustering algorithms. *The Computer Journal*, 26(4), 1983.

[16] Cecilia M. Procopiuc. Clustering problems and their applications. Department of Computer Science, Duke University.
http://www.cs.duke.edu/~magda/clustering-survey.ps.gz.

[17] Leonard J. Schulman. Clustering for edge-cost minimization. *Electronic Colloquium on Computational Complexity (ECCC)*, 6(035), 1999.

[18] Michael Steinbach, George Karypis, and Vipin Kumar. A comparison of document clustering techniques. *KDD-2000 Workshop on TextMining Workshop*, 2000.

[19] Eva Tardos and Vijay V. Vazirani. Improved bounds for the max-flow min-multicut ratio for planar and $K_{r,r}$-free graphs. *Information Processing Letters*, 47(2):77–80, 1993.

[20] V. V. Vazirani. *Approximation Algorithms.* Springer-Verlag, Berlin, 2001.

[21] M. Yannakakis, P. C. Kanellakis, S. C. Cosmadakis, and C. H. Papadimitriou. Cutting and partitioning a graph after a fixed pattern. *10th Intl. Coll. on Automata, Languages, and Programming*, 1983.

Improved Linear Time Approximation Algorithms for Weighted Matchings*

Doratha E. Drake and Stefan Hougardy

Institut für Informatik, Humboldt-Universität zu Berlin, 10099 Berlin, Germany
{drake,hougardy}@informatik.hu-berlin.de

Abstract. The weighted matching problem is to find a matching in a weighted graph that has maximum weight. The fastest known algorithm for this problem has running time $O(nm + n^2 \log n)$. Many real world problems require graphs of such large size that this running time is too costly. We present a linear time approximation algorithm for the weighted matching problem with a performance ratio of $\frac{2}{3} - \varepsilon$. This improves the previously best performance ratio of $\frac{1}{2}$.

1 Introduction

A *matching* M in a graph $G = (V, E)$ is a subset of the edges of G such that no two edges in M are adjacent. In a graph $G = (V, E)$ with edge weights given by a function $w : E \to \mathbb{R}_+$ the *weight of a matching* is defined as $w(M) := \sum_{e \in M} w(e)$. The weighted matching problem is to find a matching M in G that has maximum weight. The first polynomial time algorithm for the weighted matching problem is due to Edmonds [4]. A straightforward implementation of this algorithm requires a running time of $O(n^2 m)$, where n and m denote the number of vertices and edges in the graph. Lawler [8] and Gabow [6] improved the running time to $O(n^3)$. The fastest known algorithm to date for solving the weighted matching problem in general graphs is due to Gabow [7] and has a running time of $O(nm + n^2 \log n)$.

Many real world problems require graphs of such large size that the runtime of Gabow's algorithm is too costly. Examples of such problems are the refinement of FEM nets [9], the partitioning problem in VLSI-Design [10], and the gossiping problem in telecommunications [2]. There also exist applications were the weighted matching problem has to be solved extremely often on only moderately large graphs. An example of such an application is the virtual screening of protein databases containing the three dimensional structure of the proteins [5]. The graphs appearing in such applications only have about 10,000 edges. But the weighted matching problem has to be solved more than 100,000,000 times for a complete database scan.

Therefore, there is considerable interest in approximation algorithms for the weighted matching problem that are very fast, have ideally linear runtime, and that nevertheless produce very good results even if these results are not optimal.

---

* supported by DFG research grant 296/6-3

S. Arora et al. (Eds.): APPROX 2003+RANDOM 2003, LNCS 2764, pp. 14–23, 2003.

The quality of an approximation algorithm for solving the weighted matching problem is measured by its so-called *performance ratio*. An approximation algorithm has a performance ratio of c, if for all graphs it finds a matching with a weight of at least c times the weight of an optimal solution.

A simple approximation algorithm for the weighted matching problem with performance ratio $\frac{1}{2}$ is obtained by the following greedy approach [1]: Start with an empty matching and extend it in each step by the heaviest edge currently available. The running time of this algorithm is $O(m \log n)$ as it requires sorting the edges by decreasing weight. The first linear time $\frac{1}{2}$-approximation algorithm for the weighted matching problem was proposed by Preis [11] using the idea of locally heaviest edges. Drake and Hougardy [3] obtained a simpler linear time approximation algorithm with the same performance ratio by using a completely different approach. In this paper we improve these results by proving the existence of linear time approximation algorithms for the weighted matching problem which have approximation ratios arbitrarily close to $\frac{2}{3}$.

Main Theorem. *For each $\varepsilon > 0$ there exists a linear time approximation algorithm for the weighted matching problem with a performance ratio of $\frac{2}{3} - \varepsilon$.*

The main idea of our algorithm is to start with a maximal matching M and to increase its weight by local changes. These local changes which we call short augmentations add in each step at most two new edges to M while up to four edges of M will be removed. A graph can possess up to $\Omega(n^4)$ short augmentations. To achieve linear running time only some part of these can be looked at. For each edge of the maximal matching M our algorithm only looks at all short augmentations that involve the endpoints of this edge. This way the short augmentations considered by the algorithm are in some sense spread evenly over the graph and their number is linearly bounded.

As the short augmentations are partly overlapping it can happen that after performing one short augmentation several others are no longer available. For the performance ratio it is therefore important to be able to reject short augmentations that achieve only minor improvements in the weight of the matching. This is achieved by only taking short augmentations into considerations that gain at least some constant factor β and that additionally yield the largest possible gain from all these. Such augmentations will be called β-augmentations. In linear time it seems not to be possible to find the best β-augmentation. However we will show that in linear time a constant factor approximation of the best β-augmentation can be found.

To prove the performance ratio of our algorithm we use an amortized analysis. The idea is that the gain that is achieved by an augmentation is not realized immediately but part of it is stored in certain edges of the graph for later use. This way we are able to prove that the algorithm increases the weight of the given matching by some constant factor. By repeating the algorithm a constant number of times and choosing β sufficiently small the resulting matching will have a weight that comes arbitrarily close to $\frac{2}{3}$.

The paper is organized as follows. In Section 2 we give basic definitions. In Section 3 we define short augmentations and use these to prove the existence of the set of local improvements upon which our algorithm is based. In Section 4 we present the algorithm and prove that its performance ratio is $\frac{2}{3} - \varepsilon$ for any $\varepsilon > 0$.

2   Preliminaries

Let $G = (V, E)$ be a weighted graph with weight function $w : E \to \mathbb{R}_+$. For a subset $F \subseteq E$ the weight of F is defined as $w(F) := \sum_{f \in F} w(f)$. A matching $M \subseteq E$ is called *maximal* if no proper superset of M in E is a matching. By M_{opt} we denote a maximum weight matching in G, i.e. a matching that satisfies $w(M_{opt}) \geq w(M)$ for all other matchings M. A path or cycle is called M-*alternating* if it uses alternately edges from M and $E \setminus M$. Note that alternating cycles must contain an even number of edges. Let P be an alternating path such that if it ends in an edge not belonging to M then this endpoint of P is not covered by an edge of M. The path P is called M-*weight-augmenting* if

$$w\left(E(P) \cap M\right) \; < \; w\left(E(P) \setminus M\right) \; .$$

If P is an M-weight-augmenting path then $M \bigtriangleup P$ (the symmetric difference between M and P) is again a matching with strictly larger weight than M. The notion of M-weight-augmenting cycles is defined similarly. More generally we call an *augmentation* any process that replaces some edges of a matching M by some new edges and increases the weight of the matching.

3   Short Augmentations

A weight-augmenting path or cycle with respect to a matching M is called *short* if it contains at most two edges not belonging to M. The only weight-augmenting short cycle is by this definition an alternating cycle of length four and there exist six different types of weight-augmenting short paths. The following result shows that it is indeed enough to consider such short augmenting paths and cycles to obtain a $\frac{2}{3}$-approximation of the maximum weight matching.

Lemma 1. *For any matching M there exists a node disjoint set of weight-augmenting short paths and cycles such that augmenting along all these paths and cycles results in a matching of weight at least $\frac{2}{3} \cdot w(M_{opt})$.*

Proof. Consider the symmetric difference $M \bigtriangleup M_{opt}$. It consists of even length alternating cycles and of alternating paths. Order these paths and cycles arbitrarily and number the edges of M_{opt} in the order in which they appear in these paths and cycles. Now partition M_{opt} into three sets by taking the edge numbers modulo 3. By removing any of these three sets from $M \bigtriangleup M_{opt}$ one obtains a set of alternating paths and cycles each of which contains at most two edges of M_{opt}. Removing the lightest of these three sets shows that M can be augmented to a matching of weight at least $\frac{2}{3} \cdot w(M_{opt})$ by paths and cycles each of which contain at most two edges not in M. □

In the following we need the notion of a β-*augmentation*. For a constant $\beta > 1$ a β-augmentation of a matching M is an augmentation that has the property that the weight of the edges that are removed from M is at least by the factor β smaller than the weight of the edges that are added to M by the augmentation. The following result shows that for small enough β any matching M can be augmented by short paths and cycles each of which is a β-augmentation to a matching that has a weight close to $\frac{2}{3} \cdot w(M_{opt})$.

Lemma 2. *Let M be an arbitrary matching and $\beta > 1$ be constant. Then there exists a node disjoint set of weight-augmenting short paths and cycles each of which is a β-augmentation such that augmenting along all these paths and cycles results in a matching of weight at least $\frac{2}{3\beta} \cdot w(M_{opt})$.*

Proof. By Lemma 1 we know that there exists a node disjoint set of augmenting paths and cycles each of which contains at most two edges not in M such that augmenting along all these paths and cycles results in a matching $\tilde{M}$ of weight at least $\frac{2}{3} \cdot w(M_{opt})$. We now claim that if we take the subset of these augmenting paths and cycles that are β-augmentations, we get a matching of the desired weight.

Partition the set $\tilde{M}$ into two sets $\tilde{M}_{\geq \beta}$ and $\tilde{M}_{< \beta}$ such that $\tilde{M}_{\geq \beta}$ contains all edges of $\tilde{M}$ that are obtained by β-augmentations and let $\tilde{M}_{< \beta}$ be all other edges of $\tilde{M}$. The set M similarly can also be partitioned into two sets $M_{< \beta}$ and $M_{\geq \beta}$ according to the augmenting paths and cycles in $M \triangle \tilde{M}$ that contain these edges. By performing only the β-augmentations one obtains the matching $M_{< \beta} \cup \tilde{M}_{\geq \beta}$. The weight of this set can be bounded from below as follows:

$$w(M_{< \beta}) + w(\tilde{M}_{\geq \beta}) \geq \frac{1}{\beta} w(\tilde{M}_{< \beta}) + w(\tilde{M}_{\geq \beta}) \geq \frac{1}{\beta} w(\tilde{M}) \geq \frac{2}{3\beta} w(M_{opt}) .$$

$\square$

Let M be an arbitrary matching and let $\tilde{M}$ be a matching of weight at least $\frac{2}{3} \cdot w(M_{opt})$ such that the symmetric difference of M and $\tilde{M}$ consists of weight-augmenting short paths and cycles. The existence of $\tilde{M}$ is guaranteed by Lemma 1. For each cycle or path in $M \triangle \tilde{M}$ choose an edge in M that is adjacent to all edges of $\tilde{M}$ in this path or cycle. Call the set of all these chosen edges M^*. For each edge $e \in M^*$ denote by S_e the (at most two) edges of $\tilde{M}$ that are adjacent to e. For an arbitrary set F of edges denote by $inc(F)$ all edges in M that are incident to the endpoints of edges in F. Then $inc(S_e)$ contains at most three edges of M and $S_e \cup inc(S_e)$ is the set of edges of the path or cycle in $M \triangle \tilde{M}$ that contains the edge e.

For a given constant $\beta > 1$ we partition the set M^* into two subsets $M^*_{< \beta}$ and $M^*_{\geq \beta}$ such that $M^*_{< \beta}$ contains all edges of M^* such that $w(S_e) < \beta \cdot w(inc(S_e))$ and $M^*_{\geq \beta}$ contains all other edges of M^*.

The following result shows that if an algorithm achieves at least a constant fraction of the value $\frac{1}{\beta} \cdot w(S_e) - w(inc(S_e))$ for all $e \in M^*_{\geq \beta}$ then it will improve a given matching by a constant factor.

Lemma 3. *Let M be a matching of weight $w(M) \geq \alpha \cdot w(M_{opt})$. If the matching M' has a weight that is larger than the weight of M by at least $\varepsilon \cdot \sum_{e \in M^*_{\geq \beta}} \left(\frac{1}{\beta} \cdot w(S_e) - w(inc(S_e)) \right)$ then*

$$w(M') \geq \left(\alpha + \varepsilon \cdot \left(\frac{2}{3\beta} - \alpha \right) \right) \cdot w(M_{opt}) .$$

Proof. By the definition of $M^*_{<\beta}$ we have $w(inc(S_e)) > \frac{1}{\beta} \cdot w(S_e)$ for $e \in M^*_{<\beta}$ and $w(M) = \sum_{e \in M^*} w(inc(S_e))$. Applying these two facts we get

$$w(M') \geq w(M) + \varepsilon \cdot \sum_{e \in M^*_{\geq \beta}} \left(\frac{1}{\beta} \cdot w(S_e) - w(inc(S_e)) \right)$$

$$= (1 - \varepsilon) \cdot w(M) + \varepsilon \cdot \sum_{e \in M^*_{<\beta}} w(inc(S_e)) + \varepsilon \cdot \sum_{e \in M^*_{\geq \beta}} \frac{1}{\beta} \cdot w(S_e)$$

$$> (1 - \varepsilon) \cdot w(M) + \varepsilon \cdot \sum_{e \in M^*_{<\beta}} \frac{1}{\beta} \cdot w(S_e) + \varepsilon \cdot \sum_{e \in M^*_{\geq \beta}} \frac{1}{\beta} \cdot w(S_e)$$

$$= (1 - \varepsilon) \cdot w(M) + \frac{\varepsilon}{\beta} \cdot \sum_{e \in M^*} w(S_e)$$

$$\geq (1 - \varepsilon) \cdot \alpha \cdot w(M_{opt}) + \frac{\varepsilon}{\beta} \cdot \frac{2}{3} \cdot w(M_{opt})$$

$$= \left(\alpha + \varepsilon \cdot \left(\frac{2}{3\beta} - \alpha \right) \right) \cdot w(M_{opt})$$

$\square$

4   The Algorithm

For the algorithm we have to extend the notion of weight-augmenting short paths and cycles slightly. Let $S \subset E$ be a set of at most two non-adjacent edges such that there exists an edge e in E that is adjacent to all edges in S. Then by removing all edges from a matching M' that are adjacent with some edge in S and by adding S one obtains a new matching M''. If $w(M'') > w(M')$ we say that S is a *(short) augmenting set centered at e* with respect to M'. As S contains at most two edges there are at most four edges in M' that will be removed. Note that the sets S_e introduced in Section 3 are augmenting sets centered at e with respect to M.

For the description of the algorithm and the proof of its performance ratio we need the following additional definitions. Let M denote the maximal matching that the algorithm begins with and which is the matching that defines the set S_e as described in Section 3. Let M' denote the matching that is continuosly updated by the algorithm by means of augmentations. Let $aug(e)$ denote the set of edges that the algorithm chooses for a β-augmentation at $e \in M$. For an arbitrary set F of edges denote by $inc'(F)$ all edges in M' that are incident to the endpoints of edges in F.

Algorithm improve_matching $(G = (V, E), w : E \to \mathbb{R}_+, M)$

1   make M maximal
2    $M' := M$
3   **for** $e \in M$ **do begin**
4   **if** there exists a β-augmentation with center e
5   **then** augment M' by a good β-augmentation with center e
6   **end**
7   **return** M'

Fig. 1. Algorithm improve_matching for increasing the weight of a matching.

The algorithm, which we call improve_matching, is shown in Figure 1. Starting from a maximal matching M the algorithm visits each edge $e \in M$ exactly once. For each $e \in M$ the algorithm determines if there is any β-augmenting set centered at e in M'. If there is none then the algorithm moves on to the next edge in M. Otherwise, there is a β-augmenting set centered at e. The algorithm then tries to find the best β-augmenting set centered at e. The *gain* of an augmenting set S is defined to be $w(S) - w(inc'(S))$ which is the amount by which M' increases by augmenting S. We define the *best β-augmenting set* centered at e to be the β-augmenting set centered at e with the largest gain.

However, the algorithm is not guaranteed to find the best β-augmenting set centered at e but rather it finds a good β-augmenting set at e. We define a *good β-augmenting set* centered at e to be a β-augmenting set centered at e with a gain of at least $\frac{\beta-1}{4}$ times the gain of the best β-augmenting set centered at e. For technical reasons we assume from now on that $1 < \beta \le \frac{3}{2}$ which is no restriction as in the end β will turn out to be very close to 1.

Figure 2 shows our algorithm for finding a good β-augmentation. It takes an edge e as input and returns a good β-augmenting set centered at e if any such set exists. We need a few more definitions to describe the algorithm. For an arbitrary edge x let $w_M(x)$ be 0, if $x \notin M$ and define $w_M(x) = w(x)$ otherwise. Arbitrarily label the endnodes of e as *left* and *right*. Then any edge $a \notin M'$ that is incident to left together with $inc'(a) \setminus \{e\}$ form a *left arm* of e. The definition of a *right arm* of e is symmetrical to this. The *gain* of an arm of e that consists of $a \notin M'$ together with $inc'(a) \setminus \{e\}$ is defined as $gain_a := w_M(a) - w_M(inc'(a) \setminus \{e\})$. The gain of an arm of e that consists of just $a \notin M'$ is defined in the obvious way as just $gain_a := w_M(a)$.

We define a left arm $a \cup (inc'(a) \setminus \{e\})$ to be *allowable* if there exists a right arm $b \cup (inc'(b) \setminus \{e\})$ such that $a \cup b$ or a alone forms a β-augmenting set at e. We calculate the left allowable arms as follows: First, we calculate the greatest surplus from among the right arms, where we define the *surplus* of the right arm $b \cup (inc'(b) \setminus \{e\})$ as $w_M(b) - \beta \cdot (w_M(inc'(b) \setminus \{e\}) + w_M(e))$.

Algorithm good-β-augmentation $(G = (V, E), w : E \rightarrow \mathbb{R}_+, e \in E)$

1   find the right and left arms of e
2   determine the gains and surpluses of the left and right arms
3   *left* := largest left allowable arm and its best extension
4   *right* := largest right allowable arm and its best extension
5   **if** *left*$= \emptyset$ and *right*$= \emptyset$
6   **then return** $\emptyset$
7   **else return** max(*left*, *right*)

Fig. 2. Algorithm for finding a good β-augmentation.

One can think of the largest surplus from among all the right arms, denoted $surp_r$, as the maximum value that a right arm can loan to a left arm in order to make it part of a β-augmenting set at e. A left arm is allowable if and only if $w_M(a) - \beta \cdot w_M(inc'(a) \setminus \{e\}) + surp_r \geq 0$. The definition of a right allowable arm, the maximum left surplus $surp_l$, as well as the process for calculating these is symmetrical.

Once the algorithm has calculated the left and right allowable arms of e it chooses from among these the one with the largest gain. Without loss of generality let it be a left arm. Let $a \notin M'$ be the uncovered edge in this arm. Then the algorithm returns the best β-augmenting set centered at e that contains a.

Lemma 4. *If there exists a β-augmenting set centered at e then the algorithm good-β-augmentation (Figure 2) returns a good β-augmenting set centered at e. The running time is proportional to the sum of the degrees of the end-vertices of the edge e.*

Proof. Sketch of proof: If the largest possible gain of an arm is larger than twice the weight of e then one easily gets that the algorithm finds a β-augmentation that achieves at least $\frac{1}{3}$ of the largest possible gain. This is because then the best β-augmentation gains at most 3 times the weight of e and the algorithm finds a β-augmentation that gains at least the weight of e.

In the other case, using the fact that the algorithm finds a β-augmentation that does not share an arm with the best possible β-augmentation, one can show that $\beta - 1$ must be sufficiently small such that $\frac{4}{\beta-1}$ is big enough to scale the gain found by the algorithm. For $\beta < \frac{3}{2}$ the latter is larger. $\square$

Lemma 5. *The algorithm improve_matching improves the matching M by at least*

$$\sum_{e \in M^*_{\geq \beta}} \frac{(\beta - 1)^2}{8\beta} \cdot \left(\frac{1}{\beta} \cdot w(S_e) - w(inc(S_e)) \right).$$

Proof. Define $\delta := \frac{\beta-1}{4}$. The algorithm visits every $e \in M$ and hence every $e \in M^*_{\geq \beta}$. If it finds any β-augmenting set for e it also finds and augments a β-

augmenting set $aug(e)$ which yields at least δ of the gain of the currently best β-augmenting set at e. Even though the algorithm cannot distinguish between the edges of $M_{\geq\beta}$ and $M_{<\beta}$ or know the previously defined β-augmenting sets S_e we show by means of amortization that for each of these sets S_e it finds a constant proportion of $\frac{1}{\beta} \cdot w(S_e) - w(inc(S_e))$. The idea is that for each $e \in M_{\geq\beta}^*$ the algorithm can either find an augmenting set as good as S_e in M or the matching M' has increased by enough weight to already assign a constant proportion of this gained weight to e.

The idea of the amortized analysis is that when the algorithm augments at $e \in M_{\geq\beta}^*$ then M' either gains a constant proportion of $\frac{1}{\beta} \cdot w(S_e) - w(inc(S_e))$ right away or M' can additionally make a withdrawal of weight that has been added to M''s savings in the past in a way that brings a total win of some constant proportion of $\frac{1}{\beta} \cdot w(S_e) - w(inc(S_e))$ to M'. One builds up M''s savings as follows. For each $e \in M_{<\beta}^*$ the matching M' gets charged all of the augmentation that the algorithm finds at e and this amount gets put in savings. This is not a problem because there are no sets S_e associated with these edges anyway. If $e \in M_{\geq\beta}^*$ then M' keeps $\frac{1}{2}$ of the augmentation that the algorithm finds at e and M' gets charged the other half to savings. If this is done for all $e \in M$ then M' can later make withdrawals from savings when necessary. This is necessary when for instance one needs to augment at $e \in M_{\geq\beta}^*$ but the edges incident to S_e all have greater weight in M' than they had in M. Let $E \subset (M' \setminus M)$ denote the set of new edges incident with the nodes of S_e for some such $e \in M_{\geq\beta}^*$. Then a withdrawal from M''s savings of $\frac{1}{2} \cdot \frac{\beta-1}{\beta} \cdot w(E)$ can be made for the augmentation at e. The factor $\frac{\beta-1}{\beta}$ comes from the fact that the edges in the set E were added to M' during β-augmentations that occured in the past and so there must have been at least this much put in savings in the past. The factor $\frac{1}{2}$ comes from the fact that each edge in E has two endnodes and therefore each e in E can be involved in at most two withdrawals since each of the S_e are node disjoint.

More concretely, when the algorithm visits $e \in M_{\geq\beta}^*$ there are three possibilities for the set S_e: The first is that S_e is still β-augmenting in M' with $w(inc'(S_e)) \leq w(inc(S_e))$, the second is that S_e is still β-augmenting in M' with $w(inc'(S_e)) > w(inc(S_e))$, and the third is that S_e is no longer β-augmenting in M'.

For the first possibility for $e \in M_{\geq\beta}^*$ we have S_e is still β-augmenting with $w(inc'(S_e)) \leq w(inc(S_e))$ when the algorithm visits it. Since the algorithm always finds a β-augmentation $aug(e)$ of at least δ the gain of the largest β-augmentation at e in M' it follows that after M' has been charged $\frac{1}{2}$ of the augmentation found at e the amount of weight that M' increases by is

$$\frac{1}{2}(w(aug(e)) - w(inc'(aug(e)))) \geq \frac{\delta}{2}(w(S_e) - w(inc'(S_e)))$$

$$\geq \frac{\delta}{2}(w(S_e) - w(inc(S_e)))$$

$$\geq \frac{\delta}{2}(\frac{1}{\beta} \cdot w(S_e) - w(inc(S_e))).$$

For the second possibility for $e \in M^*_{\geq \beta}$ we have that S_e is still β-augmenting with $w(inc'(S_e)) > w(inc(S_e))$ when the algorithm visits it. Let A denote the set $inc'(S_e) \setminus inc(S_e)$. The set A contains only *new* edges, i.e., edges that were in augmentations, therefore M''s has increased in the past by at least $\frac{\beta-1}{\beta} \cdot w(A)$ of which at least half can be withdrawn by M'. This together with the augmentation that the algorithm will find at e, one half of which M' gets to keep, means that M''s total win at e is at least

$$
\begin{aligned}
&\frac{1}{2}(w(aug(e)) - w(inc'(aug(e)))) + \frac{1}{2}\frac{\beta-1}{\beta} \cdot w(A) \\
\geq\ &\frac{\delta}{2}(w(S_e) - w(inc'(S_e))) + \frac{1}{2}\frac{\beta-1}{\beta}(w(inc'(S_e)) - w(inc(S_e))) \\
\geq\ &\frac{\delta}{2}\frac{\beta-1}{\beta}(w(S_e) - w(inc'(S_e)) + w(inc'(S_e)) - w(inc(S_e))) \\
\geq\ &\frac{\delta}{2}\frac{\beta-1}{\beta}(\frac{1}{\beta}w(S_e) - w(inc(S_e))).
\end{aligned}
$$

For the third and final possibility for $e \in M^*_{\geq \beta}$ we have that S_e is no longer β-augmenting when the algorithm visits it, i.e., $w(inc'(S_e)) > \frac{1}{\beta} \cdot w(S_e)$. Therefore, the set of edges $A = inc'(S_e) \setminus inc(S_e)$ has weight $w(A) > \frac{1}{\beta} \cdot w(S_e) - w(inc(S_e))$. Then M''s savings must have increased by at least $\frac{\beta-1}{\beta} \cdot w(A)$ of which at least $\frac{1}{2}$ can be withdrawn by M'. So independently of wether the algorithm finds a β-augmenting set at e in M', M' gets a total win in this step of at least

$$
\frac{1}{2}\frac{\beta-1}{\beta}w(A) \geq \frac{1}{2}\frac{\beta-1}{\beta}(\frac{1}{\beta}w(S_e) - w(inc(S_e))).
$$

The minimum weight that M' increases by at each $e \in M^*_{\geq \beta}$ over all three cases is $\frac{\delta}{2}\frac{\beta-1}{\beta}(\frac{1}{\beta}w(S_e) - w(inc(S_e)))$ which proves the lemma since we defined δ as $\frac{\beta-1}{4}$.

$\square$

Theorem 1. *If M is any matching with $w(M) \geq \alpha \cdot w(M_{opt})$ then after applying the algorithm improve_matching one obtains a matching M' with weight at least*

$$
w(M') \geq \left(\alpha + \frac{(\beta-1)^2}{8\beta}\left(\frac{2}{3\beta} - \alpha\right)\right) \cdot w(M_{opt})
$$

Proof. This is an immediate consequence of Lemma 3 and Lemma 5. $\square$

We are now able to prove the main theorem.

Main Theorem. *For each $\varepsilon > 0$ there exists a linear time approximation algorithm for the weighted matching problem with a performance ratio of $\frac{2}{3} - \varepsilon$.*

Proof. Theorem 1 shows that by repeating algorithm improve_matching one gets a matching with weight arbitrarily close to $\frac{2}{3\beta} \cdot w(M_{opt})$. Now by choosing $\beta > 1$ small enough one gets a matching with weight arbitrarily close to $\frac{2}{3} \cdot w(M_{opt})$. Note that β and the number of repeats of algorithm improve_matching are constants depending on ε. As the algorithm improve_matching has linear running time the total running time stays linear. □

References

1. D. Avis, A Survey of Heuristics for the Weighted Matching Problem, Networks, Vol. 13 (1983), 475–493
2. R. Beier, J.F. Sibeyn, A Powerful Heuristic for Telephone Gossiping, Proc. 7th Colloquium on Structural Information and Communication Complexity, Carleton Scientific (2000), 17–35
3. D.E. Drake, S. Hougardy, A Simple Approximation Algorithm for the Weighted Matching Problem, Information Processing Letters 85 (2003), 211–213
4. J. Edmonds, Maximum matching and a polyhedron with 0,1-vertices, J. Res. Nat. Bur. Standards 69B (1965), 125–130
5. C. Frömmel, C. Gille, A. Goede, C. Gröpl, S. Hougardy, T. Nierhoff, R. Preißner, M. Thimm, Accelerating screening of 3D protein data with a graph theoretical approach, to appear in Bioinformatics
6. H.N. Gabow, An efficient implementation of Edmond's algorithm for maximum matching on graphs, Journal of the ACM 23 (1976), 221–234
7. H.N. Gabow, Data Structures for Weighted Matching and Nearest Common Ancestors with Linking, SODA 1990, 434–443
8. E.L. Lawler, Combinatorial Optimization: Networks and Matroids, Holt, Rinehart and Winston, New York, 1976
9. R.H. Möhring, M. Müller-Hannemann, Complexity and Modeling Aspects of Mesh Refinement into Quadrilaterals, Algorithmica 26 (2000), 148–171
10. B. Monien, R. Preis, R. Diekmann, Quality Matching and Local Improvement for Multilevel Graph-Partitioning, Parallel Computing, 26(12), 2000, 1609–1634
11. R. Preis, Linear Time 1/2-Approximation Algorithm for Maximum Weighted Matching in General Graphs, Symposium on Theoretical Aspects of Computer Science, STACS 99, C. Meinel, S. Tison (eds.), Springer, LNCS 1563, 1999, 259–269

Covering Graphs Using Trees and Stars[*]

G. Even[1], N. Garg[2], J. Könemann[3], R. Ravi[3], and A. Sinha[3]

[1] Tel-Aviv University, Tel-Aviv, Israel.
guy@eng.tau.ac.il
[2] Indian Institute of Technology, Delhi, India.
naveen@cse.iitd.ernet.in
[3] GSIA, Carnegie Mellon University, Pittsburgh, USA.
jkonemann@acm.org, ravi@cmu.edu, asinha@andrew.cmu.edu

Abstract. A tree cover of a graph G is defined as a collection of trees such that their union includes all the vertices of G. The cost of a tree cover is the weight of the maximum weight tree in the tree cover. Given a positive integer k, the k-tree cover problem is to compute a minimum cost tree cover which has no more than k trees. Star covers are defined analogously. Additionally, we may also be provided with a set of k vertices which are to serve as roots of the trees or stars. In this paper, we provide constant factor approximation algorithms for finding tree and star covers of graphs, in the rooted and un-rooted versions.

1 Introduction

This paper was motivated by the following "Nurse station location" problem. A hospital wanted to locate k nurses in its coverage area. Each nurse would be assigned a certain set of patients, who she would visit in her morning rounds. The objective is to figure out where to locate the nurse stations and how to assign patients to nurses so that the last completion time is minimized.

This problem is equivalent to covering a metric graph with no more than k tours, so that the maximum length of a tour is minimized. Since minimum spanning trees are constant factor approximations to traveling salesperson tours, we look at covering the graph with no more than k spanning trees, so that the maximum weight of a tree is minimized. A variant is when the nurse has to return to her station before visiting each patient, in order to pick up equipment and supplies necessary for that patient. In that case, we get the problem of covering the graph with stars. If the hospital has already built its nursing stations and only wants to assign patients to nurses, we get the rooted versions of these problems. The problems are defined formally in the next section.

---

[*] J. Könemann, R. Ravi and A. Sinha are supported by the National Science Foundation under grant No. 0105548 and the ALADDIN Center under NSF grant No. CCR-0122581. A. Sinha is also supported by a Carnegie Bosch Institute Fellowship.

S. Arora et al. (Eds.): APPROX 2003+RANDOM 2003, LNCS 2764, pp. 24–35, 2003.

Previous and related work. The problems studied in this paper are closely related to those studied by Arkin, Hassin, and Levin [1]. The problems they deal with include covering the nodes of a graph or a subset of the edges of a graph by paths, walks, or stars. Most of their approximation algorithms deal with minimizing the number of covering objects (e.g. paths) subject to a constraint on the cost of each covering object. They also consider unrooted versions of k-path covers and k-walk covers. The algorithms in [1] do not seem to extend to rooted versions.

These problems fall in the general class of "vehicle routing" problems (see [16] for a recent survey). In the k-traveling salesperson problem, a feasible solution consists of k tours that cover the nodes, where the tours share that same depot (i.e. starting and ending point). The objective is to minimize the total length of tours. The k-traveling salesperson problem was first approximated to a constant by Frederickson, Hecht and Kim [8] (see also [11]). Recently, Fakcharoenphol, Harrelson and Rao [7] provided a constant-factor approximation algorithm for the k-traveling repairman problem, where the objective is to minimize the average waiting time of the customers.

The problems we study can also be viewed as members of the family of "clustering" problems. We are partitioning the vertices of the graph into clusters, where the weight of a cluster is the weight of the cheapest spanning tree or star in the cluster. Other versions of clustering studied recently include the cases where the weight of a cluster is the cost of a clique on it [2,10], the maximum cluster diameter (also called k-center) [6,12] and the sum of radii of clusters [3].

Our results and techniques. For both the rooted and un-rooted versions of k-tree cover, we get polynomial time approximation algorithms with performance ratio 4. Both algorithms can be made strongly polynomial with a slight loss in the approximation guarantee, which becomes $4+\epsilon$. The algorithms are combinatorial, and rely on a matching construct to prove the approximation guarantee.

We use LP rounding to provide a $(4, 4)$ bi-criteria approximation algorithm for un-rooted k-star cover. That is, our algorithm outputs a solution which covers the graph with no more than $4k$ stars, and the cost of the solution is no more than four times the cost of an optimal solution which uses no more than k stars. Finally, we show that the rooted version of k-star cover reduces to a scheduling problem studied by Shmoys and Tardos [15]. This immediately yields a 2-approximation algorithm for this problem.

For the unrooted k-star problem, Levin [13] suggested an improvement of our $(4, 4)$ bi-criteria approximation to a $(3, 3)$ bi-criteria algorithm; this improvement is based on the minimum star cover approximation algorithm from [1].

Organization. We define the four versions of the problem in the next section. In Section 3, we prove that all four problems are NP-hard. We provide constant factor approximation algorithms for the rooted and un-rooted versions of k-tree cover in Section 4. We deal with k-star cover in Section 5. We conclude with some open questions in Section 6.

2   Problem Definition

k-tree cover. Let $G = (V, E)$ denote an undirected graph with positive integral edge weights $w : E \to I\!\!N^+$. A *tree cover* of a graph $G = (V, E)$ is a set $\mathcal{T}$ of trees $\{T_i\}_i$ such that $V = \bigcup_{i=1}^{k} V(T_i)$. The *cost* of a tree cover $\mathcal{T}$ is $\max_{T_i \in \mathcal{T}} w(T_i)$.

Note that trees in a tree cover may share nodes and even edges. The goal in the *min-max k-tree cover* problem is to find a minimum cost tree cover consisting of at most k trees.

Rooted k-tree cover. Let $R \subset V$ denote a set of *roots*. An *R-rooted tree cover* of a graph $G = (V, E)$ is a tree cover $\mathcal{T}$, where each tree $T_i \in \mathcal{T}$ has a distinct root in R.

As before, trees in an R-rooted tree cover may share nodes and edges. In particular, the root of T_i may be in T_j, but the roots of T_j and T_i must be distinct. Given an edge weighted graph G and a set R of roots, the *min-max R-rooted tree cover* problem is to find a minimum cost R-rooted tree cover of G.

Star covers. Star cover problems are defined over complete graphs (i.e. finite metric spaces). The goal is to cover the vertices of the graph with stars so that the maximum weight of a star is minimized.

A *star cover* is a cover of the vertex set of a graph by stars. The *cost* of a star cover is the weight of the heaviest star in the cover. In the *min-max k-star cover* problem, the goal is to find a minimum cost star cover using at most k stars. Let $R = \{r_i\}_i$ denote a set of roots. In the *min-max R-centered star cover* problem the goal to find a minimum cost star cover $\mathcal{S} = \{S_r\}_{r \in R}$ such that, for every $r \in R$, the center of S_r is r.

3   Hardness

In this section, we show that all four problems are NP-complete. We begin by showing the NP-completeness of R-centered star cover, and then extend the result to the other three problems.

We show the NP-completeness of R-centered star cover by reducing BIN-PACK to it. An instance of BIN-PACK consists of (i) a set U of elements, where the size of an element $u \in U$ is s_u, (ii) k bins, and (iii) a positive bin capacity B. The problem is to determine if there is a partition of U into k parts $U_1, \ldots, U_k$ such that for every $i = 1, \ldots, k$, we have $\sum_{u \in U_i} s_u \le B$. This was shown to be NP-hard in [9].

Theorem 1. *The min-max R-centered star cover problem is NP-complete.*

Proof. Given an instance $\Pi = \langle U, \{s_u\}_u, k, B \rangle$ of BIN-PACK, we transform it to an instance of R-centered star cover as follows. We create a complete bipartite graph $G(\Pi)$ with a vertex set $R \cup U$, where R is a set of k new nodes $R = \{r_1, \ldots, r_k\}$. For every r_i and every $u \in U$, the weight of an edge $e = (r_i, u)$ is set to $w(e) = s_u$. We complete $G(\Pi)$ into a metric space (i.e. complete graph)

Algorithm 1 Rooted-Tree-Cover(G, R, B) - Compute an R-rooted tree cover of G with cost at most $4B$.

1: Remove all edges of weight greater than B.
2: $M \leftarrow$ MST of graph obtained from G by contracting roots in R to a single node.
3: $\{T_i\}_i \leftarrow$ forest obtained from M by un-contracting roots in R.
4: Edge-decompose each tree T_i into trees $\{S_j^i\}_j + L_i$ such that $w(S_i^j) \in [B, 2B)$, for every j, and $w(L_i) < B$.
5: Try to match the trees $\{S_j^i\}_{i,j}$ to roots, subject to the constraint that a tree S_i^j may be matched only to roots of distance at most B from it.
6: If not all trees are matched, then **return** fail: "B is too low".
7: If every tree is matched, then **return** success: set of trees where each tree consists of S_j^i, its matched root r, and the leftover tree L (if any) that contains the root r.

$K(\Pi)$ by taking the metric completion of the edge weighted graph $G(\Pi)$. We designate R to be the set of roots, and ask if there is an R-centered star cover of $K(\Pi)$ of cost no more than B.

It is immediate that every bin packing induces an R-centered star cover of the same cost. Conversely, every R-centered star cover induces a partition of U, where a bin size equals the weight of the corresponding star. Since an R-centered star cover of cost B gives a solution of cost B for BIN-PACK, the theorem follows.

The following theorem can be proved by converting in polynomial time an optimal solution to any of the three other problems to an R-star cover at the same cost.

Theorem 2. *The following problems are NP-complete: min-max R-tree cover, min-max k-tree cover, and min-max k-star cover.*

4   Clustering into Trees

4.1 R-rooted Tree Cover

Algorithm. In this section we present a 4-approximation algorithm for the min-max R-rooted tree cover problem. A strongly polynomial version of this algorithm has an approximation ratio of $(4 + \varepsilon)$.

The approximation algorithm is based on Algorithm Rooted-Tree-Cover, which is given (i) a graph $G = (V, E)$ with edge weights $w(e)$, (ii) a set R of k roots, and (iii) a bound B on the weight of each tree. Algorithm Rooted-Tree-Cover either returns with a proof that B is too small (i.e., $B < B^*$, the minimum cost of a tree cover) or finds an R-rooted tree cover of cost at most $4B$. By applying binary search, a 4-approximation algorithm is obtained. In Section 4.1 we discuss a how to derive a strongly polynomial algorithm.

A listing outlining Algorithm Rooted-Tree-Cover appears in Algorithm 1. We now explain each step in detail. The algorithm begins by removing all edges of weight greater than B, since they obviously cannot be used. If as a result of deleting heavy edges there exists a node that is no longer connected to R, then obviously $B \leq B^*$. To keep the description simple, we assume that the graph remains connected even after the heavy edges are deleted. In Line 2, the roots in R are contracted to a single node, and the algorithm computes a minimum spanning tree (MST) in the contracted graph. In Line 3, the MST is broken into a set $\{T_i\}_i$ of k disjoint trees by un-contracting the nodes in R. Note that, by construction, every tree T_i is rooted at a root of R. In Line 4, the edge set of every tree T_i is decomposed into subtrees $\{S_j^i\}_j + L_i$. The subtrees may share nodes but not edges. The weight of every sub-tree S_j^i is in the range $[B, 2B)$ and there is perhaps a leftover tree L_i whose weight is less than B. We elaborate below how this edge decomposition is performed. In Line 5, a bi-partite graph is constructed as follows. One side of the vertex set is R and the other side consists of nodes representing the trees $\{S_j^i\}_{i,j}$. An edge connects a root r and a tree S_j^i if the distance between S_j^i and r is at most B. A maximum matching is then computed in this bi-partite graph. The algorithm considers now two cases: If the maximum matching does not match all the sub-trees, then in Line 6, the algorithm reports a failure by returning the statement that B is too small. If the maximum matching matches all the sub-trees, then the algorithm returns the set of trees where each tree consists of a subtree S_j^i, the root r matched to the subtree S_j^i, a shortest path from the root r to S_j^i, and the leftover tree L (if any) that contains the root r.

We now elaborate on how the edge set of every tree is decomposed in Line 4. Consider a tree T_i rooted at r. For a node v let T_v denote the rooted subtree hanging from v. Consider an edge $e = (u, v)$ where u is the parent of v. The subtree T_e is the subtree that contains three parts u, T_v, and the edge (u, v). The weight $w(T_e)$ of a subtree T_e is the sum of the edge weights in T_e. Given the threshold value B, depending on $w(T_e)$, a subtree T_e is defined as *light, medium* or *heavy* as follows. If $w(T_e) \geq 2B$, then T_e is heavy. If $w(T_e) < B$, then T_e is light. If $w(T_e) \in [B, 2B)$, then T_e is medium. The decomposition algorithm proceeds by splitting away subtrees. Recall that subtrees may share nodes, hence the definition of splitting T' away from T_i means: (i) designate T' as a new part, (ii) remove the edges of T' from T_i, and (iii) let T_i now contain only nodes and edges that are still connected to the root of T_i. Note that, when $T_{(u,v)}$ is split away from T_i, the node u remains a node in T_i so $T_{(u,v)}$ and the remaining tree share the node u.

One can always split away medium subtrees from the (remaining) tree. Since such medium weight subtrees are split away whenever possible, we now focus on the case that subtrees are either light or heavy. If every subtree is either heavy or light, let v denote a heavy node, all the children of which are light. We bunch edges $e_1, e_2, \ldots$ emanating from v to children of v until the first time the cumulative weight of the trees hanging from these edges exceeds B. We then split away the subtree $\bigcup_i T_{e_i}$ (note that this tree is a medium subtree since $w(T_{e_i}) < B$, for every i). The decomposition stops as soon as the weight of the remaining tree is less than B. If upon termination the edge set of T_i is not empty, then T_i is declared as a leftover tree L_i. Note that in this case the root of the leftover tree L_i is r (where $r \in R$ is the root of the tree T_i) .

Correctness and Approximation Ratio. In this section we prove two lemmas: Lemma 1 proves the correctness of the algorithm and Lemma 2 proves its approximation ratio. Let B^* be the minimum cost of a tree cover of G.

Lemma 1. *If Algo. Rooted-Tree-Cover returns "B is too low", then $B^* > B$.*

Proof. We prove the contrapositive, namely, if $B \geq B^*$ then there exists a matching in the bi-partite graph that matches every subtree in $\{S_j^i\}_{i,j}$ to a root in R. The existence of such a matching is equivalent by Hall's Marriage Theorem [5] to the condition that, for every subset $\mathcal{S}$ of trees from $\{S_j^i\}_{i,j}$, the neighbor set $N(\mathcal{S})$ of $\mathcal{S}$ satisfies $|N(\mathcal{S})| \geq |\mathcal{S}|$.

Consider a subset $\mathcal{S}$ of trees from $\{S_j^i\}_{i,j}$. Every tree $S \in \mathcal{S}$ satisfies $w(S) \in [B, 2B)$. Hence, $w(\mathcal{S}) \geq B \cdot |\mathcal{S}|$.

Consider an optimal R-rooted tree cover $\mathcal{T}^* = \{T_1^*, \ldots, T_k^*\}$. Let $\mathcal{T}^*(\mathcal{S})$ denote the subset of trees of $\mathcal{T}$ that are stabbed by trees from $\mathcal{S}$. Namely, $T_i^* \in \mathcal{T}^*(\mathcal{S})$ iff there exists a tree $S \in \mathcal{S}$ such that $S \cap T_i^*$ is non-empty. Note that there is an edge in the bi-partite graph between a tree S_j^i and r if the tree T_ℓ^* rooted at r intersects the tree S_j^i. Hence $|N(\mathcal{S})| \geq |\mathcal{T}^*(\mathcal{S})|$ and it suffices to prove that $|\mathcal{T}^*(\mathcal{S})| \geq |\mathcal{S}|$. Note that the weight of $\mathcal{T}^*(\mathcal{S})$ satisfies $w(\mathcal{T}^*(\mathcal{S})) \leq B^* \cdot |\mathcal{T}^*(\mathcal{S})|$.

Every node in $\bigcup \mathcal{S}$ is connected by edges in $\bigcup \mathcal{T}^*(\mathcal{S})$ to a root. Recall that every edge in S_j^i is also an edge in the MST M (from Line 2). Let M' denote the subgraph obtained from the MST M by deleting edges in $\bigcup \mathcal{S}$ and then adding edges in $\bigcup \mathcal{T}^*(\mathcal{S})$. Every vertex is connected in M' to a root, hence, the subgraph M' is connected if the roots are contracted. It follows that $w(M') \geq w(M)$, and hence, $w(\mathcal{T}^*(\mathcal{S})) \geq w(\mathcal{S})$. We conclude that $B^* \cdot |\mathcal{T}^*(\mathcal{S})| \geq w(\mathcal{T}^*(\mathcal{S})) \geq w(\mathcal{S}) \geq B \cdot |\mathcal{S}|$. Since $B^* \leq B$, it follows that $|\mathcal{T}^*(\mathcal{S})| \geq |\mathcal{S}|$. Hence, Hall's condition holds, and the lemma follows.

The following lemma proves that the approximation ratio is 4.

Lemma 2. *When successful, Algorithm Rooted-Tree-Cover finds an R-rooted tree cover of cost at most $4B$.*

Proof. By construction, each tree returned by the algorithm has a distinct root from R and every node belongs to at least one tree. The weight of each tree equals the weight of the tree S_j^i (which is bounded by $2B$), the weight of the path from the root r to a node in S_j^i (which is bounded by B), and the weight of the leftover tree (which is bounded by B). It follows that the weight of every tree is less than $4B$, and the lemma follows.

Note that a path from a root r to a subtree S_j^i may contain edges and nodes that also belong to other trees. Hence, when successful, Algorithm Rooted-Tree-Cover covers the graph with trees, but these trees are not disjoint.

Algorithm 2 Tree-Cover(G, k, B) - Compute an k-tree cover of G with cost at most $4B$.

1: Remove all edges of weight greater than B. Let $\{G_i\}_i$ denote the connected components after deleting heavy edges.
2: $MST_i \leftarrow$ MST of G_i.
3: $k_i \leftarrow \lfloor \frac{w(MST_i)}{2B} \rfloor$.
4: If $\sum_i (k_i + 1) > k$ then **Return** fail: "B is too low".
5: Edge-decompose each tree MST_i into at most $(k_i + 1)$ trees $\{S_j^i\}_j + L_i$ such that $w(S_i^j) \in [2B, 4B)$, for every j, and $w(L_i) < 2B$.
6: **Return** success: set of trees $\{S_j^i\}_{i,j} \cup \{L_i\}_i$.

Strongly Polynomial Algorithm. Let $n = |V|$ and consider an $\varepsilon > 0$. Our goal is to find a $(4 + \varepsilon)$-approximation algorithm that is polynomial in n and in $\log \frac{1}{\varepsilon}$. Sort the edge weights, let $w_1 < w_2 < \cdots < w_m$ denote the sorted edge weights. Obviously $B^* < n \cdot w_m$. If Algorithm Rooted-Tree-Cover reports that $B < B^*$ for $B = w_m$, then the weight of all edges of weight at most $w_m/(n^2/\varepsilon)$ is less than $\varepsilon \cdot B^*$. Hence, we may contract all these edges, and consider only the remaining edges of weight at least $\varepsilon \cdot w_m/n^2$. Now binary search within the range $[\varepsilon \cdot w_m/n^2, n \cdot w_m]$ is strongly polynomial.

If Algorithm Rooted-Tree-Cover does not fail with $B = w_m$ we do the following. Let i denote an index such that (i) Algorithm Rooted-Tree-Cover reports $B < B^*$ for $B = w_i$, and (ii) Algorithm Rooted-Tree-Cover finds an R-rooted tree cover of cost $4 \cdot w_{i+1}$. Hence $B^* \in (w_i, 4 \cdot w_{i+1}]$. If $w_{i+1}/w_i \leq \frac{n^2}{\varepsilon}$, binary search within the range $[w_i, w_{i+1}]$ is strongly polynomial. Otherwise, let $w' = \frac{n^2}{\varepsilon} \cdot w_i$. Run Algorithm Rooted-Tree-Cover with $B = w'$. If the algorithm finds an R-rooted tree cover of cost $4w'$, binary search within the range $[w_i, w']$ is strongly polynomial. If the algorithm reports that $w' < B^*$, the weight of all edges of weight at most w_i is bounded by $n^2 \cdot w_i \leq \varepsilon \cdot B^*$. Hence, we may contract all these edges, and consider only the remaining edges of weight at most $4 \cdot w_{i+1}$. Now binary search within the range $[w_{i+1}, 4 \cdot w_{i+1}]$ is strongly polynomial.

By combining Lemmas 1, 2, and the above discussion we conclude with the following theorem. Note that if edge weights are polynomial, then a 4-approximation algorithm follows.

Theorem 3. *For every ε, there is a $(4 + \varepsilon)$-approximation algorithm for min-max rooted tree cover that runs in time polynomial in the size of G and $\log(\frac{1}{\varepsilon})$.*

4.2 k-tree Cover

In this section we present a 4-approximation algorithm for the k-tree cover problem. A strongly polynomial version of this algorithm has an approximation ratio of $(4 + \varepsilon)$.

Algorithm. A listing of Algorithm Tree-Cover appears as Algorithm 2. The input consists of (i) a graph $G = (V, E)$ with positive integral edge weights $w(e)$,

(ii) a bound k on the number of trees allowed in the cover, and (iii) a bound B on the weight of each tree in the cover. The algorithm returns either "fail" (meaning that B is too small), or "success" with a tree cover the cost of which is bounded by $4B$.

As in Algorithm Rooted-Tree-Cover, Algorithm Tree-Cover begins by removing edges of weight bigger than B. The removal of heavy edges may render G unconnected; we denote the connected components by $\{G_i\}_i$. In Line 2, a minimum spanning tree MST_i is computed for each component G_i. In Line 3, an estimate k_i of the number of trees needed to cover G_i is computed. In Line 4, the algorithm returns with "fail" if the estimates are too small. This means that the algorithm has a proof that the cost of an optimal k-tree cover of G is greater than B. In Line 5, each tree MST_i is edge decomposed to at most $(k_i + 1)$ subtrees. Each subtree is of cost at most $4B$. The edge-decomposition procedure is the same procedure that is used in Line 4 in Algorithm Rooted-Tree-Cover (with the threshold $2B$ instead of B). In Line 6, a tree cover consisting of at most k trees is returned. The cost of the returned tree cover is at most $4B$.

Note that the edge-decomposition procedure decomposes MST_i into at most $k_i + 1$ trees. By setting a threshold of $2B$ it follows that the weight of each tree S_j^i is at least $2B$ and at most $4B$. It follows that the number of trees $\{S_j^i\}_j$ obtained when decomposing MST_i is at most k_i. Together with the the leftover tree L_i (if it exists) we obtain at most $k_i + 1$ trees.

Correctness and Approximation Ratio. Let B^* denote the cost of a minmax k-tree cover of G. Let $T^* = \{T_1^*, \ldots, T_k^*\}$ denote an optimal k-tree cover. If $B^* \leq B$, then T^* uses edges of weight no greater than B. Let k_i^* denote the number of trees in T^* that contain nodes of G_i.

Lemma 3. *If $B^* \leq B$ then $k_i + 1 \leq k_i^*$, for every i.*

Proof. For simplicity, let $T_1^*, \ldots T_{k_i^*}^*$ denote the trees that cover G_i. By adding at most $k_i^* - 1$ edges, one can connect these k_i^* trees to obtain a tree that spans G_i. Since the cost of each such edge is at most B, we obtain: $\sum_{j=1}^{k_i^*} w(T_i^*) + (k_i^* - 1) \cdot B \geq w(MST_i)$. Since $w(T_i^*) \leq B$, we obtain $k_i^* \geq \frac{w(MST_i)}{2B} + \frac{1}{2}$. The lemma follows because $k_i \leq \frac{w(MST_i)}{2B}$.

Lemma 3 immediately implies the following lemma.

Lemma 4. *If Algorithm Tree-Cover returns "B is too low", then $B^* > B$.*

We conclude with the following theorem. Note that a 4-approximation algorithm is obtained if the edge weights are polynomial.

Theorem 4. *For every ε, there is a $(4 + \varepsilon)$-approximation algorithm for minmax tree cover that runs in time polynomial in the size of the graph and in $\log(\frac{1}{\varepsilon})$.*

Proof. When $B \geq B^*$, Lemma 4 implies that Algorithm 2 is successful and a k-tree cover of cost at most $4B$ is computed. A strongly-polynomial binary search along the lines of Section 4.1 completes the proof.

5   Clustering into Stars

In this section we discuss the un-rooted and rooted k-star cover problem. Here, we are given an undirected complete graph $G = (V, E)$, a metric c on the edges of E and a parameter $k > 0$. In the rooted version, we are also given a set of root nodes R with $|R| = k$. For a set $S \subseteq V$ and a vertex $r \in V$, let $c(S, r)$ be the cost of the star rooted at r and spanning S, i.e. $c(S, r) = \sum_{v \in S} c_{rv}$.

In the un-rooted version, we are supposed to partition the vertex set into k subsets $S_1, \ldots, S_k$ and place k roots $r_1, \ldots, r_k$ such that $\max_{1 \leq i \leq k} c(S_i, r_i)$ is as small as possible. In the rooted version, $r_1, \ldots, r_k$ are given.

5.1   Un-rooted k-star Cover

We use linear programming techniques similar to those used in facility location problems [4,14] to solve this problem. We first give a natural integer programming formulation. We have a variable y_i for each $i \in V$ that indicates the number of stars that are rooted at node i. For each pair of nodes i and j, variable x_{ij} has value one if node j is in a star rooted at node i and zero otherwise. As in the previous section, we guess the maximum star cost B of an optimum solution, and try to minimize the number of stars needed to cover the graph.

$$\min \qquad \sum_{i \in V} y_i \qquad\qquad\qquad (IP_B)$$

$$\text{s.t} \qquad \sum_{i \in V : c_{ij} \leq B} x_{ij} \geq 1 \qquad \forall j \in V \qquad (1)$$

$$x_{ij} \leq y_i \qquad \forall i, j \in V \qquad (2)$$

$$\sum_{j \in V} x_{ij} c_{ij} \leq B \cdot y_i \qquad \forall i \in V \qquad (3)$$

$$x_{ij} \in \{0, 1\}, y_i \in \mathbb{N} \qquad \forall i, j \in V \qquad (4)$$

Constraints (1) ensure that each node $j \in V$ is assigned to a root; (2) makes sure that node i is marked as a root if a node j is assigned to it; (3) bounds the cost of each of the star rooted at node i by B; and finally (4) sets integrality constraints. We denote the linear programming relaxation of (IP_B) by (LP). Let (x, y) be a solution to (LP). The following observation is immediate.

Lemma 5. *Suppose there exists a solution to the un-rooted k-star cover problem with value B. Then (LP_B) has a solution (x, y) such that $\sum_{i \in V} y_i \leq k$.*

Our algorithm now rounds an optimal fractional solution (x, y) of (LP_B) to an integer solution $(\hat{x}, y')$, such that the total cost of the solution is no more than $4B$ and $\sum_i y'_i \leq 4k$. The algorithm begins with the process of *filtering*, where a new fractional solution $(\bar{x}, \bar{y})$ which costs not much more than (x, y) but has the property that $\bar{x}_{ij}$ is positive only for (i, j) which are close together. This allows us to re-assign vertices to roots as we round up $\bar{y}$ to an integer solution and set some $\bar{y}$ variables to zero. The process and a proof of its performance guarantee are described in detail in the next lemma.

Lemma 6. *Suppose that (LP$_B$) has a solution (x, y) such that $\sum_{i \in V} y_i \leq k$. Then we can find in polynomial time a 4k-star cover of cost at most 4B.*

Proof. Define the fractional assignment cost c_j as $c_j = \sum_{i \in V} x_{ij} c_{ij}$ for each $j \in V$. Now, define

$$\bar{x}_{ij} = \begin{cases} \min\{1, 2x_{ij}\} & : \quad \text{if} \quad c_{ij} \leq 2c_j \\ 0 & : \quad \text{otherwise} \end{cases}$$

and $\bar{y}_i = 2 \cdot y_i$ for all $i, j \in V$. It is not hard to see that $(\bar{x}, \bar{y})$ is a feasible solution for (LP$_B$), and $\sum_{i \in V} \bar{y}_i \leq 2k$. By scaling y, we can assume w.l.o.g. that $\sum_{i \in V} \bar{y}_i = 2k$.

We now define 0-1 variables $\hat{x}_{ij}$ as follows. Let C denote the set of "unassigned" nodes and R the set of "opened" roots. We initialize $C \leftarrow V$ and $R \leftarrow \emptyset$. As long as C is non-empty, pick $v \in C$ that attains the minimum c_v value among all nodes remaining in C, i.e. pick v such that $c_v = \min_{u \in C} c_u$.

We add v to R. For node v, let $\mathcal{F}_v$ be the set of roots that v connects to, i.e. $\mathcal{F}_v = \{i \in V : \bar{x}_{iv} > 0\}$ and let $\mathcal{N}_v$ be the set of nodes in C that are served by roots in $\mathcal{F}_v$, that is, $\mathcal{N}_v = \{j \in C : \exists i \in \mathcal{F}_v \text{ s.t. } \bar{x}_{ij} > 0\}$. We now assign all nodes in $\mathcal{N}_v$ to v, i.e. for all $j \in \mathcal{N}_v$, let

$$\hat{x}_{lj} = \begin{cases} 1 & : \quad \text{if} \quad l = v \\ 0 & : \quad \text{otherwise.} \end{cases}$$

Finally, we remove $\mathcal{N}_v$ from C and continue, until $C = \emptyset$.

For every root $i \in R$, let $\hat{y}_i = \sum_{\{j : \hat{x}_{ij} = 1\}} 2c_j / B$. If $i \notin R$, then set $\hat{y}_i = 0$. We now claim that: (i) $\sum_i \hat{y}_i \leq 2k$. (ii) The cost of the star rooted at $v \in R$ is bounded by $2B \cdot \hat{y}_v$.

Firstly, (i) follows because of the following inequalities:

$$\sum_{i \in R} \hat{y}_i = \sum_{i \in R} \sum_{\{j : \hat{x}_{ij} = 1\}} \frac{2c_j}{B} \leq \frac{2}{B} \cdot B \cdot \sum_i y_i \leq 2k.$$

Now, observe that the reason for assigning node j to the root v must have been a (fractional solution) root i that serves both v and j, i.e. $\bar{x}_{ij} > 0$ and $\bar{x}_{iv} > 0$. But this means that $c_{jv} \leq c_{ij} + c_{iv} \leq 2c_j + 2c_v \leq 4c_j$ where the last inequality follows from our choice of v.

This proves (ii), since the cost of the star rooted at v satisfies:

$$\sum_{\{j : \hat{x}_{vj} = 1\}} c_{jv} \leq \sum_{\{j : \hat{x}_{vj} = 1\}} 4c_j = 2B \cdot \hat{y}_v.$$

We now show that $|R| \leq 2k$. Consider any two vertices $u, v \in R$. By definition of our procedure of adding vertices to R, we must have $\mathcal{F}_u \cap \mathcal{F}_v = \emptyset$. However,

$\sum_i \bar{x}_{iv} \geq 1$ means $\sum_{i \in \mathcal{F}_v} \bar{y}_i \geq 1$ for all $v \in R$. Since we also have $\sum_{i \in V} \bar{y}_i \leq 2k$, we must have $|R| \leq 2k$.

Finally, we round the $\hat{y}_i$ variables to integer variables y_i', defined as $y_i' = \lceil \hat{y}_i \rceil$. Since we have $|R| \leq 2k$ and $\sum_i \hat{y}_i \leq 2k$, we obtain $\sum_i y_i' \leq 4k$. We open y_i' star centers at node i, thus creating at most $4k$ stars in total.

Using the fact that in (IP_B) we only had $x_{ij} > 0$ for those (i,j) pairs where $c_{ij} \leq B$, it follows that whenever $y_i' > 1$, we can assign the nodes served by i to distinct stars centered at node i such that each star has total cost no more than $4B$. This is because (i) we must have $c_v \leq B$ for all $v \in V$, meaning that $\hat{x}_{iv} > 0$ only if $c_{iv} \leq 2B$, and (ii) we have already guaranteed that the cost of the star rooted at v in the solution given by $\hat{y}$ is bounded above by $2B \cdot \hat{y}_v$.

Thus we finish with an integral solution y' and an integral assignment of nodes to stars given by $\hat{x}$, such that there are no more than $4k$ stars in total and each star has cost no more than $4B$.

Lemmas 5 and 6 yield the following theorem. This can also be converted into a strongly polynomial algorithm by applying the methods used in Section 4.1.

Theorem 5. *There is a polynomial-time algorithm for the un-rooted k-star cover problem that partitions the set of nodes into $4k$ stars each of which has cost bounded by $4B$, where B is the value of an optimum solution.*

5.2 Rooted k-star Cover

In the rooted version of k-star cover we are given a root set R of cardinality k in addition to the usual problem parameters and we are supposed to use the roots in R. Notice that this problem is equivalent to the following scheduling problem: We are given k machines $M_1, \ldots, M_k$ (one for each root in R) and n jobs $\{J_v\}_{v \in V}$. For each job–machine pair (M_i, J_j) we have a processing time c_{ij}. The objective now is to assign each job to a unique machine. Let J_i be the jobs assigned to machine M_i. We want to minimize the make-span, i.e. $\max_{1 \leq i \leq k} \sum_{j \in J_i} c_{ij}$. It is easy to see that this problem is equivalent to the rooted k-star packing problem. A 2-approximation due to Shmoys and Tardos [15] implies the following theorem:

Theorem 6. *There is a polynomial-time 2-approximation for the rooted k-star cover problem.*

6 Open Questions

The more obvious integer programming formulation for k-star cover would minimize B subject to opening no more than k roots. However, we were unable to prove constant factor upper bounds on the integrality gap of that formulation. It would be interesting to see if that formulation, or some other technique, yields a constant factor approximation algorithm for k-star cover which obeys the constraint that no more than k stars are used exactly.

Our algorithms for k-tree cover immediately yield constant factor approximation algorithms for the "Nurse station location" problem (which one might call the k-tour cover problem) that motivated this research. However, it may be possible to obtain improved approximation factors by attacking the k-tour cover problem directly, instead of going via k-tree cover.

Acknowledgments

We would like to thank Asaf Levin for sending us a copy of [1] and for his improved bi-criteria approximation algorithm for the unrooted k-star problem.

References

1. E. Arkin, R. Hassin and A. Levin. Approximations for minimum and min-max vehicle routing problems. *Manuscript*, 2003.
2. Y. Bartal, M. Charikar and D. Raz. Approximating min-sum k-clustering in metric spaces. In *Proceedings of the 33rd Annual ACM Symposium on Theory of Computing*, 11-20, 2001.
3. M. Charikar and R. Panigrahy. Clustering to minimize the sum of cluster diameters. In *Proceedings of the 33rd Annual ACM Symposium on Theory of Computing*, 1-10, 2001.
4. F. Chudak and D. Shmoys. Improved approximation algorithms for a capacitated facility location problem. In *Proceedings of the 10th Annual ACM-SIAM Symposium on Discrete Algorithms*, 875-876, 1999.
5. R. Diestel. *Graph Theory*, Springer-Verlag, Berlin, 2000.
6. M. Dyer and A. Frieze. A simple heuristic for the p-center problem. *Operations Research Letters*, 3(6):285-288, 1985.
7. J. Fakcharoenphol, C. Harrelson and S. Rao. The k-traveling repairman problem. In *Proceedings of the 10th Annual ACM-SIAM Symposium on Discrete Algorithms*, 655-664, 2003.
8. G.N. Frederickson, M.S. Hecht and C.E. Kim. Approximation algorithms for some routing problems. *SIAM J. Computing* 7:178-193, 1978.
9. M. Garey and D. Johnson. *Computers and Intractability: A Guide to the Theory of NP-Completeness*, W.H. Freeman, San Francisco, 1979.
10. N. Guttman-Beck and R. Hassin. Approximation algorithms for min-sum p-clustering. *Discrete Applied Mathematics*, 89:125-142, 1998.
11. M. Haimovich, A. Rinnooy Kan and L. Stougie. *Vehicle Routing: Methods and Studies*, Elsevier, 1988.
12. D. Hochbaum and D. Shmoys. A best possible approximation algorithm for the k-center problem. *Mathematics of Operations Research*, 10(2):180-184, 1985.
13. A. Levin, private communication, May 2003.
14. D. Shmoys, E. Tardos and K. Aardal. Approximation algorithms for facility location problems. In *Proceedings of the 29th Annual ACM Symposium on Theory of Computing*, 265-274, 1997.
15. D. Shmoys and E. Tardos. An approximation algorithm for the generalized assignment problem. *Mathematical Programming A*, 62:461-474, 1993.
16. P. Toth and D. Vigo (editors). *The Vehicle Routing Problem*, SIAM monographs on discrete mathematics and applications, 2002.

An Improved Decomposition Theorem for Graphs Excluding a Fixed Minor

Jittat Fakcharoenphol[1]* and Kunal Talwar[2]**

[1] Kasetsart University,
Bangkok, Thailand.
jtf@ku.ac.th
[2] Computer Science Division,
University of California, Berkeley
kunal@cs.berkeley.edu

Abstract. Given a graph G and a parameter δ, we want to decompose the graph into clusters of diameter δ without cutting too many edges. For any graph that excludes a $K_{r,r}$ minor, Klein, Plotkin and Rao [15] showed that this can be done while cutting only $O(r^3/\delta)$ fraction of the edges. This implies a bound on multicommodity max-flow min-cut ratio for such graphs. This result as well as the decomposition theorem have found numerous applications to approximation algorithms and metric embeddings for such graphs.

In this paper, we improve the above decomposition results from $O(r^3)$ to $O(r^2)$. This shows that for graphs excluding any minor of size r, the multicommodity max-flow min-cut ratio is at most $O(r^2)$ (for the uniform demand case). This also improves the performance guarantees of several applications of the decomposition theorem.

1 Introduction

A natural generalization of the s-t flow problem is the multicommodity flow problem, where we want to simultaneously route several commodities. Each commodity has a source and a sink, and the goal is to route the flows so that the total flow on any edge does not exceed its capacity. An optimization version of this problem is the *concurrent flow* problem, first defined by Shahrokhi and Matula [32], where we wish to maximize the *throughput* λ, such that we can feasibly route a λ fraction of each demand.

The sparsity of a cut $(S, \overline{S})$ is defined as $c(S, \overline{S})/d(S, \overline{S})$, where $c(S, \overline{S})$ is the sum of capacities of edges between S to $\overline{S}$ and $d(S, \overline{S})$ is the total demand from some source(sink) in S to a sink(source) in $\overline{S}$. The sparsity of any cut gives a upper bound on the maximum throughput. For the single commodity case, the max-flow min-cut theorem of Ford and Fulkerson [9] and of Elias, Feinstein and

* Reseach done while the author was a graduate student at UC Berkeley. Research partially supported by a DPST scholarship and NSF grant CCR-0105533.
** Research partially supported by NSF via grants CCR-0121555 and CCR-0105533.

S. Arora et al. (Eds.): APPROX 2003+RANDOM 2003, LNCS 2764, pp. 36–46, 2003.

Shannon [8], says that the maximum flow equals the value of the sparsest cut, and also gives an algorithm for finding the minimum cut.

The seminal work of Leighton and Rao [18] first considered approximate max-flow min-cut theorems. They showed that for the case of uniform demands, the ratio of sparsest cut to the maximum throughput in any graph is at most $O(\log n)$. Their proof also gives an algorithm to find a cut of sparsity no more than $O(\log n)$ times the maximum throughput (and hence at most $O(\log n)$ times the sparsest cut). This approximation algorithm is a basic subroutine for approximation algorithms for a variety of NP-hard problems.

For arbitrary demands, such an approximate max-flow min-cut theorem was discovered by Klein, Rao, Agrawal and Ravi [16], who showed an upper bound of $O(\log C \log D)$ where C is the sum of all capacities and D is the sum of all demands. This ratio has since been improved and the best currently known bound is $O(\log k)$, where k is the number of commodities, due to Linial, London and Rabinovich [19], and Aumann and Rabani [2] (see the related work section for details). For arbitrary graphs, this is the best(upto constants) that one can do, since an expander graph gives a matching lower bound.

Klein, Plotkin and Rao [15] considered restricted families of graphs, and showed for graphs excluding a minor of size r, the gap is $O(r^3)$ for the uniform demand case and $O(r^3 \log k)$ for the general case. The latter result was improved to $O(r^3 \sqrt{\log k})$ by Rao [26]. In particular, this showed that for planar graphs, which exclude K_5 and $K_{3,3}$ minors, the max-flow min-cut gap is $O(1)$ for the uniform case. Both the aforementioned results use a decomposition lemma proved in [15], which says that given a parameter δ, one can decompose a graph excluding a $K_{r,r}$ minor into clusters of diameter δ, while cutting only $O(r^3/\delta)$ fraction of the edges[3]. Note that any such decomposition of a path graph must cut an $O(\frac{1}{\delta})$ fraction of the edges, and thus the overhead for graphs excluding $K_{r,r}$ was shown to be $O(r^3)$. Not surprisingly, this decomposition lemma has found several other applications to approximation algorithms, distributed computing and embeddings results for such graphs.

In this paper we make some progress towards finding the right relation between the size of the forbidden minor and the overhead of such a decomposition. We show that for any graph excluding a K_r minor, we can find a decomposition into clusters of diameter δ while cutting only $O(r^2/\delta)$ fraction of the edges. This shows that the max-flow min-cut gap for such graphs is $O(r^2)$ for the uniform demands case and $O(r^2 \sqrt{\log n})$ for the general case. It also improves the performance guarantees of approximation algorithms and embeddings results for such graphs.

What is the right order of magnitude of the overhead of such a decomposition? An expander graph gives a lower bound of $\Omega(\log r)$, the upper bound we show is $O(r^2)$. Moreover, can we bound this overhead in terms of some other topological/metric properties of the graph? We leave open these intriguing questions.

---

[3] The second result actually requires the decomposition to have an additional "padding" property, details of which are deferred to the technical sections.

Related Work

As described above, Klein et.al. [16] gave the first non trivial upper bound of $O(\log C \log D)$ for multicommodity max-flow min-cut ratio for arbitrary demands. This was improved to $O(\log k^*)$ through the works of Tragoudas [34], Garg, Vazirani and Yannakakis [11], Plotkin and Tardos [23], Aumann and Rabani [2], Linial, London and Rabinovich [19], and Günlük [12] (k^* here is the size of the smallest vertex cover of the demand graph).

For several special classes of graphs, exact max-flow min-cut theorems have been proved, for example, by Hu [14], Rothschild and Whinston [29], Dinits(see [1]), Seymour [31], Lomonosov [20], Seymour [30] and Okamura and Seymour [22]. See [10] for more on this vein of work.

Network decomposition theorems like this one, are known for other classes of graphs as well. For general graphs, it is known that it suffices to cut an $O(\log n/\delta)$ fraction of the edges to decompose it into clusters of diameter δ, and this is the best one can do for general graphs. For graphs induced by real normed spaces $\mathbf{R}_p^d$, Charikar et.al. [7] show that such decompositions exist with an overhead of $O(d^{\frac{1}{p}})$ for $1 \leq p \leq 2$ and $O(d^{1-\frac{1}{p}})$ for $p > 2$, and that this is tight.

The characterization of planar graphs in terms of forbidden minors is due to Kuratowski [17]. Robertson and Seymour [28] showed that similar charcterizations exist for graphs of genus g for any g. In particular it is known that graphs of genus g exclude $K_{\Omega(\sqrt{g})}$ minor.

The approximate max-flow min-cut theorems have found numerous applications such as Oblivious routing, Data management, small area VLSI layout, efficient simulations of one interconnection network by another, etc. For more details on oblivious routing the reader is referred to the papers by Räcke [25], Azar et.al. [3], Bienkowski, Korzeniowski and Räcke [5], and Harrelson, Hildrum and Rao [13]. Data management applications have been looked at by Maggs et.al.[21]. The reader is referred to Bhatt and Leighton [4] for VLSI layout applications.

The decomposition theorem itself has found applications to approximation algorithms for various NP-hard problems. We mention a few of these applications here. Tardos and Vazirani [33] showed that the decomposition theorem implied an $O(r^3)$ bound on the max (total) flow-min multicut gap and an approximation algorithm for minimum multicut in graphs excluding a $K_{r,r}$ minor.

Rao and Richa [27] gave $O(r^3 \log \log n)$-approximation algorithms for minimum linear arrangement and minimum containing interval graph on graphs excluding K_r minor. Calinescu, Karloff and Rabani [6] gave an $O(r^3)$-approximation algorithm for the 0-extension problem on such graphs and Feige and Krauthgamer gave an $O(r^3 \log n)$-approximation algorithm to minimum bisection on such graphs.

A slight modification of these decompositions have also been used in the area of metric embeddings. Rao [26] showed that graphs excluding K_r minors can be embedded into l_2 with distortion $O(r^3 \sqrt{\log n})$. Moreover these embeddings preserve not only distances but also volumes. Recently, Rabinovich [24] showed how to embed a metric excluding K_r into a line with *average distortion* $O(r^3)$. For graphs with tree width r, they further improved the embedding to $O(\log r)$

and left open the question of the correct order for graphs excluding K_r minor. Our results improve the r^3 in all the above applications to r^2.

A Note on Techniques

The techniques used in this paper borrow generously from those used by Klein, Plotkin and Rao [15]. They showed that if their algorithm of repeatedly shattering BFS trees $O(r)$ times produced a cluster of large diameter, then they could construct a $K_{r,r}$ minor, consisting of r well spaced points in the large diameter cluster and the r roots of the BFS trees. We note that the roots of the BFS trees used were chosen arbitrarily.

Instead, we are somewhat more careful in our choice of the roots. We make sure that the roots of the BFS trees constructed are mutually far apart; this allows us to construct disjoint paths connecting these roots. This allows us to get a better guarantee on the diameter of the clusters.

2 Preliminaries

Let H and G be graphs. Suppose that for every vertex v of H, G contains a connected subgraph $\mathcal{A}(v)$ and for every edge (u, v) in H, there is an edge $\mathcal{E}(uv)$ connecting $\mathcal{A}(u)$ and $\mathcal{A}(v)$ in G. If the $\mathcal{A}(v)$'s are pairwise disjoint, we say that G contains an H-*minor* and call $\cup_v \mathcal{A}(v)$ *an H-minor of G*. We refer to the $\mathcal{A}(v)$'s as *supernodes* and $\mathcal{E}(uv)$'s as *superedges*.

We denote by K_h the complete graph on h nodes. Note that if G contains a K_h minor, it contains every minor on h vertices. Thus if G excludes any minor of size h, it excludes K_h. In particular, excluding a $K_{r,r}$ minor implies excluding a K_{2r} minor. Moreover, a $K_{r,r}$ contains a K_r minor. Thus upto a factor of 2, excluding a K_r minor and excluding a $K_{r,r}$ minor are equivalent.

Given a graph $G = (V, E)$, we can define a natural distance measure on V: $d_G(u, v)$ is the length of the shortest path from u to v. For a subset V' of V, the *weak* diameter of V' is defined to be $\max_{u,v \in V'}\{d_G(u, v)\}$. In this paper, the term diameter will always refer to weak diameter.

A δ-decomposition π of $G = (V, E)$ is a partition of V into subsets $V_1, V_2, \ldots,$ V_k such that each cluster V_i (defined as $\{v \in V : \pi(v) = i\}$) has (weak) diameter at most δ. An edge $e = (u, v)$ is said to be cut by this decomposition if u and v lie in different V_i's.

Let Π be a set of δ-decompositions of G and let $\mathcal{D}$ be a distribution over Π. We say $(\Pi, \mathcal{D})$ is α-padded if for any vertex v, and any $c < \frac{1}{2}$, the probability that v is at distance less than $c\delta$ from any cluster boundary is at most $2c\alpha$. More formally, for a partition π, let $d(v, \pi) = \min_{u:\pi(u) \neq \pi(v)} d(u, v)$. Then we say that $(\Pi, \mathcal{D})$ is α-padded if $\Pr_{\pi \in (\Pi, \mathcal{D})}[d(v, \pi) \leq c\delta] \leq 2c\alpha$. A probabilisitic version of the KPR decomposition was shown to be $O(r^3)$-padded in [26]. We shall show that our decomposition is $O(r^2)$-padded.

For ease of notation in the rest of the paper, we shall give an algorithm to construct an $O(r\delta)$-decomposition of the graph, which cuts $O(r/\delta)$ fraction of the edges, and is $O(r)$-padded. The result claimed in the introduction can of course be derived by scaling δ by a factor of $O(r)$.

3   The Decomposition Procedure

We decompose the graph recursively $r - 2$ times. At each level i, given a cluster G_i, we do the following. We pick, if possible, an *appropriate* node (explained in the next paragraph) a_i in G_i and construct a breadth first search tree rooted at a_i. We say a vertex v is at *level l* if its distance in G_i, from a_i is l. We partition the edges of G_i into δ classes. For $k = 0, 1, \ldots, \delta - 1$, the kth class consists of edges between nodes at level $j\delta + k$ and $j\delta + k + 1$ for some integer $j \geq 0$. We pick an integer $k \in \{0, \ldots, \delta - 1\}$ uniformly at random, and cut the edges in the kth class. We recurse on the resulting clusters.

By *appropriate* above, we mean a node which is at least distance $4r\delta$ far from each of roots of the breadth-first search trees in the higher levels of recursion. In case there is no such node in cluster G_i, we *shatter* the cluster in a different way - each cluster consisting of vertices close to one of the previous level roots.

Finally, we further shatter each resulting cluster G_{r-1} into at most $r - 1$ pieces by cutting out clusters of inappropriate nodes; for each of the centers $a_1, \ldots, a_{r-2}$, we cut out a set of vertices close to a_i to form a separate cluster. We redefine G_{r-1} to be the remaining set of nodes C'. The above procedures describe the set of edges that are cut; the final clusters are defined by the connected components of the remaining graph.

Figure 1 show the pseudocode of the procedures. We start by calling the procedure Decompose$(G_1 = G, 1, \{\})$.

4   Proof of the Decomposition Procedure

We first show that the decomposition constructed has the two properties that we needed.

Lemma 1. *The expected number of edges that are cut by the above procedure is $O(r|E(G)|/\delta)$.*

Proof. Note that we have at most r levels of recursion, and at most r cuts made in any shatter procedure. Thus at most $2r$ cuts potentially involve any particular edge. In each call to decompose or shatter, a fixed edge in the cluster has a probability at most $1/\delta$ of being cut (since it is at exactly one level, and we choose one of δ levels u.a.r.). Thus, any fixed edge has a probability at most $2r/\delta$ of being cut. The claim follows by linearity of expectation.

Lemma 2. *The decomposition produced is $2r$-padded.*

Proof. From the argument above, each cluster is produced as a result of at most $2r$ random cuts. Fix a vertex v and let Y_i be a random variable denoting its

Algorithm Decompose($G_i, i, p = \{a_1, \ldots, a_{i-1}\}$)

1.   **if** there exists $v \in G_i$ such that $d_G(a_j, v) \geq 4r\delta$ for all $1 \leq j \leq i - 1$ **then**

1.1 $a_i \leftarrow v$.

1.2   Create a BFS tree T_i in G_i rooted at a_i.

1.3   **if** T_i contains less than $\delta + 1$ level **then**

1.3.1   **stop.**

1.4   **for** $k = 0, 1, \ldots, \delta - 1$ **do**

1.4.1   Define the k-th cut S^k to be the set of edges between nodes
                             at level $j\delta + k$ and $j\delta + k + 1$ in T_i, for some $j \geq 0$.

1.5   Pick a k randomly in $0, 1, \ldots, \delta - 1$. Let $S = S_k$.

1.6   Cut all edges in S.

1.7   **for** each component G' in $G_i - S$ **do**

1.7.1   **if** $i < r - 2$ **then**

1.7.1.1   Decompose($G', i + 1, \{a_1, \ldots, a_{i-1}, a_i\}$).

1.7.2   **else**

1.7.2.1   Shatter($G', i, \{a_1, \ldots, a_{i-1}, a_i\}$).

2.   **else**

2.1   Shatter($G_i, i - 1, p$).

Procedure Shatter($C, k, p = \{a_1, \ldots, a_k\}$)

1. $C' \leftarrow C$.

2.   **for** $i = 1, \ldots, k$ **do**

2.1 $C_i \leftarrow$ all nodes v in C' such that $d_G(v, a_i) \leq 4r\delta$.

2.2   Create a breadth-first search tree T_i from nodes in C_i.

2.3   Let T_i' be the first $\delta + 1$ levels of T_i.

2.4   **if** T_i' covers all C' **then**

2.4.1 $C' \leftarrow \emptyset$.

2.5   **else**

2.5.1   Let j be chosen randomly in $0, 1, \ldots, \delta - 1$

2.5.2   Let T_i'' be a subtree of T_i up to level j.

2.5.2   Cut all edges at level j.

2.5.3 $C' \leftarrow C' - (C_i \cup T_i'')$.

Fig. 1. The decomposition procedures.

distance from the boundary of the ith cut. Clearly, $d(v, \pi) = \min_i Y_i$. Moreover, the ith cut was chosen uniformly at random from δ equispaced cuts, and thus Y_i is uniformly distributed in $[1, \delta/2]$. Hence $\Pr[Y_i \leq c\delta] \leq 2c$ for any $c \leq 1/2$. The claim then follows by a simple union bound.

Having established the required properties of the probabilistic decomposition, we now proceed to show that it is indeed an $O(r\delta)$-decomposition. Note that our decomposition consists of two kinds of clusters - those consisting of vertices close to some root, formed by some call to procedure shatter, and those formed by the procedure decompose. We first show that clusters of the first type have small diameters.

Lemma 3. *The procedure* Shatter *cuts out clusters each of weak diameter at most* $(8r + 2)\delta$.

Proof. For each $j = 1, \ldots, i-1$, we define the set C_j to be the set of all vertices in G_i which are at distance at most $4r\delta$ from a_j. The procedure cuts out cluster T_j'' formed by taking the set of vertices in C_j closer than some randomly chosen threshold $t \le \delta$ to a_j. Consider any pair of nodes u and v in the same connected component T_j'' in the resulting graph. It must be the case that there is some a_i such that the distance from u and v to a_i is at most $(4r + 1)\delta$ in G_1. Therefore by triangle inequality, the weak diameter of each such component is at most $(8r + 2)\delta$.

We now consider the remaining case. We wish to show that if the graph excludes a K_r minor, then the diameter of each such cluster resulting from our decomposition algorithm is small. We shall show the contrapositive - if the resulting decomposition has some cluster with large diameter, we shall show how to construct a K_r minor in the graph. Let G_{r-1} be a cluster of large diameter and let a_{r-1} and a_r be two vertices in G_{r-1} which are at least distance $4r\delta$ apart. We shall construct a K_r minor, containing a supernode centered at each a_i, for $i = 1, 2, \ldots, r$. We shall use the paths in the bfs trees to find superedges.

Lemma 4. *Suppose that a cluster* G_{r-1} *output by our algorithm has diameter* $4r\delta$. *Then* G_1 *contains a* K_r *minor.*

Proof. As above, denote by a_{r-1} and a_r two nodes in G_{r-1} at distance $4r\delta$ from each other. Note that by our construction, every pair of a_i and a_j is at least distance $4r\delta$ apart.

We shall show how to construct a K_r minor in G_1. We do so by reverse induction - we give a procedure which, for $b = r - 2, r - 3, \ldots, 1$, constructs a K_{r-b}-minor in G_{b+1}.

Recall that G_{i+1} consists of δ consecutive layers in the bfs tree T_i rooted at a_i. An *ancestor-path* of v in T_i is the path in T_i from v to the root a_i of T_i. We shall construct the minor using suitable ancestor-paths in T_i's.

Given a K_b-minor in G such that starting at each supernode $\mathcal{A}(g)$ there is path P_g, we say that the paths $\{P_g\}$ are *tails* if each path P_g is disjoint from the other paths and also from all supernodes except $\mathcal{A}(g)$. We shall refer to P_g's ending node (outside of $\mathcal{A}(g)$) as the *tip* of the tail P_g and denote it by $tip(P_g)$.

Klein, Plotkin and Rao [15] show how to construct the minor inductively by also constructing tails which are ancestor-paths of T_b and special nodes (which they called *middle nodes*) on the tails which are far apart, and using them to further construct disjoint components of the minor. We use a similar approach.

We shall construct a K_{r-b}-minor in G_{b+1}. In addition, we construct $r - b + 1$ tails $\{P_i\}$ which are ancestor-paths of T_b of length exactly 4δ such that for each tail P_i, a middle node h_i of P_i is at distance $4b\delta$ from the other middle nodes h_j's. Moreover, we require that every middle node is at distance at least $4b\delta$ from the root a_b of T_b. This shall be our (reverse) inductive claim.

For the basis step, when $b = r - 2$, let P be the shortest path from a_{r-1} to a_r in G_{r-1} (since G_{r-1} is connected, such a path exists). We construct a K_2-minor from the path P. We let $\mathcal{A}(a_r)$ be a path of length $4\delta - 1$ on P starting from a_r. The other supernode $\mathcal{A}(a_{r-1})$ is then $P - \mathcal{A}(a_r)$. We construct the tails by taking P_j to be the ancestor-paths in $\mathcal{T}_{r-2}$ of length 4δ from a_j, for $j \in \{r, r-1\}$. It can be checked that they are proper tails and the middle nodes in these tails are at distance at least $4(r-2)\delta$ from each other. Also the middle nodes in these tails are at distance at least $4(r-2)\delta$ from a_{r-2}.

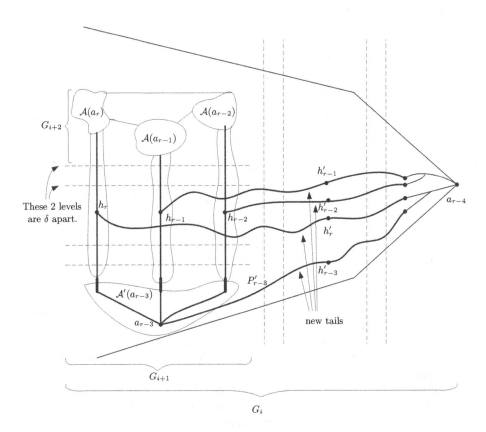

Fig. 2. The inductive step.

We now show the inductive step. Assuming that the claim is true for $b = i+1$, we want to show that the claim is true for $b = i$, i.e., G_{i+1} contains K_{r-i} as a minor and a new set of tails with the required properties.

We first construct the minor. For $j > i+1$, we create supernodes $\mathcal{A}'(a_j)$ from the supernodes of K_{r-i-1} as follows. We let $\mathcal{A}'(a_j)$ be $\mathcal{A}(a_j) \cup (P_j - \{tip(P_j)\})$. From the inductive assumption, these supernodes are disjoint. This gives us

$r - i - 1$ supernodes. We let $\mathcal{A}'(a_{i+1})$ be a union of all ancestor-paths in T_{i+1} starting from the tip of all the tails $\{P_j\}$.

We must show that $\mathcal{A}'(a_{i+1})$ is disjoint from all other new supernodes. Since we create tails of length 4δ from the ancestor-paths in T_{i+2}, the end nodes of the tails lie outside the subgraph G_{i+2}; therefore, the supernodes $\mathcal{A}(a_j)$, lying inside G_{i+2}, and $\mathcal{A}'(a_{i+1})$ are disjoint. Also, the tails $\{P_j\}$ and $\mathcal{A}'(a_{i+1})$ are disjoint by construction. Moreover, the last edges on the paths P_j give us the required additional superedges. This shows that G_{i+1} contains a K_{r-i}-minor.

To finish the inductive claim, we need to construct the tails with the desired properties. For each middle node h_j, let the tail P_j' be the ancestor-paths in T_i from h_j of length 4δ. These tails are mutually disjoint because h_j's are at distance at least $4(i+1)\delta$ from each other in G_i. We also create another tail P_i' starting from a_i in the same way. It is straightforward to verify that the new middle nodes $\{h_j'\}$ are at the right distance of each other.

We must also show that the tail P_j' are disjoint from all $\mathcal{A}'(a_k)$ where $k \neq j$. Consider any node v in $\mathcal{A}'(a_k)$. From the choice of h_j, the levels of v and h_j in T_{i+1} differ by more than δ. This implies that v does not lie on the ancestor-paths of h_j in T_i for any j[4] (since G_{i+1} consists of at most δ consecutive layers of T_i, there is a path of length at most δ from h_j to any T_i-ancestor (say w) of h_j lying in G_{i+1}. Thus h_j and w would be within δ layers of each other in T_{i+1} and hence v is different from w). Thus for any $j \neq k$, P_j' is disjoint from $\mathcal{A}'(a_k)$. To show that P_j' does not cross any tails P_k, we note that the distance between h_j and h_k is more than 6δ. Finally, since a_i is at distance $4(i+1)\delta$ from all the middle nodes h_j, the path P_i' is also a proper tail.

It only remains to show that the middles node h_j''s are at distance at least $4i\delta$ from a_{i-1}. From our construction a_{i-1} is at distance at least $4r\delta$ from a_j, where $j > i$. We know inductively that the new middle node h_j' are at distance at most $2(r-i)\delta$ from a_j. By triangle inequality then, the distance from h_j' and a_{i-1} is at least $4r\delta - 2(r-i)\delta \geq 4i\delta$. This completes the inductive argument.

Thus, when $b = 1$, the induction claim says that G_2 contains a K_{r-1}-minor and the tails with the appropriate properties. We can construct a K_r-minor in G_1 as in the inductive step. This completes the proof of Lemma 4.

From the above lemmas, we have the main theorem.

Theorem 1. *Given a graph G and parameters δ and r, we can either find a K_r minor in G or find a $O(r)$-padded $O(r\delta)$-probabilistic decomposition of the G which expects to cut at most $O(mr/\delta)$ edges.*

We can also generalize this procedure for graphs with distances and weights on the edges. Moreover, if the padding property is not required, we can easily derandomize the algorithm by picking the best cut at each step.

---

[4] This is exactly the "moat" argument in [15].

5   Acknowledgement

We would like to thank Satish Rao for several helpful discussions. We would also like to thank the anonymous referees for several helpful comments.

References

1. G. Adel'son-Vel'ski, E. Dinits, and A. Karzanov. *Flow Algorithms*. Nauka, Moscow, 1975. In Russian.
2. Y. Aumann and Y. Rabani. An $O(\log k)$ approximate min-cut max-flow theorem and approximation algorithm. *SIAM J. Comput.*, 27(1):291–301, 1998.
3. Y. Azar, E. Cohen, A. Fiat, H. Kaplan, and H. Räcke. Optimal oblivious routing in polynomial time. In *Proceedings of the thirty-fifth annual ACM symposium on Theory of computing*, 2003.
4. S. N. Bhatt and F. T. Leighton. A framework for solving VLSI graph layout problems. *Journal of Computer and System Sciences*, 28(2):300–343, Apr. 1984.
5. M. Bienkowski, M. Korzeniowski, and H. Räcke. A practical algorithm for constructing oblivious routing schemes. In *Fifteenth ACM Symposium on Parallelism in Algorithms and Architectures*, June 2003.
6. G. Calinescu, H. Karloff, and Y. Rabani. Approximation algorithms for the 0-Extension problem. In *Proceedings of the Twelfth Annual ACM-SIAM Symposium on Discrete Algorithms (SODA-01)*, pages 8–16, New York, Jan. 7–9 2001. ACM Press.
7. M. Charikar, C. Chekuri, A. Goel, S. Guha, and S. Plotkin. Approximating a finite metric by a small number of tree metrics. In IEEE, editor, *39th Annual Symposium on Foundations of Computer Science: proceedings: November 8–11, 1998, Palo Alto, California*, pages 379–388, 1109 Spring Street, Suite 300, Silver Spring, MD 20910, USA, 1998. IEEE Computer Society Press.
8. P. Elias, A. Feinstein, and C. E. Shannon. A note on the maximum flow through a network. *IEEE Trans. Inform. Th.*, IT-2:117–119, 1956.
9. L. R. Ford and D. R. Fulkerson. *Flows in Networks*. Princeton Univ. Press, Princeton, NJ, 1962.
10. A. Frank. Packing paths, circuits, and cuts - a survey. In B. Korte, L. Lovász, H.-J. Prömel, and A. Schrijver, editors, *Paths, Flows and VLSI-Layouts*, pages 47–100. Springer Verlag, 1990.
11. N. Garg, V. V. Vazirani, and M. Yannakakis. Approximate max-flow min-(multi)cut theorems and their applications. In *Proceedings of the twenty-fifth annual ACM symposium on Theory of computing*, pages 698–707. ACM Press, 1993.
12. O. Günlük. A new min-cut max-flow ratio for multicommodity flows. *Lecture Notes in Computer Science: Integer Programming and Combinatorial Optimization*, 2337:54–66, 2002.
13. C. Harrelson, K. Hildrum, and S. Rao. A polynomial-time tree decomposition to minimize congestion. In *Symposium on Parallel Algorithms and Architectures*, 2003.
14. T. Hu. Multicommodity network flows. *Operations Research*, 11:344–360, 1963.
15. P. Klein, S. A. Plotkin, and S. Rao. Excluded minors, network decomposition, and multicommodity flow. In *Proceedings of the twenty-fifth annual ACM symposium on Theory of computing*, pages 682–690. ACM Press, 1993.

16. P. N. Klein, S. Rao, A. Agrawal, and R. Ravi. An approximate max-flow min-cut relation for unidirected multicommodity flow, with applications. *Combinatorica*, 15(2):187–202, 1995.

17. K. Kuratowski. Sue le problème des courbes gauches en topologie. *Fund. Math.*, 15:217–283, 1930.

18. T. Leighton and S. Rao. An approximate max-flow min-cut theorem for uniform multicommodity flow problems with applications to approximation algorithms. In *29th Annual Symposium on Foundations of Computer Science*, pages 422–431, White Plains, New York, 24–26 Oct. 1988. IEEE.

19. N. Linial, E. London, and Y. Rabinovich. The geometry of graphs and some of its algorithmic applications. *COMBINAT: Combinatorica*, 15, 1995.

20. M. V. Lomonosov. Combinatorial approaches to multiflow problems. *Discrete Applied Math.*, 11:1–94, 1985.

21. B. M. Maggs, F. M. auf der Heide, , B. Vöcking, and M. Westermann. Exploiting locality for data management in systems of limited bandwidth. In *38th Annual Symposium on Foundations of Computer Science*, pages 284–293, Miami Beach, Florida, 20–22 Oct. 1997. IEEE.

22. H. Okamura and P. Seymour. Multicommodity flows in planar graphs. *Journal of Combinatorial Theory, Series B*, 31:75–81, 1981.

23. S. A. Plotkin and É. Tardos. Improved bounds on the max-flow min-cut ratio for multicommodity flows. In *ACM Symposium on Theory of Computing*, pages 691–697, 1993.

24. Y. Rabinovich. On average distortion of embedding metrics into l_1 and into the line yuri rabinovich. In *Proceedings of the thirty-fifth annual ACM symposium on Theory of computing*, 2003.

25. H. Räcke. Minimizing congestion in general networks. In *Proceedings of the 43rd Annual Symposium on the Foundations of Comuter Science*, pages 43–52, Nov. 2002.

26. S. Rao. Small distortion and volume preserving embeddings for planar and euclidean metrics. In *Proceedings of the fifteenth annual symposium on Computational geometry*, pages 300–306. ACM Press, 1999.

27. S. Rao and A. W. Richa. New approximation techniques for some ordering problems. In *Proceedings of the Ninth Annual ACM-SIAM Symposium on Discrete Algorithms*, pages 211–218, San Francisco, California, 25–27 Jan. 1998.

28. N. Robertson and P. D. Seymour. Graph minors. VIII. a Kuratowski theorem for general surfaces. *Journal of Combinatorial Theory Series B*, 48(2):255–288, 1990.

29. B. Rothschild and A. Whinston. On two commodity network flows. *Operations Res.*, 14:377–387, 1966.

30. P. Seymour. Matroids and multicommodity flows. *European Journal of Combinatorics*, 2:257–290, 1981.

31. P. D. Seymour. Four-terminus flows. *Networks*, 10:79–86, 1980.

32. F. Shahrokhi and D. W. Matula. The maximum concurrent flow problem. *Journal of the ACM (JACM)*, 37(2):318–334, 1990.

33. É. Tardos and V. Vazirani. Improved bounds for the max-flow min-multicut ratio for planar and $k_{r,r}$-free graphs. *Information Processing Letters*, 47:77–80, August 1993.

34. S. Tragoudas. *VLSI partitioning approximation algorithms based on multicommodity flow and other techniques*. PhD thesis, University of Texas, Dallas, 1991.

Approximation Algorithms for Channel Allocation Problems in Broadcast Networks

Rajiv Gandhi[1]*, Samir Khuller[2]**, Aravind Srinivasan[2]***, and Nan Wang[1]†

[1] Department of Computer Science,
University of Maryland, College Park, MD 20742.
`gandhi@cs.umd.edu, nwang@cs.umd.edu.`
[2] Department of Computer Science and Institute for Advanced Computer Studies,
University of Maryland, College Park, MD 20742.
`samir@cs.umd.edu, srin@cs.umd.edu.`

Abstract. We study two packing problems that arise in the area of dissemination-based information systems; a second theme is the study of *distributed* approximation algorithms. The problems considered have the property that the space occupied by a collection of objects together could be significantly less than the sum of the sizes of the individual objects. In the *Channel Allocation Problem*, there are users who request subsets of items. There are a fixed number of channels that can carry an arbitrary amount of information. Each user must get all of the requested items from one channel, i.e., all the data items of each request must be broadcast on some channel. The load on any channel is the number of items that are broadcast on that channel; the objective is to minimize the maximum load on any channel. We present approximation algorithms for this problem and also show that the problem is MAX-SNP hard. The second problem is the *Edge Partitioning Problem* addressed by Goldschmidt, Hochbaum, Levin, and Olinick (*Networks, 41:13-23, 2003*). Each channel here can deliver information to at most k users, and we aim to minimize the total load on all channels. We present an $O(n^{1/3})$–approximation algorithm and also show that the algorithm can be made fully distributed with the same approximation guarantee; we also generalize to the case of hypergraphs.

1 Introduction

We develop approximation algorithms for certain packing problems arising in broadcast systems; these have the property that the objects to be packed "overlap". In other words, the space occupied by a collection of objects together could be significantly less than the sum of the sizes of the individual objects. This is

---

* Research supported by NSF Award CCR-9820965.
** Research supported by NSF Award CCR-9820965 and an NSF CAREER Award CCR-9501355.
*** Research supported in part by NSF Award CCR-0208005.
† Research supported by NSF Award CCR-0208005.

S. Arora et al. (Eds.): APPROX 2003+RANDOM 2003, LNCS 2764, pp. 47–58, 2003.

in contrast with traditional packing problems in which the objects to be packed are disjoint. A second theme of our work is that some of our algorithms can also be made completely distributed and implemented to run in polylogarithmic time, with only a constant-factor loss in the approximation guarantee. We study problems that arise in the area of dissemination-based information systems [1,2,11,12,23]. Such systems are used in application domains such as public-safety systems, election-result servers and stock tickers [3]. One characteristic of dissemination-based applications is that there is a high degree of overlap in the user needs. Since many user-requests in such applications are similar, it would be a waste of resources to transmit the information to each user separately. For users with similar requests, if their requests are grouped and transmitted only once then this wastage of bandwidth could be avoided. On the negative side, the grouped data may contain information that would be irrelevant for some users. Hence, the users would have to process the broadcast information to obtain the data that they want. Thus, there is a trade-off between reducing the bandwidth used by grouping the requests and the amount of processing of the broadcast data that the clients need to do to obtain the data that they requested. In our model, there is a transmitter such as a satellite that broadcasts information on a fixed number of physical multicast channels. Each user is assigned to some channel on which the user gets his/her requested data. Our work deals with satisfying the client requests in a timely manner, while minimizing the amount of bandwidth used.

Problems and Results. The first problem, Channel Allocation, can be defined as follows. There is a set of topics (e.g., news, sports events, stock-market updates), as well as a set of users. Each user requests a subset of items (topics). There are a fixed number of channels that can each carry an arbitrary amount of information. Each user must get all of the requested items from one channel, i.e., all the data items of each request must be broadcast on some channel. The load on any channel is the number of items that are broadcast on that channel, and the goal is to minimize the maximum load on any channel. Formally, we are given: (i) a set of topics $T = \{t_1, t_2, \ldots, t_n\}$, (ii) a collection of user-requests $R = \{R_1, R_2, \ldots, R_m\}$, where $R_i \subseteq T$ for all i, and $\max_i |R_i|$ is a constant w; and (iii) a positive integer k denoting the number of channels. Our goal is to construct a family $C = \{C_1, C_2, \ldots, C_k\}, C_i \subseteq T$, such that for each set $R_i \in R$, there exists a C_j such that $R_i \subseteq C_j$. For all j, C_j constitutes the set of topics on channel j. If $R_i \subseteq C_j$ then we say that request R_i is satisfied by channel j. The load on channel j is the number of topics placed on it: i.e., $|C_j|$. The objective function is to minimize the maximum load on any channel, i.e., to minimize $\max_j |C_j|$. We will denote this problem as CHA.

The second problem, Edge-Partitioning (EP), basically arises by bounding the number of requests that any channel can handle, in CHA. The setting is the same as in CHA, with the additional constraint that each R_i must be assigned to some channel C_j for which $R_i \subseteq C_j$ holds; furthermore, the number of requests (i.e., users) assigned to a channel should be at most k. Subject to these constraints, the objective is to minimize $\sum_j |C_j|$. This problem was studied by Goldschmidt *et al.* [14] for the special case of $w = 2$, in the context of optical network design. (That is, given a graph G, we seek to cover the edges by subgraphs containing at most k edges each, and we aim to minimize the total number of vertices in the chosen subgraphs.) The work of [14] considers the case $w = 2$, and presents an $O(\sqrt{k})$–approximation algorithm.

We give an $O(n^{\frac{w-1}{w+1}}(\lg n)^{\frac{1}{w}})$–approximation algorithm for CHA; this is obtained by taking the better of a random construction and the output of a suitable set-cover problem. We also show that the problem is MAX-SNP hard for all $w \geq 4$; thus, a polynomial time approximation scheme for the problem would imply that $P = NP$. For the case $w = 2$, CHA is the following graph problem: cover all the edges of a given graph by a given number of subgraphs, minimizing the maximum number of vertices in these subgraphs. Here, we obtain an $O(n^{1/3-\epsilon})$–approximation algorithm for some positive constant ϵ. We also show that the problem is NP-hard for $w = 2$, even when there are only two channels.

For EP, we obtain an $O(w \cdot n^{\frac{w-1}{w+1}})$–approximation algorithm, by taking the better of a simple approach and a greedy algorithm. Recall that an $O(\sqrt{k})$–approximation algorithm was developed in [14] for the case $w = 2$; in this case, our bound of $O(n^{1/3})$ is incomparable with $O(\sqrt{k})$ (note that k can take on values from 1 up to m, the number of edges in the graph). We then present an alternative approach with the same approximation guarantee for the case $w = 2$, with the help of certain tail bounds for sums of correlated random variables [17,18,22]. We show that this can be implemented as a *polylogarithmic time, distributed* algorithm, where each arriving user only communicates with the servers handling the topics that the user is interested in. This brings us to the next main theme of this paper: that of *distributed* approximation algorithms. Given the emergence of various contexts where distributed agents (e.g., in the Internet) make decisions using only local information, it is natural to ask whether the notion of approximation algorithms can be brought to bear fruitfully in such contexts. Not many polylogarithmic-time distributed approximation algorithms are known: the few that we are aware of include [15,19,9]. We hope that the intriguing mix of approximation and the constraint of locality will be understood further by research in distributed approximation algorithms.

Related Work. A problem related to the ones we study is the well-known Dense k-Subgraph problem (DkS): given a graph G, select a subset of k vertices whose induced subgraph has the maximum number of edges. In the language of CHA, we have $w = 2$ and one channel with capacity k; we wish to satisfy the maximum number of user requests. This problem is NP-hard, and an $O(n^a)$–approximate solution for some $a < \frac{1}{3}$ was given by Feige *et al.* [10]. The problem is not even known to be MAX-SNP hard. Also, Daskin *et al.* [8] discuss the following related *printed circuit board (PCB) assembly problem*. In this problem we have a list of PCBs and a list of different component types required by each PCB. The machine that produces the PCBs can hold only a fixed number of different component types, and can be loaded any number of times. The goal here is to minimize the sum over all component types, of the number of times each component type is loaded. The users correspond to the PCBs, the items correspond to the different component types required by a PCB and the channel corresponds to the machine. In other words, the channel capacity is fixed, any number of channels could be used and the objective is to minimize the sum of the channel loads. They show that the problem is NP-hard. For the general version of the problem in which each component type (item) and PCB (user) is associated with a cost, they provide a heuristic solution. They also provide a branch-and-bound algorithm that can optimally solve small to moderate sized instances of the problem.

Due to the lack of space, many proofs are deferred to the full version.

2   The Channel Allocation Problem

2.1   Algorithm

Our approach employs two different algorithms and chooses a solution of lower cost from the two solutions obtained. As we will see, these two algorithms perform "well" on different sets of inputs that cover the entire spectrum of inputs.

The first algorithm is the following simple randomized algorithm. *Independently* place each topic on each channel, $i, 1 \leq i \leq k$, with a probability p which will be determined later. We will show that with a sufficiently high probability we obtain a feasible solution whose cost is close to its expected cost. This probability can be boosted by repeating the random process.

The second algorithm uses the greedy set cover algorithm [6,20,21] on the set cover instance, I, that is constructed as follows. The elements of the instance, I, are the requests in R. Let t be some fixed large constant. For all $i, 1 \leq i \leq \binom{t}{w}$, consider all $\binom{m}{i}$ combinations of i elements. For each combination, Z, let S_z be the set of requests corresponding to the elements in Z and let T_z be the topics obtained by taking the union of the requests in S_z. The combination Z forms a set in I iff $|T_z| \leq t$. The size of our set cover instance, $|I| = \sum_{j=1}^{t} \binom{|T|}{j} \leq \sum_{j=1}^{t} |T|^j = O(|T|^t) = O(n^t) = O(m^t)$. Let $M \doteq \max_{S_z \in I} \{|S_z|\} = O(t^w)$ be the size of the largest set in I. Since t and w are constants, $|I|$ is polynomially bounded and M is a constant. Now we use the greedy set cover algorithm on I to obtain a set cover for R. For each set S_z chosen by the set cover algorithm we create a new channel. The topics in T_z constitute this channel and hence the requests in S_z are satisfied by this channel. The set cover covers all requests in R. This solution may be infeasible as it may use more than k channels. By using Lemma 1 we can convert it into a feasible solution using k channels.

We will now analyze our algorithm. Note that we can obtain solutions with good approximation guarantees trivially for the following values of w and k. If $w = 1$ we can get an optimal solution of cost $\lceil m/k \rceil$. If $k < 2 \ln m$, we get a $2 \ln m$ approximation guarantee, since for any k we can obtain a k-approximate solution by placing all topics on each of the k channels. If $k > (\frac{n}{\ln n})^w$, we can partition the requests into groups of size $\lceil (\ln n)^w \rceil$ and place each group on a separate channel. This is a feasible solution as there are at most n^w requests. The cost of our solution is at most $O(w(\ln n)^w)$, thus giving an approximation guarantee of $O((\ln n)^w)$. For the rest of the analysis we will assume that $w \geq 2$ and $2\ln m \leq k \leq (\frac{n}{\ln n})^w$.

Let (X, y) solution to CHA denote allocating y channels such that the load on each of the channels is at most X.

Lemma 1. *If (L, k'), where $k' > k$, is a solution to CHA then there exists a feasible solution $(\lceil \frac{k'}{k} \rceil L, k)$.*

Lemma 2. *With a sufficiently high probability, the randomized algorithm gives a $(O(n(\frac{\lg n}{k})^{\frac{1}{w}}), k)$ solution.*

Lemma 3. *The set cover approach gives a $(O((L_{OPT})^w), k)$ solution.*

Theorem 1. *There is a polynomial-time algorithm for CHA that gives an $O(n^{\frac{w-1}{w+1}}(\lg n)^{\frac{1}{w}})$-approximate solution.*

2.2   Hardness of Approximation

We will prove that CHA is MAX-SNP hard via a reduction from 3-Dimensional Matching problem (3DM) which can be formally stated as follows.

3-Dimensional Matching (3DM): Given three disjoint set of elements, X, Y, and Z, such that $|X| = |Y| = |Z| = q$, and a set of triples, C, each triple containing one element from X, Y, and Z. The goal is to find the maximum number of pairwise disjoint triples.

For clarity, we prove Theorem 2 for all $w \geq 10$. We can show that CHA is MAX-SNP hard for all $w \geq 4$ by replacing each request, D, of size 10 in the reduction by $\binom{10}{4}$ requests of size 4, where each new request is a subset of D.

Theorem 2. *Unless $P = NP$, for some fixed $\epsilon > 0$, the channel allocation problem is hard to approximate to within a factor of $(1 + \epsilon)$.*

Proof. Let 3DM(I) denote the cost of an optimal solution to instance I. Similar definitions apply to CHA. 3DM is NP-complete [13]. We will prove the following.

$$I \in \text{3DM} \Longrightarrow \text{CHA}(f(I)) \leq 12 \tag{1}$$

$$I \notin \text{3DM} \Longrightarrow \text{CHA}(f(I)) \geq 13 \tag{2}$$

The function f shows that an approximation algorithm for CHA yielding a solution of cost lower than $\frac{13}{12}OPT$ would imply that P=NP. Our reduction is inspired by the NP-hardness reduction from 3DM to PARTITION INTO TRIANGLES [13].

Consider a 3DM instance I. For notational convenience we will drop the parameter I while using the symbols, for e.g., we will use C instead of $C(I)$ to denote the set of triples in I. We now describe the function f that converts I

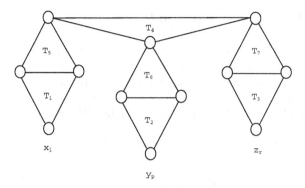

Fig. 1. Gadget corresponding to each triple.

into a CHA instance, $f(I)$, as follows. The CHA instance that we construct will have *big* requests, *small* requests and *dummy* requests. We start by describing the *big* requests. There are $(3q + 9|C|)$ big requests, one for each of the $3q$ elements in I and 9 for each triple in C. The big requests are mutually disjoint. Each big request, B, has 4 topics, $t_i^B, 1 \leq i \leq 4$. We will now describe the *small* requests. For each triple, $C_j \in C$, we will construct the gadget shown in Figure 1. Each gadget consists of 12 big requests (mentioned earlier), 9 of which are unique to the gadget and the other 3 big requests corresponding to the elements in C_j are shared between the gadgets corresponding to the triples containing the elements. Each edge connecting two big requests U and V represents 16 small requests, $\{t_i^U\} \cup \{t_j^V\}$, for all combinations of i, j for $1 \leq i, j \leq 4$. Thus each small request has size 2 and contains one topic from each of the two big requests. We also have $(144|C| - 48q)$ *dummy* requests of size 10 each. The dummy requests are mutually disjoint and disjoint from all other requests. This completes our description of requests. The set of topics is the union of the big requests and dummy requests. The total number of channels is $4q + 3(|C| - q) + 96q + 144(|C| - q) = q + 3|C| + 144|C| - 48q$.

Before we prove (1) and (2), let us define some notation. Consider a gadget representing a triple $C_j \in C$. Let $T_i^j, 1 \leq i \leq 7$, denote the set of 12 corresponding to the big requests that form the triangle T_i^j as seen in Figure 1. For notational convenience, we will drop the superscript j. Note that T_i denotes a set of topics as well as a triangle. The reference will be clear from the context in which it is used. A channel satisfying a triangle T_i would mean that the set of topics, T_i, is placed on the channel and hence the 3 big requests that form the vertices of the triangle and 48 small requests represented by the edges of the triangle are satisfied.

Claim. If $3DM(I) = q$ then $CHA(f(I)) \leq 12$.

Claim. If $3DM(I) < q$ then $CHA(f(I)) \geq 13$.

3   CHA Instances with Small Set-Size

In this section we consider the case of CHA instances when user requests are of size at most 2. In this case the user requests can be modeled as a graph in which the vertices represent the topics and the edges represent the user requests, i.e., an edge (i, j) would represent a user requesting topics i and j. The goal is to allocate channels while minimizing $\max_{1 \leq i \leq k} L_i$. We can show:

Theorem 3. *CHA is NP-hard when each request is of size two and there are two channels.*

We next give an approximation algorithm for CHA. Our algorithm uses the solution for the the Dense k-Subgraph problem (DkS) described in Section 1. Specifically, we use the approximation algorithm $\text{DkS}(G, k)$ due to [10]. Con-
Algorithm: Guess the optimal load by trying out all possible values. Con-sider a guess L. Invoke $\text{DkS}(G, L)$, which returns an approximate solution for the densest subgraph on L vertices. Place these L vertices returned by $\text{DkS}(G, L)$ onto a new channel. Remove all the covered edges from G. If any edges remain uncovered invoke DkS again.

It is not hard to show that we get an $O(\rho \lg n)$-approximate solution here, where ρ is the approximation guarantee of $\text{DkS}(G, k)$. Thus we have

Theorem 4. *For a certain constant $a < 1/3$, there is an $O(n^a \ln n)$-approximation algorithm for CHA.*

4   The Edge-Partitioning Problem

We now present approximation algorithms for EP: a sequential, deterministic algorithm in Section 4.1, and a distributed, randomized one in Section 4.2. We will throughout use hypergraph covering terminology: given a hypergraph $H = (V, E)$ with n vertices and m edges (each having a fixed number w of vertices), we wish to partition the edges into sets of at most k edges each, in order to minimize the sum of the total number of vertices in each set ("each set" here means "each block of the partition").

4.1   A Deterministic Algorithm

We now present a deterministic $O(w \cdot n^{\frac{w-1}{w+1}})$–approximation algorithm; see The-orem 5. Recall that the *degree* of a vertex in a hypergraph is the number of edges incident to it (i.e., contain it). Let $H = (V, E)$ be the given hypergraph. We start by considering the following greedy algorithm:

EDGE PARTITION($H = (V, E), k$)
1    $F \leftarrow \emptyset$
2   **While** $|E| > k$ **do**
3   Remove the isolated vertices from V

```
4          H' = (V', E') ← H = (V, E)
5          L ← ∅
6          While |E'| > k do
7               u ← a lowest degree vertex in H'
8               L ← {edges in E' that are incident to u}
9               V' ← V' \ {u}
10              E' ← E' \ L
11         End
12         R ← E' ∪ L
13         Arbitrarily remove some edges from R to make |R| = k
14         F ← F ∪ {R} (i.e., R is the set of edges assigned to a new channel)
15         H ← H\R
16    End
17    F ← F ∪ {E}
```

Lemma 4. *For each iteration of the outer* **while** *loop, the number of vertices in R is at most $w \left(\frac{k}{m'}\right)^{\frac{1}{w}} n' + 1 \simeq w \left(\frac{k}{m'}\right)^{\frac{1}{w}} n'$, where $n' = |V'|$, $m' = |E'|$ for the $H' = (V', E')$ being used in that iteration.*

Lemma 5. *The total number of vertices in the edge partition is at most $\frac{wn}{(1-1/w)} \left(\frac{m}{k}\right)^{1-1/w}$.*

Lemma 6. *The optimal solution has at least $\max\{n, \frac{w/e}{k^{1-1/w}} m\} \geq n^{1/w} \cdot \frac{(w/e)^{1-1/w}}{k^{(1-1/w)^2}} m^{1-1/w}$ vertices.*

Lemma 7. *From Lemmas 5 and 6, the approximation ratio of our algorithm is at most $\frac{w^2}{w-1} \left(\frac{en}{wk^{1/w}}\right)^{1-1/w}$.*

Note that in the case of graphs, i.e., $w = 2$, the approximation ratio of our algorithm is at most $4\sqrt{\frac{en}{2\sqrt{k}}}$. Also note that the constant factor of this ratio can be improved in the analysis for $w = 2$. The algorithm of [14] works for $w = 2$, and their approximation ratio for $w = 2$ is about $\sqrt{\frac{k}{2}}$.

Lemma 8. *By partitioning E into m parts such that each part consists of exactly one edge, we obtain a trivial algorithm whose approximation ratio is at most $ek^{1-1/w}$.*

Theorem 5. *By running the first algorithm and the trivial algorithm and taking the best solution, we obtain an algorithm with approximation ratio at most $2w \cdot n^{\frac{w-1}{w+1}}$. The running time of the composite algorithm is $O(\frac{m}{k}(m+n))$.*

4.2   A Distributed Algorithm for the Graph Case

We now present a randomized distributed $O(n^{1/3})$–approximation algorithm for
the case where the given hypergraph H is a graph $G = (V, E)$. Recall that
in the present case where $w = 2$, each user basically requests two topics. We
consider a fully distributed model where each broadcast channel has a server
running it, and where each topic also has its own distribution server. A topic-
distribution server can communicate with a channel server, if the former wants
its topic broadcast on that channel. Each arriving user communicates only with
the two topic-distribution servers of interest to it; thus, the model is distributed
in the sense that the users need not have any knowledge about each other. By
interpreting the topics as vertices and as the two topics of interest to a user
as an edge, we thus equivalently get the following familiar distributed point-
to-point model. Each vertex in the graph $G = (V, E)$ has a processor which
can communicate with its neighbors, as well as with the servers handling the
channels. Each processor knows the values of n (which is a static parameter –
the number of topics) and k. We now wish to assign each edge to one channel
(from among an arbitrary number of channels), such that each channel has at
most k edges assigned to it. (The two processors at the end-point of an edge
co-operatively decide which channel that edge gets assigned to.) The goal is to
minimize the sum, over all channels i, of the total number of vertices that use
i (a vertex v uses i iff some edge incident to v is assigned to i). Computation
proceeds in *rounds*: in each round, every node communicates with its neighbors,
and updates its internal state. The running time of an algorithm is the number
of rounds, and hence locality is the main constraint in this model; we aim for
polylogarithmic-time algorithms.

   We further distinguish two models: *strong* and *weak*. In the weak model, if a
channel has more than k edges that attempt to get assigned to it, the channel
sends back a "No" message to the end-points of these edges, after which the
end-points can retry. In the strong model, even such attempts are disallowed,
and if we ever attempt to send more than k edges to a channel, the system
enters a "Failed" state. Such a strongly constrained model is less realistic than
the weak model – in practice, a channel can typically report that it is getting
overloaded, without crashing. However, we also study the strong model and show
that if all nodes know the value of m (which can be obtained if each incoming
user "registers" with a central server which broadcasts the value of m to all
servers), then we can develop an $O(n^{1/3})$–approximation algorithm even for the
strong model. (There is a positive probability of entering the Failed state in
our algorithm for the strong model – indeed, this seems inevitable – but this
probability can be made as small as n^{-c} for any desired constant c.) In the weak
model, the processors need not know the value of m.

The algorithm. We first assume the strong model, where the value of m is
known to all nodes; we will finally show how to translate our results to the
weak model. As in Section 4.1, there is the "trivial algorithm" (which places at
most k edges arbitrarily on each channel) whose total objective function value
is at most $2m$. The trivial algorithm can be easily implemented in the strong

model with a contention-resolution type algorithm, where each edge chooses to be assigned to each channel independently with a suitable probability p. Briefly, if $k \geq \log^2 n$, we take, say, $4(m/k) \log n$ channels and $p = k/(2m)$; each edge tries each channel with probability p, and goes to the first one on which its trial came up 1. If $k < \log^2 n$, we just take $p = n^{-c}$ and take $(4/p) \log n$ channels, for a suitable constant c. It is easy to show that with high probability, we get a feasible solution with the desired objective function value of $O(m)$. Like in Section 4.1, our focus is on showing how to construct a feasible solution with objective function value $O(n\sqrt{m/k})$; taking the better of this solution and that of the trivial algorithm, will yield an $O(n^{1/3})$–approximation. For the rest of this discussion, we assume $k \geq \log^4 n$, say; if k is smaller, the above trivial algorithm already results in a poylog(n) approximation.

The heart of our algorithm is the following: a preprocessing step followed by a random-selection step. Define $\bar{d} = \lceil \frac{2m}{n} \rceil$, and let $deg(v)$ be the current degree of v. The preprocessing step is as follows; it basically ensures that the maximum degree is not much more than the average degree. Each $v \in V$ makes $\lceil \frac{deg(v)}{\bar{d}} \rceil$ virtual copies of itself; it then distributes its $deg(v)$ incident edges to these copies, so that no copy gets more than $\bar{d}$ edges. Thus we get a new graph with m edges, and maximum degree $\bar{d}$. It is easy to see that the new number of vertices is at most $2n$. So, we have a graph with number of vertices in the range $[n, 2n]$, which has m edges and maximum degree at most $2m/n$. Now, the random-selection step is as follows. Choose am/k new channels, where a is a suitable constant. Each vertex then independently goes into each of these channels with probability $p = \sqrt{k/(2m)}$. (More precisely, the choices for all virtual copies of an original vertex v, are all made independently by v.) An edge is assigned to a channel iff both of its end-points choose to go into that channel; if an edge gets assigned to more than one channel, it chooses one arbitrarily.

The above preprocessing and random-selection constitute the main iteration of the algorithm. Note that the expected number of edges on any channel is $k/2$, and that for any edge, the probability that it was assigned to at least one of the channels is $1 - (1 - k/(2m))^{am/k} \approx 1 - e^{-a/2}$. The expected total load on the channels is $a(m/k) \cdot np = an\sqrt{m/(2k)}$. If everything happens according to expectation, we would have covered a constant fraction $b \sim 1 - e^{-a/2}$ of the edges, at a total cost of $\Theta(n\sqrt{m/k})$. We can then try to iterate this argument on the residual graph, leading to a toal cost of

$$\Theta\left(\sum_{i \geq 0} n\sqrt{m(1-b)^i/k}\right) = \Theta(n\sqrt{m/k}); \tag{3}$$

furthermore, the running time is basically the number of iterations, which would be $O(\log m) = O(\log n)$ with high probability.

The above idea on total cost can be carried through by using the Chernoff-Hoeffding bounds [5,16]. However, bounding the number of edges assigned to a channel is harder, due to correlations; moreover, the correlation among the edges is in the "wrong" direction, as far as proving a concentration of measure is concerned. This is where our preprocessing step helps; intuitively, since it eliminates high-degree vertices, the correlation among the edges is lessened. First of all, to lower-bound the number of edges assigned to any channel, we use Janson's inequality for lower-tail bounds [17]. Fix a particular channel. Let $X_e = 1$ be the event that edge e is assigned to that channel, and $X_e = 0$ otherwise. Define $e \sim f$ iff the edges e and f are different, and have a common end-point. Then, $\Delta \doteq \sum_{(e,f):\ e \sim f} \Pr[X_e = X_f = 1]$ can be bounded by $O(n(\bar{d})^2 p^3) = O(k^{3/2})$; this is because there are $O(n(\bar{d})^2)$ pairs (e,f) such that $e \sim f$, and because $\Pr[X_e = X_f = 1] = p^3$ for any such pair. Thus, by Janson's inequality, the probability that at most $k/4$ edges get assigned to that channel is at most $e^{-\Omega(\mu/(2+\Delta/\mu))}$, which is $e^{-\Omega(\sqrt{k})}$. Since we have assumed that $k \geq \log^4 n$, this is negligibly small. Next, in order to follow the constraint of the strong model, we also need to show that at most k edges get assigned to the channel. Such upper-tail bounds are usually harder, but the recent tail bounds of [18,22] can be shown to help; they help show that the probability of more than k edges getting assigned to a channel is once again at most $e^{-\Omega(\sqrt{k})}$. (The fact that our preprocessing significantly reduces the maximum degree, once again plays a key role.)

The above brief sketch shows that "everything relevant happens nearly according to expectation", with high probability. The nodes no longer know the exact value of m after one or more iterations, but choose an estimate slightly larger than expectation, and repeat. We can now use the argument following (3) to claim our performance bounds, and this concludes our brief discussion of the main ideas. Finally, for the weak model, we do not know the value of m, but guess it by repeated doubling. More precisely, we first run the above protocol for the strong model assuming $m = 2$; for each surviving edge, its end-points then run the above protocol for $m = 4$, and so on. When we finally hit the correct value of m, we will terminate with high probability. Since the cost function in (3) is proportional to $\sqrt{m}$, our final cost now is just a constant times that of (3) with high probability; the running time remains polylogarithmic.

Acknowledgments. We thank Michael Franklin, Guy Kortsarz, Vincenzo Liberatore, Christine Piatko, I-Jeng Wang and An Zhu for useful discussions. The third author thanks the Institute for Mathematics and its Applications at the University of Minnesota for its pleasant hospitality during his visit in April 2003; part of this author's writing took place during this visit. We also thank the AP-PROX 2003 referees for their useful comments.

References

1. S. Acharya, R. Alonso, M. Franklin and S. Zdonik. Broadcast Disks: Data management for asymmetric communication environments. Proc. ACM SIGMOD International Conference on Management of Data, San Jose, CA., 1995.
2. S. Acharya, M. Franklin and S. Zdonik. Balancing push and pull for data broadcast. Proc. ACM SIGMOD International Conference on Management of Data, Tuscon, AZ., 1997.

3. D. Aksoy, M. Altinel, R. Bose, U. Cetintemel, M. Franklin, J. Wang and S. Zdonik. Research in Data Broadcast and Dissemination. *International Conference on Advanced Multimedia Content Processing (AMCP)*, Osaka, Japan, 1998.
4. R. Bhatia. Approximation Algorithms for Scheduling Problems. Ph.D. Thesis, University of Maryland at College Park, 1998.
5. H. Chernoff. A measure of asymptotic efficiency for tests of a hypothesis based on the sum of observations. *Annals of Mathematical Statistics*, 23:493-509, 1952.
6. V. Chvátal. A greedy heuristic for the set-covering problem. *Math. of Oper. Res.* Vol. 4, 3, 233-235, 1979.
7. A. Crespo, O. Buyukkokten and H. Garcia-Molina. Efficient Query Processing in a Multicast Environment. *Proceedings of the 16th International Conference on Data Engineering (ICDE)*, San Diego, 2000.
8. M. S. Daskin, O. Maimon, A. Shtub and D. Braha. Grouping components in printed circuit board assembly with limited component staging capacity and single card setup: problem characteristics and solution procedures. *International Journal of Production Research*, 35, 1617-1638, 1997.
9. D. Dubhashi, A. Mei, A. Panconesi, J. Radhakrishnan, and A. Srinivasan. Fast Distributed Algorithms for (Weakly) Connected Dominating Sets and Linear-Size Skeletons. *Proc. ACM-SIAM Symposium on Discrete Algorithms*, pages 717–724, 2003.
10. U. Feige, G. Kortsarz and D. Peleg. The Dense k-Subgraph Problem. *Algorithmica* 29, 410-421, 2001.
11. M. Franklin and S. Zdonik. A framework for scalable dissemination-based systems. *Proc. Object Oriented Programming Systems, Languages and Applications, OOPSLA*, 1997.
12. M. Franklin and S. Zdonik. "Data in your face": push technology in perspective. *Proceedings of ACM SIGMOD International Conference on Management of Data*, 1998.
13. M. R. Garey and D. S. Johnson. *Computers and Intractability: A Guide to the Theory of NP-Completeness*. W. H. Freeman and Company, 1979.
14. O. Goldschmidt, D. Hochbaum, A. Levin and E. Olinick. The SONET Edge-Partition Problem. *Networks*, 41:13-23, 2003.
15. D. A. Grable and A. Panconesi. Nearly optimal distributed edge coloring in $O(\log \log n)$ rounds. *Random Structures & Algorithms*, 10:385–405, 1997.
16. W. Hoeffding. Probability inequalities for sums of bounded random variables. *American Statistical Association Journal*, 58:13-30, 1963.
17. S. Janson. Poisson approximations for large deviations. *Random Structures & Algorithms*, 1:221–230, 1990.
18. S. Janson and A. Ruciński. The deletion method for upper tail estimates. Technical Report 2000:28, Department of Mathematics, Uppsala University, Sweden, 2000.
19. L. Jia, R. Rajaraman and T. Suel. An Efficient Distributed Algorithm for Constructing Small Dominating Sets. *Proc. ACM Symposium on Principles of Distributed Computing*, pages 33–42, 2001.
20. D. S. Johnson. Approximation Algorithms for Combinatorial Problems. *Journal of Computer and System Sciences*, 9:256–278, 1974.
21. L. Lovász. On the ratio of optimal integral and fractional covers. *Discrete Mathematics*, 13:383–390, 1975.
22. V. H. Vu. Concentration of non-Lipschitz functions and applications. *Random Structures & Algorithms*, 20:262–316, 2002.
23. J. Wong. Broadcast Delivery. In *Proc. of the IEEE*, 76(12), 1988.

Asymmetry in k-Center Variants

Inge Li Gørtz[1][*] and Anthony Wirth[2][**]

[1] Theory Department, The IT University of Copenhagen, Denmark.
inge@it-c.dk.
[2] Department of Computer Science, Princeton University.
awirth@cs.princeton.edu.

Abstract. This paper explores three concepts: the k-center problem, some of its variants, and asymmetry. The k-center problem is a fundamental clustering problem, similar to the k-median problem. Variants of k-center may more accurately model real-life problems than the original formulation. Asymmetry is a significant impediment to approximation in many graph problems, such as k-center, facility location, k-median and the TSP.

We demonstrate an $O(\log^* n)$-approximation algorithm for the asymmetric *weighted* k-center problem. Here, the vertices have weights and we are given a total budget for opening centers. In the *p-neighbor* variant each vertex must have p (unweighted) centers nearby: we give an $O(\log^* k)$-bicriteria algorithm using $2k$ centers, for small p.

Finally, the following three versions of the asymmetric k-center problem we show to be inapproximable: *priority k-center*, *k-supplier*, and *outliers with forbidden centers*.

1 Introduction

Imagine you have a delivery service. You want to place your delivery hubs at locations that minimize the maximum distance between customers and their nearest hubs. This is the *k-center problem*—a type of clustering problem that is similar to the facility location and k-median problems. The motivation for the *asymmetric* k-center problem, in our example, is that traffic patterns or one-way streets might cause the travel time from one point to another to differ depending on the direction of travel. Traditionally, the k-center problem was solved in the context of a metric; in this paper we retain the triangle inequality, but abandon the symmetry.

Symmetry is a vital concept in graph approximation algorithms. Very recently, the k-center problem was shown to be $\Omega(\log^* n)$ hard to approximate [6, 7], even though the symmetric version has a factor 2 approximation. Moreover, facility location and k-median both have constant factor algorithms in the

---

[*] Part of this work was performed while visiting Princeton University.

[**] Supported by a Gordon Wu Fellowship, a DIMACS Summer Research Fellowship, and NSF ITR grant CCR-0205594.

S. Arora et al. (Eds.): APPROX 2003+RANDOM 2003, LNCS 2764, pp. 59–70, 2003.

symmetric case, but are provably $\Omega(\log n)$ hard to approximate without symmetry [1]. The traveling salesman problem is a little better, in that no $\Omega(\log n)$ hardness is known, but without symmetry no algorithm better than $O(\log n)$ has been found either.

Definition 1 (k-Center). *Given $G = (V, E)$, a complete graph with nonnegative (but possibly infinite) edge costs and a positive integer k, find a set S of k vertices, called* centers, *with minimum covering radius. The covering radius of a set S is the minimum distance R such that every vertex in V is within distance R of some vertex in S.*

Kariv and Hakimi [11] showed that the k-center problem is NP-hard. Without the triangle inequality the problem is NP-hard to approximate; we henceforth assume that the edge costs satisfy the triangle inequality.

The asymmetric k-center problem has proven to be much more difficult to understand than its symmetric counterpart. Hsu and Nemhauser [10] showed that the k-center problem cannot be approximated within a factor of $(2 - \epsilon)$ unless $P = NP$. In 1985 Hochbaum and Shmoys [8] provided a (best possible) factor 2 algorithm for the symmetric k-center problem. In 1996 Panigrahy and Vishwanathan [16, 13] gave the first approximation algorithm for the asymmetric problem, with factor $O(\log^* n)$. Archer [2] proposed two $O(\log^* k)$ algorithms based on many of the ideas in [13]. We now know [6, 7] that these algorithms are asympotitcally the best possible.

Variants of the k-Center Problem A number of variants of the k-center problem have been explored in the context of symmetric graphs. Perhaps some delivery hubs are more expensive to establish than others: instead of a restriction on the number of centers we can use, each vertex has a weight and we have a budget W, that limits the total weight of centers. Hochbaum and Shmoys [9] produced a factor 3 algorithm for this *weighted k-center problem*. This has recently been shown to be tight [6].

Hochbaum and Shmoys [9] also studied the *k-supplier problem* where the vertex set is segregated into suppliers and customers. Only supplier vertices can be centers and only the customer vertices need to be covered. Hochbaum and Shmoys gave a 3-approximation algorithm and showed that this is the best possible.

Khuller et al. [12] investigated the *p-neighbor k-center problem* where each vertex must have p centers nearby. This problem is motivated by need to account for facility failures: even if up to $p - 1$ facilities fail, every demand point has a functioning facility nearby. They gave a 3-approximation algorithm for all p, and a best possible 2-approximation algorithm when $p < 4$, noting that the case where p is small is "perhaps the practically interesting case".

Perhaps some demand points are more important than others. Plesnik [14] studied the *priority k-center problem*, in which the effective distance to a demand point is increased in proportion to its specified priority. Plesnik approximates the symmetric version within a factor of 2.

Charikar et al. [4] note that a disadvantage of the standard k-center formulation is that a few distant clients, *outliers*, can force centers to be located in isolated places. They suggest a variant of the problem, the k-center problem with *outliers and forbidden centers*, where a small subset of clients can be denied service, and some points are forbidden from being centers. Charikar et al. gave a (best possible) 3-approximation algorithm for the symmetric version of this problem.

Bhatia et al. [3] considered a network model, such as a city street network, in which the traversal time change as the day progresses. This is known as the *k-center problem with dynamic distances*: we wish to assign the centers such that the objective criteria are met at all times.

Results and Organization

Table 1 gives an overview of the best known results for the various k-center problems. In this paper we explore asymmetric variants that are not yet in the literature.

Table 1. An overview of the approximation results for k-center variants. †β is the maximum ratio of an edge's greatest length to its shortest length. ‡This is a bicriteria algorithm using $k(1 + 3/(\nu + 1))$ centers. §For $p < 4$. ¶This is a bicriteria algorithm using $2k$ centers, for $p \leq n/k$

Problem	Symmetric		Asymmetric	
k-center	2	[8]	$O(\log^* k)$	[2]
k-center with dynamic distances	$1 + \beta$ †	[3]	$O(\log^* n + \nu)$ ‡	[3]
weighted k-center	3	[9]	$O(\log^* n)$	**Here**
p-neighbor k-center	3 (2 §)	[5]	$O(\log^* k)$ ¶	**Here**
priority k-center	2	[14]	**Inapproximable**	**Here**
k-center with outliers and forbidden centers	3	[4]	**Inapproximable**	**Here**
k-suppliers	3	[9]	**Inapproximable**	**Here**

Section 2 contains the definitions and notation required to develop the results. In Section 3 we briefly review the algorithms of Panigrahy and Vishwanathan [13], and Archer [2]. The techniques used in the standard k-center problem are often applicable to the variants.

Our first result, in Section 4, is an $O(\log^* n)$-approximation for the asymmetric weighted k-center problem. In Section 5 we develop an $O(\log^* k)$ approximation for the asymmetric p-neighbor k-center problem, for $p \leq n/k$. As noted by Khuller et al. [12], the case where p is small is the most interesting case in practice. This a bicriteria algorithm, allowing an increase to $2k$ centers, but it can be turned into an $O(\log k)$-approximation algorithm using only k centers. Turning to hardness, we show that the asymmetric versions of the k-center problem with outliers (and forbidden centers), the priority k-center problem, and the k-supplier problem are NP-hard to approximate (Section 6).

2   Definitions

The input to the asymmetric k-center problem is a distance function d on *ordered* pairs of vertices—distances are allowed to be infinite—and a bound k on the number of centers.

Definition 2. *Vertex c covers vertex v within r, or c r-covers v, if $d_{cv} \leq r$. We extend this definition to a set C and a set A if for every $a \in A$ there is a $c \in C$ such that c covers a within r. Often we abbreviate "1-covers" to "covers".*

Most of the algorithms do not in fact operate on graphs with edge costs. Rather, they consider restricted graphs, in which only those edges with distance lower than some threshold are included, and the edges have unit cost. Hochbaum and Shmoys [9] refer to these as bottleneck graphs. Since the optimal value of the covering radius must be one of the n^2 distance values, many algorithms essentially run through a sequence of restricted graphs of every possible threshold radius in ascending order. This can be thought of as *guessing* the optimal radius R_{OPT}. The approach works because the algorithm either returns a solution, within the specified factor of the current threshold radius, or it fails, in which case R_{OPT} must be greater than the current radius.

Definition 3 (Restricted Graph G_r). *For $r > 0$, define the restricted graph G_r of the graph $G = (V, E)$ to be the graph $G_r = (V, E_r)$, where $E_r = \{(i, j) : d_{ij} \leq r\}$ and all edges have unit cost.*

Most of the following definitions apply to *restricted* graphs.

Definition 4 (Power of Graphs). *The tth power of a graph $G = (V, E)$ is the graph $G^t = (V, E^t)$, $t > 1$, where E^t is the set of edges between distinct vertices that have a path of at most t edges between them in G.*

Definition 5. *For $i \in \mathbb{N}$ define $\Gamma_i^+(v) = \{u \in G \mid (v, u) \in E^i\} \cup \{v\}$, and $\Gamma_i^-(v) = \{u \in G \mid (u, v) \in E^i\} \cup \{v\}$, i.e. in the restricted graph there is a path of length at most i from v to u, respectively u to v.*

Notice that in a symmetric graph $\Gamma_i^+(v) = \Gamma_i^-(v)$. We extend this notation to sets so that $\Gamma_i^+(S) = \{u \in G \mid u \in \Gamma_i^+(v) \text{ for some } v \in S\}$, with $\Gamma_i^-(S)$ defined similarly. We use $\Gamma^+(v)$ and $\Gamma^-(v)$ instead of $\Gamma_1^+(v)$ and $\Gamma_1^-(v)$.

Definition 6. *For $i \in \mathbb{N}$ define $\Upsilon_i^+(v) = \Gamma_i^+(v) \setminus \Gamma_{i-1}^+(v)$, and $\Upsilon_i^-(v) = \Gamma_i^-(v) \setminus \Gamma_{i-1}^-(v)$, i.e., the nodes for which the path distance from v is exactly i, and the nodes for which the path distance to v is exactly i, respectively.*

For a set S, the extension follows the pattern $\Upsilon_i^+(S) = \Gamma_i^+(S) \setminus \Gamma_{i-1}^+(S)$. We use $\Upsilon^+(v)$ and $\Upsilon^-(v)$ instead of $\Upsilon_1^+(v)$ and $\Upsilon_1^-(v)$.

Definition 7 (Center Capturing Vertex (CCV)). *A vertex v is a* center capturing vertex *(CCV) if $\Gamma^-(v) \subseteq \Gamma^+(v)$, i.e., v covers every vertex that covers v.*

In the graph $G_{R_{OPT}}$ the optimum center that covers v must lie in $\Gamma^-(v)$. For a CCV v, it lies in $\Gamma^+(v)$, hence the name. In symmetric graphs all vertices are CCVs and this property leads to the standard 2-approximation.

Definition 8 (Dominating Set). *Given a graph $G = (V, E)$, and a weight function $w : V \to \mathbb{Q}^+$ on the vertices, find a minimum weight subset $D \subseteq V$ such that every vertex $v \in V$ is covered by D, i.e., $v \in \Gamma^+(D)$ for all $v \in V$.*

Definition 9 (Set Cover). *Given a universe $\mathcal{U}$ of n elements, a collection $\mathcal{S} = \{S_1, \ldots, S_k\}$ of subsets of $\mathcal{U}$, and a weight function $w : \mathcal{S} \to \mathbb{Q}^+$, find a minimum weight sub-collection of $\mathcal{S}$ that includes all elements of $\mathcal{U}$.*

The *Max Coverage* problem, on an instance $\langle \mathcal{U}, \mathcal{S}, k \rangle$, is similar to the Set Cover problem: instead of trying to *minimize* the number of sets used we have a *bound* on the number of sets we can use, and the problem is then to maximize the number of elements covered. The Dominating Set, Set Cover, and Max Coverage problems are all NP-complete.

3   Asymmetric k-Center Review

The $O(\log^* n)$ algorithm of Panigrahy and Vishwanathan [13] has two phases, the *halve* phase, sometimes called the *reduce* phase, and the *augment* phase. As described above, the algorithm guesses R_{OPT}, and works in the restricted graph $G_{R_{OPT}}$. In the halve phase we find a CCV v, include it in the set of centers, mark every vertex in $\Gamma_2^+(v)$ as covered, and repeat until no CCVs remain unmarked. The CCV property ensures that, as each CCV is found, the rest of the graph can be covered with one fewer center. Hence if k'' CCVs are obtained, the unmarked portion of the graph can be covered with $k' = k - k''$ centers. The authors then prove that this unmarked portion, CCV-free, can be covered with only $k'/2$ centers if we use radius 5 instead of 1. That is to say, $k'/2$ centers suffice in the graph $G_{R_{OPT}}^5$.

The k-center problem in the restricted graph is identical to the dominating set problem. This is a special case of set cover in which the sets are the Γ^+ terms. In the augment phase, the algorithm recursively uses the greedy set cover procedure. Since the optimal cover uses at most $k'/2$ centers, the first cover has size at most $\frac{k'}{2} \log \frac{2n}{k'}$.

The centers in this first cover are themselves covered, using the greedy set cover procedure, then the centers in the second cover, and so forth. After $O(\log^* n)$ iterations the algorithm finds a set of at most k' vertices that, together with the CCVs, $O(\log^* n)$-covers the unmarked portion, since the optimal solution has $k'/2$ centers. Combining these with the k'' CCVs, we have k centers covering the whole graph.

Archer [2] presents two $O(\log^* k)$ algorithms, both building on the work in [13]. The algorithm more directly connected with the earlier work nevertheless has two fundamental differences. Firstly, in the reduce phase Archer shows that the CCV-free portion of the graph can be covered with $2k'/3$ centers and radius 3. Secondly, he constructs a set cover-like linear program and solves the relaxation to get a total of k' *fractional* centers that cover the unmarked vertices. From these fractional centers, he obtains a 2-cover of the unmarked vertices with $k' \log k'$ (integral) centers. These are the seed for the augment phase, which thus produces a solution with an $O(\log^* k')$ approximation to the optimum radius.

During the preparation of the final version of this manuscript, it was announced that the asymmetric k-center problem is hard to approximate better than $\Omega(\log^* n)$ [6, 7], closing the gap with the upper bound.

4   Asymmetric Weighted k-Center

Recall the application in which the costs of delivery hubs vary. In this situation, rather than having a restriction on the number of centers used, each vertex has a weight and we have a budget W that restricts the total weight of centers used.

Definition 10 (Weighted k-Center). *Given a weight function on the vertices, $w : V \to \mathbb{Q}^+$, and a bound $W \in \mathbb{Q}^+$, the problem is to find $S \subseteq V$ of total weight at most W, so that S covers V with minimum radius.*

Hochbaum and Shmoys [9] gave a 3-approximation algorithm for the symmetric weighted version, applying their approach for bottleneck problems. We propose an $O(\log^* n)$-approximation for the asymmetric version, based on Panigrahy and Vishwanathan's technique for the unweighted problem. Note that in light of the hardness result just announced [6, 7], this algorithm is asymptotically optimal. Another variant has both the k and the W restrictions, but we will not expand on that problem here.

First a brief sketch of the algorithm, which works with restricted graphs. In the reduce phase, having found a CCV, v, we pick the lightest vertex u in $\Gamma^-(v)$ (which might be v itself) as a center in our solution. Then mark everything in $\Gamma_3^+(u)$ as covered, and continue looking for CCVs. We can show that there exists a 7-cover of the unmarked vertices with total weight less than half optimum. Finally we recursively apply a greedy procedure for weighted elements $O(\log^* n)$ times, similar to the one used for Set Cover. The total weight of centers in our solution set is at most W.

The following lemma about digraphs is the key to our reduce phase and is analogous to Lemma 4 in [13] and Lemma 16 in [2].

Lemma 1 (Cover of Half the Graph's Weight). *Let $G = (V, E)$ be a digraph with weighted vertices, but unit edge costs. Then there is a subset $S \subseteq V$, $w(S) \leq w(V)/2$, such that every vertex with positive indegree is reachable in at most 3 steps from some vertex in S.*

Proof. To construct the set S repeat the following, to the extent possible: *Select a vertex with positive outdegree, but if possible select one with indegree zero*. Let v be the selected vertex and compare sets $\{v\}$ and $\Gamma^+(v) \setminus \{v\}$: add the set of smaller weight to S and remove $\Gamma^+(v)$ from G.

It is clear that the weight of S is no more than half the weight of V. We must now show that S 3-covers all non-orphan vertices—we call x a parent of y if $x \in \Gamma^-(y)$.

The children of v are clearly 1-covered. Assume v is not in S (trivial otherwise): if v was an orphan initially then ignore it. If v is an orphan when selected, then some parent must have been removed by the selection of a grandparent, so it is 2-covered.

So v has at least one parent when it is selected, implying there are no orphan vertices at that time. Therefore the sets of parents of v, S_1, grandparents of v, S_2, and great-grandparents of v, S_3, are not empty. Although these sets might not be pairwise disjoint, if they contained any of v's children, then v would be 3-covered.

After v is removed, there are three possibilities for S_2: (i) Some vertex in S_3 is selected, removing part of S_2; (ii) Some vertex in S_2 is selected and removed; (iii) Some vertex in S_1 is selected, possibly making some S_2 vertices childless. One of these events *must* happen, since S_1 and S_2 are non-empty. As a consequence, v is 3-covered. □

Henceforth call the vertices that have not yet been covered/marked *active*. Using Lemma 1 we can show that after removing the CCVs from the graph, we can cover the active set with half the weight of an optimum cover if we are allowed to use distance 7 instead of 1.

Lemma 2 (Cover of Half Optimal Weight). *Consider a subset $A \subseteq V$ that has a cover consisting of vertices of total weight W, but no CCVs. Assume there exists a set C_1 that 3-covers exactly $V \setminus A$. Then there exists a set of vertices S of total weight $W/2$ that, together with C_1, 7-cover A.*

Proof. Let U be the subset of optimal centers that cover A. We call $u \in U$ a *near* center if it can be reached in 4 steps from C_1, and a *far* center otherwise. Since C_1 5-covers all of the nodes covered by near centers, it suffices to choose S to 6-cover the far centers, so that S will 7-cover all the nodes they cover.

Define an auxiliary graph H on the (optimal) centers U as follows. There is an edge from x to y in H if and only if x 2-covers y in G (and $x \neq y$). The idea is to show that any far center has positive indegree in H. As a result, Lemma 1 shows there exists a set $S \in U$ with $|S| \leq W/2$ such that S 3-covers the far centers in H, and thus 6-covers them in G.

Let x be any far center. Since A contains no CCVs, there exists y such that y covers x, but x does not cover y. Since $x \notin \Gamma_4^+(C_1)$, $y \notin \Gamma_3^+(C_1)$, and thus $y \in A$ (since everything not 3-covered by C_1 is in A). Thus there exists a center $z \in U$, which is not x, but might be y, that covers y and therefore 2-covers x. Hence x has positive indegree in the graph H. □

As we foreshadowed, we will use the greedy heuristic to complete the algorithm. We now analyze the performance of this heuristic in the context of the dominating set problem in node-weighted graphs. All vertices V are available as potential members of the dominating set (i.e. centers), but we need only dominate the active vertices A. The heuristic is to select the most *efficient* vertex: the one that maximizes $w(A(v))/w(v)$, where $A(v) \equiv A \cap \Gamma^+(v)$.

Lemma 3 (Greedy Algorithm in Weighted Dominating Set). *Let $G = (V, E)$, $w : V \to \mathbb{Q}^+$ be an instance of the dominating set problem in which a set A is to be dominated. Also, let w^* be the weight of an optimum solution for this instance. The greedy algorithm gives an approximation guarantee of*

$$2 + \ln \frac{w(A)}{w^*} = O\left(\log \frac{w(A)}{w^*}\right) .$$

Proof. In every application of the greedy selection there must be some vertex $v \in V$ for which

$$\frac{w(A(v))}{w(v)} \geq \frac{w(A)}{w^*} \quad \Rightarrow \quad \frac{w(A(v))}{w(A)} \geq \frac{w(v)}{w^*} \tag{1}$$

otherwise no optimum solution of weight w^* would exist. This is certainly true of the most efficient vertex v, so make it a center and mark all that it covers, leaving A' uncovered. Now,

$$w(A') = w(A) - w(A(v)) \leq w(A)\left(1 - \frac{w(v)}{w^*}\right) \leq w(A)\exp\left(-\frac{w(v)}{w^*}\right)$$

After j steps, the remaining active vertices, A^j, satisfy

$$w(A^j) \leq w(A^0) \prod_{i=1}^{j} \exp\left(-\frac{w(v_i)}{w^*}\right) , \tag{2}$$

where v_i is the ith center picked (greedily) and A^0 is the original active set.

Assume that after some number of steps, say j, there are still some active elements, but the upper bound in (2) drops below w^*. That is to say,

$$\sum_{i=1}^{j} w(v_i) \geq w^* \ln(w(A^0)/w^*) .$$

Before we picked the vertex v_j we had

$$\sum_{i=1}^{j-1} w(v_i) \leq w^* \ln(w(A^0)/w^*) , \quad \text{and so,} \quad \sum_{i=1}^{j} w(v_i) \leq w^* + w^* \ln(w(A^0)/w^*) ,$$

because (1) tells us that $w(v_i)$ is no greater than w^*. To cover the remainder, A^j, we just use A^j itself, at a cost of at most w^*. Hence the total weight of the solution is at most $w^*(2 + \ln(w(A^0)/w^*))$.

On the other hand, if the upper bound on $w(A^j)$ never drops below w^* before A^j becomes empty, then we have a solution of weight at most $w^* \ln(w(A^0)/w^*)$. $\square$

We now show that this tradeoff between covering radius and optimal cover size leads to an $O(\log^* n)$ approximation.

Lemma 4 (Recursive Set Cover). *Given $A \subseteq V$, such that A has a cover of weight W, and a set $C_1 \in V$ that covers $V \setminus A$, we can find in polynomial time a set of vertices of total weight at most $2W$ that, together with C_1, cover A (and hence V) within a radius of $O(\log^* n)$.*

Proof. Our first attempt at a solution, S_0, is all vertices of weight no more than W: only these vertices could be in the optimum center set. Their total weight is at most nW. Since C_1 covers $S_0 \setminus A$, consider $A_0 = S_0 \cap A$, which has a cover of size W. Lemma 3 shows that the greedy algorithm results in a set S_1 that covers A_0, and has weight

$$w(S_1) \leq O\left(W \log \frac{Wn}{W}\right) = O(W \log n) .$$

Set C_1 covers $S_1 \setminus A$, so we need only consider $A_1 = S_1 \cap A$, and so forth. At the ith iteration we have: $w(S_i) \leq O(W \log(w(S_{i-1})/W))$ and hence by induction at most $O(W \log^{(i)} n)$. Thus after $\log^* n$ iterations the weight of our solution set falls to $2W$. □

All the algorithmic tools can be assembled to form an approximation algorithm.

Theorem 1 (Approximation of Weighted k-Center). *We can approximate the weighted k-center problem within factor $O(\log^* n)$ in polynomial time.*

Proof. Guess the optimum radius, R_{OPT}, and work in the restricted graph $G_{R_{\text{OPT}}}$. Initially, the active set A is V. Repeat the following as many times as possible: Pick CCV v in A, add the lightest vertex u in $\Gamma^-(v)$ to our solution set of centers and, remove the set $\Gamma_3^+(u)$ from A. Since v is covered by an optimum center in $\Gamma^+(v)$, u is no heavier than this optimum center, and $\Gamma_3^+(u)$ includes everything covered by the optimum center.

Let C_1 be the centers chosen in this first phase. We know the remainder of the graph, A, has a cover of total weight $W' = W - w(C_1)$, because of our choices based on CCV and weight.

Lemma 2 shows that we can cover the remaining uncovered vertices with weight no more than $W'/2$ if we use distance 7. So let the active set A be $V \setminus \Gamma_7^+(C_1)$, and recursively apply the greedy algorithm as described in the proof of Lemma 4 on the graph $G_{R_{\text{OPT}}}^7$. As a result, we have a set of size W' that covers A within radius $O(\log^* n)$. □

5   Asymmetric p-Neighbor k-Center

Imagine that we wish to locate k facilities at such that the maximum distance of a demand point from its pth-closest facility is minimized. As a consequence, failures in $p - 1$ facilities do not bring down the network.

Definition 11 (Asymmetric p-Neighbor k-Center Problem). *Let $d_p(S, v)$ denote the distance from the pth closest vertex in S to v. The problem is to find a subset S of at most k vertices that minimizes*

$$\max_{v \in V \setminus S} d_p(S, v) .$$

We show that we can approximate the asymmetric p-neighbor k-center problem within a factor of $O(\log^* k)$ if we allow ourselves to use $2k$ centers. Our algorithm is restricted to the case $p \leq n/k$, but this is reasonable as p should not be too large [12].

We use the same techniques as usual, including restricted graphs, but in the augment phase we use the greedy algorithm for the *Constrained Set Multicover* problem [15]. That is, each element, e, needs to be covered r_e times, but each set can be picked at most once. The p-neighbor k-center problem has $r_e = p$ for all e. We say that an element e is *alive* if it occurs in fewer than p sets chosen so far. The greedy heuristic is to pick the set that covers the most live elements. It can be shown that this algorithm achieves an approximation factor of $H_n = O(\log n)$ [15]. However the following result is more appropriate to our application.

Lemma 5 (Greedy Constrained Set Multicover). *Let k be the optimum solution to the Constrained Set Multicover problem. The greedy algorithm gives approximation guarantee $O(\log(np/k))$.*

Proof. The same kind of averaging argument used for standard Set Cover shows that the greedy choice of a set reduces the total number of unmarked element copies by a factor $1 - 1/k$. So after i steps the number of copies of elements yet to be covered is $np(1 - 1/k)^i \leq np(e^{-1/k})^i$. Hence after $k \ln(np/k)$ steps the number of uncovered copies of elements is at most k. A naive cover of these last k element copies leads to the total number of sets being $k + k \ln(np/k)$. $\square$

If $p \leq n/k$ this greedy algorithm gives an approximation factor of $O(\log(n/k))$. Applying the standard recursive approach in [13], which works in the p-neighbor case, we can achieve an $O(\log n)$ approximation with k centers, or $O(\log^* n)$ with $2k$ centers. We can lower the approximation guarantee to $O(\log^* k)$, with $2k$ centers, using Archer's LP-based priming. First solve the LP for the constrained set multicover problem. In the solution each vertex is covered by an amount p of fractional centers, out of a total of k. We can now use the greedy set cover algorithm to get an initial set of $k^2 \ln k$ centers that 2-covers every vertex in the active set with at least p centers. Repeatedly applying the greedy procedure for constrained set multicover, this time for $(\log^* k + 1)$ iterations, we get $2k$ centers that cover all active vertices within $O(\log^* k)$. Alternatively, we could carry out $O(\log k)$ iterations and stick to just k centers.

6   Inapproximability Results

In this section we give inapproximability results for the asymmetric versions of the k-center problem with outliers, the priority k-center problem, and the

k-supplier problem. These problems all admit constant factor approximation algorithms in the symmetric case.

Asymmetric k-Center with Outliers

Definition 12 (k-Center with Outliers and Forbidden Centers). *Find a set $S \subseteq C$, where C is the set of vertices allowed to be centers, such that $|S| \leq k$ and S covers at least p nodes, with minimum radius.*

Theorem 2. *For any polynomial time computable function $\alpha(n)$, the asymmetric k-center problem with outliers (and forbidden centers) cannot be approximated within a factor of $\alpha(n)$, unless $P = NP$.*

Proof. We reduce instance $\langle U, \mathcal{S}, k \rangle$ of Max Coverage to our problem. Construct vertex sets A and B so that for each set $S \in \mathcal{S}$ there is $v_S \in A$, and for each element $e \in U$ there is $v_e \in B$. From every vertex $v_S \in A$, create an edge of unit length to vertex $v_e \in B$ if $e \in S$.

Let $p = |B| + k$, so that if we find k centers that cover p vertices within any finite distance, we *must* have found k vertices in A that cover all $|B|$ vertices. Hence we have solved the instance of Max Coverage which is an NP-complete problem.   □

Note that the proof never relied on the fact that the B vertices were forbidden from being centers (setting p to $|B| + k$ ensured this).

Asymmetric Priority k-Center

Definition 13 (Priority k-Center). *Given a priority function $p : V \rightarrow \mathbb{Q}^+$ on the vertices, find $S \subseteq V$, $|S| \leq k$, that minimizes R so that for every $v \in V$ there exists a center $c \in S$ for which $p_v d_{cv} \leq R$.*

Theorem 3. *For any polynomial time computable function $\alpha(n)$, the asymmetric k-center problem with priorities cannot be approximated within a factor of $\alpha(n)$, unless $P = NP$.*

Proof. The construction of the sets A and B is the similar to the proof of Theorem 2, except that we reduce from Set Cover. This time make the set A a complete digraph, with edges of length ℓ, as well as the unit length set-element edges from A to B. Give the nodes in set A priority 1 and the nodes in set B priority ℓ. An optimal solution to the priority k-center problem is k centers in A and a radius of ℓ, which covers every vertex. This implies that the k centers cover (in the Set Cover sense) all the elements in B. If $k' < k$ centers were chosen from A and $k - k'$ centers were chosen from B instead, we could trivially convert this to a solution choosing k centers from A.

Any non-optimal solution requires a radius of at least $\ell^2 + \ell$, as this would involve covering some B vertex by stepping from an A center through another A vertex. Therefore any algorithm with approximation guarantee $\ell + 1 - \varepsilon$ or better would solve Set Cover. We can make ℓ any function we like and the result follows.   □

Asymmetric k-Supplier

Definition 14 (k-Supplier). *Given a set of suppliers Σ and a set of customers C, find a subset $S \subseteq \Sigma$ that minimizes R such that S covers C within R.*

Theorem 4. *For any polynomial time computable function $\alpha(n)$, the asymmetric k-supplier problem cannot be approximated within a factor of $\alpha(n)$, unless $P = NP$.*

Proof. By a reduction from the Max Coverage problem similar to the proof of Theorem 2. □

Acknowledgements

The authors would like to thank Moses Charikar and the reviewers.

References

[1] A. Archer. Inapproximability of the asymmetric facility location and k-median problems. Unpublished manuscript available at www.orie.cornell.edu/~aarcher/Research, 2000.

[2] A. Archer. Two $O(\log^* k)$-approximation algorithms for the asymmetric k-center problem. In K. Aardal and B. Gerads, editors, *IPCO*, volume 2081 of *Lecture Notes in Computer Science*, pages 1–14. Springer-Verlag, 2001.

[3] R. Bhatia, S. Guha, S. Khuller, and Y. Sussmann. Facility location with dynamic distance function. In *Scand. Workshop on Alg. Th. (SWAT)*, pages 23–34, 1998.

[4] M. Charikar, S. Khuller, D. Mount, and G. Narasimhan. Algorithms for facility location problems with outliers. In *Proc. 12th SODA*, pages 642–51, 2001.

[5] S. Chaudhuri, N. Garg, and R. Ravi. The p-neighbor k-center problem. *Info. Proc. Lett.*, 65:131–4, 1998.

[6] J. Chuzhoy, S. Guha, S. Khanna, and S. Naor. Asymmetric k-center is $\log^* n$-hard to approximate. Technical Report 03-038, Elec. Coll. Comp. Complexity, 2003.

[7] E. Halperin, G. Kortsarz, and R. Krauthgamer. Tight lower bounds for the asymmetric k-center problem. Technical Report 03-035, Elec. Coll. Comp. Complexity, 2003.

[8] D. Hochbaum and D. Shmoys. A best possible approximation algorithm for the k-center problem. *Math. Oper. Res.*, 10:180–4, 1985.

[9] D. Hochbaum and D. Shmoys. A unified approach to approximation algorithms for bottleneck problems. *JACM*, 33:533–50, 1986.

[10] W. Hsu and G. Nemhauser. Easy and hard bottelneck location problems. *Disc. Appl. Math.*, 1:209–16, 1979.

[11] O. Kariv and S. Hakimi. An algorithmic approach to network location problems. I. The p-centers. *SIAM J. Appl. Math.*, 37:513–38, 1979.

[12] S. Khuller, R. Pless, and Y. Sussmann. Fault tolerant k-center problems. *Theor. Comp. Sci. (TCS)*, 242:237–45, 2000.

[13] R. Panigrahy and S. Vishwanathan. An $O(\log^* n)$ approximation algorithm for the asymmetric p-center problem. *J. Algorithms*, 27:259–68, 1998.

[14] J. Plesnik. A heuristic for the p-center problem in graphs. *Disc. Appl. Math.*, 17:263–268, 1987.

[15] V. Vazirani. *Approximation Algorithms*. Springer-Verlag, 2001.

[16] S. Vishwanathan. An $O(\log^* n)$ approximation algorithm for the asymmetric p-center problem. In *Proc. 7th SODA*, pages 1–5, 1996.

An FPTAS for Quickest Multicommodity Flows with Inflow-Dependent Transit Times[*]

Alex Hall[1], Katharina Langkau[2], and Martin Skutella[3]

[1] Institut TIK, Gloriastrasse 35, ETH Zentrum, 8092 Zurich, Switzerland,
hall@tik.ee.ethz.ch
[2] Institut für Mathematik, TU Berlin, Straße des 17. Juni 136, 10623 Berlin, Germany,
langkau@math.tu-berlin.de
[3] Max-Planck Institut für Informatik, Stuhlsatzenhausweg 85, 66123 Saarbrücken, Germany,
skutella@mpi-sb.mpg.de

Abstract. Given a network with capacities and transit times on the arcs, the quickest flow problem asks for a 'flow over time' that satisfies given demands within minimal time. In the setting of flows over time, flow on arcs may vary over time and the transit time of an arc is the time it takes for flow to travel through this arc. In most real-world applications (such as, e.g., road traffic, communication networks, production systems, etc.), transit times are not fixed but depend on the current flow situation in the network. We consider the model where the transit time of an arc is given as a nondecreasing function of the rate of inflow into the arc. We prove that the quickest s-t-flow problem is NP-hard in this setting and give various approximation results, including an FPTAS for the quickest multicommodity flow problem with bounded cost.

1 Introduction

Flows over time have been introduced more than forty years ago by Ford and Fulkerson [6, 7]. Given a directed graph with capacities and transit times on the arcs, a source node s, a sink node t, and a time horizon T, they consider the problem of sending the maximum possible amount of flow from s to t within T time units. A flow over time specifies a flow rate for each arc at each point in time. The capacity of an arc is an upper bound on this flow rate, i.e., on the amount of flow that can be sent into the arc during each unit of time. Flow on an arc progresses at a constant speed which is determined by its transit time.

Known results for flows over time with constant transit times. Ford and Fulkerson show that the maximum s-t-flow over time problem can be solved by essentially one static min-cost flow computation in the given network, where transit times are interpreted as costs. An arbitrary path decomposition of such a static min-cost flow can be turned into

---

[*] Extended abstract; information on the full version of the paper can be obtained via the authors' WWW-pages. This work was supported in part by the joint Berlin/Zurich graduate program Combinatorics, Geometry, and Computation (CGC) financed by ETH Zurich and the German Science Foundation grant GRK 588/2 and by the EU Thematic Network APPOL II, Approximation and Online Algorithms, IST-2001-30012.

S. Arora et al. (Eds.): APPROX 2003+RANDOM 2003, LNCS 2764, pp. 71–82, 2003.

a flow over time by sending flow at the given flow rate into each path as long as there is enough time left for the flow on a path to arrive at the sink before time T. A flow featuring this structure is called 'temporally repeated'.

A problem closely related to the maximum s-t-flow over time problem is the quickest s-t-flow problem. Here, the flow value (or 'demand') is fixed and the task is to find a flow over time with minimal time horizon T. Clearly, this problem can be solved in polynomial time by incorporating the algorithm of Ford and Fulkerson into a binary search framework. Burkard, Dlaska, and Klinz [2] give a strongly polynomial algorithm for the quickest s-t-flow problem which is based on the parametric search method of Megiddo [15]. Hoppe and Tardos [10, 11] study the quickest transshipment problem which, given supplies and demands at the nodes, asks for a flow over time that zeroes all supplies and demands within minimal time. They give a polynomial time algorithm which is, however, based on a submodular function minimization routine.

The latter fact already indicates that flow over time problems are, in general, considerably harder than their static counterparts in classical network flow theory. The best evidence for this allegation is maybe provided by a surprising result of Klinz and Woeginger [12]. They show that computing a quickest s-t-flow of minimum cost in a network with cost coefficients on the arcs is already NP-hard in series-parallel networks. Moreover, it is even strongly NP-hard to find a quickest temporally repeated s-t-flow of minimum cost. Only recently, Hall, Hippler, and Skutella [8] showed that computing quickest multicommodity flows is NP-hard, even on series-parallel networks.

On the other hand, Ford and Fulkerson [6, 7] introduce the concept of time-expanded networks which allows to solve many flow over time problems in pseudopolynomial time. The node set of a time-expanded network consists of several copies of the node set of the underlying graph building a 'time layer'. The number of time layers is equal to the integral time horizon T and thus pseudopolynomial in the input size. Copies of an arc of the underlying graph join copies of its end-nodes in time layers whose distances equal the transit time of that arc. Ford and Fulkerson observe that a flow over time in the given graph corresponds to a static flow in the time-expanded network, and vice versa. Thus, many flow over time problems can be solved by static flow computations in the time-expanded network.

Fleischer and Skutella [4] come up with so-called 'condensed' time-expanded networks which are of polynomial size and can be used to compute provably good multicommodity flows over time with costs in polynomial time. In particular, they present a fully polynomial time approximation scheme (FPTAS) for the quickest multicommodity flow problem with bounded cost [4, 5]. Using completely different techniques, they also show that 2-approximate temporally repeated flows can be obtained from a static, length-bounded flow computation in the given graph [4]. The advantage of the latter solutions is that they have a very simple structure and also do not use storage of flow at intermediate nodes.

Flow-dependent transit times. So far we have considered the setting of flows over time where transit times of arcs are fixed. In many practical applications, however, the latter assumption is not realistic since transit times vary with the flow situation on an arc. We refer to [1, 16, 17] for an overview and further references. Usually, the correlation of the transit time and the flow situation on an arc is highly complex. It is a major challenge to come up with a mathematical model that, on the one hand, captures the real behavior as realistically as possible and, on the other hand, can be solved efficiently even on large networks.

Köhler and Skutella [14] consider a model where, at any moment in time, the actual speed of flow on an arc depends on the current amount of flow on the arc. Under this assumption, they give a 2-approximation algorithm for the quickest s-t-flow problem and show that no polynomial time approximation scheme (PTAS) exists, unless P=NP. A simpler model is studied by Carey and Subrahmanian [3]. They assume that the transit time on an arc only depends on the current rate of inflow into the arc and propose a time-expanded network whose arcs somehow to reflect this behavior. Köhler, Langkau, and Skutella [13] give a 2-approximation algorithm for the quickest s-t-flow problem in the setting of inflow-dependent transit times. The algorithm uses the algorithm of Ford and Fulkerson [6, 7] on a so-called 'bow graph' with fixed transit times on the arcs. In the bow graph, every arc of the original graph is replaced by a bunch of arcs corresponding to different transit times. The quickest flow problem in the bow graph is a relaxation of the quickest flow problem with inflow-dependent transit times.

Contribution of this paper. While, for the special case of constant transit times, quickest s-t-flows can be computed in polynomial time [2, 6, 7], we show in Section 6 that the problem becomes NP-hard if we allow inflow-dependent transit times. In Section 4, we generalize the 2-approximation result given in [13] to the setting with costs and multiple commodities. Our approach is based on a new and stronger relaxation of the quickest flow problem, which we introduce in Section 3. This relaxation is defined in a bow graph similar to the one introduced in [13], but it uses additional 'coupling constraints' between flow values on different copies of one arc in the original graph. In particular, this relaxation can no longer be solved by standard network flow algorithms but requires general linear programming techniques. Nevertheless, as shown in Section 4, the approximation technique based on length-bounded static flows presented in [4] can be generalized to yield provably good solutions to our bow graph relaxation. Moreover, we prove that such a solution to the relaxation can be turned into a feasible multicommodity flow over time with inflow-dependent transit times and bounded cost.

The main result of this paper is a fully polynomial time approximation scheme for the quickest multicommodity flow problem with bounded cost and inflow-dependent transit times (see Section 5). It again uses the new bow graph relaxation introduced in Section 3 and generalizes the approach based on condensed time-expanded networks from [5]. Interestingly, the time-expanded version of our bow graph relaxation essentially coincides with the modified time-expanded graph considered by Carey and Subrahmanian [3].

Due to space limitations, we omit most proofs in this extended abstract.

2 Preliminaries

We are considering network flow problems in a directed graph $G = (V, E)$ with $n :=$ $|V|$ nodes and $m := |E|$ arcs. Each arc $e \in E$ has associated with it a positive capacity u_e and a nonnegative, nondecreasing transit time function $\tau_e : [0, u_e] \to \mathbb{R}^+$. There is a set of commodities $K = \{1, \ldots, k\}$; every commodity $i \in K$ is defined by a source-sink pair[4] $(s_i, t_i) \in V \times V$. The objective is to send a prespecified amount of flow $d_i > 0$, called the demand, from s_i to t_i. Finally, each arc e has associated cost coefficients $c_{e,i}$, for $i \in K$, where $c_{e,i}$ is interpreted as the cost (per flow unit) for sending flow of commodity i through the arc. For an arc $e = (v, w) \in E$, we use the notation head$(e) := w$ and tail$(e) := v$.

Flows over time with constant transit times. A *(multicommodity) flow over time* f in G with time horizon T is given by Lebesgue-measurable functions $f_{e,i} : [0, T) \to \mathbb{R}^+$, where $f_{e,i}(\theta)$ is the rate of flow (per time unit) of commodity i entering arc e at time θ. In order to simplify notation, we sometimes use $f_{e,i}(\theta)$ for $\theta \notin [0, T)$, implicitly assuming that $f_{e,i}(\theta) = 0$ in this case. The capacity u_e is an upper bound on the rate of flow entering arc e at any moment of time, i.e., $f_e(\theta) \le u_e$ for all $\theta \in [0, T)$ and $e \in E$. Here, $f_e(\theta) := \sum_{i \in K} f_{e,i}(\theta)$ is the total rate at which flow is entering arc e at time θ.

In the original setting of flows over time, the transit time function τ_e of arc e is assumed to be constant. Then, the flow $f_{e,i}(\theta)$ of commodity i entering arc e at time θ arrives at head(e) at time $\theta + \tau_e$. All arcs must be empty from time T on, i.e., $f_{e,i}(\theta) = 0$ for $\theta \ge T - \tau_e$. To generalize the notion of *flow conservation*, we define $D_{v,i}^-(\xi) := \sum_{e \in \delta^-(v)} \int_{\tau_e}^\xi f_{e,i}(\theta - \tau_e)\, d\theta$ to be the total inflow of commodity $i \in K$ into node v until time $\xi \in [0, T]$. Similarly, $D_{v,i}^+(\xi) := \sum_{e \in \delta^+(v)} \int_0^\xi f_{e,i}(\theta)\, d\theta$ is the corresponding outflow. We consider the model with storage of flow at intermediate nodes. That is, flow entering a node can be held back for some time before it is sent onward. To rule out deficit at any node, we require $D_{v,i}^-(\xi) - D_{v,i}^+(\xi) \ge 0$, for all $\xi \in [0, T), i \in K$, and $v \in V \backslash \{s_i\}$. Moreover, flow must not remain in any node other than the sinks at time T. Therefore, we require that equality holds for every $i \in K$, $v \in V \backslash \{s_i, t_i\}$, at time $\xi = T$. The flow over time f satisfies the multicommodity demands if $D_{t_i,i}^-(T) - D_{t_i,i}^+(T) = d_i$, for any commodity $i \in K$. The cost of a flow over time f is defined as $c(f) := \sum_{e \in E} \sum_{i \in K} c_{e,i} \int_0^T f_{e,i}(\theta) d\theta$.

Time-expanded graphs. Many flow over time problems can be solved by static flow algorithms in time-expanded graphs [6, 7]. Given a graph $G = (V, E)$ with integral transit times on the arcs and an integral time horizon T, the T-*time-expanded graph* of G, denoted G^T, is obtained by creating T copies of V, labeled V_0 through V_{T-1}, with the θth copy of node v denoted $v(\theta)$, $\theta = 0, \dots, T - 1$. For every arc $e = (v, w) \in E$ and $\theta = 0, \dots, T - 1 - \tau_e$, there is an arc $e(\theta)$ from $v(\theta)$ to $w(\theta + \tau_e)$ with the same capacity and costs as arc e. In addition, there is an infinite capacity *holdover arc* from $v(\theta)$ to $v(\theta + 1)$, for all $v \in V$ and $\theta = 0, \dots, T - 2$, which models the possibility to hold flow at node v during the time interval $[\theta, \theta + 1)$.

Any static flow in this time-expanded network corresponds to a flow over time of equal cost: interpret the flow on arc $e(\theta)$ as the flow through arc $e = (v, w)$ that starts at node v in the time interval $[\theta, \theta + 1)$. Similarly, any flow over time completing by time T corresponds to a static flow in G^T of the same value and cost obtained by mapping the total flow starting on e in time interval $[\theta, \theta + 1)$ to flow on arc $e(\theta)$. Thus, we may

---

[4] To simplify notation, we restrict to the case of only one source and one sink for each commodity. However, our results can be directly generalized to the case of several sources and sinks with given supplies and demands for each commodity.

solve a flow over time problem by solving the corresponding static flow problem in the time-expanded network.

One drawback of this approach is that the size of G^T depends linearly on T, so that if T is not bounded by a polynomial in the input size, this is not a polynomial-time method. However, the following useful observation can be found in [4]: If all transit times are multiples of some large number $\Delta > 0$, then instead of using the T-time-expanded graph, we may rescale time and use a Δ-*condensed time-expanded graph* that contains only $\lceil T/\Delta \rceil$ copies of V. Since in this setting every arc corresponds to a time interval of length Δ, capacities are multiplied by Δ. For more details we refer to [4].

Flows with inflow-dependent transit times. In the original setting of flows over time discussed above, it is assumed that transit times are fixed throughout, so that flow on arc e progresses at a uniform speed. In the following, we will consider the more general model of *inflow-dependent* transit times. Here, the transit time of an arc may vary with the current amount of flow using this arc. Each arc e has an associated non-negative transit time function τ_e, which determines the time it takes for flow to traverse arc e. Flow of commodity i entering arc e at time θ at rate $f_{e,i}(\theta)$ arrives at head(e) at time $\theta + \tau_e(f_e(\theta))$. We will later need the following simple observation which follows from the fact that flow can be stored at intermediate nodes.

Observation 1. *For every arc $e \in E$, let $\tau_e : [0, u_e] \to \mathbb{R}^+$ and $\tau'_e : [0, u_e] \to \mathbb{R}^+$ be transit time functions on arc e such that $\tau'_e(x) \leq \tau_e(x)$ for all $x \in [0, u_e]$. Then, a flow over time with inflow-dependent transit times $(\tau_e)_{e \in E}$ and time horizon T also yields a flow over time with inflow-dependent transit times $(\tau'_e)_{e \in E}$ and time horizon T.*

3   The Bow Graph

In this section, we will define a so-called bow graph that is very similar to the one defined in [13]. Let us for the moment assume that all transit time functions are piecewise constant, non-decreasing, and left-continuous. This transit time function of arc e is denoted by τ_e^s. It is given by breakpoints $0 = x_0 < x_1 < \cdots < x_\ell$ and corresponding transit times $\tau_1 < \cdots < \tau_\ell$. Flow entering arc e at rate $x \in (x_{i-1}, x_i]$ needs τ_i time to traverse arc e. Later we will use the fact that general transit time functions can be approximated by such step functions within arbitrary precision.

The bow graph, denoted $G^B = (V^B, E^B)$, is defined on the same node set as G, i.e., $V^B := V$, and is obtained by creating several copies of an arc, one for every possible transit time on this arc. Thus, arc e is replaced by ℓ parallel *bow arcs* $a_1, \ldots, a_\ell$. The transit time of bow arc a_i is τ_i and its capacity is x_i, $i = 1, \ldots, \ell$. We will denote the set of bow arcs corresponding to arc $e \in E$ by E_e^B, and refer to E_e^B as the *expansion* of arc e. The cost coefficients of every arc $a \in E_e^B$ are identical to those of e, i.e., $c_{a,i} := c_{e,i}$, for $i \in K$.

3.1   A Relaxation of Inflow-Dependent Transit Times

We will now discuss the relationship between flows over time with inflow-dependent transit times in G and flows over time in the bow graph G^B. Any flow over time f in G

with inflow-dependent transit times $(\tau_e^s)_{e \in E}$ and time horizon T can be interpreted as a flow over time f^B in G^B (with constant transit times) with the same time horizon T: If flow is entering arc $e \in E$ at time θ with flow rate $f_e(\theta)$, then, in the bow graph, this flow is sent onto the bow arc $a \in E_e^B$ representing the transit time $\tau_e^s(f_e(\theta))$.

Unfortunately, an arbitrary flow over time f^B in G^B does not correspond to a flow over time f with inflow-dependent transit times $(\tau_e^s)_{e \in E}$ in G. In addition, we have to require the following property: For every original arc $e \in E$ and at every point in time θ, the flow f^B sends flow into at most one bow arc $a \in E_e^B$. A flow over time in G^B fulfilling this property is called *inflow-preserving*.

Observation 2. *Every inflow-preserving flow over time f^B in G^B with time horizon T corresponds to a flow over time f in G with inflow-dependent transit times $(\tau_e^s)_{e \in E}$ and time horizon T, and vice versa.*

Notice that the set of inflow-preserving flows over time is not convex. In particular, it is difficult to compute inflow-preserving flows directly. Therefore, we also consider a relaxed notion which can be interpreted as a convexification of inflow-preserving flows: For any arc $a \in E^B$, let $\lambda_a(\theta) := f_a^B(\theta)/u_a$ denote the *per capacity inflow rate* into arc a at time θ. Then, a flow over time f^B in G^B with time horizon T is called *weakly inflow-preserving* if $\sum_{a \in E_e^B} \lambda_a(\theta) \leq 1$ for all $e \in E$ and $\theta \in [0, T)$. Since every inflow-preserving flow over time is also weakly inflow-preserving, it follows from Observations 1 and 2 that weakly inflow-preserving flows over time in G^B constitute a relaxation of flows over time with inflow-dependent transit times in G:

Observation 3. *For every arc $e \in E$, let $\tau_e^s : [0, u_e] \to \mathbb{R}^+$ and $\tau_e : [0, u_e] \to \mathbb{R}^+$ be transit time functions on arc e such that τ_e^s is a step function with $\tau_e^s(x) \leq \tau_e(x)$ for all $x \in [0, u_e]$. Then, every flow over time with inflow-dependent transit times $(\tau_e)_{e \in E}$ and time horizon T in G yields a (weakly) inflow-preserving flow over time with time horizon T in G^B.*

The basic idea of the approximation algorithms presented in this paper is to compute weakly inflow-preserving flows over time in an appropriate bow graph and turn these into flows over time in G with inflow-dependent transit times. The following lemma and its corollary make this approach work. Consider the expansion of a single arc $e \in E$ to bow arcs $E_e^B = \{a_1, \ldots, a_\ell\}$.

Lemma 1. *Let f^B be a weakly inflow-preserving flow over time with time horizon T in E_e^B and $\delta > 0$. Then, f^B can be turned into an inflow-preserving flow over time $\hat{f}^B$ in E_e^B such that every (infinitesimal) unit of flow in $\hat{f}^B$ reaches $\mathrm{head}(e)$ at most δ time units later than it does in f^B.*

Proof. For every bow arc a_i, $i = 1, \ldots, \ell$, we set up a buffer b_i in $\mathrm{tail}(e)$ for temporary storage of flow. The buffer b_i is collecting all flow in f^B which is about to be shipped through bow arc a_i. It can output this flow in a first-in-first-out manner, i.e., flow units must enter and leave the buffer in the same order. Buffer b_i has only two output modes. Either it is *closed*, then no flow is leaving the buffer, or it is *open* and flow is leaving the buffer at constant rate u_{a_i}, immediately entering arc a_i. In our modified solution $\hat{f}^B$, at every point in time at most one of the buffers b_i, $i = 1, \ldots, \ell$, will be open. This guaranties that $\hat{f}^B$ is inflow-preserving.

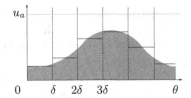   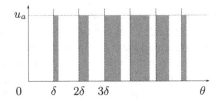

Fig. 1. Original flow rate on bow arc a and modified flow rate produced by buffering in tail(a).

As above, let $\lambda_a(\theta) := f_a^B(\theta)/u_a$ be the per capacity inflow rate of f^B on arc $a \in E_e^B$ at time θ. We partition the time horizon into intervals of length $\tilde{\delta}$, where $\tilde{\delta} := \delta/2$. Let $\lambda_{a,j}$ be the average per capacity inflow rate on arc $a \in E_e^B$ during time interval $[(j-1)\,\tilde{\delta}, j\,\tilde{\delta})$, $j = 1, \ldots, \lceil T/\tilde{\delta} \rceil$. We define the modified flow $\hat{f}^B$ as follows: During the first $\tilde{\delta}$-round, all buffers are closed. During each following $\tilde{\delta}$-round, we open the buffers in a 'round robin' fashion. More precisely, during time interval $[j\,\tilde{\delta}, (j+1)\,\tilde{\delta})$, we first open buffer b_1 for $\lambda_{a_1,j}\tilde{\delta}$ time, then buffer b_2 for $\lambda_{a_2,j}\tilde{\delta}$ time, and so on. Since f^B is weakly inflow-preserving, $\sum_{i=1}^{\ell} \lambda_{a_i,j} \leq 1$ holds and the last buffer is closed again before the end of this $\tilde{\delta}$-round. Figure 1 illustrates how the buffer changes the original inflow rate of a single bow arc a.

We show that the buffers are never empty while they are open. Consider bow arc a_i. During the interval $[(j-1)\,\tilde{\delta}, j\,\tilde{\delta})$, the flow f^B sends $\tilde{\delta}\lambda_{a_i,j}u_{a_i}$ units of flow into bow arc a_i. This is exactly the amount of flow that the corresponding buffer b_i is sending out during the succeeding interval $[j\,\tilde{\delta}, (j+1)\,\tilde{\delta})$. Hence buffer b_i is never emptied and, in particular, every unit of flow is delayed for at most $2\tilde{\delta} = \delta$ time. Note that throughout these modifications no flow is rerouted. We only make use of storage in nodes. Therefore, the cost of f^B remains unchanged. $\qquad\square$

For $\delta > 0$, we call a flow over time f^B in G^B δ-resting if, for every node $v \in V \setminus \{s_1, \ldots, s_k\}$, all flow arriving at v is stored there for at least δ time units before it moves on. A weakly inflow-preserving flow over time f^B in G^B which is δ-resting can easily be interpreted as an inflow-preserving flow over time $\hat{f}^B$: Consider a single arc $e \in E$ and its expansion E_e^B. Applying Lemma 1, the flow over time f^B restricted to E_e^B can be modified to an inflow-preserving flow over time such that every unit of flow is delayed by at most δ. The resting property of f^B makes up for this delay and ensures that every such flow unit can continue its way on time. Applying Observation 2, the flow $\hat{f}^B$ can then be interpreted as a flow over time f in G with inflow-dependent transit times $(\tau_e^s)_{e \in E}$.

Corollary 1. *Let f^B be a weakly inflow-preserving flow over time in G^B with time horizon T which is δ-resting. Then, f^B can be turned into a flow over time f in G with inflow-dependent transit times $(\tau_e^s)_{e \in E}$ and with the same time horizon and the same cost as f^B. Moreover, the flow over time f is given by piecewise constant functions $(f_e)_{e \in E}$ such that the number of breakpoints of f_e is bounded by $2\,|E_e^B|\,\lceil T/\delta \rceil$.*

4   A $(2 + \varepsilon)$-Approximation Algorithm for Quickest Flows

In this section we present a fairly simple $(2+\varepsilon)$-approximation algorithm for the quickest multicommodity flow problem with inflow-dependent transit times. The algorithm consists of the following three main steps. First, the original transit times $(\tau_e)_{e \in E}$ are replaced by lower step functions $(\tau_e^s)_{e \in E}$ and the corresponding bow graph G^B is constructed. Then, an appropriately modified version of the $(2 + \varepsilon)$-approximation algorithm presented in [4] is applied yielding a weakly inflow-preserving flow over time in G^B. Finally, the output is turned into a feasible solution to the original problem. The bow graph G^B is defined in the first step according to step functions fulfilling the requirements stated in the following observation. We will later specify the parameters $\delta, \eta > 0$ such that the size of the resulting bow graph is polynomial in the input size and $1/\varepsilon$.

Observation 4. *Let $\delta, \eta > 0$. For every non-negative, non-decreasing, and left-continuous function $\tau : [0, u] \to \mathbb{R}^+$, there exists a step function $\tau^s : [0, u] \to \mathbb{R}^+$, with*

(i) *$\tau^s(x) \leq \tau(x) \leq (1 + \eta)\,\tau^s(x) + \delta$ for every $x \in [0, u]$,*
(ii) *the number of breakpoints of τ^s is bounded by $\lceil \log_{1+\eta}(\tau(u)/\delta) \rceil + 1$.*

4.1 $(2 + \varepsilon)$-Approximate Quickest Weakly Inflow-Preserving Flows

Fleischer and Skutella [4] propose a $(2 + \varepsilon)$-approximation algorithm for the quickest multicommodity flow problem with bounded cost and constant transit times. The method is based on an approximate length-bounded static flow computation. The same approach can be applied to the problem of finding a quickest weakly inflow-preserving multicommodity flow over time with bounded cost in the bow graph.

Let f^B be an optimal solution to this problem with minimal time horizon T. As suggested in [4], we consider the static multicommodity flow x^B in G^B which results from averaging the flow f^B on every arc $a \in E^B$ over the time interval $[0, T)$. As proven in [4], this static flow (i) satisfies a fraction of $1/T$ of the demands covered by the flow over time f^B, (ii) has cost $c(x^B) = c(f^B)/T$, and (iii) is T-length-bounded. The latter property means that the flow of every commodity $i \in K$ can be decomposed into a sum of flows on s_i-t_i-paths such that the length $\tau(P) := \sum_{a \in P} \tau_a$ of any such path P is at most T. Since f^B is weakly inflow-preserving, so is x^B, i.e., its *per capacity flow values* $\lambda_a := x_a^B/u_a$, $a \in E^B$, satisfy $\sum_{a \in E_e^B} \lambda_a \leq 1$ for every arc $e \in E$. We refer to this property as property (iv).

Any static flow x in G^B meeting requirements (i) – (iv) can be turned into a weakly inflow-preserving flow over time g in G^B meeting the same demands at the same cost as f^B within time $2T$: Send flow into every s_i-t_i-path P given by the length-bounded path decomposition of x at the corresponding flow rate $x_{P,i}$ for exactly T time units; wait for at most another T time units until all flow has arrived at its destination. Since $g_a(\theta)/u_a$ is always upper-bounded by x_a/u_a, it follows from property (iv) that g is weakly-inflow preserving. Thus, g is a 2-approximate solution to the problem under consideration.

Unfortunately, computing the T-length-bounded flow x is NP-hard, even for the special case of a single commodity [9]. Yet, as discussed in [4], the T-length-bounded multicommodity flow problem can be approximated within arbitrary precision in polynomial time by slightly relaxing the length bound T. It is easy to generalize this observation to length-bounded, weakly inflow-preserving flows. This finally yields a $(2+\varepsilon)$-approximate solution.

Lemma 2. *Assume that there exists a weakly inflow-preserving multicommodity flow over time with time horizon T and cost at most C. Then, for every $\varepsilon > 0$, a weakly inflow-preserving multicommodity flow over time with time horizon at most $(2 + \varepsilon) T$ and cost at most C can be computed in time polynomial in the input size and $1/\varepsilon$.*

If all transit time functions τ_e are constant, the $(2 + \varepsilon)$-approximation algorithm in Lemma 2 and the one presented in [4] basically coincide. In [4], an example is given which shows that the performance guarantee of both algorithms is not better than 2.

4.2 $(2 + \varepsilon)$-Approximate Quickest Flows with Inflow-Dependent Transit Times

So far, we have presented an algorithm to compute a $(2 + \varepsilon)$-approximate solution to the quickest multicommodity flow problem in the relaxed model of weakly inflow-preserving flows over time. Such a solution has a simple structure, namely it is generated from a path decomposition of a static flow in the bow graph. We will use this property to turn such a flow into a solution to the original problem. Throughout this modification we will make sure that the time horizon only increases by a small factor.

Let f^B be a weakly inflow-preserving multicommodity flow over time with time horizon T^B in G^B, which is generated from a static flow x^B as described in the last section. In particular, x^B is weakly inflow-preserving and has a length-bounded path decomposition. Let $\mathcal{P}_i$ denote the set of s_i-t_i-paths from the length-bounded path decomposition of x^B and $\mathcal{P} := \cup_{i=1}^{k} \mathcal{P}_i$.

Lemma 3. *The flow over time f^B can be turned into a flow over time f in G with inflow-dependent transit times $(\tau_e)_{e \in E}$ and time horizon T, where T is bounded from above by $(1 + \eta)T^B + 2n\delta$.*

We are now ready to state the main result of this section.

Theorem 1. *For the quickest multicommodity flow problem with inflow-dependent transit times and bounded cost, there exists a polynomial time algorithm that, for any $\varepsilon > 0$, finds a solution of the same cost as optimal with time horizon at most $2 + \varepsilon$ times the optimal time horizon T^*.*

5   An FPTAS for Quickest Flows

In this section we present an FPTAS for the quickest multicommodity flow problem with inflow-dependent transit times and bounded cost. We use ideas similar to the ones employed in [5] for the problem with fixed transit times. The FPTAS is based on a static weakly inflow-preserving flow computation in a condensed time-expanded bow graph.

Theorem 2. *There is an FPTAS for the quickest multicommodity flow problem with inflow-dependent transit times and bounded cost.*

5.1   The Algorithm

To state our algorithm and prove its correctness we define the following two bow graphs: Given $G = (V, E)$ with transit time functions $(\tau_e)_{e \in E}$ and a time horizon T, let $G^{\downarrow}$ denote the *lower bow graph* constructed from the lower step functions $\tau_e^{\downarrow}(x) := \lfloor \tau_e(x)/\Delta \rfloor \Delta$, for $e \in E$, $x \in [0, u_e]$. Here, $\Delta := \varepsilon^2 T/n$ for a given small constant $\varepsilon > 0$ (we assume that n/ε^2 is integral such that T is a multiple of Δ). That is, $\tau_e(x)$ is rounded down to the nearest multiple of Δ. By choice of Δ, the size of $G^{\downarrow}$ is polynomially bounded since we can delete all arcs with transit times greater than T. The second graph is the 2Δ-*lengthened bow graph*, denoted by $G^{\uparrow\uparrow}$, which is constructed from $G^{\downarrow}$ by lengthening the transit time of each arc by 2Δ. The corresponding transit time step functions are given by $\tau_e^{\uparrow\uparrow}(x) := \tau_e^{\downarrow}(x) + 2\Delta$, for $e \in E$, $x \in [0, u_e]$.

Let the *fan graph* $G^F = (V^F, E^F)$ be the Δ-condensed time-expansion of $G^{\uparrow\uparrow}$ for time horizon T (see Section 2). Each arc $e \in E$ is represented in the bow graph $G^{\uparrow\uparrow}$ by its expansion $E_e^{\uparrow\uparrow}$. Thus, the fan graph contains, for each time $\theta \in S := \{0, \Delta, \dots, T - \Delta\}$, a 'fan' of arcs $E_e^F(\theta) := \{a(\theta) : a \in E_e^{\uparrow\uparrow}, \theta + \tau_a \in S\}$, where $a(\theta) = (v(\theta), w(\theta + \tau_a))$. For a static flow x in G^F, we define $\lambda_a(\theta) := x_{a(\theta)}/u_{a(\theta)}$ to be the per capacity inflow value on arc $a(\theta) \in E^F$. With these definitions, the concept of (weakly) inflow-preserving flows directly carries over to static flows in G^F. Moreover, the problem of computing a weakly inflow-preserving static flow in G^F can easily be formulated as a linear program. Take a standard network flow formulation and add an extra constraint for each fan in G^F. In particular, such a flow can be computed in polynomial time. Note that any (weakly) inflow-preserving *static flow* in G^F directly corresponds to a (weakly) inflow-preserving *flow over time* in $G^{\uparrow\uparrow}$, as described in Section 1.

Let T^* denote the time horizon of a quickest flow with inflow-dependent transit times in G. We can now give an overview of our algorithm:

FPTAS FOR QUICKEST FLOWS WITH INFLOW-DEPENDENT TRANSIT TIMES

1. Guess T such that $T^* \le T \le (1 + \mathcal{O}(\varepsilon))T^*$. This is done via geometric mean binary search, starting with good upper and lower bounds, obtained, e.g., with help of the $(2 + \varepsilon)$-approximation in Section 4.
2. Construct the fan graph G^F for time horizon T and compute a weakly inflow-preserving *static* multicommodity flow satisfying all demands at minimum cost.
3. Interpret this static flow as a weakly inflow-preserving *flow over time* in $G^{\uparrow\uparrow}$. Modify this flow to make it inflow-preserving and, from this, derive a flow over time in G with inflow-dependent transit times and time horizon at most T.

We now proceed as follows: First we discuss issues related to the running time of the algorithm and detail how step 3 is implemented. Then, in the next section, we prove that a static flow in G^F with the properties claimed in step 2 actually exists.

The upper and lower bounds obtained from the $(2 + \varepsilon)$-approximation in step 1 are within a constant factor of each other. Thus, the estimate T can be found within $\mathcal{O}(\log(1/\varepsilon))$ geometric mean binary search steps. The fan graph G^F constructed in step 2 contains $\mathcal{O}(n^2/\varepsilon^2)$ nodes and $\mathcal{O}(mn^2/\varepsilon^4)$ arcs; note that each fan contains

$\mathcal{O}(n/\varepsilon^2)$ arcs, potentially one for each layer of G^F. Therefore, the static flow in G^F can be computed in polynomial time. We now go into the details of step 3. As mentioned before, interpreting the static flow in G^F as a weakly inflow-preserving flow over time in $G^{\uparrow\uparrow}$ is done in the canonical way, as described in Section 1. If we now shorten all arcs of $G^{\uparrow\uparrow}$ by Δ (we refer to the resulting bow graph as $G^{\uparrow}$), we obtain a weakly inflow-preserving flow over time in $G^{\uparrow}$ which is Δ-resting. Applying Corollary 1, we derive an inflow-preserving flow over time in $G^{\uparrow}$. Finally, by Observation 1, we get a flow over time in G with inflow-dependent transit times $(\tau_e)_{e \in E}$ with time horizon at most T. Clearly, step 3 can be done in polynomial time.

5.2   Transforming a Flow over Time in G to a Static Flow in G^F

In this section we prove that our algorithm actually is an FPTAS by showing that a feasible flow as claimed in step 2 exists. To this end, we transform a quickest flow in G with inflow-dependent transit times to a weakly inflow-preserving static flow in G^F and thereby lengthen the time horizon by at most a factor of $1 + \mathcal{O}(\varepsilon)$. This transformation is done in several steps which are illustrated in the following diagram:

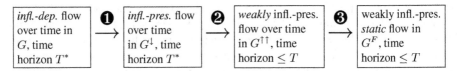

With Observation 3, step ❶ is easy to see. For step ❸, flow in $G^{\uparrow\uparrow}$ is mapped to G^F as described in Section 1: the total flow entering arc $a \in E^{\uparrow\uparrow}$ in the interval $[\theta, \theta + \Delta)$ is assigned to $a(\theta) \in E^F$, for $\theta \in S$. Clearly, if the flow was (weakly) inflow-preserving in $G^{\uparrow\uparrow}$, it will be weakly inflow-preserving in G^F, too. Step ❷ is the most interesting but also the most intricate one. It is done similarly to [5] by carefully averaging flow to derive an 'almost feasible' flow, then subsequently sending less to obtain a feasible flow and finally increasing the time horizon to meet the demands (we refer to [5] for details). We can adopt this method since the transit times in bow graphs $G^{\uparrow}$ and $G^{\uparrow\uparrow}$ are constant. However, in contrast to [5], our flows must have the additional property of being weakly inflow-preserving.

Lemma 4. *A (weakly) inflow-preserving flow over time f in $G^{\uparrow}$ with time horizon T^* can be transformed into a* weakly *inflow-preserving flow over time in $G^{\uparrow\uparrow}$ with time horizon at most $T := (1 + \mathcal{O}(\varepsilon))T^*$ and the same cost as f.*

This concludes the proof of Theorem 2.

6   Complexity

Theorem 3. *The quickest s-t-flow problem with inflow-dependent transit times, with or without storage of flow at intermediate nodes, is NP-hard in the strong sense.*

The proof uses a reduction from the well-known NP-complete problem 3-PARTITION.

References

[1] J. E. Aronson. A survey of dynamic network flows. *Annals of Operations Research*, 20:1–66, 1989.

[2] R. E. Burkard, K. Dlaska, and B. Klinz. The quickest flow problem. *ZOR — Methods and Models of Operations Research*, 37:31–58, 1993.

[3] M. Carey and E. Subrahmanian. An approach to modelling time-varying flows on congested networks. *Transportation Research B*, 34:157–183, 2000.

[4] L. Fleischer and M. Skutella. The quickest multicommodity flow problem. In W. J. Cook and A. S. Schulz, editors, *Integer Programming and Combinatorial Optimization*, volume 2337 of *Lecture Notes in Computer Science*, pages 36–53. Springer, Berlin, 2002.

[5] L. Fleischer and M. Skutella. Minimum cost flows over time without intermediate storage. In *Proceedings of the 14th Annual ACM–SIAM Symposium on Discrete Algorithms*, pages 66–75, Baltimore, MD, 2003.

[6] L. R. Ford and D. R. Fulkerson. Constructing maximal dynamic flows from static flows. *Operations Research*, 6:419–433, 1958.

[7] L. R. Ford and D. R. Fulkerson. *Flows in Networks*. Princeton University Press, Princeton, NJ, 1962.

[8] A. Hall, S. Hippler, and M. Skutella. Multicommodity flows over time: Efficient algorithms and complexity. In *Proceedings of the 30th International Colloquium on Automata, Languages and Programming (ICALP)*, Eindhoven, The Netherlands, 2003. To appear.

[9] G. Handler and I. Zang. A dual algorithm for the constrained shortest path problem. *Networks*, 10:293–310, 1980.

[10] B. Hoppe. *Efficient dynamic network flow algorithms*. PhD thesis, Cornell University, 1995.

[11] B. Hoppe and É. Tardos. The quickest transshipment problem. *Mathematics of Operations Research*, 25:36–62, 2000.

[12] B. Klinz and G. J. Woeginger. Minimum cost dynamic flows: The series-parallel case. In E. Balas and J. Clausen, editors, *Integer Programming and Combinatorial Optimization*, volume 920 of *Lecture Notes in Computer Science*, pages 329–343. Springer, Berlin, 1995.

[13] E. Köhler, K. Langkau, and M. Skutella. Time-expanded graphs for flow-dependent transit times. In *Proceedings of the 10th Annual European Symposium on Algorithms (ESA)*, volume 2461 of *Lecture Notes in Computer Science*, pages 599–611. Springer, Berlin, 2002.

[14] E. Köhler and M. Skutella. Flows over time with load-dependent transit times. In *Proceedings of the 13th Annual ACM–SIAM Symposium on Discrete Algorithms*, pages 174–183, San Francisco, CA, 2002.

[15] N. Megiddo. Combinatorial optimization with rational objective functions. *Mathematics of Operations Research*, 4:414–424, 1979.

[16] W. B. Powell, P. Jaillet, and A. Odoni. Stochastic and dynamic networks and routing. In M. O. Ball, T. L. Magnanti, C. L. Monma, and G. L. Nemhauser, editors, *Network Routing*, volume 8 of *Handbooks in Operations Research and Management Science*, chapter 3, pages 141–295. North–Holland, Amsterdam, The Netherlands, 1995.

[17] B. Ran and D. E. Boyce. *Modelling Dynamic Transportation Networks*. Springer, Berlin, 1996.

On the Complexity of Approximating
k-Dimensional Matching[*]

Elad Hazan[1], Shmuel Safra[2,3], and Oded Schwartz[2]

[1] Department of Computer Science, Princeton University, Princeton NJ, 08544 USA
ehazan@princeton.edu
[2] School of Computer Science, Tel Aviv University, Tel Aviv 69978, Israel
{safra,odedsc}@post.tau.ac.il
[3] School of Mathematics, Tel Aviv University, Tel Aviv 69978, Israel

Abstract. We study the complexity of bounded variants of graph problems, mainly the problem of k-Dimensional Matching *(k-DM)*, namely,
the problem of finding a maximal matching in a k-partite k-uniform balanced hyper-graph. We prove that k-DM cannot be efficiently approximated to within a factor of $O(\frac{k}{\ln k})$ unless $P = NP$. This improves the
previous factor of $\frac{k}{2^{O(\sqrt{\ln k})}}$ by Trevisan [Tre01]. For low k values we prove
NP-hardness factors of $\frac{54}{53}-\varepsilon$, $\frac{30}{29}-\varepsilon$ and $\frac{23}{22}-\varepsilon$ for 4-DM, 5-DM and 6-DM
respectively. These results extend to the problem of k-Set-Packing and
the problem of Maximum Independent-Set in $(k + 1)$-claw-free graphs.

1 Introduction

Bounded variants of optimization problems are often easier to approximate than
the general, unbounded problems. The Independent-Set problem illustrates this
well: it cannot be approximated to within $O(N^{1-\varepsilon})$ unless $P = NP$ [Hås99],
nevertheless, once the input graph has a bounded degree d, much better approximations exist (e.g., a $\frac{d \log \log d}{\log d}$ approximation by [Vis96]).

We next examine some bounded variants of the set-packing *(SP)* problem
and try to illustrate the connection between the bounded parameters (e.g, sets
size, occurrences of elements) and the complexity of the bounded problem.

In the problem of SP, the input is a family of sets $S_1, ..., S_N$, and the objective
is to find a maximal packing, namely a maximal number of pairwise disjoint sets
from the family. This problem is often phrased in terms of Hyper-graphs: we have
a vertex v_x for each element x and a hyper-edge e_S for each set S of the family
(containing all vertices v_x which correspond the elements x in the set S). The
objective is to find a maximal matching. Alternatively one can formulate this
problem using the dual-graph: a vertex v_S for each set S and a hyper-edge e_x for
each element (v_S is contained in all edges e_x such that $x \in S$). The objective is
to find a maximal independent set (namely, a maximal number of vertices, such
that no two of them are contained in the same edge).

---

[*] Research supported in part by the Fund for Basic Research Administered by the
Israel Academy of Sciences, and a Bikura grant.

S. Arora et al. (Eds.): APPROX 2003+RANDOM 2003, LNCS 2764, pp. 83–97, 2003.
© Springer-Verlag Berlin Heidelberg 2003

The general problem of SP has been extensively studied (for example [Wig83], [BYM84], [BH92], [Hås99]). Quite tight approximation algorithms and inapproximability factors are known for this problem. Håstad [Hås99] proved that Set-Packing cannot be approximated to within $O(N^{1-\varepsilon})$ unless $NP \subseteq ZPP$ (for every $\varepsilon > 0$, where N is the number of sets). The best approximation algorithm achieves an approximation ratio of $O(\frac{N}{\log^2 N})$ [BH92]. In contrast, the case of bounded variants of this problem seems to be of a different nature.

1.1 Bounded Variants of Set-Packing

For bounded variant it seems natural to think of SP using hyper-graph notions. One may think of two natural bounds: the size of the edges (size of the sets) and the degree of the vertices (number of occurrences of each element). For example, k-Set-Packing *(k-SP)* is this problem where the size of the hyper-edges is bounded by k. If we also bound the degree of the vertices by two this becomes the problem of maximum independent-set in k bounded degree graphs *(k-IS)* (recall the dual-graph defined above).

Another natural bound is the colorability of the input graph. Consider the problem of 3-Dimensional Matching *(3-DM)*. It is a variant of $3 - SP$ where the vertices of the input hyper-graph are a union of three disjoint sets, $V = V_1 \uplus V_2 \uplus V_3$, and each hyper-edge contains exactly one vertex from each set, namely, $E \subseteq V_1 \times V_2 \times V_3$. In other words, the vertices of the hyper-graph can be colored using 3 colors, so that each hyper-edge does not contain the same color twice. A graph having this property is called *3-strongly-colorable* (in general - k-strongly-colorable). Thus the color-bounded version of k-SP, namely the problem of k-DM, is

Definition 1 (k-DM). *k-Dimensional Matching*
Input: *A k-uniform k-strongly colorable hyper-graph $H = (V^1, ..., V^k, E)$.*
Problem: *Find a matching of maximal size in H .*

These bounded variants of SP are known to admit approximation algorithms better than their general versions, the quality of the approximation being a function of the bounds. An extensive body of algorithmic work has been devoted to these restricted problems (for example, [HS89]), but matching inapproximability results have only recently been explored (notably by Trevisan [Tre01]).

With some abuse of notations, one can say that hardness of approximation factor of SP is a monotonous increasing function in each of the bounded parameters: the edges size, the vertices degree and the colorability (of edges and vertices). For example, inapproximability factor for graphs of degree bounded by 3 holds for graphs with degree bounded by 4. We next try to overview what is known regarding the complexity of this problem as a function of these bounds.

1.2 Previous Results

2-DM is known to be solvable in polynomial time, say by a reduction to network flow problems [Pap94]. Polynomial time algorithms are also known for graphs that are not bipartite [Edm65].

In contrast, for all $k \geq 3$, k-DM is NP-hard [Kar72, Pap94]. Furthermore, for $k = 3$, the problem is known to be APX-hard [Kan91].

For large k values, we are usually interested in the asymptotic dependence of the approximation ratio (and inapproximability factor) on k. Currently, the best polynomial time approximation algorithm for k-SP achieves an approximation ratio of $\frac{k}{2}$ [HS89]. This is, to date, the best approximation algorithm for k-DM as well.

Alon et al [AFWD95] proved that for suitably large k, k-IS is NP-hard to approximate to within $k^c - \varepsilon$ for some $c > 0$. This was later improved to the currently best asymptotical inapproximability factor [Tre01] of $\frac{k}{2^{O(\sqrt{\ln k})}}$. All hardness factors for k-IS hold in fact for k-DM as well (by a simple reduction).

The best known approximation algorithm for k-IS achieves an approximation ratio of $O(k \log \log k / \log k)$ [Vis96]. For low bound values instances of k-IS, the best approximation algorithm achieves an approximation ratio of $(k + 3)/5$ for $k \geq 3$ (see [BF94, BF95]). For $k = 3, 5$ [BK99, BK03] showed inapproximability factors of $\frac{1676}{1675}$ and $\frac{332}{331}$ respectively.

1.3   Our Contribution

We improve the inapproximability factor for the variant k-DM, and show:

Theorem 1 (Asymptotic Hardness). *It is NP-hard to approximate k-DM to within* $O\left(\frac{k}{\ln k}\right)$

In addition, we show inapproximability factors for 4-DM, 5-DM and 6-DM:

Theorem 2 (Hardness for Low Bound Values). *For every* $\varepsilon > 0$ *it is NP-hard to approximate 4-DM, 5-DM and 6-DM to within* $\frac{54}{53} - \varepsilon$, $\frac{30}{29} - \varepsilon$ *and* $\frac{23}{22} - \varepsilon$ *respectively.*

These results extend to k-SP and Independent-Set in $k + 1$-claw-free graphs *(k + 1-ISCFG)* (see [Hal98] for definition of $k + 1$-ISCFG and reduction from k-SP). They do not hold, however, for k-IS. The table below summarizes known upper and lower bounds.

Recently there have been noteworthy developments namely by [CC02, BK03, CC03]. Inapproximability factors of $\frac{95}{94} - \varepsilon$ for 3-IS and 3-DM and of $\frac{48}{47} - \varepsilon$ for 4-IS and 4-DM were shown by [CC03].

1.4   Outline

Some preliminaries are given in section 2. Section 3 presents the notion of hypergraph-dispersers. Section 4 contains the proof of the asymptotic hardness of approximation for k-SP. Section 5 extends the proof to hold for k-DM. The proof for the low-values inapproximability factors will be given in the full version. The existence of a good hyper-disperser is proved in section 6. The optimality of its parameter is shown in section 6.1. Section 7 contains a discussion on the implications of our results, the techniques used and some open problems.

Problem	Approximation Ratio	Prev. Inapproximability	Our Inapproximability
k-DM k + 1-ISCFG k-SP	$\frac{k}{2}$ [HS89]	$\frac{k}{2^{O(\sqrt{\ln k})}}$ [Tre01]	$O\left(\frac{k}{\ln k}\right)$
4-DM, 4-SP 5-ISCFG	2 [HS89]	$\frac{74}{73} - \varepsilon$ [BK99]	$\frac{54}{53} - \varepsilon$
5-DM, 5-SP 6-ISCFG	$\frac{5}{2}$ [HS89]	$\frac{68}{67} - \varepsilon$ [BK99]	$\frac{30}{29} - \varepsilon$
6-DM, 6-SP 7-ISCFG	$\frac{6}{2}$ [HS89]	$\frac{68}{67} - \varepsilon$ [BK99]	$\frac{23}{22} - \varepsilon$

Table 1. Approximation ratios versus inapproximability factors for k-DM and related problems

2 Preliminaries

In order to prove inapproximability of a maximization problem, one usually defines a corresponding gap problem.

Definition 2 (Gap problems). *Let A be a maximization problem.* **gap-A-**$[a, b]$ *is the following decision problem:*
Given an input instance, decide whether

- *there exists a solution of fractional size at least b, or*
- *every solution of the given instance is of fractional size smaller than a.*

If the size of the solution resides between these values, then any output suffices.

Clearly, for any maximization problem, if gap-A-$[a, b]$ is NP-hard, than it is NP-hard to approximate A to within any factor smaller than $\frac{b}{a}$.
Our main result in this paper is derived by a reduction from the following problem.

Definition 3 (Linear Equations). **MAX-3-LIN-q** *is the following optimization problem:*
Input: *A set Φ of linear equations over $GF(q)$, each depending on 3 variables.*
Problem: *Find an assignment that satisfies the maximum number of equations.*

The following central theorem stemmed from extensive research that formulated in the celebrated PCP theorem (see [ALM+92, AS92]):

Theorem 3 (Håstad [Hås97]). *gap-MAX-3-LIN-q-$[\frac{1}{q} + \varepsilon, 1 - \varepsilon]$ is NP-Hard for every $q \in \mathbb{N}$ and $\varepsilon > 0$. Furthermore, the result holds for instances of MAX-3-LIN-q in which the number of occurrences of each variable is a constant (depending on ε only), chosen from two possible values, and in which no variable appears more than once in a single equation.*

We denote an instance of MAX-3-LIN-q by $\Phi = \{\varphi_1, ..., \varphi_m\}$. Φ is over the set of variables $X = \{x_1, ..., x_n\}$. Let $\Phi(x)$ be the set of all equations in Φ depending on x. Denote by $Sat(\Phi, A)$ the set of all equation of Φ satisfied by an assignment A. If A is an assignment to an equation $\varphi \in \Phi(x)$, we denote by $A[\varphi]_{|x}$ the corresponding assignment to x.

We next explain the reduction from Linear equations to out problem. The reduction gives an inapproximability factor for k-SP. We later amend it to hold for k-DM too.

3   Hyper Dispersers

The following definition is a generalization of disperser graphs. For definitions and results regarding dispersers see [RTS00].

Definition 4 ($((q, \delta)$-Hyper-Graph Edge-Disperser). *We call a hyper graph $H = (V, E)$ a (q, δ)-Hyper-Graph Edge-Disperser if there exists a partition of its edges:*

$$E = E_1 \uplus ... \uplus E_q \quad , \quad |E_1| = ... = |E_q|$$

such that every large matching M of H is (almost) concentrated in one part of the edges. Formally, there exists i so that

$$|M \setminus E_i| \leq \delta |E|$$

Lemma 1. *For every $q > 1$ and $t > 1$ there exists a hyper-graph $H = (V, E)$ such that*

- $V = [t] \times [d]$, whereas $d = \Theta(q \ln q)$.
- H is $(q, \frac{1}{q^2})$-hyper-edge-disperser
- H is d uniform, d-strongly-colorable.
- H is q regular, q-strongly-edge-colorable.

We denote this graph by $(t, q) - \mathcal{D}$.

This lemma is in section 6.

4   Proof of the Asymptotic Inapproximability Factor for k-SP

This section provides a deterministic polynomial time reduction from MAX-3-LIN-q to k-SP.

4.1   The Construction

Let $\Phi = \{\varphi_1, ..., \varphi_n\}$ be an instance of MAX-3-LIN-q over the sets of variables X and Y, where each variable $x \in X$ and $y \in Y$ occurs a constant number of times c_X and c_Y respectively (recall Theorem 3). We now describe how to deterministically construct, in polynomial time, an instance of k-SP - the hyper-graph $H_\Phi = (V, E)$.

Let $\mathcal{D}_X$ be a $(c_X, q) - \mathcal{D}$ and $\mathcal{D}_Y$ be a $(c_Y, q) - \mathcal{D}$ (which exist by lemma 1). For every variable $x \in X$ (and $y \in Y$) we have a copy $\mathcal{D}_x$ of $\mathcal{D}_X$ (or $\mathcal{D}_y$ of $\mathcal{D}_Y$). The vertices of H_Φ are the union of the vertices of all these hyper-disperses. Formally,

$$V \quad = \quad X \times [c_X] \times [q] \quad \cup \quad Y \times [c_Y] \times [q]$$

namely,

$$V = \{v_{x,\varphi,i} \mid x \in X \cup Y, \varphi \in \Phi(x), i \in [d]\}$$

The Edges of H_Φ. We have an edge for each equation $\varphi \in \Phi$ and a satisfying assignment to it. Consider an equation $\varphi = x+y+z = a \mod q$, and a satisfying assignment A to that equation (note that there are q^2 such assignments, as assigning the first two variables, determines the third). The corresponding edge, $e_{\varphi,A}$, is composed of three edges, one from the hyper-graph $\mathcal{D}_x$, one from $\mathcal{D}_y$ and the last from $\mathcal{D}_z$. Formally:

$$e_{\varphi,A} = e_{x,\varphi,A_{|x}} \cup e_{y,\varphi,A_{|y}} \cup e_{z,\varphi,A_{|z}}$$

Where $A_{|x}$ is the restrictions of the assignment A to the variable x, and $e_{x,\varphi,A_{|x}}$ is the edge $e[\varphi, A_{|x}]$ of $\mathcal{D}_x$ (and similarly for y and z). The edges of H_Φ are

$$E = \{e_{\varphi,A} \mid \varphi \in \Phi, A \text{ is a satisfying assignment to } \varphi\}$$

Clearly, the cardinality of $e_{\varphi,A}$ is $3d$ (and note that each of the three composing edges participates in creating q edges). This concludes the construction.

Notice that the construction is indeed deterministic, as each variable occurs a constant number of times (see Theorem 3). Hence, the size of $\mathcal{D}_X$ and $\mathcal{D}_Y$ is constant and its existence (see lemma 1) suffices, as one can enumerate all possible hyper-graphs, and verify their properties.

Claim. [Completeness] If there is an assignment to Φ which satisfies $1 - \varepsilon$ of its equations, then there is a matching in H_Φ of size $\left(\frac{1-\varepsilon}{q^2}\right)|E|$.

Proof. Let A be an assignment that satisfies $1 - \varepsilon$ of the equations. Consider the matching $M \subseteq E$ comprised of all edges corresponding to A, namely

$$M = \{e_{\varphi,A(\varphi)} \mid \varphi \in Sat(\Phi, A)\}$$

Trivially, $|M| = \left(\frac{1-\varepsilon}{q^2}\right)|E|$, as we took one edge corresponding to each satisfied equation. These edges are indeed a matching since for each variable, only edges corresponding to a single assignment to that variable are taken.

Lemma 2. *[Soundness] If every assignment to Φ satisfies at most $\frac{1}{q}+\varepsilon$ fraction of its equations, then every matching in H_Φ is of size $O\left(\frac{1}{q^3}|E|\right)$.*

Proof. Denote by E_x the edges of H_Φ corresponding to equations φ containing the variable x, namely,

$$E_x = \{e_{\varphi,A} \mid \varphi \in \Phi(x), e_{\varphi,A} \in E\}$$

Denote by $E_{x=a}$ the subset of E_x corresponding to an assignment of a to x, that is,

$$E_{x=a} = \{e_{\varphi,A} \mid e_{\varphi,A} \in E_x, A_{|x} = a\}$$

Let M be a matching of maximal size in H_Φ. Let A_{maj} be the most popular assignment. That is, for every $x \in X \cup Y$ choosing the assignment of x to be such that it corresponds to maximal number of edges. Formally, choose

$$A_{maj}(x) \in [q] \text{ s.t. } |E_{x=a} \cap M| \text{ is maximized}$$

Let M_{maj} be the set of edges in M that agree with A_{maj}, and M_{min} be all the other edges in M, namely

$$M_{maj} = \{e_{\varphi,A_{maj}}\}_{\varphi \in \Phi}$$

$$M_{min} = M \setminus M_{maj}$$

As $|Sat(\Phi, A_{maj})| \leq \frac{1}{q} + \varepsilon$, we have $|M_{maj}| < (\frac{1}{q} + \varepsilon)\frac{E}{q^2}$.
From the disperser-properties of $\mathcal{D}_X$ and $\mathcal{D}_Y$ (derived from lemma 1) we know that for every $x \in X \cup Y$

$$\sum_{a \neq A_{maj}(x)} |M_{min} \cap E_{x=a}| \leq \frac{1}{q^2} E(\mathcal{D}_x)$$

This means that

$$\sum_{a \neq A_{maj}(x)} |M_{min} \cap E_{x=a}| \leq \frac{1}{q^3}|E_x|$$

as every edge of $\mathcal{D}_x$ is a subset of q hyper edges in E_x, but only one of such q edges can be taken to M as they share vertices (recall that M is a matching). Therefore,

$$|M_{min}| \leq \sum_{x \in X \cup Y, a \neq A_{maj}(x)} |M_{min} \cap E_{x=a}| \leq \frac{1}{q^3} \sum_{x \in X \cup Y} |E_x| = \frac{3}{q^3}|E|$$

and thus

$$|M| = |M_{min}| + |M_{maj}| \leq (\frac{4}{q^3} + \varepsilon)|E|$$

By claim 4.1 and lemma 2 we showed that Gap-k-SP-$\left[\frac{4}{q^3} + \varepsilon, \frac{1}{q^2} - \varepsilon\right]$ is NP-hard. Since each edge is of size $k = 3d = \Theta(q \log q)$ it is NP-hard to approximate k-SP to within $O(\frac{\ln k}{k})$.

5   Extending the Proof for k-DM

The proof for k-DM follows the steps of the proof for k-SP. The difference being that we use three dispersers for each variable (instead of one) - a different disperser for each location in the equations. Denote by $\Phi(x, l)$ the subset of $\Phi(x)$ where x is the l'th variable in the equation (clearly $l \in [3]$). Note that w.l.o.g. we may assume that for every $x \in X, \Phi(x, 1) = \Phi(x, 2) = \Phi(x, 3)$ (as we can take three copies of each equation, and shift the location of the variables).

Let $\mathcal{D}_X \equiv (c_X/3, q) - \mathcal{D}$ and $\mathcal{D}_Y \equiv (c_Y/3, q) - \mathcal{D}$ (as stated in lemma 1). For every variable $x \in X$ (or $y \in Y$) and position $l \in [3]$, we have a copy $\mathcal{D}_{x,l}$ of $\mathcal{D}_X$ (or $\mathcal{D}_{y,l}$ of $\mathcal{D}_Y$).

$$V \;\; = \;\; X \times V(\mathcal{D}_X) \times [3] \;\; \cup \;\; Y \times V(\mathcal{D}_Y) \times [3]$$

namely,

$$V = \{v_{x,\varphi,i} \mid x \in X \cup Y, \varphi \in \Phi(x), i \in [d]\}$$

where the index $i \in [q]$ is given by a strong-coloring of the edges with q colors (recall that such a coloring exists as $(t, q) - \mathcal{D}$ is q-strongly colorable).

The Edges of H_Φ. We have an edge for each equation $\varphi \in \Phi$ and a satisfying assignment to it. Consider an equation $\varphi = x+y+z = a \mod q$, and a satisfying assignment A to that equation. The corresponding edge, $e_{\varphi,A}$, is composed of three edges, one from the hyper-graph $\mathcal{D}_{x,1}$, one from $\mathcal{D}_{y,2}$ and the last from $\mathcal{D}_{z,3}$. Formally:

$$e_{\varphi,A} = e_{x,\varphi,A_{|x}} \cup e_{y,\varphi,A_{|y}} \cup e_{z,\varphi,A_{|z}}$$

Where $e_{x,\varphi,A_{|x}}$ is the edge $e[\varphi, A_{|x}]$ of $\mathcal{D}_{x,1}$, $e_{y,\varphi,A_{|y}}$ is the edge $e[\varphi, A_{|y}]$ of $\mathcal{D}_{y,2}$ and $e_{z,\varphi,A_{|z}}$ is the edge $e[\varphi, A_{|z}]$ of $\mathcal{D}_{z,3}$. The edges of H_Φ are

$$E = \{e_{\varphi,A} \mid \varphi \in \Phi, A \text{ is a satisfying assignment to } \varphi\}$$

This concludes the construction for k-DM. We next show that the graph constructed is indeed a k-DM instance:

Proposition 1. *H_Φ is 3d-strongly-colorable.*

Proof. We show how to partition V into $3d$ independent sets of equal size. Let the sets be $P_{l,i}$ whereas $i \in [d]$ and $l \in [3]$:

$$P_{l,i} = \{v_{x,\varphi,i} \mid x \in X \cup Y, \varphi \in \Phi(x, l)\}$$

$P_{l,i}$ is clearly a partition of the vertices, as each vertex belongs to a single part. We now explain why each part is an independent set. Let $P_{l,i}$ be an arbitrary part, and let $e_{\varphi,A} \in E$ be an arbitrary edge, where $\varphi \equiv x + y + z = a \mod q$:

$$e_{\varphi,A} = e_{x,\varphi,A[\varphi]_{|x}} \cup e_{y,\varphi,A[\varphi]_{|y}} \cup e_{z,\varphi,A[\varphi]_{|z}}$$

$P_{l,i} \cap e_{\varphi,A}$ may contain vertices corresponding only to one of the variables x, y, z, since it contains variables corresponding to a single location (first, second or third). Let that variable be, w.l.o.g, x. The edge $e_{x,\varphi,A[\varphi]_{|x}}$ contains exactly one vertex from each of the d parts, as the graph $D_{x,1}$ is d-partite. Therefore, the set $P_{l,i} \cap e_{\varphi,A}$ contains exactly one vertex. Since $|P_{l,i} \cap e_{\varphi,A}| = 1$ for every edge and every set $P_{l,i}$, the graph H_Φ is $3d$-partite-balanced.

The completeness claim for k-SP (claim 4.1) holds here too. The soundness lemma for k-SP holds with minor changes:

Lemma 3. *[Soundness] If every assignment to Φ satisfies at most $\frac{1}{q} + \varepsilon$ fraction of its equations, then every matching in G is of size $O\left(\frac{1}{q^3}E\right)$.*

Proof. We repeat the soundness proof of k-SP but the definition of the most-popular assignment is slightly different, and takes into account the three different dispersers per variable.

Denote by $E_{x,l}$ the edges of H_Φ corresponding to equations φ containing the variable x in location l, namely,

$$E_{x,l} = \{e_{\varphi,A} \mid \varphi \in \Phi(x,l), A \in [q^2]\}$$

Denote by $E_{x=a,l}$ the subset of $E_{x,l}$ corresponding to an assignment of a to x, that is,

$$E_{x=a,l} = \{e_{\varphi,A} \mid \varphi \in \Phi(x,l), A[\varphi]_{|x} = a\}$$

Let M be a matching of maximal size. Let A_{maj} be the most popular of most popular assignment. That is, for every $x \in X \cup Y$ choose the location (of equations of edges of M) in which x appears maximal number of time :

$$\hat{l}(x) \in [3] \text{ s.t. } |E_{x,\hat{l}(x)} \cap M| \text{ is maximized} \tag{1}$$

Then choose an assignment for x such that it corresponds to maximal number of those edges. Formally, choose

$$A_{maj}(x) \in [q] \text{ s.t. } |E_{x=a,\hat{l}(x)} \cap M| \text{ is maximized}$$

As before, let M_{maj} be the set of edges in M that agree with A_{maj}, and M_{min} be all the other edges in M, namely

$$M_{maj} = \{e_{\varphi,A_{maj}}\}_{\varphi \in \Phi}$$

$$M_{min} = M \setminus M_{maj}$$

For the exact same reasons as in the k-SP proof, we have

$$|M_{maj}| < \left(\frac{1}{q} + \varepsilon\right)\frac{|E|}{q^2} \tag{2}$$

and for every x,

$$\sum_{a \neq A_{maj}(x)} |M_{min} \cap E_{x=a,\hat{l}(x)}| \leq \frac{1}{q^3}|E_{x,\hat{l}(x)}| \tag{3}$$

Therefore,

$$|M| = \sum_{x,l} |M \cap E_{x,l}|$$

$$\leq \sum_{x,l} |M_{maj} \cap E_{x,l}| + \sum_{x,l,a \neq A_{maj}(x)} |M_{min} \cap E_{x=a,l}|$$

by (1) we have

$$\leq 3 \cdot \sum_{x} |M_{maj} \cap E_{x,\hat{l}(x)}| + 3 \cdot \sum_{x,a \neq A_{maj}(x)} |M_{min} \cap E_{x=a,\hat{l}(x)}|$$

$$\leq 3 \cdot |M_{maj}| + 3 \cdot \sum_{x,a \neq A_{maj}(x)} |M_{min} \cap E_{x=a,\hat{l}(x)}|$$

thus by (2) and (3)

$$< 3(\frac{1}{q} + \varepsilon)\frac{|E|}{q^2} + \frac{3}{q^3} \sum_{x} |E_{x,\hat{l}(x)}|$$

$$= (\frac{12}{q^3} + 3\varepsilon)|E|$$

By claim 4.1 and lemma 3 we showed that Gap-k-DM-$\left[\frac{12}{q^3} + 3\varepsilon, \frac{1}{q^2} - \varepsilon\right]$ is NP-hard, thus it is NP-hard to approximate k-DM to within $O(\frac{\ln k}{k})$.

6   Hyper-Dispersers

In this section, we prove lemma 1. As stated before, these are generalizations of disperser graphs. In section 6.1, we prove that these are the best (up to a constant) parameters for a hyper-disperser one can hope to achieve.

Lemma 1 *For every $q > 1$ and $t > 1$ there exists a hyper-graph $H = (V, E)$ such that*

- *$V = [t] \times [d]$, whereas $d = \Theta(q \ln q)$.*
- *H is $(q, \frac{1}{q^2})$-hyper-edge-disperser*
- *H is d uniform, d-strongly-colorable.*
- *H is q regular, q-strongly-edge-colorable.*

We denote this graph by $(t, q) - \mathcal{D}$.

Proof. Let

$$V = [t] \times [d]$$

and denote $V_i = [t] \times \{i\}$.
We next randomly construct the edges of the hyper-graph, so that it is d-uniform, q-regular. Let S_t be all permutation over t elements, and let

$$\Pi_{i_1,i_2} \in_R S_t , \ (i_1, i_2) \in [q] \times [d]$$

(that is, qd permutations, chosen uniformly from S_t). Define

$$e[i,j] = \{ \ (\Pi_{j,1}(i),1), \ (\Pi_{j,2}(i),2) \ , ..., \ (\Pi_{j,d}(i),d) \ \} \tag{4}$$

and let

$$E = \{e[i,j] \mid (i,j) \in [t] \times [q]\}$$

Hence $|E| = tq$. Define a partition of the edges as follows: $E_j = \{e[i,j] \mid i \in [t]\}$. Thus $|E_1| = ... = |E_q| = t$ and each set of edges E_j covers every vertex exactly once. Therefore, H is q strongly-edge-colorable. On the other hand, every edge contains exactly one vertex from each set of vertices V_i. Thus H is d-strongly-colorable.

We next show that with high probability H has the disperser property, namely, every matching M of H is concentrated on a single part of the edges, except for maybe $\frac{1}{q^2}|E| = \frac{t}{q}$ edges of M. Denote by P the probability that H does *not* have the disperser property. Let $\mathcal{M}$ be the family of all subsets $M \subseteq E$ of interest, that is, a family (of *not* concentrated subsets of edges) that ought to be inspected in order to determine whether H has the disperser property, namely,

$$\mathcal{M} = \{M \mid M \subseteq E, |M| = \frac{t}{q} + \frac{t}{q^2}, \exists i, |M \setminus E_i| = \frac{t}{q}\}$$

(note that if a set M is a matching, so is any subset of M, hence it suffices to check H for all subsets $M \in \mathcal{M}$). Denote by $\Pr[M]$ the probability (over the random choice of H) that M is a matching. By union bound,

$$P = \Pr_H[\exists M \in \mathcal{M}, \ M \text{ is a matching }] \leq$$

$$\leq \sum_{M \in \mathcal{M}} \Pr[M] \leq |\mathcal{M}| \Pr[\hat{M}] \tag{5}$$

where $\hat{M} \in \mathcal{M}$ is the set which maximizes $\Pr[\hat{M}]$. Clearly,

$$|\mathcal{M}| \leq q \binom{(q-1)t}{\frac{t}{q}} \binom{t}{\frac{t}{q^2}} \leq q(eq^2)^{\frac{t}{q}} (eq^2)^{\frac{t}{q^2}} \leq (eq)^{\frac{3t}{q}} \tag{6}$$

We next bound $\Pr[\hat{M}]$. Let $M_i = \hat{M} \cap E_i$. Let $B_{i,j}$ be the event that the sets of edges M_i and M_j do not share a vertex, and $A_i = \cap_{j<i}B_{i,j}$. Then

$$\Pr[\hat{M}] = \Pr\left[\bigcap_i A_i\right] = \prod_i \Pr\left[A_i \mid \bigcap_{l<i} A_l\right]$$

Note however, that the event A_i is independent of the event $\cap_{l<i} A_l$ as A_i is determined by (the independently chosen permutations) $\{\Pi_{i,j} \mid j \in [d]\}$, whereas $\cap_{l<i} A_l$ is determined by the permutations $\{\Pi_{l,j} \mid l < i, j \in [d]\}$. Thus

$$\Pr[\hat{M}] = \prod_i \Pr[A_i] \tag{7}$$

Let $C_{i,j}$ be the event that there is no collision (common vertex) of M_i and $\bigcup_{l<i} M_l$ on the subset of vertices V_j (clearly $A_i = \bigcap_{j\in[d]} C_{i,j}$). Hence, as for $j_1 \neq j_2$, C_{i,j_1} and C_{i,j_2} are determined by independent sets of permutations (recall (4)) we have

$$\Pr[A_i] = \prod_{j\in[d]} \Pr[C_{i,j}] = (\Pr[C_{i,1}])^d \le (1 - \frac{|M_i|}{t})^{d|\bigcup_{l<i} M_l \cap V_1|} = (1 - \frac{|M_i|}{t})^{d\sum_{l<i} |M_l|}$$

where the sum in the exponent of the rightmost expression is by assuming no collisions between edges of $\bigcup_{l<i} M_l$ on V_j (which is implied by $\bigcap_{l<i} A_l$). Thus by equation (7) we have (as $1 - x \le e^{-x}$):

$$\Pr[\hat{M}] \le \prod_i \left(1 - \frac{|M_i|}{t}\right)^{d\sum_{j<i} |M_j|} \le e^{-\frac{d}{t}\sum_{i=2}^{q}(|M_i|\sum_{j=1}^{i-1} |M_j|)}$$

Under the constraint that $\hat{M} \in \mathcal{M}$ the sum $\sum_{i=2}^{q}(|M_i|\sum_{j=1}^{i-1} |M_j|)$ is minimized for $|M_2| = |M_3| = \frac{t}{2q}$ hence

$$\Pr[\hat{M}] \le e^{-\frac{dt}{4q^2}} \tag{8}$$

Therefore by equations (5),(6),(8),

$$P \le (eq)^{\frac{3t}{q}} e^{-\frac{dt}{4q^2}}$$

Any d which guarantees that $(eq)^{\frac{3t}{q}} e^{-\frac{dt}{4q^2}} \ll 1$ suffices (for example $d \ge 20q \ln q$) as $P < 1$, thus there exists H with the disperser properties.

6.1   Optimality of Hyper-Disperser Construction

We now turn to see why the hyper-disperser from lemma 1 has optimal parameters. We base our observation on a lemma from [RTS00]:

Definition 5. *A bipartite graph $G = (V_1, V_2, E)$ is called a δ-disperser if for every $U_1 \subseteq V_1, U_2 \subseteq V_2, |U_1|, |U_2| \ge \delta|V_1| = \delta|V_2|$, the subset $U_1 \cup U_2$ is not an independent set.*

Lemma 4. *Every bipartite d-regular $\frac{1}{k}$-disperser must satisfy $d = \Omega(k \ln k)$.*

Proposition 2. *Every d-uniform q-strongly-edge-colorable q-regular d-strongly colorable $(q, \frac{1}{q^2})$-hyper-edge-disperser must satisfy $d = \Omega(q \ln q)$.*

Proof. We prove that in case there exists such a hyper-graph which satisfies $d = o(q \ln q)$, then there exists a bipartite $o(q \ln q)$-regular $\frac{1}{q}$-disperser, in contrast to lemma 4. We transform a d-partite d-uniform q-regular q-strongly-edge colorable

$(q, \frac{1}{q^2})$-hyper-disperser $H = (V_H, E_1, E_2, ..., E_q)$ into a bipartite d-regular $\frac{1}{q}$-disperser $G = (V_1, V_2, E_G)$ in the following way. Let

$$V_1 = E_1$$

$$V_2 = E_2$$

$$E_G = \{(e_1, e_2) \mid e_1 \cap e_2 \neq \phi\}$$

Obviously G is a bipartite d-regular graph (we allow multi-edges). In addition, suppose two sets of fractional sizes:

$$S_1 = \frac{1}{q}V_1, S_2 = \frac{1}{q}V_2$$

are an independent set in G. Then the corresponding sets of edges in H are disjoint and are of fractional size $\frac{2}{q^2}$, thus contradicting the fact that H is a $(q, \frac{1}{q^2})$-hyper-disperser.

7 Discussion

An interesting property of our construction (for both asymptotic and low bound values results) is the *almost perfect completeness*. This property refers to the fact that the matching proved to exist in the completeness claim 4.1 is an almost perfect matching, that is, it covers $1 - \varepsilon$ of the vertices. Knowing the location of a gap is interesting by itself and may proof useful (in particular if it is extreme on either the completeness or the soundness parameters, see for example [Pet94]). In fact, applying our reduction on other PCP variants instead of Max-3-Lin-q (e.g. parallel repetition of 3-SAT) yields perfect completeness for k-DM (but with weaker hardness factors).

The ratio between the asymptotic inapproximability factor presented herein for k-DM and k-SP, and the tightest approximation algorithm known, was reduced to $O(\ln k)$. The open question of where in the range, from $\frac{k}{2}$ to $O(\frac{k}{\ln k})$ is the approximability threshold is interesting by itself, as well as its implications to the difference between k-DM and k-IS. The current asymptotic inapproximability factor of $O(\frac{k}{\ln k})$ for k-DM approaches the tightest approximation ratio known for k-IS, namely $O(k \log \log k / \log k)$ [Vis96]. Thus, a small improvement in either the approximation ratio or the inapproximability factor will show these problems to be of inherently different complexity.

An improvement in the low bound values hardness factor for k-DM may also separate these problems. The tightest known approximation algorithm for low bound values of k-IS achieves an approximation ratio of $(k + 3)/5$ for $k \geq 3$ [BF94, BF95]. Thus, improving the low bound values factors up to $\frac{6}{5} + \varepsilon$ for 3-DM or $\frac{7}{5} + \varepsilon$ for 4-DM, suffices for separating these problems.

8 Acknowledgements

We would like to thank Adi Akavia and Dana Moshkovitz for their helpful and insightful comments.

References

[AFWD95] N. Alon, U. Fiege, A. Wigderson, and D.Zuckerman. Derandomized graph products. *Computational Complexity*, 5:60–75, 1995.

[ALM+92] S. Arora, C. Lund, R. Motwani, M. Sudan, and M. Szegedy. Proof verification and intractability of approximation problems. In *Proc. 33rd IEEE Symp. on Foundations of Computer Science*, pages 13–22, 1992.

[AS92] S. Arora and S. Safra. Probabilistic checking of proofs: A new characterization of NP. In *Proc. 33rd IEEE Symp. on Foundations of Computer Science*, pages 2–13, 1992.

[BF94] P. Berman and M. Furer. Approximating maximum independent set in bounded degree graphs. *SODA*, pages 365–371, 1994.

[BF95] P. Berman and T. Fujito. On the approximation properties of independent set problem in degree 3 graphs. *WADS*, pages 449–460, 1995.

[BH92] R. Boppana and Magnus M. Halldorsson. Approximating maximum independent sets by excluding subgraphs. *Bit 32*, pages 180–196, 1992.

[BK99] P. Berman and M. Karpinski. On some tighter inapproximability results. *DIMACS Technical Report 99-23*, 1999.

[BK03] Piotr Berman and Marek Karpinski. Improved approximation lower bounds on small occurrence optimization. *ECCC TR03-008*, 2003.

[BYM84] R. Bar-Yehuda and S. Moran. On approximation problems related to the independent set and vertex cover problems. *Discrete Applied Mathematics*, 9:1–10, 1984.

[CC02] M. Chlebik and J. Chlebikova. Approximation hardness for small occurrence instances of NP-hard problems. *ECCC TR02-073*, 2002.

[CC03] M. Chlebik and J. Chlebikova. Inapproximability results for bounded variants of optimization problems. *ECCC TR03-026*, 2003.

[Edm65] J. Edmonds. Paths, trees and flowers. *Canadian Journal of Mathematics*, 17:449–467, 1965.

[Hal98] Magnus M. Halldorsson. Approximations of independent sets in graphs. *APPROX*, 1998.

[Hås97] Johan Håstad. Some optimal inapproximability results. In *Proceedings of the Twenty-Ninth Annual ACM Symposium on Theory of Computing*, pages 1–10, El Paso, Texas, 4–6 May 1997.

[Hås99] Johan Håstad. Clique is hard to approximate within $n^{1-\epsilon}$. *Acta Math.*, 182(1):105–142, 1999.

[HS89] C. A. J. Hurkens and A. Schrijver. On the size of systems of sets every t of which have an sdr, with an application to the worst-case ratio of heuristics for packing problems. *SIAM Journal Discrete Math*, 2:68–72, 1989.

[Kan91] V. Kann. Maximum bounded 3-dimensional matching is MAXSNP-complete. *Information Processing Letters*, 37:27–35, 1991.

[Kar72] R. M. Karp. Reducibility among combinatorial problems. *Complexity of Computer Computations*, pages 83–103, 1972.

[Pap94] C. Papadimitriou. *Computational Complexity*. Addison Wesley, 1994.

[Pet94] E. Petrank. The hardness of approximation - gap location. *Israel Symposium on Theory of Computing Systems*, 1994.

[RTS00] J. Radhakrishnan and A. Ta-Shma. Bounds for dispersers, extractors, and depth-two superconcentrators. *SIAM Journal on Discrete Mathematics*, 13:2–24, 2000.

[Tre01]   L. Trevisan. Non-approximability results for optimization problems on bounded degree instances. *In Proc. of the 33rd ACM STOC*, 2001.

[Vis96]   S. Vishwanathan. Personal communication to m. halldorsson cited in [Hal98]. 1996.

[Wig83]   A. Wigderson. Improving the performance guarantee for approximate graph coloring. *Journal of the Association for Computing Machinery*, 30(4):729–735, 1983.

Approximating Market Equilibria

Kamal Jain[1], Mohammad Mahdian[2], and Amin Saberi[3]

[1] Microsoft Research, One Microsoft Way, Redmond, WA 98052.
kamalj@microsoft.com
[2] Laboratory for Computer Science, MIT, Cambridge, MA 02139, USA.
mahdian@theory.lcs.mit.edu
[3] College of Computing, Georgia Tech, Atlanta, GA 30332, USA.
saberi@cc.gatech.edu

Abstract. In this paper we consider the classic problem of finding the market equilibrium prices under linear utility functions. A notion of *approximate market equilibrium* was proposed by Deng, Papadimitriou and Safra [5]. Using this notion, we present the first fully polynomial-time approximation scheme for finding a market equilibrium price vector. The main tool in our algorithm is the polynomial-time algorithm of Devanur et al. [6] for a variant of the problem in which there is a clear demarcation between buyers and sellers. Their algorithm is used as a subroutine in our algorithm.

1 Introduction

The behavior of a complex marketplace with multiple goods, buyers, and sellers, can only be understood by analyzing the system in its entirety. In practice, such markets tend toward a delicate balance of supply and demand as determined by the agents' fortunes and utilities. The study of this equilibrium situation is known as general equilibrium theory, and was first formulated by Léon Walras in 1874 [12]. In the Walrasian model, the market consists of a set of agents, each with an initial endowment of goods, and a function describing the utility each one will derive from any allocation. The initial allocation could be sub-optimal, and the task of exchanging goods to mutually increase the utilities might be fairly complicated. A functioning market accomplishes this exchange by determining appropriate prices for the goods. Given these prices, all agents independently maximize their own utility by selling their endowments and buying the best bundle of goods they can afford. This new allocation will be an equilibrium allocation if the total demand for every good equals its supply. The prices that induce this equilibrium are called the market-clearing prices, and the equilibrium itself is called a market equilibrium.

Much work has been devoted to establishing the existence of market equilibria [1, 11]. This difficult problem is approached by placing different assumptions on the endowment and utility functions of the agents. The seminal work of Arrow and Debreu [1] proves the existence of market equilibria in the quite general setting of concave utility functions by applying Kakutani's fixed point theorem.

S. Arora et al. (Eds.): APPROX 2003+RANDOM 2003, LNCS 2764, pp. 98–108, 2003.

This generality comes at a high price: the proof is non-constructive and so does not give an algorithm to compute the equilibrium prices. Yet computing these prices can be of considerable importance for predicting the market. For example, in order to determine the effects of a change in a tariff, we must be able to compute the equilibrium prices before and after the tariff change. Equilibrium prices also have applications in computer science. Kelly and Vazirani [8] show that the rate control for elastic traffic in a network can be reduced to a market equilibrium problem.

Despite the impressive progress in computing equilibrium prices [2, 3, 4, 10], especially the seminal work of Scarf [10], polynomial-time algorithms have evaded researchers. In the special case of linear utility functions, Deng, Papadimitriou, and Safra [5] (see also [9]) provide a polynomial-time algorithm when the number of goods or agents is bounded. Devanur et al. [6] obtain a polynomial-time algorithm via a primal-dual-type approach when there is a demarcation between sellers and buyers. However, the question of existence of a polynomial-time algorithm for the general case is still open. In this paper, we present the first fully polynomial-time approximation scheme for this problem. Since the market equilibrium problem is not an optimization problem, we need to clarify what we mean by an *approximate market equilibrium*. For this, we use a definition proposed by Deng, Papadimitriou, and Safra [5]. According this definition, an approximate market equilibrium is a price vector for which there is an allocation of goods to the agents that approximately clears the market and each agent is approximately maximally happy with the allocation (subject to her budget constraint). The precise definition is presented in Section 2.

In the market equilibrium problem, all agents are buyers as well as sellers. The algorithm of Devanur et al. [6] works only when the buyers and sellers are different. The reason is that their algorithm requires that the buyers' budgets be known beforehand. The main idea of our algorithm is to overcome this difficulty by running in iterations, and letting the budget of an agent in the current iteration be the revenue she generated in the previous iteration. The algorithm of Devanur et al. [6] requires an initial setting of prices in which no good is undersold. We satisfy this requirement by adding a dummy buyer who has enough money to buy the residual goods.

The rest of this paper is organized as follows. In Section 2, we precisely define the model, our assumptions, and the notion of approximate market equilibria. In Section 3, we present our algorithms. In Section 4, we prove that one of our algorithms computes an approximate equilibrium in polynomial time. Finally, in Section 5 we conclude with a summary of our results and a discussion of remaining open questions.

2   Definitions and Preliminaries

Consider a market consisting of n agents trading m types of divisible goods. Initially, each agent i has an endowment $w^i \in \mathbb{R}^m$ of goods (i.e., w^i_j indicates the amount of good j that agent i initially has). We assume, without loss of generality, that in total there is one unit of each type of good in the initial endowments (i.e., $\sum_{i=1}^{n} w_j^i = 1$ for every good j). Also, each agent i has a utility function $u_i : \mathbb{R}^m \mapsto \mathbb{R}^+$. That is, if $x \in \mathbb{R}^m$ is a vector that specifies how much of each good agent i has, then $u_i(x)$ indicates the utility (or happiness) of agent i. If a price of p_j^* dollars is set for one unit of good j, then agent i can sell her endowment for a total of $\sum_{j=1}^{m} p_j^* w_j^i$ dollars. Using this money, she can buy a *bundle* $x \in \mathbb{R}^m$ of goods. Since each agent is trying to maximize her utility, the bundle x is a solution to the following maximization program.

$$\text{maximize } u_i(x)$$
$$\text{subject to } \sum_{j=1}^{m} p_j^* x_j \leq \sum_{j=1}^{m} p_j^* w_j^i. \tag{1}$$

Such a solution x is called an *optimal bundle* for agent i. If the function u_i is strictly concave (i.e., for every $x \neq x' \in \mathbb{R}^m$, $u_i((x+x')/2) > (u_i(x)+u_i(x'))/2)$, then there is a unique optimal bundle for agent i. The Arrow-Debreu theorem states the following.

Theorem A (Arrow and Debreu [1]). *Consider the above setting and assume that u_i's are strictly concave. Then there is a price vector $\mathbf{p}^*$ such that if each agent buys the optimal bundle with respect to $\mathbf{p}^*$, then the market clears. In other words, if $x^i \in \mathbb{R}^m$ is the optimal bundle for agent i with respect to $\mathbf{p}^*$, then for every good j, $\sum_{i=1}^{n} x_j^i = \sum_{i=1}^{n} w_j^i$.*

If the utility functions are concave but not strictly concave (e.g., if they are linear), then the optimal bundle is not necessarily unique. In this case, the Arrow-Debreu theorem says that there is a price vector $\mathbf{p}^*$ and a bundle x^i for each agent i, such that x^i is *an* optimal bundle for i with respect to $\mathbf{p}^*$, and if for every i, agent i buys the bundle x^i, then the market clears.

The proof of the Arrow-Debreu theorem is existential and uses a fixed point theorem. Therefore, a natural question is whether one can efficiently compute the equilibrium prices that are guaranteed to exist by the Arrow-Debreu theorem. This problem is still widely open.

In [6], Devanur et al. present a polynomial-time algorithm that computes the market-clearing prices in a market with the following conditions:

1. All utility functions are linear, i.e., $u_i(x) = \sum_{j=1}^{m} u_{ij} x_j$ for non-negative constants u_{ij}.
2. There is a distinction between buyers and sellers in the market. More precisely, there are m sellers, each having one unit of a different type of good, and n buyers in the market. Each buyer i has a given budget e_i, and wants to buy a certain amount of each good to maximize her utility, subject to her budget constraint. We will call a market with this property a *dichotomous market*.

In this paper, we refer to the algorithm of Devanur et al. [6] as the *DPSV algorithm*. The idea of the DPSV algorithm is to start from a price vector $\mathbf{p}^0$ satisfying an invariant stated below, and keep increasing the prices subject to not violating the invariant, until we converge to the equilibrium prices. In order to introduce the invariant, we first define the concept of the *equality subgraph*.

Let $\mathbf{p}$ be a price vector. For each agent i, let $\alpha_i = \max_j \{u_{ij}/p_j\}$ (α_i is agent i's *bang per buck*). The equality subgraph $N(\mathbf{p})$ is a network whose vertex set consists of a source s, a vertex a_j for each good j, a vertex b_i for each buyer i, and a sink t. Let A and B denote the sets of a_j's and b_i's, respectively. There is an edge from s to each $a_j \in A$ of capacity p_j (the price of j), and an edge from each $b_i \in B$ to t of capacity e_i (the budget of i). Also, for each buyer i and good j, if $\alpha_i = u_{ij}/p_j$, then we put an edge from a_j to b_i of infinite capacity. This edge is called an *equality edge*. Notice that by this definition a bundle is optimal for buyer i with respect to the prices $\mathbf{p}$ if and only if its total price is equal to the budget of i, and it only contains goods that have an equality edge to b_i in $N(\mathbf{p})$.

Now, we can state the invariant of the DPSV algorithm.

Invariant 1 *The prices $\mathbf{p}$ are such that $(s, A \cup B \cup t)$ is a min-cut in $N(\mathbf{p})$.*

For a price vector $\mathbf{p}$ and a subset S of goods, we define $\Gamma_p(S)$ as the set of buyers i such that $N(\mathbf{p})$ contains an edge from a_j to b_i for some $j \in S$. In other words, $\Gamma_p(S)$ is the set of buyers who are interested in at least one of the goods in S at price $\mathbf{p}$. For any $S \subseteq A$, the *money* of S (denoted by $m_p(S)$) is the sum of the prices of the goods in S. Similarly, the money of a subset S of B (denoted by $m_e(S)$) is the sum of the budgets of the buyers in S.

By the above definition, it is straightforward to see that Invariant 1 is equivalent to the following.

Invariant 2 *The prices $\mathbf{p}$ are such that for every $S \subseteq A$, we have $m_p(S) \leq m_e(\Gamma_p(S))$.*

Since the DPSV algorithm starts with an arbitrary price vector satisfying the invariant, and only increases the prices until it reaches the equilibrium, therefore it proves the following stronger statement. We will use this observation in the analysis of our algorithm.

Theorem B (Devanur et al. [6]). *Let $\mathbf{p}^0$ be a price vector satisfying Invariant 1. Then there is a market-clearing price vector $\mathbf{p}^*$ such that $p_j^* \geq p_j^0$ for every good j. Furthermore, $\mathbf{p}^*$ can be computed in polynomial time.*

In this paper, we present an algorithm that computes an *approximate market equilibrium* in the setting of the Arrow-Debreu theorem (where there is no dichotomy between buyers and sellers) assuming that the utility functions are linear. This "approximately" answers an open question of [6].

Since the market equilibrium problem is not an optimization problem, we need to clarify what we mean by an *approximate market equilibrium*. Deng et al. [5] presented the following natural definition for the notion of approximate market equilibria.

Definition 1. *An ε-approximate equilibrium for a market is a price vector $\mathbf{p}^*$ and a bundle x^i for each agent i such that*

- *The market approximately clears, i.e., for every good j, $(1 - \varepsilon) \sum_{i=1}^{n} w_j^i \leq \sum_{i=1}^{n} x_j^i \leq \sum_{i=1}^{n} w_j^i$.*
- *For all i, the utility $\sum_{j=1}^{m} u_{ij} x_j^i$ of agent i is at least $(1 - \varepsilon)$ times the value of the optimum solution of the maximization program (1).*

3   The Algorithm

In this section, we present two algorithms for computing market-clearing prices in a market with m types of goods and n agents, each having an initial endowment w^i of goods and a linear utility function $u_i(x) = \sum_{j=1}^{m} u_{ij} x_j$. The first algorithm is similar in nature to the DPSV algorithm, and is based on the simple approach of increasing the price of *oversold* items until we reach an equilibrium. Unable to analyze the running time of this algorithm, we present a modification of this algorithm which we will prove, using Theorem B, that computes an approximate equilibrium in polynomial time.

Before we present the algorithm, we define the equality subgraph corresponding to the price vector $\mathbf{p}$. The definition is similar to the definition of the equality subgraph in a dichotomous market presented in Section 2, except here the budget of each buyer is a function of prices. More precisely, the equality subgraph has m vertices in the first part A, n vertices in the second part B, equality edges between A and B as defined in Section 2, an edge of capacity p_j from the source s to the vertex $a_j \in A$, and an edge of capacity $\sum_{j=1}^{m} p_j w_j^i$ from the vertex $b_i \in B$ to the sink t. We will denote this equality subgraph by $N'(\mathbf{p})$ to avoid confusion with the equality subgraph for dichotomous markets defined in Section 2. The money of a set (denoted by $m_p(S)$) is defined as in Section 2, using $\sum_{j=1}^{m} p_j w_j^i$ as the budget of buyer i.

For a set $S \subseteq A$, we define the *deficiency* of S (denoted by $\mathrm{def}_p(S)$) as $m_p(S) - m_p(\Gamma_p(S))$. The *maximum deficiency* of the price vector $\mathbf{p}$ (denoted by $\mathrm{maxdef}(\mathbf{p})$) is the maximum value of $\mathrm{def}_p(S)$ over all $S \subseteq A$. The following fact is easy to observe.

Proposition 1. *Assume $\mathbf{p}$ is a price vector and the budgets defined above are non-zero. Let $\{s\} \cup S \cup T$ be the s-side of the minimum st-cut in $N'(\mathbf{p})$. Then $T = \Gamma_p(S)$, and the deficiency of the set S is equal to the maximum deficiency of $\mathbf{p}$.*

We call a set S with $\mathrm{def}(S) = \mathrm{maxdef}(\mathbf{p})$ a *maximally deficient set* with respect to $\mathbf{p}$. By the above fact, finding a maximally deficient set is equivalent to finding a minimum st-cut in $N'(\mathbf{p})$.

We are now ready to state our first algorithm.

Algorithm 1

1. Start from an arbitrary price vector, say $\mathbf{p}^0 = (1, 1, \ldots, 1)$.

2. Find the largest maximally deficient set S. Let $D = \text{def}(S)$. If $D = 0$ then stop.

3. Remove all equality edges between $A \setminus S$ and $\Gamma_p(S)$ from $N'(\mathbf{p})$.

4. Increase the prices of the goods in $A \setminus S$ continuously and at the same rate (i.e., multiply these prices by a factor δ initially equal to 1, and increase δ continuously), until one of the following events occur:

   (a) A new equality edge is added to $N'(\mathbf{p})$.

   (b) For a set $S' \not\subseteq S$, the deficiency of S' becomes equal to D.

   In either case, continue from Step 2. If none of the above events happens for any value of $\delta > 1$, then proceed to the next step.

5. Set the prices of the goods in S to zero, remove these goods from the set of goods, and start again from Step 2.

Step 4 in the above algorithm can be implemented using binary search over values of δ or using a parametric network flow algorithm [7] to find the first event that occurs. Notice that Step 5 in the above algorithm is only for taking care of (pathological) cases where in the equilibrium some of the prices are zero. If, for example, we assume that each agent has a non-zero utility for each good (i.e., $u_{ij} > 0$ for every i, j), then we will not need this step.

The intuition behind Algorithm 1 is simple: it is easy to observe that if the maximum deficiency of the initial price vector $\mathbf{p}^0$ is D^0, then the algorithm never lets the maximum deficiency of $\mathbf{p}$ to increase beyond D^0. On the other hand, the algorithm keeps increasing the total price of all goods. Therefore, the ratio of the maximum deficiency to the total prices will converge to zero. However, since in each step we might only slightly increase the prices, we were unable to prove any polynomial upper bound on the running time of Algorithm 1. Instead, we will change the algorithm to use the DPSV algorithm as a subroutine in each iteration. This enables us to prove a polynomial bound on the time it takes until the algorithm reaches an approximate equilibrium.

Algorithm 2

1. Start from an arbitrary price vector, say $\mathbf{p} := (1, 1, \ldots, 1)$.

2. Let $D := \text{maxdef}(\mathbf{p})$.

3. Construct an instance M_p of a dichotomous market as follows: There are m types of goods and $n + 1$ buyers. For $i = 1, \ldots, n$, the utility of buyer i for the goods is the same as the utility of the corresponding agent in the original instance. Also, the budget of buyer i is $e_i := \sum_{j=1}^{m} p_j w_j^i$. The $(n+1)$'th

buyer has a budget of $e_{n+1} := D$, and its utility for good j is equal to p_j (i.e., at price $\mathbf{p}$, buyer $n + 1$ is equally interested in all goods).

4. Run the DPSV algorithm on the instance M_p starting from the price vector $\mathbf{p}$. Let $\mathbf{p}'$ denote the output of this algorithm.

5. For every agent i, let $e'_i := \sum_{j=1}^m p'_j w^i_j$ be the budget of i with respect to $\mathbf{p}'$. If $e'_i/e_i \leq 1 + \varepsilon$ for $every$ agent i, then output $\mathbf{p}'$ and stop.

6. Let $\mathbf{p} := \mathbf{p}'$. Go to Step 2.

We will show in the next section that after at most polynomially many iterations, Algorithm 2 finds an ε-approximate market equilibrium.

4   Analysis

In this section we will prove that Algorithm 2 is correct (i.e., it computes an ε-approximate market equilibrium) and terminates in polynomial time. We start with the following simple lemma, which shows that the price vector $\mathbf{p}$ satisfies Invariant 2 of the DPSV algorithm on the instance M_p, and therefore in Step 4 of Algorithm 2 we can run the DPSV algorithm with the initial price vector $\mathbf{p}$.

Lemma 1. *In Step 4 of Algorithm 2, the price vector $\mathbf{p}$ satisfies Invariant 2 of the DPSV algorithm on the instance M_p .*

Proof. It is enough to notice that by the definition, at the price $\mathbf{p}$, the buyer $n + 1$ is interested in all goods. Therefore, adding this buyer to the set of buyers decreases the deficiency of every set by the budget of buyer $n + 1$, which is D. Therefore, after adding buyer $n + 1$, the maximum deficiency is non-positive. Thus, $\mathbf{p}$ satisfies Invariant 2 on the instance M_p. □

The following lemma shows that when Algorithm 2 stops in Step 5, it must have found an ε-approximate market equilibrium.

Lemma 2. *Assume Algorithm 2 terminates and outputs the price vector $\mathbf{p}^* := \mathbf{p}'$. Then there exist a bundle x^i for each agent i such that*

- *The market clears, i.e., for every good j, $\sum_{i=1}^n x^i_j = \sum_{i=1}^n w^i_j$.*
- *For all i, the utility $\sum_{j=1}^m u_{ij} x^i_j$ of agent i is at least $(1 - \varepsilon)$ times the value of the optimum solution of the maximization program (1).*

Therefore, the price vector $\mathbf{p}^*$ together with the allocation x is an ε-approximate market equilibrium.

Proof. Consider the instance M_p constructed in the last iteration of the algorithm, and the equality subgraph $N(\mathbf{p}')$ for this instance. Find a maximum flow from s to t in this network, and let y^i_j denote the amount of flow from the a_j to b_i divided by p'_j. Thus, the total amount of flow entering the vertex b_i is

$\sum_j p'_j y^i_j$. Therefore, since $\mathbf{p}'$ is a market-clearing price in M_p, we must have $\sum_j p'_j y^i_j = e_i$ for every i. By Theorem B we have $\mathbf{p}' \geq \mathbf{p}$ and therefore $e'_i \geq e_i$ for every i. This shows that the allocation y^i does not violate the budget constraint of agents. Also, by the termination condition of the algorithm, we have $e_i \geq e'_i/(1+\varepsilon) \geq (1-\varepsilon)e'_i$. Thus, $\sum_j p'_j y^i_j \geq (1-\varepsilon)e'_i$. That is, every agent uses at least a $(1-\varepsilon)$ fraction of her budget. Since utility functions are linear, we know that the solution of the maximization program 1 is precisely the budget of agent i times the *bang per buck* for agent i. By the definition of the equality subgraph, the agent only buys goods that have the highest bang per buck for her. Therefore, the utility that agent i has for the allocation y^i is at least a $(1-\varepsilon)$ fraction of her optimal bundle. Thus, the allocation y^i satisfies the second condition.

In order to satisfy the first condition, we change the allocation y^i as follows: by the *principle of conservation of money* the total extra money that the agents have after buying the bundles y^i is equal to the total price of the unsold goods. We distribute these goods among the agents arbitrarily, so that all goods are sold (i.e., the market clears). Let x^i's denote the resulting allocations. Since by doing so we do not decrease the utility of any agent, therefore the allocation x^i satisfies both conditions of the lemma. $\qquad\square$

Lemmas 1 and 2 together prove that Algorithm 2 is correct. Now, we only need to show that it terminates after polynomially many iterations. This is based on the fact that the price vector $\mathbf{p}$ in Algorithm 2 satisfies the following invariant.

Lemma 3. *Algorithm 2 never increases the maximum deficiency of the price vector* $\mathbf{p}$.

Proof. We need to show that the maximum deficiency of the price vector $\mathbf{p}'$ computed in Step 4 is not more than D (the maximum deficiency of $\mathbf{p}$). Since the output $\mathbf{p}'$ of the DPSV algorithm must satisfy Invariant 2, we have $m_{p'}(S) \leq m_e(\Gamma'_{p'}(S))$ for every set S of goods in M_p, where $\Gamma'_{p'}(S)$ denotes the set of buyers that have an equality edge from the goods in S in the equality subgraph $N(\mathbf{p}')$ for the instance M_p (we use Γ' instead of Γ to indicate the presence of the dummy buyer $n+1$), and $m_e(\Gamma'_{p'}(S))$ is computed using the budgets $e_i := \sum_{j=1}^m p_j w^i_j$. Therefore, if we remove the buyer $n+1$ from this instance, we still have

$$m_{p'}(S) - m_e(\Gamma'_{p'}(S) \setminus \{n+1\}) \leq D \qquad (2)$$

for every set S.

On the other hand, by Lemma 2 and Theorem B, the price vector $\mathbf{p}'$ must satisfy $p'_j \geq p_j$ for every good j. Therefore, we have

$$m_e(\Gamma'_{p'}(S) \setminus \{n+1\}) = \sum_{i \in \Gamma'_{p'}(S) \setminus \{n+1\}} e_i$$

$$= \sum_{i \in \Gamma_{p'}(S)} \sum_{j=1}^m p_j w^i_j$$

$$\leq \sum_{i \in \Gamma_{p'}(S)} \sum_{j=1}^{m} p'_j w^i_j$$

$$= m_{p'}(\Gamma_{p'}(S)). \tag{3}$$

By Equations 2 and 3, we have

$$\text{def}_{p'}(S) = m_{p'}(S) - m_{p'}(\Gamma_{p'}(S)) \leq m_{p'}(S) - m_e(\Gamma'_{p'}(S) \setminus \{n+1\}) \leq D.$$

This completes the proof of the lemma.   □

We are now ready to analyze the running time of Algorithm 3.

Lemma 4. *Let $e_{\min} := \min_i \sum_j w^i_j$ be the minimum budget e_i in the first itera-tion of the algorithm. Then Algorithm 2 terminates after at most $O(\frac{n}{\varepsilon} \log(\frac{m}{\varepsilon e_{\min}}))$ iterations.*

Proof. By Theorem B we have $\mathbf{p}' \geq \mathbf{p}$ and therefore $e'_i \geq e_i$ for every i. On the other hand, we have

$$\sum_i e'_i = \sum_j p'_j = \sum_j p_j + D = \sum_i e_i + D.$$

Therefore, for every i,

$$e'_i - e_i \leq D. \tag{4}$$

Let D^0 denote the maximum deficiency of the original price vector $(1, 1, \ldots, 1)$. By Lemma 3, the value of D in Algorithm 3 is always less than or equal to D^0. Also, $D^0 \leq m$ by definition. Therefore, by Equation (4), we have $e'_i - e_i \leq m$. Thus,

$$\frac{e'_i}{e_i} \leq 1 + \frac{m}{e_i}.$$

By the above inequality, the event $\frac{e'_i}{e_i} > 1 + \varepsilon$ can happen only if $\frac{m}{e_i} > \varepsilon$ or $e_i < m/\varepsilon$. However, if this event happens in some iteration, then the value of e_i in the next iteration (which is the same as the value of e'_i in the current iteration) will grow by a factor of $1 + \varepsilon$. This means that after $k = O(\frac{1}{\varepsilon} \log(\frac{m}{\varepsilon e_{\min}}))$ occurrences of the event $\frac{e'_i}{e_i} > 1 + \varepsilon$, the value of e_i will be at least $e_{\min}(1 + \varepsilon)^k > \frac{m}{\varepsilon}$, and therefore by the above observation the event $\frac{e'_i}{e_i} > 1 + \varepsilon$ cannot happen anymore. On the other hand, in every iteration in which the algorithm does not stop, this event must happen for at least one i. Thus, after at most $O(\frac{n}{\varepsilon} \log(\frac{m}{\varepsilon e_{\min}}))$ iterations the algorithm stops.   □

Lemmas 2 and 4 together with the observation that $\log(1/e_{\min})$ is upper bounded by a polynomial in the size of input imply our main result.

Theorem 1. *For every $\varepsilon > 0$, Algorithm 2 computes an ε-approximate market equilibrium in time polynomial in $1/\varepsilon$ and the size of the input.*

Remark 1. Using Lemma 3 and the fact that in each iteration $\sum_j p'_j = \sum_j p_j + D$, it is straightforward to show the ratio of the maximum deficiency to the total price of goods $(\mathrm{maxdef}(\mathbf{p})/\sum_j p_j)$ in the rth iteration of Algorithm 2 is at most $1/r$. Therefore, if instead of the requirements of Definition 1 we only need the relative maximum deficiency to be less than ε, it is enough to run Algorithm 2 for $1/\varepsilon$ iterations.

5   Conclusions

In this paper we presented a polynomial-time approximation scheme for computing an approximate market equilibrium for a general market with linear utilities. The main problem that remains open is to obtain a polynomial-time algorithm for computing the exact equilibrium. We introduced Algorithm 1 as a candidate for such an algorithm, but have been unable to analyze the running time of this algorithm. The problem of analyzing the running time of Algorithm 1 is similar in nature to the question left open in [6] on the running time of their basic algorithm. It is conjectured by Goemans that the basic algorithm of [6] runs in strongly polynomial time. A solution to this conjecture might be the first step toward analyzing the running time of Algorithm 1.

Another interesting open question is to generalize the result of this paper or [6] to the case of strictly concave utility functions. In the Arrow-Debreu setting, strictly concave utility functions are more interesting than linear utility functions, since if the utilities are strictly concave, all optimal bundles are uniquely determined from the prices. Even for special classes of strictly concave utility functions, we do not know how to compute market-clearing prices efficiently.

Throughout this paper, we assumed that we know the initial endowment and the utility of the participating agents. It would be interesting to consider scenarios where the agents are allowed to behave strategically in announcing their initial endowment or utility function.

Acknowledgments. The second author would like to thank Michel Goemans for helpful discussions. Parts of this research was done while the second and third authors were visiting Microsoft Research. We would like to thank the theory group at Microsoft Research for their hospitality. We would like to thank Nicole Immorlica for her comments on a draft of this paper.

References

[1] K. Arrow and G. Debreu. Existence of an equilibrium for a competitive economy. *Econometrica*, 22:265–290, 1954.

[2] K. J. Arrow, H. D. Block, and L. Hurwicz. On the stability of competitive equilibrium II. *Econometrica*, 27:82–109, 1959.

[3] K. J. Arrow and L. Hurwicz. On the stability of competitive equilibrium I. *Econometrica*, 26:522–52, 1958.

[4] W. C. Brainard and H. E. Scarf. How to compute equilibrium prices in 1891. *Cowles Foundation Discussion Paper 1270*, 2000.

[5] Xiaotie Deng, Christos Papadimitriou, and Shmuel Safra. On the complexity of equilibria. In *Proceedings of ACM Symposium on Theory of Computing*, 2002.

[6] Nikhil R. Devanur, Christos H. Papadimitriou, Amin Saberi, and Vijay V. Vazirani. Market equilibrium via a primal-dual-type algorithm. In *The 43rd Annual IEEE Symposium on Foundations of Computer Science*, 2002.

[7] Giorgio Gallo, Michael D. Grigoriadis, and Robert E. Tarjan. A fast parametric maximum flow algorithm and applications. *SIAM J. Comput.*, 18(1):30–55, 1989.

[8] F. P. Kelly and V. V. Vazirani. Rate control as a market equilibrium. In preparation.

[9] Christos H. Papadimitriou. Algorithms, games, and the internet. In *Proceedings of ACM Symposium on Theory of Computing*, 2001.

[10] H. Scarf. *The Computation of Economic Equilibria (with collaboration of T. Hansen)*. Cowles Foundation Monograph No. 24., New Haven: Yale University Press, 1973.

[11] A. Wald. On some systems of equations of mathematical economics. *Zeitschrift für Nationalökonomie*, Vol. 7, 1936. Translated, 1951, Econometrica 19(4), pp. 368-403.

[12] L. Walras. *Éléments d'économie politique pure; ou, Théorie de la richesse sociale (Elements of Pure Economics, or the theory of social wealth)*. Lausanne, Paris, 1874. (1899, 4th ed.; 1926, rev ed., 1954, Engl. transl.).

Approximating the Degree-Bounded Minimum Diameter Spanning Tree Problem

Jochen Könemann[1]*, Asaf Levin[2], and Amitabh Sinha[3]**

[1] GSIA, Carnegie Mellon University, Pittsburgh PA 15213, USA.
jochen@cmu.edu
[2] Department of Statistics and Operations Research, Tel-Aviv University,
Tel-Aviv 69978, Israel.
levinas@post.tau.ac.il
[3] GSIA, Carnegie Mellon University, Pittsburgh PA 15213, USA.
asinha@andrew.cmu.edu

Abstract. We consider the problem of finding a minimum diameter spanning tree with maximum node degree B in a complete undirected edge-weighted graph. We prove that the problem is NP-complete, and provide an $O(\sqrt{\log_B n})$-approximation algorithm for the problem. Our algorithm is purely combinatorial, and relies on a combination of *filtering* and *divide and conquer*.

1 Introduction

The importance of algorithms for designing efficient networks in today's interconnected world can hardly be overstated. The operative word here is "efficient", and indeed, there are many (often conflicting) ways to measure the efficiency of a network. Suppose a telecommunication company is building a communication network. While budgeting constraints may require the company to minimize total cost, there are also quality of service and technological constraints which may require the network to have low diameter and low degree.

Low diameter is essential to ensure that any pair of nodes can communicate fast. It is also useful to force reliability constraints, as explained in the following (see also [8] and [13]): Assume that an edge e fails with probability $1-p_e$, and that all failures occur independently. Then, the probability that a path $e_1, e_2, \ldots, e_k$ is operational is $p_{e_1} \times p_{e_2} \times \cdots \times p_{e_k}$. Given a certain threshold value for the desired reliability, there is a corresponding parameter D such that the diameter of the network defined by edge length $(|\log p_e|)_{e \in E}$ is required to be at most D. Therefore, the reliability constraint is transformed into a diameter constraint.

Degree-constraints appear naturally in graph-theoretic abstractions of communication network design problems. As an example, consider the so called *IP*

---

* This material is based upon work supported by the National Science Foundation under Grant No. 0105548.
** This material is based upon work supported by the National Science Foundation under Grant No. 0105548 and a Carnegie Bosch Institute Fellowship.

multicast [4,5] problem where we would like to disseminate centrally stored information from a server node to a set of client hosts. The standard solution is to compute a tree in the given graph that spans the server node and all client nodes. We then send data packets from the root along each of its incident edges in the tree. An internal node forwards incoming information to its descendants in the tree. The degree of a node in this tree is proportional to the amount of work that the node has to do and it is hence natural to aspire to compute spanning trees of low maximum degree (see also [1,2,3]).

Our work is motivated by precisely these considerations. We proceed by defining our problem.

1.1   Problem Definition

Formally, we consider the following BOUNDED DEGREE MINIMUM DIAMETER SPANNING TREE PROBLEM (BDST): given an undirected complete graph $G = (V, E)$ whose edges are endowed with a metric length function $\{l_e\}_{e \in E}$ and a parameter $B \geq 2$, we want to find a spanning tree T of G of maximum node-degree at most B. At the same time we want to minimize the *diameter* of T, i.e. we would like to minimize

$$\Delta(T) := \max_{u,v \in V} \mathtt{dist}_l^T(u, v)$$

where $\mathtt{dist}_l^T(u, v)$ denotes the l-length of the unique u, v-path in T.

Let the *height* of a tree T rooted at node r be the maximum number of edges on any r, v-path, where v is a leaf node in T and denote it by $\mathtt{height}(T)$. We also use n and m to denote $|V|$ and $|E|$, respectively.

For $B = 2$, BDST can be approximated within a constant using approximation algorithms for the *Traveling Salesperson* problem. In this paper we consider the case $B \geq 3$.

1.2   Results and Paper Outline

Our main result is an $O(\sqrt{\log_B n})$ approximation algorithm for BDST. The algorithm is described and analyzed in Section 2. There are two main ideas in the algorithm. First, we break up the graph into clusters of low diameter. For each cluster, we compute a balanced $(B-1)$-ary tree. We then compute a global tree over the clusters, and show that the resulting tree has low diameter.

Our algorithm is the first known sub-logarithmic approximation for this problem. An $O(\log_B n)$ approximation is trivial; any complete balanced $(B-1)$-ary spanning tree of the graph will do. We also prove the NP-completeness of BDST in Section 3. We conclude the paper with some open questions.

1.3   Related Work

Ravi [12] considered the MINIMUM POISE SPANNING TREE PROBLEM defined as follows: given an unweighted graph $G = (V, E)$, we want to find a spanning tree

$T = (V, E_T)$ such that the sum of the maximum degree of a node in T and the diameter of T is minimized. In order to provide an approximation algorithm for this problem, he presented an $(O(\log n), O(\log^2 n))$-bicriteria approximation for the BDST problem with a length metric defined by the distances in an unweighted graph G, with the restriction that we can only use edges from G.

The MINIMUM DIAMETER SPANNING TREE PROBLEM is the following: given an undirected graph $G = (V, E)$ and length function defined over its edge set $\{l_e\}_{e \in E}$, we want to find a spanning tree of G of minimum diameter. This problem is equivalent to finding the shortest paths tree from the absolute 1-center of G (see [9]), and therefore, is solvable in $O(mn + n^2 \log n)$ time.

The MINIMUM DEGREE SPANNING TREE PROBLEM is the following (see [6]): given an undirected graph $G = (V, E)$, we want to find a spanning tree of G whose maximum node-degree is minimized. Fürer and Raghavachari [6] provided a polynomial time algorithm which computes a spanning tree with maximum degree at most $\Delta^* + 1$ where Δ^* denotes the smallest possible maximum degree of any spanning tree of the input graph. The algorithm in [6] extends also to Steiner trees.

Könemann and Ravi [10,11] studied the MINIMUM-COST DEGREE BOUNDED SPANNING TREE problem, where in addition to an undirected graph and non-negative edge-costs we are also given a parameter $B_v > 1$ for each node $v \in V$. The objective is to find a minimum-cost spanning tree where the degree of each vertex $v \in V$ is at most B_v. The authors show how to compute a tree T where each node v has degree $O(B_v + \log(n))$ and the cost of T is $O(\text{opt})$ where opt is the minimum cost of any tree obeying all degree-bounds exactly.

2   Algorithm and Analysis

2.1   Overview

The main idea behind our algorithm is *filtering*. Let $\alpha > 0$ be a threshold, where distances more than α are called *long* and distances less than α are *short*. We partition the node set of G into clusters such that the diameter of each cluster is short, but the number of clusters is also low. We do this by filtering the node set so that we retain one representative node for each cluster, and define an artificial degree bound for this representative node to account for the degree capacity of the entire cluster.

We obtain our performance guarantee from the following two observations. Since the number of clusters is small, any balanced tree which spans the representatives has a small number of long edges. And since each cluster has small diameter, the overhead added to any path by the expansion of the representative nodes into trees spanning the clusters is also small. The rest of this paper shows that such a threshold exists and yields our claimed performance guarantee.

Algorithm 1 GlobTree($\{(v_i, B_i)\}_{i=1}^l$): Compute a tree T on the nodes $\{v_i\}_{i=1}^l$ such that node v_i has node degree at most B_i for all i.

1: Assume $B_1 \geq \ldots \geq B_l$.
2: $T \leftarrow \emptyset$.
3: $d_i \leftarrow B_i$ for all $1 \leq i \leq l$
4: **for** $i = 2$ to l **do**
5:     Let $1 \leq j \leq i$ be smallest with $d_j > 0$.
6:     Add edge (v_j, v_i) to T.
7:     $d_j \leftarrow d_j - 1$.
8:     $d_i \leftarrow d_i - 1$.
9: **end for**
10: **return** Tree T with root v_1.

2.2   Algorithm

Given an appropriately chosen threshold α, the first step of our algorithm is to find representatives $R = \{v_1, \ldots, v_l\} \subseteq V$ and a partition of V into pairwise disjoint sets:

$$V = V_1 \cup \ldots \cup V_l \tag{1}$$

such that $v_i \in V_i$ and $\mathrm{dist}_l(v_i, u) \leq 3 \cdot \alpha$ for all $1 \leq i \leq l$ and for all $u \in V_i$. Roughly speaking, we then construct a low-degree and low-diameter tree on the nodes of R. This tree determines the global structure of our solution. In addition we construct low-diameter degree-B-bounded trees for the nodes of each set V_i, $1 \leq i \leq l$. We finish by replacing the nodes from R in the global solution by the respective spanning trees.

In the following we assume that we have a guess for the optimum diameter Δ. This is justified since the diameter of an optimum tree is within the interval $[\max_{e \in E} l_e, n \cdot \max_{e \in E} l_e]$ and we can perform a binary search in order to find a proper approximate guess (i.e., a guess within twice the optimum diameter).

We now detail the process of finding the partition from (1). We proceed in iterations: in iteration $1 \leq i \leq l$, we compute the set V_i and its representative v_i. For ease of notation, we use U_i^γ to denote the set of nodes that are at a distance of at least γ from the first $i - 1$ representatives $\{v_1, \ldots, v_{i-1}\}$. In order to define these sets formally, let $\mathrm{cov}_\gamma(v, U) = \{u \in U : \mathrm{dist}_l(v, u) \leq \gamma\}$ be the set of nodes in U that are within a distance of γ of vertex v (we also say that v γ-covers the nodes in $\mathrm{cov}_\gamma(v, U)$). Then, we let $U_1^\gamma = V$ for all $\gamma > 0$. For $i > 1$ we define $U_i^\gamma = V \setminus \bigcup_{1 \leq j \leq i-1} \mathrm{cov}_\gamma(v_j, V)$.

Let α be a given contraction threshold. In iteration i, we then pick vertex $v_i \in U_i^\alpha$ that α-covers most nodes in $U_i^{3\alpha}$, i.e. we let $v_i = \mathrm{argmax}_{v \in U_i^\alpha} |\mathrm{cov}_\alpha(v, U_i^{3\alpha})|$, and $V_i = \mathrm{cov}_{3\alpha}(v_i, U_i^{3\alpha})$.

The algorithm stops as soon as all nodes in V are within a distance of at most 3α from some representative. We assume that this happens after l iterations. We have $U_{l+1}^{3\alpha} = \emptyset$ and $U_i^{3\alpha} \neq \emptyset$ for all $1 \leq i \leq l$.

Algorithm 2 BDST(G, Δ): Compute a degree B tree of diameter no more than $O(\sqrt{\log_B n})\Delta$.

1: $\alpha \leftarrow \Delta/\sqrt{\log_B n}$.
2: $i \leftarrow 0$.
3: $U_1^{3\alpha} \leftarrow V$.
4: **while** $U_i^{3\alpha} \neq \emptyset$ **do**
5:     $v_i \leftarrow \text{argmax}_{v \in U_i^\alpha}|\text{cov}_\alpha(v, U_i^{3\alpha})|$.
6:     $V_i \leftarrow \text{cov}_{3\alpha}(v_i, U_i^{3\alpha})$.
7:     $i \leftarrow i + 1$.
8: **end while**
9: Reorder $\{v_1, v_2, \ldots, v_l\}$ so that $|V_1| \geq |V_2| \geq \ldots \geq |V_l|$.
10: Compute B_i as defined in (2).
11: $T^g \leftarrow \textbf{GlobTree}(\{(v_i, B_i)\}_{i=1}^l)$.
12: **for** $i = 1$ to l **do**
13:     $T_i \leftarrow$ Tree spanning V_i of degree at most B and minimum height.
14:     Replace v_i by T_i, and distribute the edges in T^g incident on v_i over the nodes of T_i so that the maximum degree of any node in T_i is at most B.
15: **end for**
16: **return** Resulting tree T^{apx}.

In order to compute the final tree, we go through two main steps. Assume that we have reordered the sets $\{V_i\}$ such that $|V_1| \geq |V_2| \geq \ldots \geq |V_l|$.

Global structure For each $1 \leq i \leq l$, let the degree bound of node v_i be

$$B_i = \begin{cases} |V_i| \cdot (B - 2) + 2 & : \quad i = 1 \\ |V_i| \cdot (B - 2) + 1 & : \quad \text{otherwise.} \end{cases} \tag{2}$$

We then compute a tree $T^g = \textbf{GlobTree}(\{(v_i, B_i)\}_{i=1}^l)$ on the nodes $\{v_1, \ldots, v_l\}$ of low diameter. See Algorithm 1 for the details.

Local structure For each $1 \leq i \leq l$ we construct a tree T_i spanning the nodes of V_i of minimum height such that the degree of each node is at most B.

Finally, we compute the final tree T^{apx} by taking the global tree T^g which is rooted at v_1 and replacing each node v_i by the tree T_i. We distribute the edges that are incident to v_i in T^g over the nodes of T_i evenly, such that the maximum degree of any node of T_i is as small as possible.

A listing outline of the algorithm is shown in Algorithm 2. Its output is always a tree of degree no more than B. We do a binary search over Δ to obtain a tree of minimum diameter. In the following, we analyze the performance of the algorithm, assuming the correct value for Δ is fed to Algorithm 2.

2.3   Performance Ratio

Theorem 1. *Suppose that there is a tree T^* with maximum node-degree B and diameter Δ. Then Algorithm BDST(G, Δ) produces a tree T^{apx} with maximum node-degree B and diameter $O(\sqrt{\log_B n} \cdot \Delta)$.*

Theorem 1 is the main result we are trying to prove. We prove it at the end of this section, using a sequence of lemmas which follow.

Lemma 1. *The maximum degree of* $T^{\mathtt{apx}}$ *is no more than* B.

Proof. For any i, the degree of vertex v_i in the global tree T^g is bounded by $(|V_i| - 2) \cdot B + 2$ for all $1 \leq i \leq l$. Also, tree T_i has $|V_i|$ nodes each with capacity B and there are exactly $|V_i| - 1$ edges in T_i. Hence the total available capacity of the nodes in V_i for edges outside T_i is $|V_i| \cdot (B - 2) + 2$. Hence, there is a way of distributing the edges of T^g that are incident to node v_i over all nodes of T_i such that the maximum degree in $T^{\mathtt{apx}}$ is at most B.

We now prove that $T^{\mathtt{apx}}$ has diameter $O(\sqrt{\log_B n} \cdot \Delta)$. In the following, we say that an edge $uv \in E$ is *short* if $u, v \in V_i$ for some $1 \leq i \leq l$, and uv is *long* otherwise. Our proof of the diameter bound has two parts: the first part shows that the maximum number of long edges on any root,leaf-path in $T^{\mathtt{apx}}$ is $O(\sqrt{\log_B n})$. The second part shows that there are $O(\log_B n)$ short edges on any root,leaf-path in $T^{\mathtt{apx}}$. This suffices, because the length of any edge in our input graph is at most Δ and the length of a short edge in G is at most 6α (using triangle inequality).

First, we prove that any root,leaf-path contains at most $O(\sqrt{\log_B n})$ long edges. We begin by creating a partition of V using T^*'s structure. We root T^* at v_1^* (chosen arbitrarily), and let V_1^* be the set of nodes $u \in V$ such that the unique (v_1^*, u) path in T^* has length at most α. We let $S^* = \{v_1^*\}$, and let the set of *uncovered nodes* be $U = V \setminus V_1^*$ initially.

We continue until there are no uncovered nodes remaining. In iteration $i > 1$, let $v_i^* \in U$ be an uncovered node of smallest height in T^* (i.e. v_i^*'s parent in T^* is already covered). We then say that a node u is *covered* by v_i^* if u is a descendant of v_i^* and the length of the path from v_i^* to u in T^* is at most α. We let V_i^* be the set of nodes in U that are covered by v_i^*. We remove V_i^* from U and repeat.

Assume that the final partition has sets $V_1^*, \ldots, V_q^*$ and representatives $v_1^*, \ldots, v_q^*$. Since the subtree $T^*[V_i^*]$ of T^* induced by the nodes of V_i^* is connected, a counting argument shows that the nodes of V_i^* have at most

$$B_i^* = \begin{cases} |V_i^*| \cdot (B - 2) + 2 & : \quad i = 1 \\ |V_i^*| \cdot (B - 2) + 1 & : \quad \text{otherwise.} \end{cases} \tag{3}$$

children from $V \setminus V_i^*$ in T^*. Order the sets such that $|V_1^*| \geq \ldots \geq |V_q^*|$ and let T be the tree produced by **GlobTree**$(\{v_i^*\}_{i=1}^q, \{B_i^*\}_{i=1}^q)$.

Definition 1. $\{(\overline{v}_i, \overline{V}_i)\}_{i=1}^p$ *is called a* proper collection *of* V *for a given node set* V *if the following conditions hold:*

1. $\overline{V}_i \subset V$ *and* $\overline{v}_i \in \overline{V}_i$ *for all* $1 \leq i \leq p$.
2. $\overline{V}_i \cap \overline{V}_j = \emptyset$ *for all* $1 \leq i < j \leq p$.
3. $|\overline{V}_1| \geq \ldots \geq |\overline{V}_p|$.
4. $\mathtt{dist}_l(\overline{v}_i, u) \leq \alpha$ *for all* $1 \leq i \leq p$ *and for all* $u \in \overline{V}_i$.

The following lemma is useful in order to prove that the height of the global tree T^g is at most that of T.

Lemma 2. *Let $\{(V_i, v_i)\}_{i=1}^l$ be a partition of V together with a corresponding set of representatives created by Steps 1-8 of Algorithm 2. Let $\{(\overline{v}_i, \overline{V}_i)\}_{i=1}^l$ be a proper collection of G as defined in Definition 1. We then must have*

$$\sum_{i=1}^{j} |V_i| \geq \sum_{i=1}^{j} |\overline{V}_i| \tag{4}$$

for all $1 \leq j \leq \max\{l, p\}$.

Proof. We prove the lemma by induction over $|V| = n$. For $n = 1$ the lemma is trivially satisfied since in this case $V_1 = \overline{V}_1 = V$. For $n > 1$, assume that the lemma holds for all node sets with at most $n - 1$ nodes.

Assume now, for the sake of contradiction, that the lemma does not hold. Let j be the minimum index such that $\sum_{i=1}^{j} |V_i| < \sum_{i=1}^{j} |\overline{V}_i|$. Then, there must exist an index $j_0 \leq j$ such that

$$\overline{V}_{j_0} \not\subseteq \bigcup_{1 \leq i \leq j} V_i$$

and hence $\overline{V}_{j_0} \not\subseteq \mathrm{cov}_{3\alpha}(v_i, V)$ for all $1 \leq i \leq j$. Notice that this implies $\overline{v}_{j_0} \in U_j^\alpha$.

Now consider the application of the induction hypothesis for the set of nodes $V' = V \setminus \overline{V}_{j_0}$. Since $\overline{V}_{j_0} \cap \mathrm{cov}_\alpha(v_i, V) = \emptyset \; \forall i$, the application of our algorithm with V' yields the exact same set of the first $j - 1$ representatives $v_1, v_2, \ldots, v_{j-1}$ and the corresponding subsets $\{V_i \setminus \overline{V}_{j_0}\}_{i=1}^{j-1}$ of V'. Note that $\{\overline{V}_i\}_{i=1}^{p} \setminus \{\overline{V}_{j_0}\}$ is a proper collection of V'. Therefore, by the induction hypothesis, we conclude that

$$\sum_{i=1}^{j-1} |V_i \setminus \overline{V}_{j_0}| \geq \sum_{i=1}^{j} |\overline{V}_i| - |\overline{V}_{j_0}|. \tag{5}$$

Let us now lower-bound the difference $\sum_{i=1}^{j} |V_i| - \sum_{i=1}^{j-1} |V_i \setminus \overline{V}_{j_0}|$.

This difference can be expressed as the sum of two terms: the size of the set V_j and the increase of the sizes of the biggest $j - 1$ sets of our partition. Hence, we obtain

$$\sum_{i=1}^{j} |V_i| \geq \sum_{i=1}^{j-1} |V_i \setminus \overline{V}_{j_0}| + \left| \overline{V}_{j_0} \cap \bigcup_{1 \leq i < j} \mathrm{cov}_{3\alpha}(v_i, V) \right| + |V_j|. \tag{6}$$

Observe that in the j-th iteration of our algorithm we could have chosen $\overline{v}_{j_0}$ as a representative instead of v_j since $\overline{v}_{j_0} \in U_j^\alpha$. Therefore, we must have that

$$|V_j| = |\mathrm{cov}_\alpha(v_j, U_j^{3\alpha})| \geq |\mathrm{cov}_\alpha(\overline{v}_{j_0}, U_j^{3\alpha})|.$$

Using (5) together with (6) and noting that

$$\left| \overline{V}_{j_0} \cap \bigcup_{1 \leq i < j} \text{cov}_{3\alpha}(v_i, V) \right| + |\text{cov}_{\alpha}(\overline{v}_{j_0}, U_j^{3\alpha})| \geq |\overline{V}_{j_0}| \tag{7}$$

finally yields

$$\sum_{i=1}^{j} |V_i| \geq \sum_{i=1}^{j} |\overline{V}_i|.$$

This contradicts our assumption, and the lemma follows.

Corollary 1. *Let $\{V_i\}_{i=1}^{l}$ be the partition of V generated by Step 9 of Algorithm 2, and $\{V_i^*\}_{i=1}^{l}$ be the partition of V generated from the optimum tree. For all $1 \leq j \leq \max\{l, p\}$, we have:*

$$\sum_{i=1}^{j} |V_i| \geq \sum_{i=1}^{j} |V_i^*| \tag{8}$$

Proof. The statement in (8) clearly holds for the partition $\{V_i'\}_{i=1}^{l}$ generated by steps 1–8 of Algorithm 2, noting that $\{(v_i^*, V_i^*)\}_{i=1}^{q}$ is a proper collection of V as defined in Definition 1.

The corollary follows by observing that reordering the sets of the partition by non-increasing size increases the left hand side of (8) and does not change the right hand side.

We can now prove that the height of the global tree T^g is at most the height of the tree T.

Lemma 3. *When T is constructed from T^* by $\textbf{GlobTree}(\{v_i^*\}_{i=1}^{q}, \{B_i^*\}_{i=1}^{q})$, we have $\texttt{height}(T^g) \leq \texttt{height}(T)$.*

Proof. We say that the *level* of node v of T is the number of edges in the unique path from the root of T to v. We now claim that the level of node v_i in T^g is at most the level of node v_i^* in T for all $1 \leq i \leq l$. We use induction over i to prove the claim.

The claim is clear for $i = 1$. For $i > 1$, assume that $\texttt{GlobTree}$ connects node v_i^* to node v_p^* for some $p < i$. It follows from (2),(3) and Corollary 1 that

$$\sum_{j=1}^{p} B_j^* \leq \sum_{j=1}^{p} B_j$$

and hence there must exist a $1 \leq p' \leq p$ such that $d_{p'} > 0$ in $\texttt{GlobTree}$ at the time when node v_i is connected. By the induction hypothesis, we know that the level of $v_{p'}$ in T^g is at most that of node $v_{p'}^*$ in T. It follows from the definition of $\texttt{GlobTree}$ that the level of $v_{p'}^*$ is at most the level of v_p^*. Hence, the level of v_i in T^g is at most the level of v_i^* in T and this finishes the induction.

Observe that the height of T^g is equivalent to the level of node v_l in T^g, and that the height of T equals the level of v_q^* in T. This implies the lemma.

Lemma 4. *Let T^g be a tree returned by* `GlobTree(`$\{(v_i, B_i)\}_{i=1}^l$`)`. *Then, T^g must be a tree of minimum height among all trees that satisfy the given degree constraints.*

Proof. Given a tree T, we define the following total order of the nodes in T. The order is a breadth-first-search order, with the refinement that the nodes of each level are ordered in non-increasing order of their corresponding sets V_i. In particular, the nodes of T are ordered $v_1, v_2, \ldots, v_l$ such that $i < j$ if $\texttt{level}(v_i) < \texttt{level}(v_j)$ or $\texttt{level}(v_i) = \texttt{level}(v_j)$ and $|V_i| \geq |V_j|$. By construction of T^g, we have that if $i < j$ in the total order of the nodes in T^g, then $|V_i| \geq |V_j|$, regardless of their levels. Moreover, every tree of minimum height for which this holds must have the same height as $\texttt{height}(T^g)$.

Assume for the sake of contradiction that there is a tree T' such that $\deg_{T'}(v_i) \leq B_i$ for all $1 \leq i \leq l$ and $\texttt{height}(T') < \texttt{height}(T^g)$. Let $v_1', \ldots, v_l'$ be the total order induced by T', as defined above. By the observation in the preceding paragraph, for some $i < j$, we have $|V_i'| < |V_j'|$. We call this an *inversion*, and without loss of generality, assume that T' is a tree with the fewest number of inversions among all trees that satisfy the degree constraints and have height less than $\texttt{height}(T^g)$.

We show that we can reduce the number of inversions in T' without increasing the tree's height. This contradicts the inversion-minimality of T'.

Let $\langle v_i', v_j' \rangle$ be an inversion in T'. We swap labels: relabel node v_i' as v_j' and relabel v_j' as v_i'. The resulting tree may now violate the degree constraints at node v_i'. We counter this by moving a sufficient number of v_i''s children to v_j'. This does not increase $\texttt{height}(T')$, and reduces the number of inversions in T', which is a contradiction.

Lemma 5. *Any root,leaf-path in $T^{\texttt{apx}}$ has at most $\sqrt{\log_B n}$ long edges.*

Proof. Let d^* denote the maximum number of long edges on any root,leaf-path in T^*. It follows from Lemma 4 that $\texttt{height}(T) \leq d^*$ and hence, together with Lemma 3, we have that $\texttt{height}(T^g) \leq d^*$.

By the construction of the partition $V_1^*, \ldots, V_q^*$, we know that a root,leaf-path P in T^* that contains d^* long edges must have length at least $\alpha \cdot d^*$. Since T^* has diameter at most Δ it then follows that $d^* \leq \Delta/\alpha = \sqrt{\log_B n}$ by our choice of α.

Lemma 5 bounds from above the contribution of long edges to the diameter of $T^{\texttt{apx}}$. We bound the contribution of short edges in the next lemma. For a root,leaf-path P in $T^{\texttt{apx}}$, let $|P|_s$ denote the number of short edges in P.

Lemma 6. *Let P be an arbitrary root,leaf-path in $T^{\texttt{apx}}$. Then,*

$$|P|_s = O(\log_B n).$$

Proof. Let P_1 and P_2 be two root,leaf-paths in $T^{\texttt{apx}}$, and let P_1^g and P_2^g be their images in T^g, i.e. $P_1^g = \langle v_1^1, \ldots, v_{l_1}^1 \rangle$ and $P_2^g = \langle v_1^2, \ldots, v_{l_2}^2 \rangle$.

We define a relation $\prec$ on two root,leaf-paths as follows. We say that $P_1 \prec P_2$ if $|V_j^1| \geq |V_j^2|$ for all $1 \leq j \leq \max\{l_1, l_2\}$, with $|V_j^i| = 0$ if V_j^i does not exist. By construction of T^g, for every two paths P_1 and P_2 at least one of the following holds: $P_1 \prec P_2$ or $P_2 \prec P_1$. Moreover, if $P_1 \prec P_2$ then $l_2 \leq l_1 \leq l_2 + 1$.

Recall that T_i denotes the *local* tree that spans the nodes of V_i. For the purpose of this proof, we assume that all edges of the form (v_i, v_j) in T^g such that v_i is a parent of v_j are attached to leaf nodes in T_i. This assumption only increases the number of short edges in root,leaf-paths, and hence is valid.

Consider two paths P_1 and P_2 such that $P_1 \prec P_2$. Since each T_i is a balanced $(B-1)$-ary tree, we have $|P_2|_s \leq |P_1|_s + |P_1| \leq |P_1|_s + \log_B n$. We also have $|V_i^1| \leq |V_{i-1}^2|$ for $i > 1$, by construction of T^g. Therefore, $|P_1|_s \leq |P_2|_s + |P_1| + \mathtt{height}(T_{v_1^1}) \leq |P_2|_s + 2\log_B n$. Hence, there exists a γ such that $|P|_s \in [\gamma, \gamma + 2\log_B n]$ for all root,leaf-paths P in $T^{\mathtt{apx}}$.

Observe that on any root,leaf-path P in $T^{\mathtt{apx}}$, all but at most $O(\log_B n)$ of the short edges must be incident to nodes of degree B. This follows from the fact that T^g has $O(\log_B n)$ levels. Since there are n nodes in our graph, we must have that $\gamma = O(\log_B n)$. This finishes the proof of the lemma.

We are now ready to prove Theorem 1.

Proof. (of Theorem 1) Lemma 1 shows that $T^{\mathtt{apx}}$ has maximum degree B.

Long edges in $T^{\mathtt{apx}}$ have length no more than Δ, since the graph has a spanning tree of diameter Δ and we are assuming we have the correct guess of Δ. Hence, it follows from Lemma 5 that the contribution of long edges to the diameter of $T^{\mathtt{apx}}$ is no more than $2\Delta\sqrt{\log_B n}$.

Short edges in $T^{\mathtt{apx}}$ have length no more than $6\alpha = 6\Delta/\sqrt{\log_B n}$. Lemma 6 bounds the number of short edges in any root,leaf-path, so the total contribution of short edges to the diameter of $T^{\mathtt{apx}}$ is no more than $O(\alpha \log_B n) = O(\Delta\sqrt{\log_B n})$.

All edges are either long or short; this completes the proof of the theorem.

3   Hardness

In this section we prove that for any $B \geq 3$ the BDST problem is NP-hard. We prove the NP-completeness of BDST by reducing SET COVER to it.

Suppose we are given an instance of SET COVER $\mathcal{S}$, specified by subsets $\{S_1, \ldots, S_m\}$ of a universe $U = \{u_1, \ldots, u_n\}$, and a number C. We want to find out if there is a sub-collection of at most C subsets that covers U. We fix a parameter $B \leq C$, and convert $\mathcal{S}$ into a graph $G(\mathcal{S})$ as follows.

The graph has four kinds of nodes. It has one node for every element of U. For each set S_j, the graph has $\left\lceil \frac{|S_j|}{B-1} \right\rceil$ nodes. Before we describe the other sets of nodes, we describe the edges between these two sets. Every element-set pair (u_i, S_j) such that $u_i \in S_j$ is represented by an edge of length 1 between u_i and one of the nodes representing S_j, such that every node that represents a subset has at most $(B-1)$ such adjacent edges.

$G(\mathcal{S})$ has a set of *artificial* nodes, as follows. There is a special artificial node called the root, denoted r. For each set S_j, we build a tree of degree at most B such that the nodes representing S_j are at the leaves of this tree. The root of the tree is connected to r by a single edge of length 1. All other edges of these trees have length zero.

There are more artificial nodes to *cover* the artificial nodes described above. We build yet another degree B tree, whose leaves include all the nodes in the tree built for each set. The root of this tree is also connected to r with an edge of length 1. In this tree, all edges incident to leaves have length 1, and all other edges have length 0.

We replace the root by a minimum height $(B-1)$-ary tree with $\left\lceil \frac{C+2}{B-1} \right\rceil$ leaves and where all the inner nodes have exactly $(B-1)$ children. Let r be the root of this tree. The edges of the tree has zero length, and for every neighbor of the old root it is now a neighbor of all the leaves of this tree with edges of length 1. We denote the node set of this tree by $ROOT$.

We add another $(B-1)\left\lceil \frac{C+2}{B-1} \right\rceil - C$ *extra nodes* that are connected to the nodes of $ROOT$ with edges of length 2. Note that there are at least 2 extra nodes. The BDST instance is defined by the metric closure of the above distances.

Lemma 7. *$\mathcal{S}$ has a set cover of size C if and only if $G(\mathcal{S})$ has a degree B bounded spanning tree with diameter no more than 4.*

Proof. Given a set cover of size B, we can embed it into $G(\mathcal{S})$ in the obvious way, with the BDST being rooted at r. Since every element is covered by the set cover, there is a path of length 2 from $ROOT$ to every element of U. The edges that connect the extra nodes provide a path of length 2 to these nodes as well. The artificial tree constructed above provides a path of length 2 to all nodes which do not participate in the set cover. The degree constraints are automatically satisfied by construction. Hence, if $\mathcal{S}$ has a set cover of size C, then $G(\mathcal{S})$ has a BDST of diameter 4.

Conversely, suppose $G(\mathcal{S})$ has a BDST of degree B and diameter no more than 4. By construction of $G(\mathcal{S})$, it is impossible for a tree of diameter 3 or less to span it; hence we may assume that the diameter of the tree is exactly 4. In this case, the tree must have a node such that all other nodes in the graph are at distance no more than 2 from it. We call this the *center* of the tree. Since there are at least two extra nodes and the only nodes that are within a distance of at most 2 from the extra nodes are $ROOT$, then the center must be at $ROOT$.

If a BDST centered at $ROOT$ spans the entire graph with paths of length no more than 2, then it must reach all the elements via sets which include those elements, and also it must reach the "extra nodes" directly. The construction of $G(\mathcal{S})$ ensures that these paths induce a valid set cover, and the degree constraint ensures that this set cover has size no more than C. Hence if $G(\mathcal{S})$ has a BDST of degree B and diameter 4, then it has a set cover of size no more than C.

Since SET COVER is NP-hard [7], so is BDST. Clearly BDST is in NP, since it is easy to check whether a tree has diameter Δ and degree no more than B. We conclude as follows:

Corollary 2. *BDST is NP-complete.*

Corollary 3. *If $P \neq NP$, then there is no approximation algorithm with performance guarantee of less than $\frac{5}{4}$.*

The reduction is not approximation preserving, so even though SET COVER cannot be approximated to within a logarithmic factor, such a result is not implied for BDST.

4   Open Questions

In some situations, rather than bounding the diameter of the tree, it is required to bound the *dilation* of every pair of nodes. The dilation of a pair of nodes is defined as the ratio between their distance in the tree and their distance in the original metric. An approximation algorithm for degree bounded minimum dilation spanning trees is still open. This problem is closely related to the well-studied problem of approximating a general metric space by a tree metric.

Our algorithm crucially uses the fact that the input graph is a complete metric. In particular, our algorithm does not work if we are given an (incomplete) input graph and a metric induced by the edge-lengths of its edges (and we are enforced to use only edges from the input graph). Thus, an improvement over the bicriteria $(O(\log n), O(\log^2 n))$ approximation algorithm from [12] for this case is still open.

References

1. F. Bauer and A. Varma. Degree-Constrained Multicasting in Point-to-Point Networks. In *Proc. of the 14th Annual Joint Conference of the IEEE Computer Communications Societies (INFOCOMM'95)*, 369-376, 1995.
2. Y. Chu, S. G. Rao, S. Seshan, and H. Zhang. Enabling conferencing applications on the internet using an overlay multicast architecture. In *Proceedings of SIGCOMM*, pages 55–68, 2001.
3. W. De Zhong. A copy network with shared buffers for large-scale multicast ATM switching. *IEEE/ACM Transactions on Networking*, 1(2):157-165, 1993.
4. S. E. Deering and D. R. Cheriton. Multicast routing in datagram internetworks and extended LANs. *ACM Transactions on Computer Systems*, 8(2):85, May 1990.
5. S. Deering, D. Estrin, and D. Farinacci. An architecture for wide-area multicast routing. In *Proceedings of SIGCOMM*, 1994.
6. M. Fürer and B. Raghavachari. An NC approximation algorithm for the minimum degree spanning tree problem. In *Proc. of the 28th Annual Allerton Conference on Communication, Control and Computing*, 274-281, 1990.
7. M.R. Garey and D.S. Johnson. *Computers and Intractability: A Guide to the Theory of NP-Completeness*, Freeman, New York, 1979.
8. L. Gouveia. Multicommodity flow models for spanning trees with hop constraints. *European Journal of Operational Research*, 95:178-190, 1996.
9. R. Hassin and A. Tamir. On the minimum diameter spanning tree problem. *Information Processing Letters*, 53:109-111, 1995.

10. J. Könemann and R. Ravi. A matter of degree: Improved approximation algorithms for degree-bounded minimum spanning trees. *SIAM Journal of Computing*, 31(6):1783-1793, 2002.
11. J. Könemann and R. Ravi. Primal-dual algorithms come of age: Approximating MST's with nonuniform degree bounds. To appear in *Proc. of the 35th ACM Symposium on Theory of Computing*, 2003.
12. R. Ravi. Rapid rumor ramification: Approximating the minimum broadcast time. In *Proc. of the 35th Annual IEEE Symposium on Foundations of Computer Science*, 202-213, 1994.
13. S. Voss. The Steiner tree problem with hop constraints. *Annals of Operations Research*, 86:321-345, 1999.

On the Hardness of Approximate Multivariate Integration

Ioannis Koutis

Computer Science Department, Carnegie Mellon University
Pittsburgh, PA 15213 USA
ioannis.koutis@cs.cmu.edu

Abstract. We show that it is NP-hard to 2^{n^k}-approximate the integral of a positive, smooth, polynomial-time computable n-variate function, for any fixed integer k.

1 Introduction

Suppose $F(\cdot)$ is a real positive function defined on a cube C in Euclidean n-dimensional space $\mathbf{R}^n$. We consider the problem of approximating the integral $I(F)$ of F over C, with relative error ϵ, under the additional assumption that F satisfies a smoothness condition.

The exact integration of multivariate functions is hard, under the widely conjectured hardness of #P, given the result in [3], which implies that the exact calculation of the volume of an n-dimensional polytope is #P-complete. In view of this, we would like to address the question whether there is an algorithm that returns a value $\hat{V}$ such that $1/(1 + \epsilon) \leq I(F)/\hat{V} \leq (1 + \epsilon)$, in other words an algorithm that ϵ-approximates $I(F)$.

The first somewhat surprising answer to this question came with the major result of Dyer, Frieze and Kannan ([5]), who showed that there is a fully polynomial randomized approximation scheme (FPRAS) for the volume of an n-dimensional convex body. More precisely, they showed that the volume of an n-dimensional convex body $\mathcal{K}$ given by a weak membership oracle $\mathcal{M}$, can be ϵ-approximated with failure probability ξ, with $poly(n, \epsilon^{-1}, \log \xi^{-1})$ calls to $\mathcal{M}$. Here, $\mathcal{M}$ can be thought of as a black-box algorithm that decides whether a given point is in $\mathcal{K}$. This directly implies that there is a FPRAS for the integration of n-variate concave functions that can be evaluated in time $poly(n)$ at any point in the cube C.

Subsequently, Applegate and Kannan ([2]), extended this result to positive, smooth and nearly log-concave functions. Define

$$f(X) = \ln F(X)$$

and let c be the edge length of C, $t(n)$ be an upper bound on the time needed to evaluate F at any point in C, and α, β satisfy

$$|f(X) - f(Y)| \leq \alpha \left(\max_{i \in [1,n]} |x_i - y_i| \right) \tag{1}$$

$$f(\lambda X + (1 - \lambda)Y) \geq \lambda f(X) + (1 - \lambda)f(Y) - \beta \tag{2}$$

S. Arora et al. (Eds.): APPROX 2003+RANDOM 2003, LNCS 2764, pp. 122–128, 2003.
© Springer-Verlag Berlin Heidelberg 2003

for all $x, y \in C$ and $\lambda \in [0, 1]$. Their algorithm has running time

$$O(t(n)\frac{n^7}{\epsilon^2}c^2\alpha^2 e^{2\beta} \log \frac{n}{\xi'} \log \frac{d\alpha n}{\xi}).$$

It can be seen that α measures the smoothness of F. This gives rise to the following definition of smoothness.

Definition 1. *A function $F(\cdot)$ is called k-smooth if it satisfies $\alpha \leq n^k$. We denote by $\mathcal{S}_k$ the set of k-smooth functions, and by $\mathcal{S} = \bigcup_k \mathcal{S}_k$ the set of smooth functions .*

If $\beta = 0$, the function is log-concave (i.e. its logarithm is concave), so β can be viewed as a measure of the distance of F from log-concavity. The natural question is whether the dependence on β can be removed or somewhat alleviated. The contribution of this paper is to show that for any fixed integer k, it is NP-hard to 2^{n^k}-approximate the integral of positive smooth functions that are computable in polynomial time. In fact, we show that considerably small improvements on the dependence on β would imply unexpected (and rather indirect) algorithmic improvements for well studied NP-complete problems. Formally, we show the following.

Theorem 1. *For any fixed integer $k \geq 3$, if there is a (randomized) 2^{n^k}- approximation algorithm with time complexity $O(poly(\alpha)2^{g(\beta)})$ for the problem of integration of functions from $\mathcal{S}_{k+3}$, then there is a $O(poly(\alpha)n^{(g(n))^{k+3}})$ (randomized) algorithm for the Hamilton Path problem on graphs with n vertices.*

Corollary 1. *For any fixed integer k, it is NP-hard to 2^{n^k}-approximate the integral of polynomial-time computable functions from $\mathcal{S}$.*

We note here that, in general, only a few negative results concerning the approximability of counting problems are known. As observed in [6], the hardness of counting problems in most cases follows either from the NP-completeness of the corresponding decision problem, or from applying some "boosting" reduction which exploits an embedded NP-complete problem (see [10,6]). There appears to be a paucity of results that prove the hardness of approximate counting problems for some other more "interesting" reason. One such case is [4], which proves that there is no FPRAS for counting the number of independent sets in graphs of maximum degree $\Delta \geq 25$, unless NP=RP. As noted in [7], in view of the lack of "satisfactory" results that prove inapproximability under reasonable complexity-theoretical assumptions, research efforts have often been directed towards proving that certain restricted algorithmic approaches fail (see section 4 of [7] and the references therein).

The rest of the paper is organized as follows. In section 2 we give an overview of the proof technique, in section 3 we give the details of the proof and finally in section 4 we make some concluding comments.

2   Overview

We derive the result through a reduction from Hamilton Path (HP for short). Recall that HP is one of the first problems shown to be NP-complete (see [9]). Given a graph G (in some usual representation), HP asks whether there exists a simple path of length n, i.e. a path that goes through every vertex of G exactly once.

With every graph G, we associate a function F_G. If G has n vertices, F_G is a function of n^2 variables. The function F_G has the the following useful characteristics. It can be computed at any point x in a cube C of interest, in time polynomial in n. The parameters α, β of F_G (defined in inequalities 1,2), are polynomial in n. Also, the value of the integral of F_G depends on whether G contains a Hamilton Path or not. Specifically, if there is a HP, the integral of F_G over a cube C of constant edge size c, is lower bounded by an explicitly known quantity I_H. If not, it is upper bounded by I_{NH}, with $I_H/I_{NH} \geq 2^{n^k}$, for any fixed constant k. It follows that the integral is not 2^{n^k}-approximable.

Also, since $\beta = O(n^d)$ for some constant d (the smallest value of d we are able to exhibit in this paper is 6), an improvement of the running time of the integration algorithm to $poly(n, \epsilon^{-1}, \alpha, 2^{b^{(1/d)-e'}})$, for any $e' > 0$, would give a $2^{o(n)}$ randomized algorithm for Hamilton Path (the best currently known upper bound is $O(2^n)$, see [1]), and through the Sparsification Lemma of [8] a $2^{o(n)}$ randomized algorithm for 3-SAT, where now n is the number of variables.

3   The Proof

3.1   Definition and Properties of the Function F_G

Let G be a graph with n vertices and $\mathcal{P}$ be the set of length-n paths of G. The function $F_G(X)$ is a function of n^2 variables, $X = \{x_{11}, \ldots, x_{nn}\}$. Each path $p \in \mathcal{P}$ is associated with a term $f_p(X)$, and $F_G(X) = \sum_{p \in \mathcal{P}} f_p(X)$.

We now describe the term $f_p(X)$ for a path p. Assume an arbitrary numbering of the graph vertices with numbers in $[n]$. We consider p as an ordered set of vertices $v_1, \ldots, v_n$, where $v_i \in [n]$. We let $m = n^k$, where k is an integer constant to be discussed later. We define

$$f_p(X) = \prod_{i=1}^n g_i(X)$$

with

$$g_i(X) = \frac{\prod_{j=1}^{i-1} x_{v_i j}^m}{x_{v_i i}^m}$$

We will integrate F_G over the cube $C = [1, c]^{n^2}$, so we study its properties in this cube. Each term $f_p(X)$ is increasing in the variables appearing in the numerator and decreasing in the variables appearing in the denominator. By setting the former to c and the latter to 1, we get that the maximum value of $f_p(X)$ is $O(c^{n^2m})$. Since there are at most $n!$ paths, it follows that for any $X \in C$, $F_G(X)$ can be expressed with $O(mn^3 \log n)$ bits.

As noted in [2], the smoothness parameter α, can be upper bounded by

$$\alpha \le n^2 \max_{X \in C, x_i \in X} \left| \frac{\partial}{\partial x_i} \ln F(X) \right| = n^2 \max_{X \in C, x_i \in X} \left| \frac{\sum_{p \in \mathcal{P}} \partial f_p(X)/\partial x_i}{\sum_{p \in \mathcal{P}} f_p(X)} \right| \tag{3}$$

Let $x_i \in X$ be any variable. Since the exponent of x_i is at most nm, for all points X in C, we have

$$\frac{\partial f_p(X)}{\partial x_i} \le nm f_p(X)$$

which combined with inequality 3, gives $\alpha \le n^3 m$.

A note about the algorithm of [2] is due here. The algorithm operates on a grid imposed on C. The coordinates of the grid are multiples of $\gamma \le 1/2\alpha$. From the bound on α it follows that we are interested in evaluating F_G at points which are rationals expressible in polynomial space. From the definition of F_G, its value at any point of the grid is also a rational expressible in polynomial space.

The definition of β trivially implies that any upper bound for $f(X)$ is also an upper bound for β. From the above analysis we get $\beta \le O(mn^3 \log n)$. For a lower bound on β note that $f(X)$ can be written as $f(X) = \ln P(X) - m \sum_{i,j \in [n]} \ln x_{ij}$, where $P(X)$ is a multivariate polynomial. Since $P(X)$ is not log-concave in general, the value of β can be lower bounded from the value of β for the function $\hat{f}(X) = -m \sum_{i,j \in [n]} \ln x_{ij}$, which can be seen to be $O(mn^2)$. Thus, we get $\beta \ge mn^2$.

We finally note that $F_G(X)$ has some additional interesting properties. First, F_G has derivatives of any order, everywhere in the cube C. Also, its form is relatively simple, as it is a sum of rational multivariate polynomials. In addition, given a graph G we can easily obtain a closed form for the integral of F_G, though of exponential length.

3.2   A Polynomial Time Algorithm for the Evaluation of F_G

We give an algorithm that computes $F_G(X)$ at any point X, in n time steps. We extend the definition of the path terms, to paths of length t. Concretely, we let

$$f_p(X) = \prod_{i=1}^{t} g_i(X)$$

with

$$g_i(X) = \frac{\prod_{j=1}^{i-1} x_{v_{ij}}^m}{x_{v_i i}^m}$$

Let $\mathcal{P}_t(v)$ be the set of paths of length t that end in node v. Also, let $Q_1(v) = x_{v1}^{-m}$. Inductively, assume that just before time step t, for every $v \in V$ we have computed

$$Q_{t-1}(v) = \sum_{p \in \mathcal{P}_{t-1}(v)} f_p(X)$$

Let $N(v)$ denote the set of neighbors of node v. At step t, for each node v we compute

$$Q_t(v) = \left(x_{vt}^{-m} \prod_{j=1}^{t-1} x_{vj}^m \right) \sum_{v' \in N(v)} Q_{t-1}(v')$$

After n steps the quantities $Q_n(v)$ have been computed for all vertices $v \in V$. Then,

$$F(X) = \sum_{v \in V} Q_n(v)$$

The computation of $Q_t(V)$ requires a polynomial number of operations. Since there are n steps and n vertices, it follows that F_G can be computed with a polynomial number of operations. The points we are interested in are rationals expressible in polynomial space, and from the observations of the previous subsection, all the intermediate quantities are expressible in polynomial space. It follows that $F_G(X)$ can be evaluated exactly, at any point $X \in C$, in time polynomial in n.

3.3   Bounding the Integrals

We integrate $F(X)$ over a cube $C = [1, c]^{n^2}$. Let $dX = dx_{11} \cdot \ldots \cdot dx_{nn}$ and π be a permutation of the variable names. Since

$$\int_{X \in C} \sum_{p \in \mathcal{P}} f_p(X) dX = \sum_{p \in \mathcal{P}} \int_{X \in C} f_p(X) dX$$

we can consider the integral of each path separately. We will refer to the value of the integral of a term corresponding to a path p as the integral of p. Also, since

$$\int_{X \in C} f_p(x_{11}, \ldots, x_{nn}) dX = \int_{X \in C} f_p\left(\pi(x_{11}), \ldots, \pi(x_{nn}) \right) dX$$

we can rename the variables in any term of F. It is then easy to see that the integral of a path depends only on the structure of the path and not on the particular vertices appearing on it.

We first consider the integral of a HP. Since HP is a simple path, there are no cancellations of variables and its integral is

$$I_{HP} = \int_{X \in C} \left(\prod_{i=1}^{n} \frac{1}{x_{ii}^m} \right) \cdot \left(\prod_{1 \le i \le n,\, 1 \le j \le i-1} x_{ij}^m \right) dX =$$

$$= (m+1)^{-n(n-1)/2} (m-1)^n (1 - c^{-m+1})^n (c^{m+1} - 1)^{n(n-1)/2} (c-1)^{n(n-1)/2}$$

Let us now consider the integrals of other non-simple paths. Suppose a path p goes through $n - d$ distinct nodes. Then, the corresponding term f_p is of the form

$$f_p(X) = \left(\prod_{i=1}^{n-d} \frac{1}{x_i^m}\right) \cdot \left(\prod_{i=n-d+1}^{n(n+1)/2-t} x_i^{a_i m}\right)$$

where t and a_i are integers that depend on the structure of p. In this case, d monomials in the denominator cancel with variables in the numerator, so that

$$\sum_{i=n-d+1}^{n(n-1)/2-t} a_i = n(n-1)/2 - d$$

By integrating, we get

$$\int_{X \in C} f_p(X) \le (1 - c^{-m+1})^{n-d} (c-1)^{n(n+1)/2+t} \prod_{i=n-d+1}^{n(n-1)/2-t} c^{a_i m+1}$$

$$\le c^{n^2} c^{m(\sum_{i=n-d+1}^{n(n-1)/2-t} a_i)} = c^{n^2} c^{mn(n-1)/2} c^{-md}$$

Now suppose we are given a non-Hamiltonian graph. Since there are at most $n! \le c^{n^2}$ paths in the graph, the integral of the associated function is

$$I_{NH} \le c^{2n^2} c^{mn(n-1)/2} c^{-m}$$

On the other hand, if the given graph is Hamiltonian (and even if we consider only the integral of the HP), the integral of the associated function is

$$I_H \ge c^{-O(n^2 \log n)} c^{mn(n-1)/2}$$

which gives a large gap, namely

$$\frac{I_H}{I_{NH}} \ge c^{m-O(n^2 \log n)}$$

Recall that $m = n^k$. By taking any fixed $k \ge 3$ we get Theorem 1.

4   Conclusions

We showed that it is NP-hard to 2^{n^k}-approximate the integral of smooth positive n-variate functions, for any fixed integer k. We also argued that the currently best known integration algorithm cannot be substantially improved, unless there exist faster algorithms for Hamilton Path and 3-SAT.

Note that the 2^{n^k}-inapproximability holds for $(k+3)$-smooth functions, with $k \ge 3$. Also, in order to obtain the full range of our inapproximability result, we make use of functions that progressively become less efficiently computable. It is an interesting question whether similar inapproximability properties can be shown for classes of functions with different trade-offs between their evaluation time complexity and the value of their α, β parameters.

We feel that the most interesting open question is whether a lower bound can be proved on β, for any smooth polynomially computable function F_G which can be constructed using the techniques of this paper.

References

1. Noga Alon, Raphael Yuster, Uri Zwick: Color Coding, Journal of the ACM, **42(4)** (1995) 844-856
2. David Applegate, Ravi Kannan: Sampling and Integration of Near Log-concave Functions, Proceedings of the 23rd Annual ACM Symposium on Theory of Computing (1991) 156-163
3. Graham Brightwell, Peter Winkler: Counting Linear Extensions is #P-complete, Proceedings of the 23th Annual ACM Symposium on Theory of Computing, (1991) 175-181
4. Martin Dyer, Alan Frieze, Mark Jerrum: On Counting Independent Sets in Graphs, Proceedings of the 40nd Annual Symposium on the Foundations of Computer Science, (1999) 210-217
5. Martin Dyer, Alan Frieze, Ravi Kannan: A Random Polynomial-time Algorithm for Approximating the Volume of Convex Bodies, Journal of the ACM, **38**, (1991) 1-17
6. Martin Dyer, Leslie Ann Goldberg, Catherine Greenhill, Mark Jerrum: On the Relative Complexity of Approximate Counting Problems, Algorithmica, to appear
7. Martin Dyer, Catherine Greenhill: Random Walks on Combinatorial Objects, Surveys in Combinatorics, **267** (1999) 101-136
8. Rusell Impagliazzo, Ramamohan Paturi: Which Problems Have Strongly Exponential Complexity?, Proceedings of the 39th Annual Symposiym on the Foundations of Computer Science (1998)
9. Christos Papadimitriou: Computational Complexity, Addison-Wesley (1994)
10. Alistair Sinclair: Algorithms for Random Generation and Counting: A Markov Chain Approach, Progress in Theoretical Computer Science, (1993)

A 2-Approximation Algorithm for the Soft-Capacitated Facility Location Problem

Mohammad Mahdian[1], Yinyu Ye[2*], and Jiawei Zhang[2*]

[1] Laboratory for Computer Science, MIT, Cambridge, MA 02139, USA.
mahdian@theory.lcs.mit.edu
[2] Department of Management Science and Engineering, School of Engineering,
Stanford University, Stanford, CA 94305, USA.
{yinyu-ye,jiazhang}@stanford.edu

Abstract. This paper is divided into two parts. In the first part of this paper, we present a 2-approximation algorithm for the soft-capacitated facility location problem. This achieves the integrality gap of the natural LP relaxation of the problem. The algorithm is based on an improved analysis of an algorithm for the linear facility location problem, and a *bifactor* approximate-reduction from this problem to the soft-capacitated facility location problem. We will define and use the concept of bifactor approximate reductions to improve the approximation factor of several other variants of the facility location problem. In the second part of the paper, we present an alternative analysis of the authors' 1.52-approximation algorithm for the uncapacitated facility location problem, using a single factor-revealing LP. This answers an open question of [16]. Furthermore, this analysis, combined with a recent result of Thorup [21] shows that our algorithm can be implemented in quasi-linear time, achieving the best known approximation factor in the best possible running time.

1 Introduction

Variants of the facility location problem (FLP) have been studied extensively in the operations research and management science literatures and have received considerable attention in the area of approximation algorithms [17]. In the metric uncapacitated facility location problem (UFLP), which is the most basic facility location problem, we are given a set $\mathscr{F}$ of *facilities*, a set $\mathscr{C}$ of *cities* (a.k.a. clients), a cost f_i for opening facility $i \in \mathscr{F}$, and a connection cost c_{ij} for connecting client j to facility i. The objective is to open a subset of the facilities in $\mathscr{F}$, and connect each city to an open facility so that the total cost is minimized. We assume that the connection costs are metric, meaning that they are symmetric and satisfy the triangle inequality.

Since the first constant factor approximation algorithm due to Shmoys, Tardos and Aardal [18], a large number of approximation algorithm have been proposed for the UFLP [19, 11, 12, 15, 20, 13, 1, 3, 4, 5, 8, 13, 14], and the current

---

* Research supported in part by NSF grant DMI-0231600.

best known approximation factor is 1.52 given by Mahdian, Ye and Zhang [16]. Guha and Khuller [8] proved that it is impossible to get an approximation guarantee of 1.463 for the UFLP, unless $\mathbf{NP} \subseteq \mathrm{DTIME}[n^{O(\log \log n)}]$.

The growing interests in the UFLP rely on not only its applications in a large number of settings [7], but also the fact that the UFLP is one of the most basic models among discrete location problems. The insights gained in dealing with the UFLP may also apply to more complicated location models, and in many cases the latter can be reduced directly to the UFLP.

In this paper, we give a 2-approximation algorithm for the soft-capacitated facility location problem (SCFLP) by reducing it to the UFLP. The SCFLP is similar to the UFLP, except that there is a capacity u_i associated with each facility i, which means that if we want this facility to serve x cities, we have to open it $\lceil x/u_i \rceil$ times at a cost of $f_i \lceil x/u_i \rceil$. This problem is also known as facility location problem with integer decision variables in the operations research literature (See [2]). Chudak and Shmoys [6] gave a 3-approximation algorithm for the SCFLP with uniform capacities (i.e., $u_i = u$ for all $i \in \mathscr{F}$) using LP rounding. For non-uniform capacities, Jain and Vazirani [13] showed how to reduce this problem to the UFLP, and by solving the UFLP through a primal-dual algorithm, they obtained a 4-approximation. A local search algorithm proposed by Arya et al [1] had an approximation ratio 3.72. Following the approach of Jain and Vazirani [13], Jain, Mahdian, and Saberi [12] showed that the SCFLP can be solved within a factor of 3. This result was further improved by the authors [16] to a 2.89-approximation for the SCFLP. This is the best previously known algorithm for this problem. We improve this factor to 2, achieving the integrality gap of the natural LP relaxation of the problem. The main idea of our algorithm is to consider algorithms and reductions that have separate (not necessarily equal) approximation factors for facility and connection costs. We will define the concept of *bifactor* approximate reduction in this paper, and show how it can be used to get an approximation factor of 2 for the SCFLP. We will also generalize our algorithm to a common generalization of the SCFLP and the concave-cost FLP. The idea of using bifactor approximation algorithms and reductions can be used to improve the approximation factor of several other problems in a straightforward manner.

In the second part of this paper, we present an alternative analysis for the 1.52-approximation algorithm for the UFLP [16] using a single factor-revealing LP. This answers an open question of [16]. Furthermore, this analysis shows that the second phase of the 1.52 algorithm can be implemented in quasi-linear time. This, together with a recent result of Thorup [21], prove that our algorithm can be implemented in quasi-linear time, achieving the best known approximation factor in essentially the best possible running time.

The rest of this paper is organized as follows: In Section 2 we present the necessary definitions and notations. In Section 3 we present a lemma on the approximability of the linear-cost facility location problem. In Section 4 we define the concept of bifactor approximate reductions between facility location problems, and present an algorithm for the SCFLP and a common generalization of the SCFLP and the concave-cost FLP using the lemma proved in Section 3 and a bifactor reduction from the SCFLP to the linear-cost FLP. Then, in Section 5, we present a new analysis on the 1.52 algorithm for the UFLP and show how it leads to an implementation in quasi-linear times.

2   Definitions and Notations

In this paper, we will define reductions between various facility location problems. Many such problems can be considered as special cases of the *generalized facility location problem*, as defined below. This problem was first defined and studied in [9].

Definition 1. *In the* metric generalized facility location problem, *we are given a set $\mathscr{C}$ of n_c cities, a set $\mathscr{F}$ of n_f facilities, a connection cost c_{ij} between city j and facility i for every $i \in \mathscr{F}, j \in \mathscr{C}$, and a facility cost function $f_i : \{0, \ldots, n_c\} \mapsto \mathbb{R}^+$ for every $i \in \mathscr{F}$. Connection costs are symmetric and obey the triangle inequality. The value of $f_i(k)$ equals the cost of opening facility i, if it is used to serve k cities. A solution to the problem is a function $\phi : \mathscr{C} \to \mathscr{F}$ assigning each city to a facility. The* facility cost F_ϕ *of the solution ϕ is defined as $\sum_{i \in \mathscr{F}} f_i(|\{j : \phi(j) = i\}|)$, i.e., the total cost of opening facilities. The* connection cost *(a.k.a. service cost) C_ϕ of ϕ is $\sum_{j \in \mathscr{C}} c_{\phi(j),j}$, i.e., the total cost of connecting each city to its assigned facility. The objective is to find a solution ϕ that minimizes the sum $F_\phi + C_\phi$.*

Now we can define uncapacitated and soft-capacitated facility location problems as special cases of the generalized FLP:

Definition 2. *The* metric uncapacitated facility location problem (UFLP) *is a special case of the generalized FLP in which all facility cost functions are of the following form: for each $i \in \mathscr{F}$, $f_i(k) = 0$ if $k = 0$, and $f_i(k) = f_i$ if $k > 0$, where f_i is a constant (which is called the facility cost of i).*

Definition 3. *The* metric soft-capacitated facility location problem (SCFLP) *is a special case of the generalized FLP in which all facility cost functions are of the form $f_i(k) = f_i \lceil k/u_i \rceil$, where f_i and u_i are constants for every $i \in \mathscr{F}$. u_i is called the capacity of facility i.*

The 1.52-approximation algorithm of Mahdian, Ye, and Zhang [16] is built upon an earlier approximation algorithm of Jain, Mahdian, and Saberi [12]. We denote these two algorithms by the MYZ and the JMS algorithms, respectively. The analyses of both of these algorithms have the feature that allow the approximation factor for the facility cost to be different from the approximation factor for the connection cost, and give a way to compute the tradeoff between these two factors. The following definition captures this notion.

Definition 4. *An algorithm is called a (γ_f, γ_c)-approximation algorithm for the generalized FLP, if for every instance $\mathscr{I}$ of the generalized FLP, and for every solution SOL for $\mathscr{I}$ with facility cost F_{SOL} and connection cost C_{SOL}, the cost of the solution found by the algorithm is at most $\gamma_f F_{SOL} + \gamma_c C_{SOL}$.*

Recall the following theorem of Jain et al. [12] on the approximation factor of the JMS algorithm.

Theorem A [12]. *Let $\gamma_f \geq 1$ be fixed and $\gamma_c := \sup_k\{z_k\}$, where z_k is the solution of the following optimization program (which we call the factor-revealing LP).*

$$\text{maximize} \quad \frac{\sum_{i=1}^k \alpha_i - \gamma_f f}{\sum_{i=1}^k d_i} \tag{1}$$

$$\text{subject to} \quad \forall\, 1 \leq i < k : \ \alpha_i \leq \alpha_{i+1} \tag{2}$$

$$\forall\, 1 \leq j < i < k : \ r_{j,i} \geq r_{j,i+1} \tag{3}$$

$$\forall\, 1 \leq j < i \leq k : \ \alpha_i \leq r_{j,i} + d_i + d_j \tag{4}$$

$$\forall\, 1 \leq i \leq k : \ \sum_{j=1}^{i-1} \max(r_{j,i} - d_j, 0) + \sum_{j=i}^k \max(\alpha_i - d_j, 0) \leq f \tag{5}$$

$$\forall\, 1 \leq j \leq i \leq k : \ \alpha_j, d_j, f, r_{j,i} \geq 0 \tag{6}$$

Then the JMS algorithm is a (γ_f, γ_c)-approximation algorithm for the UFLP.

We will use the above theorem in this paper to give an alternative proof of the following theorem about the performance of the MYZ algorithm.

Theorem B [16]. *Let (γ_f, γ_c) be a pair obtained from the above factor-revealing LP. Then for every $\delta \geq 1$, there is a $(\gamma_f + \ln(\delta) + \epsilon, 1 + \frac{\gamma_c - 1}{\delta})$-approximation algorithm for the UFLP.*

3   The Linear-Cost Facility Location Problem

The *linear-cost facility location problem* is a special case of the generalized FLP in which the facility costs are of the form

$$f_i(k) = \begin{cases} 0 & k = 0 \\ a_i k + b_i & k > 0 \end{cases}$$

where a_i and b_i are nonnegative values for each $i \in \mathscr{F}$. b_i and a_i are called the setup and marginal (a.k.a. incremental) cost of facility i, respectively.

We denote an instance of the linear-cost FLP with marginal costs (a_i), setup costs (b_i), and connection costs (c_{ij}) by $LFLP(a, b, c)$. Clearly, the regular UFLP is a special case of the linear-cost FLP with $a_i = 0$, i.e., $LFLP(0, b, c)$. Furthermore, it is straightforward to see that $LFLP(a, b, c)$ is equivalent to an instance of the regular UFLP in which the marginal costs are added to the connection costs. More precisely, let $\bar{c}_{ij} = c_{ij} + a_i$ for $i \in \mathcal{F}$ and $j \in \mathcal{C}$, and consider an instance of the UFLP with facility costs (b_i) and connection costs $(\bar{c}_{ij})$. We denote this instance by $UFLP(b, c + a)$. It is easy to see that $LFLP(a, b, c)$ is equivalent to $UFLP(b, c + a)$. Thus, the linear-cost FLP can be solved using any algorithm for the UFLP, and the overall approximation ratio will be the same. However, for applications in the next section, we need bifactor approximation factors of the algorithm (as defined in Definition 4).

It is not necessarily true that applying a (γ_f, γ_c)-approximation algorithm for the UFLP on the instance $UFLP(b, a + c)$ will give a (γ_f, γ_c)-approximate solution for $LFLP(a, b, c)$. However, we will show that the JMS algorithm has this property. The following lemma, whose proof is presented in Appendix A, generalizes Theorem A to the linear-cost FLP.

Lemma 1. *Let (γ_f, γ_c) be a pair obtained from the factor-revealing LP in Theorem A. Then applying the JMS algorithm on the instance $UFLP(b, a + c)$ will give a (γ_f, γ_c)-approximate solution for $LFLP(a, b, c)$.*

The above lemma and Theorem 9 in [12] give us the following corollary, which will be used in the next section.

Corollary 1. *There is a $(1, 2)$-approximation algorithm for the linear-cost facility location problem.*

It is worth mentioning that the MYZ algorithm can also be generalized for the linear-cost FLP. The only trick is to scale up both a and b in the first phase by a factor of δ, and scale them both down in the second phase. The rest of the proof is almost the same as the proof of Lemma 1.

4   The Soft-Capacitated Facility Location Problem

In this section we will show how the soft-capacitated facility location problem can be reduced to the linear-cost FLP. In Section 4.1 we define the concept of reduction between facility location problems. We will use this concept in Sections 4.2 and 4.3 to obtain approximation algorithms for the SCFLP and a generalization of the SCFLP and the concave-cost FLP.

4.1   Reduction between Facility Location Problems

A reduction from a facility location problem $\mathcal{A}$ to another facility location problem $\mathcal{B}$ is an efficient procedure R that maps every instance $\mathcal{I}$ of $\mathcal{A}$ to an instance $R(\mathcal{I})$ of $\mathcal{B}$. This procedure is called a (σ_f, σ_c)-reduction if the following conditions hold.

1. For any instance $\mathcal{I}$ of $\mathcal{A}$ and any feasible solution for $\mathcal{I}$ with facility cost $F_{\mathcal{A}}^*$ and connection cost $C_{\mathcal{A}}^*$, there is a corresponding solution for the instance $R(\mathcal{I})$ with facility cost $F_{\mathcal{B}}^* \le \sigma_f F_{\mathcal{A}}^*$ and connection cost $C_{\mathcal{B}}^* \le \sigma_c C_{\mathcal{A}}^*$.

2. For any feasible solution for the instance $R(\mathscr{I})$, there is a corresponding feasible solution for $\mathscr{I}$ whose total cost is at most as much as the total cost of the original solution for $R(\mathscr{I})$. In other words, the facility location instance $R(\mathscr{I})$ is an over-estimate of the facility location instance $\mathscr{I}$.

Theorem 1. *If there is a (σ_f, σ_c)-reduction from a facility location problem $\mathscr{A}$ to another facility location problem $\mathscr{B}$, and a (γ_f, γ_c)-approximation algorithm for $\mathscr{B}$, then there is a $(\gamma_f \sigma_f, \gamma_c \sigma_c)$-approximation algorithm for $\mathscr{A}$.*

Proof. On an instance $\mathscr{I}$ of the problem $\mathscr{A}$, we compute $R(\mathscr{I})$, run the (γ_f, γ_c)-approximation algorithm for $\mathscr{B}$ on $R(\mathscr{I})$, and output the corresponding solution for $\mathscr{I}$. In order to see why this is a $(\gamma_f \sigma_f, \gamma_c \sigma_c)$-approximation algorithm for $\mathscr{A}$, let SOL denote an arbitrary solution for $\mathscr{I}$, ALG denote the solution that the above algorithm finds, and $F_{\mathscr{P}}^*$ and $C_{\mathscr{P}}^*$ ($F_{\mathscr{P}}^{ALG}$ and $C_{\mathscr{P}}^{ALG}$, respectively) denote the facility and connection costs of SOL (ALG, respectively) when viewed as a solution for the problem $\mathscr{P}$ ($\mathscr{P} = \mathscr{A}, \mathscr{B}$). By the definition of (σ_f, σ_c)-reductions and (γ_f, γ_c)-approximation algorithms we have

$$F_{\mathscr{A}}^{ALG} + C_{\mathscr{A}}^{ALG} \leq F_{\mathscr{B}}^{ALG} + C_{\mathscr{B}}^{ALG} \leq \gamma_f F_{\mathscr{B}}^* + \gamma_c C_{\mathscr{B}}^* \leq \gamma_f \sigma_f F_{\mathscr{A}}^* + \gamma_c \sigma_c C_{\mathscr{A}}^*,$$

which completes the proof of the lemma.   □

We will see examples of reductions in the rest of this paper.

4.2   The Soft-Capacitated Facility Location Problem

In this subsection, we give a 2-approximation algorithm for the soft-capacitated FLP by reducing it to the linear-cost FLP.

Theorem 2. *There is a 2-approximation algorithm for the soft-capacitated facility location problem.*

Proof. We use the following reduction: Construct an instance of the linear-cost FLP, where we have the same sets of facilities and clients. The connection costs remain the same. However, the facility cost of the ith facility is $(1 + \frac{k-1}{u_i})f_i$ if $k \geq 1$ and 0 if $k = 0$. Note that, for every $k \geq 1$, $\lceil \frac{k}{u_i} \rceil \leq 1 + \frac{k-1}{u_i} \leq 2 \cdot \lceil \frac{k}{u_i} \rceil$. Therefore, it is easy to see that this reduction is a $(2, 1)$-reduction. By Lemma 1, there is a $(1, 2)$-approximation algorithm for the linear-cost FLP, which together with Theorem 1 completes the proof.   □

We now illustrate that the following natural linear programming formulation of the SCFLP has an integrality gap of 2. This means that we cannot obtain a better approximation ratio using this LP relaxation as the lower bound.

$$\text{minimize} \qquad \sum_{i \in \mathscr{F}} f_i y_i + \sum_{i \in \mathscr{F}} \sum_{j \in \mathscr{C}} c_{ij} x_{ij} \qquad\qquad (7)$$

$$\text{subject to} \quad \forall i \in \mathscr{F}, j \in \mathscr{C} : \; x_{ij} \le y_i$$

$$\forall i \in \mathscr{F} : \; \sum_{j \in \mathscr{C}} x_{ij} \le u_i y_i$$

$$\forall j \in \mathscr{C} : \; \sum_{i \in \mathscr{F}} x_{ij} = 1$$

$$\forall i \in \mathscr{F}, j \in \mathscr{C} : \; x_{ij} \in \{0,1\} \qquad\qquad (8)$$

$$\forall i \in \mathscr{F} : \; y_i \text{ is a nonnegative integer} \qquad\qquad (9)$$

In a natural linear program relaxation, we replace the constraints (8) and (9) by $x_{ij} \ge 0$ and $y_i \ge 0$. Here we observe that even if we only relax constraint (9), the integrality gap is 2. Consider an instance of the SCFLP that consists of only one potential facility i, and $k \ge 2$ clients. Assume that the capacity of facility i is $k-1$, the facility cost is 1, and all connection costs are 0. Clearly, the optimal integral solution has cost 2. However, after relaxing constraint (9), the optimal fractional solution has cost $1 + \frac{1}{k-1}$. Therefore, the integrality gap between the integer program and its relaxation is $\frac{2k}{k-1}$ which tends to 2 as k tends to infinity.

4.3   The Concave Soft-Capacitated Facility Location Problem

In this subsection, we consider a common generalization of the soft-capacitated facility location problem and the concave-cost facility location problem. This problem, which we refer to as the *concave soft-capacitated FLP*, is the same as the soft-capacitated FLP except that if $r \ge 0$ copies of facility i are open, then the facility cost is $g(r)a_i$ where $g(r)$ is a given concave function of r. In other words, the concave soft-capacitated FLP is a special case of the generalized FLP in which the facility cost functions are of the form $f_i(x) = a_i g(\lceil x/u_i \rceil)$ for constants a_i, u_i and a concave function g. It is also a special case of the so-called stair-case cost facility location problem [10]. On the other hand, it is a common generalization of the soft-capacitated FLP (when $g(r) = r$) and the concave-cost FLP (when $u_i = 1$ for all i). The concave-cost FLP is a special case of the generalized FLP in which facility cost functions are required to be concave (See [9]). The main result of this subsection is the following.

Theorem 3. *The concave soft-capacitated FLP is $(\frac{g(2)}{g(1)}, 1)$-reducible to the linear-cost FLP.*

The proof of the above theorem is omitted here. The idea is to show that the concave soft-capacitated FLP is $(\frac{g(2)}{g(1)}, 1)$ reducible to the concave-cost FLP, and the latter is equivalent to the linear-cost FLP. Therefore, by Theorem 3, a good approximation algorithm for linear-cost FLP would imply a good approximation for the concave soft-capacitated FLP.

5   The Uncapacitated Facility Location Problem

In this section, we present a new analysis of the 1.52-approximation algorithm of Mahdian, Ye, and Zhang [16] for the UFLP. The analysis of the MYZ algorithm in [16] is based on combining a result of Jain et al. [12] (which is proved using factor-revealing LPs) with an analysis of a greedy augmentation procedure of Charikar et al. [3]. Here, we analyze the MYZ algorithm using a single factor-revealing LP. This gives us a new perspective on the MYZ algorithm. As a corollary, we use a recent result of Thorup [21] that the JMS algorithm can be implemented in quasi-linear time to improve the running time of the MYZ algorithm.

We begin by sketching the MYZ algorithm. The algorithm consists of two phases. In the first phase, we scale up the facility costs in the instance by a factor of δ (which will be fixed later), and then run the JMS algorithm (see [12] for a description) on the modified instance. In addition to finding a solution for the scaled instance, the JMS algorithm outputs the *share* of each city of the total cost of the solution. Let α_j denote the share of city j of the total cost (Therefore the total cost of the solution is $\sum_j \alpha_j$). The main step in the analysis of the JMS algorithm is to prove that for any collection $\mathcal{S}$ of one facility $f_{\mathcal{S}}$ with opening cost δf (f in the original instance) and k cities with connection costs $d_1, \dots, d_k$ to $f_{\mathcal{S}}$ and shares $\alpha_1, \dots, \alpha_k$ of the total cost, the values δf, d_j's, α_j's and $r_{j,i}$'s (whose definition is omitted here, since we don't need it) satisfy the inequalities (2)-(6) in Theorem A, except that the inequality (5) is replaced by

$$\forall\, 1 \leq i \leq k : \quad \sum_{j=1}^{i-1} \max(r_{j,i} - d_j, 0) + \sum_{j=i}^{k} \max(\alpha_i - d_j, 0) \leq \delta f \qquad (10)$$

In the second phase of the MYZ algorithm we reduce the scaling factor δ continuously, until it gets to 1. If at any point during this process a facility could be opened without increasing the total cost (i.e., if the opening cost of the facility equals the total amount that cities can save by switching their "service provider" to that facility), then we open the facility and connect each city to its closest open facility. The second phase of the MYZ algorithm is equivalent to a greedy augmentation procedure of [8, 3], and, in fact, a lemma from [3] is used in [16] in order to analyze the second phase. Here we analyze this phase differently. First, we modify the second phase as follows: Instead of decreasing the scaling factor continuously from δ to 1, we decrease it discretely in L steps where L is a constant. Let δ_i denote the value of the scaling factor in the i'th step. Therefore, $\delta = \delta_1 > \delta_2 > \dots > \delta_L = 1$. We will fix the values of δ_i's later. After decreasing the scaling factor from δ_{i-1} to δ_i, we consider facilities in an *arbitrary* order, and open those that can be opened without increasing the total cost. We denote this modified algorithm by MYZ_L. Clearly, if L is sufficiently large (depending on the instance), the algorithm MYZ_L computes the same solution as the MYZ algorithm.

In order to analyze the above algorithm, we need to add extra variables and inequalities to the inequalities (2), (3), (4), (10) and (6). Let $r_{j,k+i}$ denote the connection cost that city j in $\mathcal{S}$ pays after we change the scaling factor to δ_i and process all facilities as described above (Thus, $r_{j,k+1}$ is the connection cost of city j after the first phase). Therefore, by the description of the algorithm, we have

$$\forall 1 \le i \le L: \quad \sum_{j=1}^{k} \max(r_{j,k+i} - d_j, 0) \le \delta_i f, \tag{11}$$

since otherwise we could open f_S and decrease the total cost.

Now, we compute the share of the city j of the total cost of the solution that the MYZ$_L$ algorithm finds. In the first phase of the algorithm, the share of city j of the total cost is α_j. Of this amount, $r_{j,k+1}$ is spent on the connection cost, and $\alpha_j - r_{j,k+1}$ is spent on the facility costs. However, since the facility costs are scaled up by a factor of δ in the first phase, therefore the share of city j of *facility costs* in the original instance is equal to $(\alpha_j - r_{j,k+1})/\delta$. After we reduce the scaling factor from δ_i to δ_{i+1} ($i = 1, \ldots, L-1$), the connection cost of city j is reduced from $r_{j,k+i}$ to $r_{j,k+i+1}$. Therefore, in this step, the share of city j of the facility costs is $r_{j,k+i} - r_{j,k+i+1}$ with respect to the scaled instance, or $(r_{j,k+i} - r_{j,k+i+1})/\delta_{i+1}$ with respect to the original instance. Thus, at the end of the algorithm, the total share of city j of facility costs is

$$\frac{\alpha_j - r_{j,k+1}}{\delta} + \sum_{i=1}^{L-1} \frac{r_{j,k+i} - r_{j,k+i+1}}{\delta_{i+1}}.$$

We also know that the final amount that city j pays for the connection cost is $r_{j,k+L}$. Therefore, the share of the facility j of the total cost of the solution is:

$$\frac{\alpha_j - r_{j,k+1}}{\delta} + \sum_{i=1}^{L-1} \frac{r_{j,k+i} - r_{j,k+i+1}}{\delta_{i+1}} + r_{j,k+L+1} =$$

$$\frac{\alpha_j}{\delta} + \sum_{i=1}^{L-1} \left(\frac{1}{\delta_{i+1}} - \frac{1}{\delta_i} \right) r_{j,k+i}. \tag{12}$$

This, together with a *dual fitting* argument similar to [12], imply the following.

Theorem 4. *Let (ξ_f, ξ_c) be such that $\xi_f \ge 1$ and ξ_c is an upper bound on the solution of the following maximization program for every k.*

$$\text{maximize} \qquad \frac{\sum_{j=1}^{k} \left(\frac{\alpha_j}{\delta} + \sum_{i=1}^{L-1} \left(\frac{1}{\delta_{i+1}} - \frac{1}{\delta_i} \right) r_{j,k+i} \right) - \xi_f f}{\sum_{i=1}^{k} d_i} \tag{13}$$

$$\text{subject to} \qquad (2), (3), (4), (10), (11), (6)$$

Then, the MYZ$_L$ algorithm is a (ξ_f, ξ_c)-approximation algorithm for the UFLP.

In the following theorem, we analyze the factor-revealing LP (13) and rederive the main result of [16]. In order to do this, we need to set the values of δ_i's. Here, for simplicity of computations, we set δ_i to $\delta^{\frac{L-i}{L-1}}$; however, it is easy to observe that any choice of δ_i's such that the limit of $\max_i(\delta_{i+1} - \delta_i)$ as L tends to infinity is zero, will also work. The proof is omitted here.

Theorem 5. *Let (γ_f, γ_c) be a pair given by the maximization program in Theorem A, and $\delta \geq 1$ be an arbitrary number. Then for every ϵ, if L is a sufficiently large constant, the MYZ_L algorithm is a $(\gamma_f + \ln(\delta) + \epsilon, 1 + \frac{\gamma_c - 1}{\delta})$-approximation algorithm for the UFLP.*

The above analysis enables us to prove the following result.

Corollary 2. *For every $\epsilon > 0$, there is a quasi-linear time $(1.52 + \epsilon)$-approximation algorithm for the UFLP, both in the distance oracle model (where the connection costs are given by a matrix) and in the sparse graph model (where the connection costs are distances in a given graph).*

Proof Sketch. We use the MYZ_L algorithm for a large constant L. Thorup [21] shows that for every $\epsilon > 0$, the JMS algorithm can be implemented in quasi-linear time (in both distance oracle and graph models) with an approximation factor of $1.61 + \epsilon$. It is straightforward to see that his argument actually implies the stronger conclusion that the quasi-linear algorithm is a $(\gamma_f + \epsilon, \gamma_c + \epsilon)$-approximation, where (γ_f, γ_c) are given by Theorem A. This shows that the first phase of the MYZ_L algorithm can be implemented in quasi-linear time. The second phase consists of constantly many rounds. Therefore, we only need to show that each of these rounds can be implemented in quasi-linear time. This is easy to see in the distance oracle model. In the graph model, we can use the exact same argument as the one used by Thorup in the proof of Lemma 5.1 of [21]. □

Acknowledgements. We would like to thank Asaf Levin for pointing out that our analysis of the 2-approximation algorithm for the soft-capacitated facility location problem is tight.

References

[1] V. Arya, N. Garg, R. Khandekar, A. Meyerson, K. Munagala, and V. Pandit. Local search heuristics for k-median and facility location problems. In *Proceedings of 33rd ACM Symposium on Theory of Computing*, 2001.

[2] P. Bauer and R. Enders. A capacitated facility location problem with integer decision variables. In *International Symposium on Mathematical Programming (ISMP)*, 1997.

[3] M. Charikar and S. Guha. Improved combinatorial algorithms for facility location and k-median problems. In *Proceedings of the 40th Annual IEEE Symposium on Foundations of Computer Science*, pages 378–388, October 1999.

[4] F.A. Chudak. Improved approximation algorithms for uncapacited facility location. In R.E. Bixby, E.A. Boyd, and R.Z. Ríos-Mercado, editors, *Integer Programming and Combinatorial Optimization*, volume 1412 of *Lecture Notes in Computer Science*, pages 180–194. Springer, Berlin, 1998.

[5] F.A. Chudak and D. Shmoys. Improved approximation algorithms for the uncapacitated facility location problem. unpublished manuscript, 1998.

[6] F.A. Chudak and D. Shmoys. Improved approximation algorithms for the capacitated facility location problem. In *Proc. 10th Annual ACM-SIAM Symposium on Discrete Algorithms*, pages 875–876, 1999.

[7] G. Cornuejols, G.L. Nemhauser, and L.A. Wolsey. The uncapacitated facility location problem. In P. Mirchandani and R. Francis, editors, *Discrete Location Theory*, pages 119–171. John Wiley and Sons Inc., 1990.

[8] S. Guha and S. Khuller. Greedy strikes back: Improved facility location algorithms. *Journal of Algorithms*, 31:228–248, 1999.

[9] M. Hajiaghayi, M. Mahdian, and V.S. Mirrokni. The facility location problem with general cost functions. *Networks*, 42(1):42–47, August 2003.

[10] K. Holmberg. Solving the staircase cost facility location problem with decomposition and piecewise linearization. *European Journal of Operational Research*, 74:41–61, 1994.

[11] K. Jain, M. Mahdian, E. Markakis, A. Saberi, and V.V. Vazirani. Approximation algorithms for facility location via dual fitting with factor-revealing LP. to appear in Journal of the ACM, 2002.

[12] K. Jain, M. Mahdian, and A. Saberi. A new greedy approach for facility location problems. In *Proceedings of the 34st Annual ACM Symposium on Theory of Computing*, 2002.

[13] K. Jain and V.V. Vazirani. Approximation algorithms for metric facility location and k-median problems using the primal-dual schema and lagrangian relaxation. *Journal of the ACM*, 48:274–296, 2001.

[14] M.R. Korupolu, C.G. Plaxton, and R. Rajaraman. Analysis of a local search heuristic for facility location problems. In *Proceedings of the 9th Annual ACM-SIAM Symposium on Discrete Algorithms*, pages 1–10, January 1998.

[15] M. Mahdian, E. Markakis, A. Saberi, and V.V. Vazirani. A greedy facility location algorithm analyzed using dual fitting. In *Proceedings of 5th International Workshop on Randomization and Approximation Techniques in Computer Science*, volume 2129 of *Lecture Notes in Computer Science*, pages 127–137. Springer-Verlag, 2001.

[16] M. Mahdian, Y. Ye, and J. Zhang. Improved approximation algorithms for metric facility location problems. In *Proceedings of 5th International Workshop on Approximation Algorithms for Combinatorial Optimization (APPROX 2002)*, 2002.

[17] D.B. Shmoys. Approximation algorithms for facility location problems. In K. Jansen and S. Khuller, editors, *Approximation Algorithms for Combinatorial Optimization*, volume 1913 of *Lecture Notes in Computer Science*, pages 27–33. Springer, Berlin, 2000.

[18] D.B. Shmoys, E. Tardos, and K.I. Aardal. Approximation algorithms for facility location problems. In *Proceedings of the 29th Annual ACM Symposium on Theory of Computing*, pages 265–274, 1997.

[19] M. Sviridenko. An 1.582-approximation algorithm for the metric uncapacitated facility location problem. In *Proceedings of the 9th Conference on Integer Programming and Combinatorial Optimization*, 2002.

[20] M. Thorup. Quick k-median, k-center, and facility location for sparse graphs. In *Automata, Languages and Programming, 28th International Colloquium, Crete, Greece*, volume 2076 of *Lecture Notes in Computer Science*, pages 249–260, 2001.

[21] M. Thorup. Quick and good facility location. In *Proceedings of the 14th ACM-SIAM symposium on Discrete Algorithms*, 2003.

A   Proof of Lemma 1

Proof. Let SOL be an arbitrary solution for $LFLP(a, b, c)$, which can also be viewed as a solution for $UFLP(b, \bar{c})$ for $\bar{c} = c + a$. Consider a facility f that is open in SOL, and the set of clients connected to it in SOL. Let k denote the number of these clients, $f(k) = ak + b$ (for $k > 0$) be the facility cost function of f, and $\bar{d}_j$ denote the connection cost between client j and the facility f in the instance $UFLP(b, a+c)$. Therefore, $d_j = \bar{d}_j - a$ is the corresponding connection cost in the original instance $LFLP(a, b, c)$. Recall [12] the definition of α_j and r_{ij} in the factor-revealing LP of Theorem A. It is proved [12] that $\alpha_i \le r_{j,i} + \bar{d}_j + \bar{d}_i$. We strengthen this inequality as follows.

Claim. $\alpha_i \le r_{j,i} + d_j + d_i$

Proof. It is true if $\alpha_i = \alpha_j$ since it happens only if $r_{j,i} = \alpha_j$. Otherwise, consider clients i and $j(< i)$ at time $t = \alpha_i - \epsilon$. Let s be the facility j is assigned to at time t. By triangle inequality, we have

$$\bar{c}_{si} = c_{si} + a_s \le c_{sj} + d_i + d_j + a_s = \bar{c}_{sj} + d_i + d_j \le r_{j,i} + d_i + d_j.$$

On the other hand $\alpha_i \le \bar{c}_{si}$ since otherwise i could have connected to facility s at a time earlier than t.   □

It is also known [12] that

$$\sum_{j=1}^{i-1} \max(r_{j,i} - \bar{d}_j, 0) + \sum_{j=i}^{k} \max(\alpha_i - \bar{d}_j, 0) \le b.$$

Notice that $\max(a - x, 0) \ge \max(a, 0) - x$ if $x \ge 0$. Therefore, we have

$$\sum_{j=1}^{i-1} \max(r_{j,i} - d_j, 0) + \sum_{j=i}^{k} \max(\alpha_i - d_j, 0) \le b + ka. \tag{14}$$

Claim A and Inequality 14 show that the values α_j, r_{ij}, d_j, a, and b constitute a feasible solution of the following optimization program.

maximize $\dfrac{\sum_{i=1}^{k} \alpha_i - \gamma_f(ak + b)}{\sum_{i=1}^{k} d_i}$ $\tag{15}$

subject to $\forall\, 1 \le i < k : \ \alpha_i \le \alpha_{i+1}$

$\forall\, 1 \le j < i < k : \ r_{j,i} \ge r_{j,i+1}$

$\forall\, 1 \le j < i \le k : \ \alpha_i \le r_{j,i} + d_i + d_j$

$\forall\, 1 \le i \le k : \ \displaystyle\sum_{j=1}^{i-1} \max(r_{j,i} - d_j, 0) + \sum_{j=i}^{k} \max(\alpha_i - d_j, 0) \le b + ka$

$\forall\, 1 \le j \le i \le k : \ \alpha_j, d_j, a, b, r_{j,i} \ge 0$

However, it is clear that the above optimization program and the factor-revealing LP in Theorem A are equivalent. This completes the proof of this lemma.   □

Approximating Rooted Connectivity Augmentation Problems

Zeev Nutov

Open University of Israel, 16 Klausner Str., Ramat-Aviv, 61392, Israel,
nutov@openu.ac.il

Abstract. A graph is called ℓ-*connected from U to r* if there are ℓ internally disjoint paths from every node $u \in U$ to r. The *Rooted Subset Connectivity Augmentation Problem (RSCAP)* is as follows: given a graph $G = (V + r, E)$, a node subset $U \subseteq V$, and an integer k, find a smallest set F of new edges such that $G + F$ is k-connected from U to r. In this paper we consider mainly a restricted version of RSCAP in which the input graph G is already $(k-1)$-connected from U to r. For this version we give an $O(\ln |U|)$-approximation algorithm, and show that the problem cannot achieve a better approximation guarantee than the Set Cover Problem (SCP) on $|U|$ elements and with $|V| - |U|$ sets. For the general version of RSCAP we give an $O(\ln k \ln |U|)$-approximation algorithm. For $U = V$ we get the *Rooted Connectivity Augmentation Problem (RCAP)*. For directed graphs RCAP is polynomially solvable, but for undirected graphs its complexity status is not known: no polynomial algorithm is known, and it is also not known to be NP-hard. For undirected graphs with the input graph G being $(k-1)$-connected from V to r, we give an algorithm that computes a solution of size exceeding a lower bound of the optimum by at most $(k-1)/2$ edges.

1 Introduction and Notation

A graph is called ℓ-*connected from U to r* if there are ℓ internally disjoint paths from every node in U to r. In this paper we consider the following problem:
Rooted Subset Connectivity Augmentation Problem (RSCAP):
Input: A graph $G = (V + r, E)$, node subset $U \subseteq V$, and integer k.
Output: A minimum size set F of new edges such that $G + F$ is k-connected from U to r.

For G being k_0-connected from U to r we give an $O(\ln(k - k_0) \ln |U|)$-approximation algorithm for both a directed and an undirected RSCAP. On the other hand, we show that even for $k_0 = k - 1$, the directed RSCAP cannot have a better approximation ratio than the Set Cover Problem (SCP) on $|U|$ elements and with $|V| - |U|$ sets.

For $U = V$ we get the *Rooted Connectivity Augmentation Problem (RCAP)*. A generalization of RCAP when one seeks an augmenting edge set of minimum weight is polynomially solvable for directed graphs [6], but is NP-hard for undirected graphs. However, the complexity status of an undirected RCAP (where

S. Arora et al. (Eds.): APPROX 2003+RANDOM 2003, LNCS 2764, pp. 141–152, 2003.

every new edge has weight 1) is not known: no polynomial algorithm is known, and it is also not known to be NP-hard. We show an algorithm that computes a solution which size exceeds the optimum by at most $(k-1)/2$ edges.

We note that RCAP is related to the well-studied *Vertex-Connectivity Augmentation Problem (VCAP)*: given a graph G and an integer k, find a smallest set F of new edges for which the graph $G + F$ is k-(node) connected. For directed graphs, Frank and Jordán [5] showed a polynomial algorithm. The complexity status of undirected VCAP is not known, but the following algorithms were obtained. For the case of G being $(k-1)$-connected, Jordán [9,10] gave an algorithm that computes a solution which size exceeds the optimum by at most $(k-1)/2$ edges. Recently, Jordán and Jackson [7] gave an algorithm that computes a solution with an additive gap at most $((k-k_0)(k-1)+4)/2$, where k_0 is the initial connectivity of G. In [8], the same authors give an algorithm that for any fixed k computes an optimal augmenting edge set in polynomial time.

Here is some notation and preliminary statements used in the paper. An edge from u to v is denoted by uv. Given a graph, we call new edges that can be added to the graph *links*, to distinguish them from the existing edges. Let $opt(G) = opt_k(G)$ denote the size of an optimal solution to the RSCAP on input G and k. For an arbitrary edge set F and set X let $\deg_F(X)$ denote the *degree* of X with respect to F. Let $G = (V + r, E)$ be a graph. For $X \subseteq V$ we denote by $\Gamma_G(X) = \Gamma(X)$ the set $\{v \in V - X : uv \in E \text{ for some } u \in X\}$ of *neighbors* of X in V, and let $X^* = V - (X + \Gamma(X))$. Let $d_r(X)$ denote the number of edges going from X to r, and define $g(X) = d_r(X) + |\Gamma(X)|$. We say that X is ℓ-tight (or simply that X is tight, if ℓ is understood) if $g(X) = \ell$. The following statement, which applies for both directed and undirected graphs, stems from Menger's Theorem.

Proposition 1. *A graph $G = (V + r, E)$ is ℓ-connected from U to r if and only if $g(X) \geq \ell$ for all $X \subseteq V$ with $X \cap U \neq \emptyset$.*

Let $G = (V + r, E)$ be a graph and let $X, Y \subseteq V$ be arbitrary. The following "submodular" inequality which is valid for both directed and undirected graphs can be easily proved by counting the contribution of the nodes in $\Gamma(X), \Gamma(Y)$ to its sides (e.g., see [2]).

$$g(X) + g(Y) \geq g(X \cap Y) + g(X \cup Y) \tag{1}$$

2   Rooted Subset Connectivity Augmentation

Theorem 2. *For the restriction of a directed RSCAP to instances in which G is $(k-1)$-connected from U to r, there exists:*

(i) An $O(\ln |U|)$-approximation algorithm, and
(ii) A polynomial time reduction from the Set Cover Problem (SCP) on universe U with $|V| - |U|$ sets such that there is a solution of size τ to SCP if and only if there is a solution of size τ to RSCAP.

Part (ii) of the theorem says that if one finds an algorithm with an approximation guarantee better than $O(\ln |U|)$ for the above restricted version of the RSCAP, then one can get an approximation guarantee better than $O(\ln |U|)$ for the SCP on groundset U; the latter is possible only if NP-hard problems can be solved in quasipolynomial time, see [4].

To prove the theorem, we will use the following well-known formulation of the SCP; in this formulation, J is the incidence graph of sets and elements, where A is the family of sets and B is the universe.

Input: A bipartite graph $J = (A + B, I)$ without isolated nodes.
Output: A minimum size subset $D \subseteq A$ such that $\Gamma_J(D) = B$.

The proof of Theorem 2 follows. The following lemma follows from inequality (1) and Proposition 1.

Lemma 3. *Let G be ℓ-connected from U to r, and let X, Y be ℓ-tight sets such that $X \cap Y \cap U \neq \emptyset$. Then $X \cap Y$ and $X \cup Y$ are both ℓ-tight.*

Given an instance of the directed RSCAP with the input graph G being ℓ-connected from U to r, we construct an instance $J = (A + B, I)$ of the SCP as follows: B is the family of the inclusion minimal sets among the ℓ-tight sets intersecting U, $A = V$, and for $a \in A, b \in B$ we have $ab \in I$ if, and only if, the subset of V corresponding to b contains a. Note that $|B| \leq |U|$, by Lemma 3. The above construction is polynomial, since for every node $u \in U$ we can compute the unique set in B containing u (or determine that such does not exist) in polynomial time using max-flow techniques.

Let τ^* be the optimal value of the following LP-relaxation for the obtained instance of the SCP:

$$\tau^* = \min \left\{ \sum_{a \in A} x_a : \sum_{a \in \Gamma_J(b)} x_a \geq 1 \ \forall b \in B, x_a \geq 0 \ \forall a \in A \right\}.$$

By a well-known result of Lovász [11], the greedy algorithm (which repeatedly removes from J the node of maximum degree in A and all its neighbors, until B becomes empty) computes a feasible solution $D \subseteq A$ to the SCP of size at most $H(|B|)\tau^*$. By Proposition 1, $G + \{vr : v \in D\}$ is $(\ell + 1)$-connected from U to r. We claim that $|D| \leq \frac{1}{k-\ell} H(|U|)opt_k(G)$. Let F be a link set such that $G + F$ is k-connected from U to r, and let x be the vector on $A = V$ defined by $x_v = \frac{1}{k-\ell} \deg_F(v)$. Since $\deg_F(X) \geq k-\ell$ for any ℓ-tight set X of G, x is a feasible solution to the above LP-relaxation. Thus $|D| \leq H(|B|)\tau^* \leq H(|B|) \sum_{v \in V} x_v = H(|U|)\frac{1}{k-\ell}|F|$, where $H(j)$ denotes the jth harmonic number. Consequently, the algorithm finds a link set that augments G to be $(\ell+1)$-connected from U to r of size at most $\frac{1}{k-\ell} H(|U|)opt_k(G)$. Thus we have proved the following statement, which for $k_0 = k - 1$ implies part (i) of Theorem 2:

Corollary 4. *There exists an $H(|U|)H(k-k_0)$-approximation algorithm for the directed RSCAP with the input graph G being k_0-connected from U to r.*

To prove part (ii) of Theorem 2, we will show that given an instance $J = (A + B, I)$ of the SCP, one can construct in polynomial time an instance $G = (V + r, E)$ of the directed RSCAP with $k_0 = k - 1$, $V = A + B$, and $U = B$, and such that:

(a) For any solution F' for the RSCAP there exists a solution F with $|F| = |F'|$ such that every edge in F connects some node in $V - U = A$ to r.

(b) $D \subseteq A$ is a solution to the SCP on J if, and only if, $F = \{vr : v \in D\}$ is a solution to the RSCAP on G.

Note that by Proposition 1, replacing any edge xy in a directed graph which is k-connected from U to r by a new edge xr results again in a graph that is k-connected from U to r. This implies that for any feasible solution F' for a directed RSCAP there always exists a feasible solution F with $|F| = |F'|$ such that r is the head of all the edges in F.

Given an instance $J = (A + B, I)$ for the SCP, we construct an instance $G = (V + r, E)$ for a directed RSCAP by directing the edges in J from B to A, adding a new node r and $k - 1$ edges from each node in B to r, and setting $U = B$. Then G is $(k-1)$-connected from U to r, and by Proposition 1, (b) holds. Now let F' be a set of links incident to r such that $G + F'$ is k-connected from U to r. If there is $ur \in F'$ with $u \in U$, then $\Gamma_G(u) \neq \emptyset$ (since in J there are no isolated nodes), and for any $a \in \Gamma_G(u)$ the graph $G + F$ where $F = F' - ur + ar$ is k-connected from U to r. Thus for the obtained instance of the RSCAP (a) holds. This finishes the proof of Theorem 2.

For the undirected RSCAP similar results can be deduced. Using standard constructions, it is easy to prove that a ρ-approximation algorithm for the directed RSCAP implies a 2ρ-approximation algorithm for the undirected RSCAP. In particular, by Corollary 4, there exists a $2H(|U|)H(k - k_0)$-approximation algorithm for the undirected RSCAP. On the other hand, for $k_0 = k - 1$, one can show by a similar reduction that if one finds a solution of size τ to the corresponding instance of the undirected RSCAP, then one can find a solution of size at most 2τ for the SCP.

3   Undirected Rooted Connectivity Augmentation

In the rest of the paper we consider an undirected RCAP with $k_0 = k - 1$; that is we will assume that G is $(k - 1)$-connected (from V) to r, and "tight" means $(k-1)$-tight. By Lemma 3, the (inclusion) minimal tight sets are pairwise disjoint, and let $\nu = \nu(G)$ denote their number. For $T \subseteq V$, the T-components are the connected components of $G - T$, and the T-components not containing r are the sides of T. Let $b(T)$ be the number of T-components. If $|T| = k - 1$ and $b(T) \geq 3$ then T is a shredder. Let $b(G) = b_k(G) = \max\{b(T) : T \subset V, |T| = k - 1\}$. If $G + F$ is k-connected then $|F| \geq \nu(G)/2$ (since $\deg_F(X) \geq 1$ for every tight set $X \subseteq V$) and $|F| \geq b(G) - 1$ (since for any $T \subseteq V$ with $|T| = k - 1$, F must induce a connected graph on the T-components). Thus

$$opt(G) \geq \max\{\lceil \nu(G)/2 \rceil, b(G) - 1\}.$$

For $k - 1 = 0$, it is clear that any tree on the components of G is an augmenting edge set of size $b(G) - 1$. For $k - 1 = 1$ it is also easy to compute an optimal solution in polynomial time using the lower bound $\max\{\lceil \nu'(G)/2 \rceil, b(G) - 1\}$, where $\nu'(G) = \nu(G) + 1$ if there is a tight set that contains all the minimal tight sets, and $\nu'(G) = \nu(G)$ otherwise. For $k - 1 \geq 2$ we prove the following theorem:

Theorem 5. *There is a polynomial algorithm that given a graph G which is $(k - 1)$-connected to r finds a link set F of size at most $\max\{\lceil \nu(G)/2 \rceil + \lfloor (k - 1)/2 \rfloor, b(G) - 1\}$ such that $G + F$ is k-connected to r.*

We now give some preliminary statements used in the rest of the paper. The following inequality can be easily verified by counting the contribution to its sides of nodes in $\Gamma(X), \Gamma(Y)$ and the edges incident to r.

$$g(X) + g(Y) \geq g(X^* \cap Y) + g(X \cap Y^*) + 2d_r(X \cap Y) \tag{2}$$

Two disjoint subsets X, Y of V are *adjacent* if there is an edge with one end in X and the other end in Y. Using inequalities (1) and (2) it is not hard to derive the following properties of tight sets.

Lemma 6. *Let X, Y be two tight sets in G.*

   (i) *If $X \cap Y \neq \emptyset$ then $X \cap Y, X \cup Y$ are both tight.*
   (ii) *If the sets $X \cap Y^*, Y \cap X^*$ are nonempty, then they are both tight and nonadjacent and $d_r(X \cap Y) = 0$.*
   (iii) *If X, Y are disjoint and $|X| \leq |Y|$ then exactly one of the following holds: (a) $X \cap Y^*, Y \cap X^*$ are nonadjacent tight sets, or (b) $X \subseteq \Gamma(Y)$.*

3.1   Independent Families

Definition 7. *A family $\mathcal{F}$ of pairwise disjoint tight sets is* independent *if there exists a partition Π of $\mathcal{F}$ and a family $\mathcal{S}(\mathcal{F}) = \{S_\mathcal{P} : \mathcal{P} \in \Pi\}$ of pairwise disjoint tight sets such that:*

   (i) *For every $\mathcal{P} \in \Pi$ holds: $\cup\{S : S \in \mathcal{P}\} \subseteq S_\mathcal{P}$, if $|\mathcal{P}| \neq 2$ then equality holds, and if $|\mathcal{P}| \geq 3$ then $\mathcal{P}$ consists of some sides of a shredder.*
   (ii) *For any disjoint $X, Y \in \mathcal{F} \cup \mathcal{S}(\mathcal{F})$ holds: $X - \Gamma(Y), Y - \Gamma(X)$ are both nonempty if, and only if, X, Y belong to the same part in Π.*

If in addition to (i) and (ii), for any part $\mathcal{P} = \{S_i, S_j\} \in \Pi$ we have that any tight set that intersects $S_\mathcal{P}$ is contained in one of S_i, S_j then $\mathcal{F}$ is strongly independent.

Let $\mathcal{R}$ be the following relation on tight sets: $(X, Y) \in \mathcal{R}$ if $X - \Gamma(Y)$ and $Y - \Gamma(X)$ are both nonempty. Given a family $\mathcal{F}$ of tight sets, let $\mathcal{R}(\mathcal{F})$ denote the restriction of $\mathcal{R}$ to $\mathcal{F}$. Clearly, $\mathcal{R}(\mathcal{F})$ is symmetric and reflexive and, if $\mathcal{F}$ is independent, then $\mathcal{R}(\mathcal{F})$ is an equivalence, with the corresponding partition into equivalence classes Π as in the definition. It is not hard to verify that any subfamily of an independent family is also independent. Note that condition (ii) in the definition of an independent family implies $S' \subseteq \Gamma(S'')$ or $S'' \subseteq \Gamma(S')$ for any distinct $S', S'' \in \mathcal{S}(\mathcal{F})$. But Lemma 6(iii) implies a stronger statement:

Proposition 8. *If $|S''| \leq |S'|$ for distinct $S', S'' \in \mathcal{S}(\mathcal{F})$ then $S'' \subseteq \Gamma(S')$.*

We call an independent family *trivial* if the corresponding partition is trivial, that is if $\Pi = \{\mathcal{F}\}$. Let $\beta(G)$ denote the maximum cardinality of an independent family in G. Note that even trivial independent families strictly generalize shredders. Indeed, any subfamily of sides of a shredder forms a trivial independent family; thus $\beta(G) \geq b(G) - 1$. However, even trivial independent families with two sets might not correspond to a shredder, see Example 1 below. If $\mathcal{F}$ is a trivial independent family, then $|\mathcal{F}|$ can be as large as $n - k + 1$. However, as Theorem 9 below implies, a nontrivial independent family has at most $k - 1$ sets; Examples 2,3 below show that this bound is tight. For a family $\mathcal{F}$ of sets, let $||\mathcal{F}||$ denote the cardinality of the union of the sets in $\mathcal{F}$.

Theorem 9. *Let $\mathcal{F}$ be a nontrivial independent family, and let $S' = S_{\mathcal{P}'}$ be the largest set in $\mathcal{S}(\mathcal{F})$. Then $|\mathcal{P}'| + ||\Pi - \mathcal{P}'|| \leq k - 1$.*

Proof. We need the following claim:
Claim: Let Y be an ℓ-tight set, and suppose that there is a node $v \in Y$ such that there are ℓ internally disjoint paths from r to v. Then for any set $X \subseteq V$ disjoint to Y holds: $d_r(X) + |\Gamma(X) - (Y \cup \Gamma(Y))| \geq |X \cap \Gamma(Y)|$.
Proof: Consider a set of ℓ internally disjoint paths from r to v. Then $|X \cap \Gamma(Y)|$ of them contain a node from X. In each of these $|X \cap \Gamma(Y)|$ paths pick the first node whose successor is in X. Such a node is either r or in $\Gamma(X) - (Y \cup \Gamma(Y))$, so it contributes 1 to the left side of the inequality.
Note that $|S_{\mathcal{P}}| \geq ||\mathcal{P}||$ and $|S_{\mathcal{P}}| = ||\mathcal{P}||$ if $|\mathcal{P}| \neq 2$ for any $\mathcal{P} \in \Pi$. Let $S'' = S_{\mathcal{P}''}$ be the second largest set in $\mathcal{S}(\mathcal{F})$. Then $g(S'') = k - 1$, $|\Gamma(S'') \cap S'| \geq |\mathcal{P}'|$ (by condition (ii) in Definition 7), and $S'' \cap \Gamma(S') = S''$ (by Proposition 8). The statement follows by applying the claim above on $S' = Y$ and $S'' = X$:

$$g(S'') = d_r(S'') + |\Gamma(S'') - (S' \cup \Gamma(S'))| + |\Gamma(S'') \cap S'| + |\Gamma(S'') \cap \Gamma(S')| \geq$$
$$\geq |S''| + |\mathcal{P}'| + ||\Pi - \mathcal{P}' - \mathcal{P}''|| = |\mathcal{P}'| + ||\Pi - \mathcal{P}'||.$$

Examples:
1. Let u, v be two nodes of a cycle, where $r \neq u, v$ is arbitrary and $k - 1 = 2$. Then $\mathcal{F} = \{\{u\}, \{v\}\}$ is an independent family. If u, v are adjacent, then $\mathcal{F}$ is nontrivial and strongly independent. Assume that u, v are nonadjacent. Then $\mathcal{F}$ is trivial and not strongly independent. Let us consider some modifications. Let P be the path between u and v. Suppose that none of u, v is incident to r. Let u' be the neighbor of u not belonging to P and define v' in the same way. Let G be the graph obtained by connecting each of u', v' to all the internal nodes of P. If P has at least two internal nodes, then $\mathcal{F}$ is strongly independent in G. Otherwise (P has one internal node) $\mathcal{F}$ is not strongly independent, but the family $\{\{u, u'\}, \{v, v'\}\}$ is strongly independent; $\mathcal{F}$ becomes strongly independent if we add to G the links ru', rv'. Note that there are no shredders in the graphs considered.
2. Let $G = K_k$ be a complete graph on k nodes. Then $\lceil \nu(G)/2 \rceil = \lceil (k-1)/2 \rceil$, $b(G) - 1 = 0$, but $opt(G) = \beta(G) = k - 1$. Here any family of pairwise disjoint

tight sets forms a trivial independent family. Let us replace one node of G by a clique of size at least $k - 1$, connecting the edges of K_k to distinct nodes of the clique. The new graph contains a nontrivial independent family of size $k - 1$.

3. Let $G = K_{k-1,k-1}$ be a complete bipartite graph with $k - 1$ nodes on each side and parts R, S where $r \in R$. Then $\lceil \nu(G)/2 \rceil = \lceil (2k - 3)/2 \rceil = k - 1$ and $\beta(G) = b(G) + 1 = k - 1$. Indeed, $b(G) = b(S) = k - 2$, and $\beta(G) = |R| - 1 + 1 = k - 1$ since $R - r + s$ is an independent family for any $s \in S$ (so there are $k - 1$ distinct nontrivial independent families of size $\beta(G) = k - 1$ in G). Also, $opt(G) = k - 2 + \lceil (k - 1)/2 \rceil$. An optimal augmenting edge set is obtained by connecting every node in $R - r$ to r, picking a maximum matching on S, and if $k - 1$ is odd adding one more edge from the unmatched node in S to r.

4. Let r be a leaf of a tree G (so $k - 1 = 1$) with odd number $\nu + 1$ of leaves, in which every non-leaf node has degree 3. Then $b(G) - 1 = \beta(G) = 2$ but $opt(G) = \lceil (\nu + 1)/2 \rceil$.

3.2   Main Results

A tight set is a *core* if it contains a unique minimal tight set. By Lemma 6(i), the union and intersection of any two intersecting cores are also cores. Thus for every minimal core C (that is, a minimal tight set) there exists a unique maximal core S containing it. As was mentioned in Section 2, the minimal cores can be computed in polynomial time. Let $C_1, \ldots, C_\nu$ be the minimal cores of G. By Proposition 1 $G + F$ is k-connected to r if, and only if, $G + F$ has no cores; thus the graph $G + \{v_i r : v_i \in C_i, i = 1, \ldots, \nu\}$ is k-connected to r, and $opt(G) \leq \nu(G)$. We say that a link e is $(\nu, 2)$-*reducing* for G if $\nu(G + e) \leq \nu(G) - 2$. To prove Theorem 5 we use the following two theorems:

Theorem 10. *Let G be $(k - 1)$ connected to r and let $\mathcal{F}$ be a subfamily of the family of maximal cores of G. Then exactly one of the following holds:*
(i) there is a $(\nu, 2)$-reducing link for G connecting two distinct cores in $\mathcal{F}$, or
(ii) $\mathcal{F}$ is strongly independent.
Thus if $|\mathcal{F}| \geq k$, then either there exists a $(\nu, 2)$-reducing link connecting two cores in $\mathcal{F}$, or the sets in $\mathcal{F}$ are sides of the same shredder. In particular, if $\nu(G) \geq k$, then either there exists a $(\nu, 2)$-reducing link for G, or $\nu(G) = b(G) - 1$.

Theorem 11. *Let G be $(k - 1)$ connected to r. If $b(G) \geq k$, then there exists a polynomial algorithm that finds a link set F of size at most $\max\{\lceil (\nu(G) + 1)/2 \rceil, b(G) - 1\}$ such that $G + F$ is k-connected to r.*

Proof. (of Theorem 5): The algorithm is as follows:

If $b(G) \geq k$, find an augmenting link set as in Theorem 11.
Else, perform the following two steps:
1. Find and add a $(\nu, 2)$-reducing link, as long as one exists.
2. In the resulting graph, add one link from every minimal core to r.

If $b(G) \geq k$ the algorithm finds an augmenting link set as required, by Theorem 11. Suppose that $b(G) \leq k-1$, and let F_1, F_2 be the link sets added at steps 1, 2, respectively. Then the resulting graph $G + F_1 + F_2$ is k-connected to r, by Proposition 1. By Theorems 10 and 9 $|F_2| = \nu(G + F_1) \leq k - 1$. Thus

$$|F_1| + |F_2| = (\nu - |F_2|)/2 + |F_2| = \lceil \nu/2 \rceil + \lfloor |F_2|/2 \rfloor \leq \lceil \nu/2 \rceil + \lfloor (k-1)/2 \rfloor.$$

The proof of Theorems 10 and 11 follows. Let $C_1, \ldots, C_\nu$ be the minimal cores of G. For $I \subseteq \{1, \ldots, \nu\}$, let $\mathcal{S}_I$ denote the collection of tight sets containing $\cup_{i \in I} C_i$ and not containing any other minimal core. Let S_I be the union of the sets in $\mathcal{S}_I$; we set $S_I = \emptyset$ if $\mathcal{S}_I = \emptyset$. By Lemma 6(i), if $\mathcal{S}_I \neq \emptyset$ then S_I is tight, and thus it is the inclusion maximal set in $\mathcal{S}_I$. Also, for any $I' \subset I$, $S_{I'} \subset S_I$ holds. For simplicity, S_{ij} means $S_{\{i,j\}}$ and $S_i = S_{\{i\}} = S_{ii}$. Note that $S_I \cap S_J = S_{I \cap J}$ for any $I, J \subseteq \{1, \ldots, \nu\}$ with $\mathcal{S}_I, \mathcal{S}_J \neq \emptyset$. Thus we have:

Proposition 12. (i) The sets S_i are pairwise disjoint.
(ii) If $\mathcal{S}_{ip}, \mathcal{S}_{pj} \neq \emptyset$, then $S_{ip} \cap S_{pj} = S_p$.

Clearly, if there is a $(\nu, 2)$ reducing link, then its endnodes belong to distinct minimal cores C_i, C_j. Using Lemma 6 it is not hard to prove the following statement:

Proposition 13. Let C_i, C_j be minimal cores. Then the following are equivalent:

(i) There exists a $(\nu, 2)$-reducing link connecting C_i and C_j.
(ii) (A) $S_i - \Gamma(S_j)$ and $S_j - \Gamma(S_i)$ are both nonempty, and (B) $\mathcal{S}_{ij} = \emptyset$.
(iii) Any link connecting C_i and C_j is $(\nu, 2)$-reducing.

Let $\mathcal{F}$ be a subfamily of the family of maximal cores of G.

Lemma 14. If no $(\nu, 2)$-reducing link that connects two cores in $\mathcal{F}$ exists, then the relation $\mathcal{R}(\mathcal{F})$ is an equivalence.

Proof. Symmetry and reflexivity are obvious, so we need to prove transitivity. Suppose therefore that $\{S_i, S_p\}, \{S_p, S_j\} \in \mathcal{R}(\mathcal{F})$ for distinct $S_i, S_p, S_j \in \mathcal{F}$. Then $\mathcal{S}_{ip}, \mathcal{S}_{pj} \neq \emptyset$ by Proposition 13. Thus $C_i \subseteq S_{ip} \cap S_{pj}^* \subseteq S_{pj}^*$, and $S_j \subseteq S_{pj}$, by Lemma 6(ii). Thus we must have $C_i \cap \Gamma(S_j) = \emptyset$. For a similar reason, $C_j \cap \Gamma(S_i) = \emptyset$. This proves transitivity.

The following two lemmas (the proof is omitted) are used to establish that the equivalence classes of size at least three of $\mathcal{R}(\mathcal{F})$ correspond to sides of a shredder.

Lemma 15. Let A, B be disjoint nonadjacent tight sets. If $A \cup B$ is tight, then $\Gamma(A) = \Gamma(B)$ and $d_r(A) = d_r(B) = 0$.

Lemma 16. Let A, B, C be pairwise disjoint tight sets such that none of them is contained in the set of neighbors of the other, and such that the union of any two of them is tight. Then $d_r(A) = d_r(B) = d_r(C) = 0$ and $\Gamma(A) = \Gamma(B) = \Gamma(C)$.

Corollary 17. *Let $\mathcal{F}$ be a subfamily of the family of maximal cores of G such that no $(\nu, 2)$-reducing link that connects two cores in $\mathcal{F}$ exists, and let $\mathcal{P}$ be an equivalence class of $\mathcal{R}(\mathcal{F})$. Then:*
(i) If $|\mathcal{P}| \geq 3$ then $\mathcal{P}$ consists of some sides of the same shredder.
(ii) If $\mathcal{P} = \{S_i, S_j\}$ then $S_{ij} \neq \emptyset$.

Proof. Part (ii) follows from Proposition 13 and we will prove Part (i). We will show that if distinct S_i, S_j, S_p belong to the same class of $\mathcal{R}(\mathcal{F})$ then they satisfy the conditions of Lemma 16. Since S_i, S_j, S_p are distinct, they are pairwise disjoint (by Proposition 12(i)), and by the definition of $\mathcal{R}(\mathcal{F})$, none of them is contained in the set of neighbors of the other. It remains therefore to show that the union of any two of them, say $S_i \cup S_j$, is tight. By Proposition 13 and the definition of $\mathcal{R}(\mathcal{F})$, each one of the sets S_{ij}, S_{jp}, S_{pi} exists. Thus $(S_{ip} \cup S_{pj}) \cap S_{ij}$ is tight, by Lemma 6, and $(S_{ip} \cup S_{pj}) \cap S_{ij} = (S_{ip} \cap S_{ij}) \cup (S_{pj} \cap S_{ij}) = S_i \cup S_j$, where the last equation follows from Proposition 12(ii). Thus $S_i \cup S_j$ is tight.

Proof. (of Theorem 10): Let $\mathcal{F}$ be as in Theorem 10. From Proposition 13 and the definition of an independent family it follows that if case (ii) of Theorem 10 holds (that is, if $\mathcal{F}$ is strongly independent), then case (i) cannot hold. The rest of the proof shows that if case (i) does not hold, then case (ii) must hold. Suppose therefore that no $(\nu, 2)$-reducing link connecting two distinct cores in $\mathcal{F}$ exists. Then, by Lemma 14, $\mathcal{R}(\mathcal{F})$ is an equivalence, and let Π be its partition into the corresponding equivalence classes. For $\mathcal{P} \in \Pi$ let $S_{\mathcal{P}}$ be the union of the sets in $\mathcal{P}$ if $|\mathcal{P}| \neq 2$, and $S_{\mathcal{P}} = S_{ij}$ if $\mathcal{P} = \{S_i, S_j\}$. Combining this setting with Corollary 17, we conclude that condition (i) in the definition of an independent family is satisfied for $\mathcal{F}$, Π, and $\mathcal{S}$, and, moreover, if $\mathcal{F}$ is independent, then it is strongly independent. We show that condition (ii) is also satisfied. Let $X, Y \in \mathcal{F} \cup \mathcal{S}$ be disjoint with $X - \Gamma(Y) \neq \emptyset, Y - \Gamma(X) \neq \emptyset$. Then by Lemma 6(iii) $X \cap Y^*, Y \cap X^*$ are both tight. Let S_X be an arbitrary maximal core intersecting $X \cap Y^*$, and let S_Y be an arbitrary maximal core intersecting $Y \cap X^*$. Note that $S_i \subseteq S$ and $S_i \in \mathcal{F}$ for any maximal core S_i and $S \in \mathcal{F} \cup \mathcal{S}$ that intersect. Thus $S_X \subseteq X$ and $S_Y \subseteq Y$, and $S_X, S_Y \in \mathcal{F}$. However, S_X intersects $X \cap Y^*$, S_Y intersects $Y \cap X^*$, implying that $S_X - \Gamma(S_Y), S_Y - \Gamma(S_X)$ are both nonempty; therefore, S_X, S_Y belong to the same class of $\mathcal{R}(\mathcal{F})$. Since X, Y are disjoint, $X = S_X$ and $Y = S_Y$, which finishes the proof.

The proof of Theorem 10 is done. We now prove Theorem 11.

Lemma 18. *Let T be a shredder and let Y be a tight set.*
(i) If $\Gamma(Y) = T$, then Y is a union of some sides of T.
(ii) If Y intersects two distinct sides X_i, X_j of T, then $X_i, X_j \subseteq Y$.

Proof. Part (i) is obvious, and we will prove part (ii). By Lemma 6(i), the sets $Y \cap X_i, Y \cap X_j$ are tight, and their union (which is the intersection of two intersecting tight sets $Y, X_i \cup X_j$) is also tight. Moreover, $Y \cap X_i, Y \cap X_j$ are nonadjacent, since X_i, X_j are nonadjacent. Thus $\Gamma(Y \cap X_i) = \Gamma(Y \cap X_j) = T$, by Lemma 15. Part (ii) follows now from part (i).

Two intersecting sets X, Y are *crossing*, (or Y *crosses* X) if none of them contains the other.

Lemma 19. *No tight set crosses a side or the union of sides of a shredder.*

Proof. Let Y be a tight set intersecting some side X of a shredder T. By Lemma 18(ii), if Y intersects all sides of T, then it contains all of them. Assume therefore that there is a side X' of T disjoint to Y. Let $Z = X \cup Y$. Then (i) $d_r(Z \cup X') = d_r(Z)$ (since $d_r(X') = 0$); (ii) $\Gamma(Z \cup X') \subseteq \Gamma(Z)$ and $\Gamma(Z \cup X') = \Gamma(Z)$ if, and only if, Z and X' are nonadjacent (since $\Gamma(X) = \Gamma(X') = T$ and $X \subseteq Z$). Thus Z and X' are tight and nonadjacent. Moreover, $Z \cup X'$ is tight (since $Z, X \cup X'$ are intersecting and tight, and since $X \subseteq Z$). Thus $\Gamma(Z) = \Gamma(X') = T$, by Lemma 15. Consequently, Z must be a union of some sides of T, by Lemma 18(i). Now, if Y intersects a side of T distinct from X, then $X \subset Y$, by Lemma 18(i); otherwise, $Y \subseteq X$, and the proof is complete. ∎

Given a nontrivial partitition $\mathcal{W}$ of a groundset W, a link set F on W is a $\mathcal{W}$-*connecting* W-*cover* if the following three conditions hold: (a) $\deg_F(w) \geq 1$ for every $w \in W$; (b) every link in F connects distinct parts of $\mathcal{W}$; (c) F induces a connected graph on the parts of $\mathcal{W}$. Let $\max(\mathcal{W})$ denote the largest cardinality of a set in $\mathcal{W}$. The following statement can be proved by induction on $|W|$.

Lemma 20. *Let $\mathcal{W}$ be a nontrivial partition of a groundset W. Then the minimum cardinality of a $\mathcal{W}$-connecting W-cover equals $\max\{\lceil |W|/2 \rceil, \max(\mathcal{W}), |\mathcal{W}| - 1\}$, and an optimal cover can be found in polynomial time.*

Corollary 21. *Let T be a shredder with $b(T) \geq k$ and suppose that every T-component contains at most $b(T) - 1$ minimal cores. Then given T, an augmenting link set for G of size $\max\{\lceil (\nu(G) + 1)/2 \rceil, b(T) - 1\}$ can be found in polynomial time.*

Proof. Let R be the side of T that contains r, let W' be the set of minimal cores of G, and let $W = W' + r$. By Lemma 19 the inclusion in the T-components induces a partition $\mathcal{W}$ of W, and let F be a minimum cardinality $\mathcal{W}$-connecting cover of W. Note that F can be computed in polynomial time. By Lemma 19, for any tight set Y of G exactly one of the following holds:
(i) Y is properly contained in a T-component or is a union of some but not all T-components, and thus F has an edge connecting Y and Y^*, or
(ii) Y contains all T-components, and thus F has an edge connecting Y to r.
Thus $G + F$ is k-connected to r. Note that $|W| = \nu(G) + 1$, $|\mathcal{W}| = b(T)$, and $\max(\mathcal{W}) \leq b(T) - 1 = |\mathcal{W}| - 1$. Hence, by Lemma 20, $|F| = \max\{\lceil |W|/2 \rceil, |\mathcal{W}| - 1\} = \max\{\lceil (\nu(G) + 1)/2 \rceil, b(T) - 1\}$. ∎

Consider the following algorithm applied on a shredder T with $b(T) = b(G) \geq k$.
Phase 1: *While* there exists a T-component X containing $b(T)$ minimal cores
        add to G a $(\nu, 2)$-reducing link connecting two cores in X.
    End While
Phase 2: Add to G a link set as in Corollary 21.

The condition in the loop of Phase 1 ensures that a $(\nu, 2)$-reducing link connecting two cores in X exists; otherwise by Theorem 10 the maximal cores contained in X are sides of the same shredder with at least $b(T)$ sides, while T has $b(T) - 1$ sides; this contradicts the maximality of $b(T)$. Consequently, the algorithm correctly finds an augmenting link set of size at most $\max\{\lceil (\nu(G) + 1)/2 \rceil, b(T) - 1\}$, by Corollary 21.

To finish the proof of Theorem 11, it remains to show that a shredder T with $b(T) = b(G)$ can be found in polynomial time. In fact, all the shredders can be found in polynomial time (the number of shredders is at most $(2|V| - 2k + 1)/3$, see Theorem 22 below). This can be done using the algorithm of Cheriyan and Thurimella [3] who showed that a corresponding problem in a $(k - 1)$-connected graph is solvable in polynomial time.

4   Applications

Here we discuss some consequences from the previous sections, starting with deriving an upper bound on the number of shredders. Consider the family $\mathcal{L}$ obtained by picking for every shredder its sides and the union of its sides; we color the former blue and the latter red. Let U be the union of the sets in $\mathcal{L}$, and note that $|U| \leq |V| - |\Gamma(r)| \leq |V| - k + 1$. Note that $\mathcal{L}$ is laminar (that is, its members are pairwise noncrossing), by Lemma 19. It is well known that a laminar family on U has at most $2|U| - 1$ members, thus $|\mathcal{L}| \leq 2(|V| - |\Gamma(r)|) - 1$. We can represent $\mathcal{L}$ as a forest of rooted trees if we order the sets in $\mathcal{L}$ by inclusion: X is a child of Y if X is the largest set in $\mathcal{L}$ properly contained in Y. Then this forest has the following properties: (i) every set is either blue or red, but not both; (ii) the children of every red set are all blue, and there are at least two of them. Therefore, the number of red sets, which is exactly as the number of shredders in the graph, is at most half the number of blue sets. Thus we have:

Theorem 22. Let $G = (V + r, E)$ be $(k - 1)$-connected to r. Then the number of shredders in G is at most $(2|V| - 2|\Gamma(r)| - 1)/3 \leq (2|V| - 2k + 1)/3$.

An edge e of a graph H is *critical* w.r.t. a certain property if H satisfies this property but $H - e$ does not. Splitting off two edges su, sv means replacing them by a new edge uv. Using Theorem 10 it is not hard to prove the following "splitting off" theorem:

Theorem 23. Let $H = (V + r, E)$ be k-connected from $V - s$ to r, where s is a neighbor of r, such that every edge sv of H, $v \neq r$ is critical with respect to k-connectivity from $V - s$ to r. Then either (i) there exists a pair of edges su, sv with $u, v \in \Gamma_H(s)$ that can be split-off while preserving the k-connectivity from $V - s$ to r, or (ii) the family of maximal cores of $G = H - s$ is independent.

We note that Theorem 23 is related to (but is also independent of) similar theorems in [1], [9], and [2]. Provided that (A) $\deg(s) \geq k + 2$ and (B) $|V| \geq 2k$, these theorems give a characterization when there exists a pair of edges incident to s that can be split-off while preserving: "global" k-connectivity in [1] and [9,10], and k-connectivity from V to $s = r$ in [2]. Our Theorem 23 which considers a different but related setting, gives a necessary and sufficient condition without restrictions (A) and (B). However, if (A) holds, then our characterization takes a similar form to the one given in [2, Theorem 3].

Let us call a sequence $F^* = (e_1, \ldots, e_p)$ of links $(\nu, 2)$-reducing for G if e_i is $(\nu, 2)$-reducing for $G + \{e_1, \ldots, e_{i-1}\}$, $i = 1, \ldots, p$. Let $\zeta(G)$ be the maximum length of a $(\nu, 2)$-reducing link sequence for G. A link set is *basic* if every its link connects two minimal cores of G or connects a minimal core of G to r. It is easy to see that there exists a basic augmenting link of size $opt(G)$. Using Theorem 10 and Proposition 8, we can prove the following theorem:

Theorem 24. *Among all basic augmenting link sets of size $opt(G)$, let F be one with the maximal number of links incident to r. Then an arbitrary ordering of the links in F that are not incident to r is a $(\nu, 2)$-reducing sequence for G of maximal length. Thus $opt(G) = \nu(G) - \zeta(G) \geq \beta(G)$.*

Note that computing a maximum length $(\nu, 2)$-reducing sequence for G is not equivalent to finding a maximum matching in the graph induced on the minimal cores by the $(\nu, 2)$-reducing links (formally, the nodes of this graph are the minimal cores of G, and we connect two cores by an edge if and only if there exists a $(\nu, 2)$-reducing link connecting them); see Example 4 in Section 3.1.

References

1. D. Bienstock, E. F. Brickell, and C. L. Monma, On the structure of minimum-weight k-connected spanning networks, *SIAM J. Discrete Math.* 3, 1990, 320–329.
2. J. Cheriyan, T. Jordán, and Z. Nutov, On rooted node connectivity problems, *Algorithmica* 30, 2001, 353–375.
3. J. Cheriyan and R. Thurimella, Fast algorithms for k-shredders and k-node connectivity augmentation, *J. Algorithms* 33, no. 1, 1999, 15–50.
4. U. Feige, A threshold of $\ln n$ for approximating set cover, *Journal of the ACM* 45, 1998, 634–652.
5. A. Frank and T. Jordán, Minimal edge-coverings of pairs of sets, *J. Comb. Theory B* 65, 1995, 73–110.
6. A. Frank and É. Tardos, An application of submodular flows, *Linear Algebra and its Applications* 114/115, 1989, 329–348.
7. B. Jackson and T. Jordán, A near optimal algorithm for vertex connectivity augmentation, manuscript.
8. B. Jackson and T. Jordán, Independence free graphs and vertex connectivity augmentation, manuscript.
9. T. Jordán, On the optimal vertex-connectivity augmentation, *J. Comb. Theory B* 63, 1995, 8–20.
10. T. Jordán, A note on the vertex connectivity augmentation, *J. Comb. Theory B* 71 no. 2, 1997, 294–301.
11. L. Lovász, On the ratio of optimal integral and fractional covers, *Discrete Math.* 13, 1975, 383–390.

Effective Routing and Scheduling in Adversarial Queueing Networks[*]

Jay Sethuraman[1] and Chung-Piaw Teo[2]

[1] IEOR Department, Columbia University, New York, NY 10027, USA,
jay@ieor.columbia.edu
[2] Department of Decision Sciences, National University of Singapore,
Singapore 117591,
bizteocp@nus.edu.sg

1 Introduction

Motivation. Scheduling and packet-routing have emerged as important problems in modern computer and communication systems. In this paper, we consider such problems in the setting of an arbitrary synchronous, *adversarial* network. In an adversarial network, the nature of the incoming traffic is decided by an adversary, operating under a reasonable rate restriction. Such networks have attracted attention in recent years as they appear to be a convenient and useful way to model packet injections into a communication network; in addition, these networks inspire algorithm developers to design robust algorithms that provide a performance guarantee regardless of the nature of the incoming traffic. Thus, the adversarial input model provides a valuable, complementary point of view to that of the more traditional stochastic model.

Problem description. The communication network is modeled by a directed graph $G = (V, E)$ in which the nodes represent processors and the arcs (or edges) represent links between processors. Two natural models arise, depending on whether the adversary specifies a route for the packets she injects: In the *non-adaptive* (or circuit-switched) model, the algorithm is required to route a packet along the path specified by the adversary; in the *adaptive* (or packet-switched) model, the adversary specifies only the origin and destination for each packet, but does not specify a path. In this case, the algorithm is free to route a packet along any path from its origin to its destination.

Packets are injected by an adversary subject to a natural rate restriction specified in terms of two parameters r and w. For the non-adaptive model, the packets injected by the adversary (and their associated paths) should be such that in any time window of size w, the number of packets injected during this window requiring any arc must be at most $\lfloor rw \rfloor$. For the adaptive model, the analogous restriction is that the adversary *must be able to* associate paths to the

---

[*] The research of the first author was partially supported by an NSF CAREER Award and an IBM Faculty partnership award. The research of the second author was partially supported by a Fellowship from the Singapore-MIT Alliance Program.

S. Arora et al. (Eds.): APPROX 2003+RANDOM 2003, LNCS 2764, pp. 153–164, 2003.

packets injected in any time window of size w such that the number of packets requiring any arc is at most $\lfloor rw \rfloor$. This condition can be conveniently captured by an associated integer multicommodity flow problem having an optimal value at most $\lfloor rw \rfloor$.

In this paper we focus on the adaptive model, although most of our results can be extended to the non-adaptive model as well, with virtually no changes. In fact, we focus on the adaptive model in which the adversary is allowed to *split* packets and route them using multiple paths. Essentially, the restriction on the adversary translates to an associated *fractional* multicommodity flow problem having an optimal value at most rw. For this model, we consider the problem of designing effective routing/scheduling algorithms. Our main result is a simple algorithm for this problem that is *stable* (bounded number of packets in the system), with a bound on the number of packets in the system that is $O(w/(1-r))$ for any *fixed* network G. This implies a worst-case delay bound on packets that is relatively small as well. A noteworthy feature of this result is that this matches the traditional queueing-theoretic number-in-system bound, which is usually $O(1/(1-r))$. In the rest of this paper, we assume a fixed network G, and so we often omit the dependence of the bounds on the network parameters.

Related work. Adversarial networks have received a lot of attention in recent years. They were first introduced by Borodin et al. [9], and further elaborated by Andrews et al. [3,4]. Later, these were seen to be non-trivial generalizations of earlier models of Cruz [10]. The original papers of Borodin et al. [9] and Andrews et al. [3,4] contain a wealth of interesting results, but mostly on the non-adaptive case.

The models most closely related to our work were first introduced by Aiello et al. [2]. In their work, they provided an elegant extension of the restriction on the adversary, which was previously considered only for the non-adaptive case. Furthermore, they constructed a distributed protocol with the number of packets in the system being $O(w/(1-r))$. Their results were derived for the *integer* (w, r) adversary. Motivated by the observation that this restriction is not efficiently checkable, Gamarnik [12] introduced the *fractional* (w, r) adversary: here, the adversary is allowed to associate fractional paths ("flows") to the packets to satisfy the load condition. An interesting question, then, is to quantify the performance loss due to the increased power given to the adversary. Gamarnik [12] constructed an algorithm such that the number in system is $O(w^2/(1-r)^2)$; furthermore, he observed that a naive adaptation of the methods of Aiello et all. [2] can at best lead to a bound of $O(1/(1-r)^3)$.

In more recent work, Andrews et al. [5] derive *distributed* source routing and scheduling algorithms with polynomial delay bounds using a discrete-review like strategy; these delays bounds obviously translate to bounds on the number-in-system. The algorithm described in this paper can also be viewed as a source routing/scheduling algorithm, as the route for a packet is determined at its source; the queue-length bounds we prove are stronger than those implicit in [5], but our algorithm is centralized. For the special case in which there is only a single destination, stronger bounds are known [6].

Results. For the dynamic adaptive packet routing problem in an adversarial queuing network with a fractional (w, r) adversary, we design an efficient algorithm that keeps the queue-lengths bounded. Specifically, we show that the number of packets in the system at any time t, $Q(t)$, satisfies

$$Q(t) \leq \frac{m(m + 2n + m^2 n^2 + w)}{1 - r}, \qquad (1)$$

where m and n are the number of arcs and nodes in the network. This matches the known bound (as a function of w and r) for the same problem with an integer (w, r) adversary. Our results immediately imply small delay bounds for the packets as well.

Our bounds obviously apply in the special case when rates are associated with origin-destination pairs. Specifically, suppose packets for a particular origin-destination pair i, j arrive at rate r_{ij}. As long as an associated *fractional* multicommodity flow problem has optimal value at most 1, we can find a scheduling policy with the number of packets bounded by the expression (1), where r can be explicitly determined based on the r_{ij} and the network topology alone.

Our results are achieved by a combination of techniques: we use a discrete review policy, which reduces the dynamic scheduling and routing problem to a sequence of *static*, adaptive packet routing problems; using a rounding theorem due to Karp et al. [13], we reduce each of these problems to a *non-adaptive* packet scheduling problem; these packet scheduling problems can be solved effectively using algorithms due to Bertsimas and Sethuraman [8] or Sevastyanov [14,15,16].

The rest of this paper is structured as follows: in Section 2 we describe the model in more detail; Section 3 describes the scheduling/routing algorithm, and formally specifies the details in each of the steps informally outlined above.

2   Model

The model we consider is the "adversarial queueing network" model advocated by Borodin et al. [9], as modified by Aiello et al. [2]; we refer the reader to these original papers for a thorough motivation of the adversarial model. The basic model used throughout this paper can be described as follows: The communication network is modeled by a directed graph $G = (V, E)$, with $|V| = n$ and $|E| = m$; this network is populated by *packets*, which originate in some node of the network, and need to reach some other node of the network. Associated with each arc (u, v) is an infinite buffer that stores the packets requiring the arc (u, v). We assume a synchronous network, in which time is divided into *steps*, conveniently numbered by the non-negative integers, and indexed by t. Packets require unit time to traverse an arc, and each arc can process at most one packet in a time step.

Packets are injected into the network by an *adversary* operating under a restriction specified in terms of two parameters r and w. Restrictions of this sort were first considered in [9,3,4] for the non-adaptive version, and were extended in an elegant way by Aiello et al. [2] to the adaptive version as follows: Let $A_{ij}[t_1, t_2]$

be the set of packets injected into the network during the time interval $[t_1, t_2]$, with origin i and destination j, and let

$$A[t_1, t_2) = \bigcup_{i,j \in V} A_{ij}[t_1, t_2).$$

An adversary is an *integer* (w, r) adversary for some r $(0 < r < 1)$ and some integer $w \geq 1$ if and only if for any t, the adversary can associate a path to each packet in $A[t, t + w)$ such that every arc belongs to at most $\lfloor rw \rfloor$ paths. (Note that the adversary is not constrained to have a single path in her mind for the packets she injects. A packet p injected at time t will belong to w different time windows; the adversary is allowed to associate different paths to packet p at the time instants $t - w + 1, t - w + 2, \ldots, t - 1, t$.)

Consider the following *integer* multicommodity flow problem

(IMF)   Min $C(t)$
subject to:

$$\sum_{l:(i,l) \in E} x_{ij}^{il} = A_{ij}[t, t + w), \quad \forall i, j \in V,$$

$$\sum_{k:(k,j) \in E} x_{ij}^{kj} = A_{ij}[t, t + w), \quad \forall i, j \in V,$$

$$\sum_{l:(k,l) \in E} x_{ij}^{kl} = \sum_{l:(l,k) \in E} x_{ij}^{lk} \quad \forall i, j \in V, \; k \neq i, j,$$

$$C^{kl} = \sum_{i,j \in V} x_{ij}^{kl}, \quad \forall (k, l) \in E,$$

$$C(t) \geq C^{kl}, \quad \forall (k, l) \in E,$$

$$x_{ij}^{kl} \geq 0, \; \text{integer},$$

where x_{ij}^{kl} represents the number of packets that travel from node i to node j that use the arc (k, l). It is easy to see that an adversary is an integer (w, r) if and only if the optimal value, $C^*(t)$, of (IMF) is at least $\lfloor rw \rfloor$. Since the integer (w, r) adversary is defined in terms of an integer multicommodity flow problem, it is NP-complete to check whether or not an input stream generated by an adversary respects the restrictions imposed. To overcome this limitation, Gamarnik [12] considered a model in which the adversary is allowed to split packets. An adversary is a *fractional* (w, r) adversary for some r $(0 < r < 1)$ and some integer $w \geq 1$ if and only if for any t, the adversary can *fractionally* schedule (or associate *flows* with) all the packets in $A[t, t + w)$ such that the load on each arc is at most rw. Equivalently, an adversary is a fractional (w, r) adversary if and only if the linear programming relaxation of (IMF) has optimal value at most rw. The fractional (w, r) adversary is less constrained, and hence can generate input streams that are inadmissible for the integer (w, r) adversary.

For the integer (w, r) adversary, Aiello et al. [2] constructed a routing and scheduling policy for which the total number of packets in the system is

$$O\left(\frac{n^{5/2}m^{5/2}w}{1-r}\right).$$

In fact, their algorithm is *distributed* and uses only local information. Gamarnik [12] designed a *centralized* algorithm for the fractional (w, r) adversary for which the total number of packets in the system is

$$O\left(\frac{n^4m^3 + w^2m}{(1-r)^2}\right).$$

Gamarnik [12] left open the problem of designing an algorithm for which the total number of packets in the system is $O(w/(1-r))$, matching the bound of Aiello et al. [2] for the integer (w, r) adversary. Our main result is an algorithm with this performance bound. We achieve this using a combination of techniques that have proved to be useful in a host of other problems: these include a scheduling algorithm for large job shop scheduling problems due to Bertsimas and Sethuraman [8], and the rounding theorem due to Karp et al. [13].

To avoid ambiguity, we specify explicitly the sequence of events occurring at any time step: first, packets traverse arcs; next, the adversary injects new packets into the nodes; and finally, packets that reach their destination are absorbed by the corresponding node.

3   The Routing and Scheduling Algorithm

An overview of the algorithm is as follows:

(a) The *dynamic* routing and scheduling problem in adversarial networks can be (approximately) solved as a sequence of *static, adaptive* packet routing problems;
(b) Each of these adaptive packet routing problems can be (approximately) solved as a (non-adaptive) packet *scheduling* problem with a *small* number of paths;
(c) Each of these packet scheduling problems can be (approximately) solved; and
(d) the performance loss in each of these steps is relatively *negligible*.

The rest of this section is devoted to showing the details involved in each of these steps.

Reduction to static, adaptive, packet routing. The dynamic routing and scheduling problems in adversarial queueing networks can be reduced to a sequence of adaptive packet routing problems by using *discrete review* policies. In any such policy, the system is reviewed at discrete points in time, say, at

$$T_0 \equiv 0^+, T_1, T_2, \ldots, T_i, T_{i+1}, \ldots.$$

Policies differ in the way in which the review epochs are picked; we shall not expand on this point any further because our algorithm picks these review epochs in a natural way, as described below.

Suppose T_i is a review epoch chosen by our algorithm. At T_i, we solve an adaptive packet routing problem, with the inputs given by $\{A_{kl}[T_{i-1}, T_i)\}$. In other words, the packets considered by the algorithm at time T_i are *precisely* those that were injected into the network at or after the previous review epoch; these are routed to their respective destinations using a "good" adaptive packet routing algorithm. The epoch at which all of these packets are routed to their destinations defines the next review epoch T_{i+1}. Note that packets that arrived at or after T_i are ignored by the adaptive packet routing algorithm until T_{i+1}. Clearly, the review epochs chosen by are a function of the adaptive packet routing algorithm used; and the effectiveness of such a policy will critically depend on how good the adaptive packet routing algorithm actually is. We shall analyze this next.

At the epoch T_i, we shall process all the packets that arrived during the interval $[T_{i-1}, T_i)$. Let W_i be the optimal value of the associated fractional multicommodity flow problem. It is clear that every algorithm will require at least W_i units of time to process this input; specifically, in the absence of arrivals at or after T_i, no algorithm can process all of the input by time $t < T_{i-1} + W_i$.

Suppose our adaptive packet routing algorithm is able to route all of these packets to their destinations in at most $W_i + f$ steps, for some (constant) f that depends only m and n, but not on the input to the packet routing problem. (It is important that f be independent of W_i.) Thus, f is a measure of the inefficiency of the adaptive packet routing algorithm, and bears directly on the amount of "work" seen by the algorithm at the next review epoch. Given this, how large can W_{i+1} be? Clearly, W_{i+1} represents the maximum load on any arc due to arrivals in $[T_i, T_{i+1})$, which by our assumption is contained in $[T_i, T_i + W_i + f)$. Therefore,

$$W_{i+1} \leq \left\lceil \frac{(T_{i+1} - T_i)}{w} \right\rceil rw < \left(\frac{(T_{i+1} - T_i)}{w} + 1 \right) rw < r(T_{i+1} - T_i) + w, \quad (2)$$

since $r < 1$.

A recursive application of Eq. (2) implies

$$\limsup_{i \to \infty} W_i \leq \frac{f + w}{1 - r}.$$

Thus, letting $Q(t)$ denote the total number of packets in the system at time t, we have

$$Q(t) \leq m \limsup_{i \to \infty} W_i \leq \frac{m(f + w)}{1 - r}. \quad (3)$$

Thus, the dynamic routing/scheduling problem in an adversarial queueing network can be solved as a sequence of *static*, adaptive packet routing problems, as long as each of these problems is solved relatively well; in particular, the queue-length bound of Eq. (3) will hold as long as the *makespan* of the static, adaptive packet routing problem is within an *additive constant* of the associated *congestion* lower-bound.

Identifying a small set of "good" paths. Our goal now is to consider a static, adaptive packet routing algorithm. Let t be a review epoch, and let A_{ij} be the number of packets in the system with origin i and destination j at time t. Let W_t be the optimal value of the (fractional) multicommodity flow problem (IMF) defined by the packets present in the system at time t, and let $(\overline{x})$ be such a solution. Note that without loss of generality, we can assume that $A_{i,j} > 0$. Given $\overline{x}$, we can also assume that there does not exist any cycle with positive flow; hence we can decompose the solution (arc-flows) into flows along paths P_k, $k = 1, \ldots, K$, with the (fractional) flow value on path P_k being y_{P_k}, and such that

$$\sum_{k:(i,j)\in E(P_k)} y_{P_k} = \sum_{u,v\in V} \overline{x}_{u,v}^{i,j} \leq W_t,$$

and

$$\sum_{k:o(P_k)=i,d(P_k)=j} y_{P_k} = A_{i,j}.$$

In the expressions above, $o(P_k)$ and $d(P_k)$ denote the origin and destination of path P_k. We refer the reader to Ahuja et al. [1] for a discussion on flow decomposition.

Our task now is to select precisely $A_{i,j}$ paths from i to j, without affecting the congestion along any arc adversely; in other words, we need to round the fractional solution $(\overline{x})$ to an integral 0-1 solution in a suitable manner. We do this by using the following rounding algorithm of [13]:

Theorem 1. ([13]) *Let A be a real valued $s_1 \times s_2$ matrix, and y be a real-valued s_2-vector. Let b be a real valued vector such that $Ay = b$ and $\hat{t}$ be a positive real number such that, in **every column** of A, (i) the sum of all the positive entries is at most $\hat{t}$ and (ii) the sum of all the negative entries is at least $-\hat{t}$. Then we can compute an integral vector $\overline{y}$ such that for every i, either $\overline{y}_i = \lfloor y_i \rfloor$ or $\overline{y}_i = \lceil y_i \rceil$ and $A\overline{y} = \overline{b}$ where $\overline{b}_i - b_i < \hat{t}$ for all i. Furthermore, if y contains d non-zero components, the integral approximation can be obtained in time $O(s_1^3 \lg(1 + s_2/s_1) + s_1^3 + d^2 s_1 + s_1 s_2)$.*

To use Theorem 1, we first transform our linear system above to the following equivalent form:

$$\sum_{k:(i,j)\in E(P_k)} y_{P_k} \leq W_t \quad \forall\, (i,j) \in E(G)$$

$$\sum_{k:o(P_k)=i,d(P_k)=j} (-m)y_{P_k} = -mA_{i,j} \quad \forall\, i,j \in V.$$

The set of variables above is $\{y_{P_k} : k = 1, \ldots, K\}$. Note that $y_{P_k} \in [0,1]$ for all these variables. Furthermore, in this linear system, the positive column sum is bounded by the maximum length of the paths, which in turn is bounded by m, the number of arcs in the graph. The negative column sum is also bounded by $-m$. Thus, the parameter $\hat{t}$ for this linear system, in the notation of Theorem 1, can be taken to be m. Hence by Theorem 1, we can obtain in polynomial time an **integral** solution $\bar{y}$ satisfying

$$\sum_{k:(i,j)\in E(P_k)} \bar{y}_{P_k} \leq W_t + m \ \ \forall\, (i,j) \in E(G)$$

$$\sum_{k:o(P_k)=i,d(P_k)=j} (-m)\bar{y}_{P_k} < -mA_{i,j} + m \ \ \forall\, i,j \in V.$$

For each i, j, we have

$$\sum_{k:o(P_k)=i,d(P_k)=j} \bar{y}_{P_k} > A_{i,j} - 1.$$

Note the crucial role of the *strict* inequality. Thus, we have selected at least $A_{i,j}$ paths from i to j; furthermore, the congestion along every arc is bounded by $W_t + m$.

To summarize what we have achieved: starting from an arc flow solution, we used flow decomposition and an application of the rounding theorem to derive an integer solution such that the load on any arc is increased by at most m. Each "commodity" (i.e., origin-destination pair) is now routed along at most m paths. We can now reformulate this adaptive packet routing problem as a (non-adaptive) packet *scheduling* problem as follows: think of each path from i to j as a *type*, and assume that $\bar{y}_k$ packets have to be sent from i to j along path P_k. (To avoid cumbersome notation, we have dropped the dependence of $\bar{y}$ on the origin-destination pair.) In essence, we have used the rounding algorithm to compute a small set of good paths for the adaptive packet routing problem; we now pretend that the problem to be solved is really a packet scheduling problem in which an explicit path is associated with each packet; the number of packets to be routed along a given path is determined by applying the rounding algorithm on an optimal (fractional) multicommodity flow solution.

Solving the packet scheduling problem. The dynamic routing/scheduling problem on an adversarial network is now reduced to a simpler, *static*, packet *scheduling* problem. For convenience, we describe the input to this packet scheduling problem slightly differently. The packet scheduling problem consists of K *types* of packets; packets of type k require a path P_k through the network, are initially available at $o(P_k) \in V$, and need to reach $d(P_k) \in V$; there are n_k packets of type k. The objective is to find a schedule for all of these packets that minimizes makespan. Each packet requires unit time to traverse an arc; each arc can process one packet per unit time. Obviously, this is an NP-hard problem. Fortunately, we do not need to find an optimal schedule; all we need is a schedule with makespan within an additive constant of the associated congestion lower bound. Note that this additive constant could depend on m, n, K, but cannot depend on $n_1, n_2, \ldots, n_k$ themselves; this is because in the packet scheduling instances that will arise in the solution of the adversarial network will have m, n, and K will be independent of r and w, the parameters of the adversary, whereas the n_k will depend on r and w. We briefly outline two solution methods to this packet scheduling problem, and specify the corresponding bounds.

Fluid synchronization algorithm. The packet scheduling problem outlined here is a special case of the job shop scheduling problem with the makespan objective considered by Bertsimas and Sethuraman [8]. In that work, they consider a *fluid relaxation* of the job shop scheduling problem, which can be viewed as a continuous analog of the discrete job shop scheduling problem. Using an optimal solution to the fluid relaxation, they find nominal start times for each packet at each of the arcs it has to visit; these nominal start times are carefully constructed in a recursive manner, based on both the optimal fluid solution and the partial discrete schedule.

More precisely, suppose type k packets need to visit arcs $a_{k,1}, a_{k,2}, \ldots, a_{k,i_k}$ in that order. Suppose W is the maximum load on any arc. The scheduling algorithm discussed in [8] first determines the *fluid* start and completion times for each packet at each stage. The fluid start time, $FS_{k,j}(n)$, of the nth type k packet at (its) stage j (arc $a_{k,j}$) is defined to be $(n-1)W/n_k$; the corresponding fluid completion time $FC_{k,j}(n)$ is nW/n_k.

Since the fluid relaxation processes packets continuously, each type k packet is processed by *all* its stages simultaneously at a uniform rate n_k/W; for this reason, the fluid start and completion times for any packet is *independent* of its "stage," and depends only on the packet number. In trying to "round" this fluid schedule to an implementable discrete schedule, we need to overcome two difficulties: first, the fluid relaxation treats packets as continuous entities, with the effect that the same packet can be "scheduled" by multiple arcs simultaneously; and second, the fluid relaxation allows arcs to split their effort across multiple packet types, as long as the overall effort allocated by each arc is at most 1 per unit time. In other words, the fluid relaxation views both the packets and the processing resources as being infinitely divisible. (The resulting lower bound is naturally just the congestion lower bound; the dilation bound does not arise because of the continuous nature of the jobs.)

The fluid start of a given packet at a given stage may be viewed as the ideal start time of that packet at that stage, but clearly, this is an unrealistic ideal. Motivated by the question of defining a more realistic target start time for each packet at each stage, Bertsimas and Sethuraman [8] defined *nominal start times*; these are defined in terms of the fluid start and completion times as well as the partial discrete schedule. The *nominal start time*, $NS_{k,j}(n)$, of the nth type k packet at its stage j (arc $a_{k,j}$) is defined by

$$NS_{k,1}(n) = FS_{k,1}(n),$$
$$NS_{k,i}(1) = DS_{k,i-1}(1) + 1, \quad i > 1,$$
$$NS_{k,i}(n) = \max\left\{ NS_{k,i}(n-1) + \frac{W}{n_k}, \ DS_{k,i-1}(n) + 1 \right\}, \quad n, i > 1,$$

where $DS_{k,i-1}(n)$ is the start time of the nth type k packet at stage $(i-1)$ (arc $a_{k,(i-1)}$) in the discrete schedule.

Bertsimas and Sethuraman [8] proposed a simple scheduling rule (called "fluid synchronization algorithm") based on these nominal start times: whenever a node has to make a processing decision, it schedules an available packet with the earliest nominal start time. Note that whenever a packet is chosen to be scheduled at a certain node, its nominal processing time at its next stage can be calculated; so the nominal start times for every packet queued at a node will be known.

The main result of [8] adapted to this special case can be stated as follows:

Theorem 2. *Consider a (non-adaptive) packet scheduling problem with K job types and m arcs. Given initially n_k jobs of type $k = 1, 2, \ldots, K$, suppose the maximum load on any arc is W, and let W^* be the optimal makespan. Then, the fluid synchronization algorithm produces a schedule with makespan time W_{D} such that*

$$W \leq W^* \leq W_{\mathrm{D}} \leq W + n(K+2). \tag{4}$$

Sevastyanov's algorithm. In the mid-seventies, interesting approximation algorithms were derived for several shop scheduling problems. These algorithms were based on beautiful, geometric arguments, and were discovered independently by Belov and Stolin [7], Sevastyanov [14], and Fiala [11]. These methods constructed schedules for job shop scheduling problems with an additive error term that depended only on the number of machines, and the maximum processing time of a job, but not on the number of jobs. Since it is not central to this paper (and in the interest of space), we do not discuss these algorithms in detail; we refer the interested reader to the original papers cited earlier as well as the excellent survey of Sevastyanov [17]. The strongest of these results, due to Sevastyanov [15,16], provides a schedule of length at most $(n-1)(mn^2 + 2n - 3)$.

Remark. Note that depending on K, this may or may not be better than the schedule provided by the fluid synchronization algorithm. For the adaptive case, it is seen that the guarantee provided by the fluid synchronization algorithm is slightly better than the one provided by Sevastyanov's algorithm. Moreover, the fluid-based algorithm is not computationally intensive at all, and is very simple to implement. On the other hand, for the non-adaptive case, the adversary may insist that the algorithm route packets along exponentially many paths; in this case, the guarantee provided by the fluid-based method is unattractive, and Sevastyanov's method is clearly better.

The main result. Our main result is obtained by putting all of these steps together. Fix a review epoch i, with W_i being the work seen by the scheduler at this epoch. Then, step 2 results in an instance of the non-adaptive packet scheduling problem with maximum congestion at most $W_i + m$; using the fluid synchronization algorithm for this packet scheduling problem results in a schedule with length at most $W_i + m + n(K+2)$. Noting that there are at most n^2

commodities, and that each of which may use at most m paths, we conclude that the schedule computed at epoch i will have length at most $W_i + m + n^2 m^2 + 2n$. Thus, the inefficiency parameter f is at most $m + 2n + m^2 n^2$; using this in Eq. (3), we have

$$Q(t) \leq \frac{m(f + w)}{1 - r} \leq \frac{m(m + 2n + m^2 n^2 + w)}{1 - r}, \tag{5}$$

where $Q(t)$ represents the number of packets in the system at time t.

For Sevastyanov's algorithm a similar guarantee can be shown to hold. We omit the details.

Our results can now be formally stated as the following theorem.

Theorem 3. *Consider an adversarial queueing network under a fractional (w, r) adversary. If $r < 1$, then the discrete review scheduling policy constructed keeps the number of packets in the system bounded at all times. In particular, the total number of packets in the system at time t, $Q(t)$, satisfies*

$$Q(t) \leq \frac{m(m + 2n + m^2 n^2 + w)}{1 - r}.$$

$\square$

An immediate corollary is that for adversarial queueing networks in which the arrival rates for packets with origin i and destination j is r_{ij}, an algorithm for which the number in system is $O(w/(1 - r))$, can be designed, where r can be explicitly computed based on the r_{ij} using a fractional multicommodity flow formulation. Gamarnik [12] considered this model and showed that stable policies exist for this system if and only if the associated fractional multicommodity flow problem has value at most 1. (The r in the expression for the number-in-system bound is exactly the optimal solution to this multicommodity flow problem.)

Since the number in system is relatively small, one can expect the proposed algorithm to provide good delay guarantees for all the packets as well. This can be formally established using the fact than any packet stays in the system for at most two review periods. Discussion on this topic is deferred to the full version of this paper, as is the discussion of results on the non-adaptive version of the problem. At this point, we simply note that these techniques lead to excellent performance guarantees for the non-adaptive version of the problem as well.

Future work. Several outstanding questions remain; we point out two explicitly. First, we hope to consider the case $r = 1$; this seems difficult to understand, and may in fact exhibit different behavior depending on whether the adversary is fractional (w, r) or integer (w, r) restricted. Moreover, the algorithm we propose is (semi) centralized, although the queue-length information is used only at the discrete review epochs. In contrast, Aiello et al. [2] proposed a distributed algorithm for the integer (w, r) adversary. It will be interesting to design a distributed algorithm for the problem considered here. We hope to address this in future work as well.

References

1. R. K. Ahuja, T. L. Magnanti, and J. B. Orlin. *Network flows: theory, algorithms, and applications*. Prentice Hall, Englewood Cliffs, New Jersey, 1993.
2. W. Aiello, E. Kushilevitz, R. Ostrovsky, and A. Rosen. Adaptive Packet Routing for Bursty Adversarial Traffic. *Journal of Computer and System Sciences*, **60**, 482–509, 2000. Preliminary version in *Proceedings of the 30th STOC*, 359–368, 1998.
3. M. Andrews, B. Awerbuch, A. Fernandez, J. Kleinberg, T. Leighton, and Z. Liu. Universal-stability results for greedy contention-resolution protocols. *Proceedings of the 37th FOCS*, 380–389, 1996.
4. M. Andrews, B. Awerbuch, A. Fernandez, F. T. Leighton, Z. Liu, and J. Kleinberg. Universal-stability results and performance bounds for greedy contention-resolution protocols. *Journal of the ACM*, **48**(1):39–69, 2001.
5. M. Andrews, A. Fernandez, A. Goel, and L. Zhang. Source Routing and Scheduling in Packet Networks. *Proceedings of the 42nd FOCS*, 168–177, 2001.
6. B. Awerbuch, P. Berenbrink, A. Brinkmann, C. Scheideler. Simple Routing Strategies for Adversarial Systems. *Proceedings of the 42nd FOCS*, 158–167, 2001.
7. I. S. Belov and Ya. N. Stolin. An algorithm in a single path operations scheduling problem. *Mathematical Economics and Functional Analysis*, pages 248–257, 1974. (in Russian).
8. D. Bertsimas and J. Sethuraman. From fluid relaxations to practical algorithms for job shop scheduling: the makespan objective. *Mathematical Programming*, 92(1):61–102, 2002.
9. A. Borodin, J. Kleinberg, P. Raghavan, M. Sudan, and D. P. Williamson. Adversarial Queueing Theory. *Journal of the ACM*, **48**(1):13–38, 2001. Preliminary version in *Proceedings of the 28th STOC*, 376–385, 1996.
10. R. Cruz. A Calculus for network delay, part II: network analysis. *IEEE Transactions on Information Theory*, 37:132–141, 1991.
11. T. Fiala. Közelítő algorithmus a három gép problémára. *Alkalmazott Matematikai Lapok*, 3:389–398, 1977.
12. D. Gamarnik. Stability of Adaptive and Non-Adaptive Packet Routing Policies in Adversarial Queueing Networks. *Proceedings of 31st STOC*, 1999.
13. R. M. Karp, F. T. Leighton, R. L. Rivest, C. D. Thompson, U. V. Vazirani, and V. V. Vazirani. Global Wire Routing in Two-Dimensional Arrays. *Algorithmica*, 2:113–129, 1987.
14. S. V. Sevastyanov. On an asymptotic approach to some problems in scheduling theory. In *Abstracts of papers at 3rd All-Union Conference of Problems of Theoretical Cybernetics*, pages 67–69, Novosibirsk, 1974. Inst. Mat. Sibirsk. Otdel. Akad. Nauk SSSR.
15. S. V. Sevastyanov. Efficient construction of schedules close to optimal for the cases of arbitrary and alternative routes of parts. *Soviet Math. Dokl.*, 29(3):447–450, 1984.
16. S. V. Sevastyanov. Bounding algorithm for the routing problem with arbitrary paths and alternative servers. *Kibernetika*, 22(6):74–79, 1986. Translation in *Cybernetics*, 22:773–780.
17. S. V. Sevastyanov. On some geometric methods in scheduling theory: a survey. *Discrete Applied Mathematics*, 55:59–82, 1994.

Approximation Schemes for Generalized 2-Dimensional Vector Packing with Application to Data Placement

Hadas Shachnai[1]* and Tami Tamir[2]

[1] Bell Laboratories, Lucent Technologies, 600 Mountain Ave. Murray Hill, NJ 07974
[2] Department of Computer Science & Engineering, Univ. of Washington,
Box 352350, Seattle, WA 98195.
tami@cs.washington.edu

Abstract. Suppose that we have a set of items and a set of devices, each possessing two limited resources. Each item requires a given amount of the resources. Further, each item is associated with a profit and a color, and items of the same color can *share* the use of one resource. We need to allocate the resources to the most profitable (feasible) subset of items. In alternative formulation, we need to *pack* the most profitable subset of items in a set of 2-dimensional bins (knapsacks), in which the capacity in one dimension is *sharable*. Indeed, the special case where we have a single item in each color is the well-known 2-*dimensional vector packing (2DVP)* problem. Thus, the problem that we study is strongly NP-hard for a single bin, and MAX-SNP hard for multiple bins. Our problem has several important applications, including *data placement* on disks in media-on-demand systems.

We present approximation algorithms as well as optimal solutions for some instances. In some cases, our results are similar to the best known results for 2DVP. Specifically, for a single knapsack, we show that our problem is solvable in pseudo-polynomial time and develop a *polynomial time approximation scheme (PTAS)* for general instances. For a natural subclass of instances we obtain a simpler scheme. This yields the first *combinatorial* PTAS for a non-trivial subclass of instances for 2DVP. For multiple knapsacks, we develop a PTAS for a subclass of instances arising in the data placement problem. Finally, we show that when the number of distinct colors in the instance is fixed, our problem admits a PTAS, even if the items have arbitrary sizes and profits, and the bins are arbitrary.

1   Introduction

Consider the following optimization problem. Suppose that we have a set of n items and a set of N devices, each possessing a limited supply of two resources.

---

* On leave from the Department of Computer Science, Technion, Haifa 32000, Israel.
E-mail: hadas@cs.technion.ac.il

S. Arora et al. (Eds.): APPROX 2003+RANDOM 2003, LNCS 2764, pp. 165–177, 2003.

Each item requires given amounts of the resources. Further, an item is associated with a profit, that is obtained if the resources are allocated to that item, and a color; items of the same color can *share* the use of one of the resources. The goal is to allocate the resources to a subset of the items, subject to availability constraints, such that the overall profit is maximized.

Formally, suppose that the j-th device, $1 \leq j \leq N$, has V_j and C_j units from the first and second resource, respectively. Each item i, $1 \leq i \leq n$, is associated with a profit, p_i. Also, item i requires s_i units from the first resource and c_i units from the second resource. We assume that the second resource can be *shared* by some items. Specifically, the instance I is partitioned into M sets, by colors; all items of the same color k, $1 \leq k \leq M$, require the same amount, c_k ,from the second resource, and can share its use. The goal is to select a feasible most-profitable subset of items. A subset is feasible if the total allocation from the first (second) resource on the j-th device does not exceed V_j (C_j), for $1 \leq j \leq N$.

In alternative formulation, the above set of items needs to be packed into N bins (knapsacks); the j-th bin has capacity V_j and C_j compartments. Each item can be packed in any of the bins. When the first item of color k is packed in some bin, c_k compartments are allocated to this *color*; additional items of color k will be accommodated in the same set of compartments. A packing is feasible if the total size of the packed items in any bin, j, is at most V_j, and the total number of compartments allocated in bin j is at most C_j, $1 \leq j \leq N$. The goal is to pack a subset of the items of maximum total profit. Indeed, the special case where we have a single item in each color is the well-known *2-dimensional vector packing problem (2DVP)*. Thus, our problem is strongly NP-hard for a single bin [10] and MAX-SNP hard for multiple bins, already in the case where the bins are identical, and the items have unit profits, i.e., $p_i = 1 \ \forall i, 1 \leq i \leq n$ [16]. We call this problem *vector packing with a shareable dimension* (VPSD).

An important application of VPSD is *data placement* on disks in media-on-demand systems [6,12,9]. In such systems (see, e.g., [17,7]), a large database of M video program files is stored on a centralized server. Each program file, k, $1 \leq k \leq M$, is associated with a number of desired broadcasts of this file, n_k, and a size (storage requirement), c_k. The files are stored on N shared disks. Each disk, j, is characterized by (i) its storage capacity, C_j, that is, the total size of the files that can reside on it, and (ii) its load capacity, V_j, which is the number of data streams that can be read simultaneously from that disk. The files need to be placed on the disks so as to maximize the total number of requests for broadcasts that can be satisfied simultaneously. In the resulting instance of the VPSD problem, the bins represent disks, and the items are broadcast requests. To satisfy a request, some disk has to broadcast a data stream to the client. This disk must hold a copy of the requested file. Note that storage is a shared resource – that can be used by all the streams broadcasting the same data from the same disk. Different files may have different sizes, thus, we may have different c_j values. On the other hand, all the broadcast streams require the same (non-shareable) bandwidth; thus, we have $\forall i \ s_i = 1$.

Other applications of VPSD are production planning and scheduling parallel tasks (see in [13]). Of particular interest in our study is the subclass of *uniform profit/size ratio* instances of VPSD, in which for some $\alpha > 0$, $\forall i$ $p_i = \alpha s_i$. Such instances naturally arise in real systems, where client payments for service (i.e., item profits) are proportional to the amounts of resources consumed (i.e., the items sizes).

1.1   Our Results

In the following we summarize our main results.

Single knapsack We show (in Section 2) that VPSD can be optimally solved in pseudo-polynomial time. We then develop an LP based PTAS for general instances. For the subclass of uniform profit/size ratio instances, we develop (in Section 3) a simpler approximation scheme, that is based on extension of a PTAS proposed in [11] for the classical knapsack problem. By this, we obtain the first *combinatorial* PTAS for a non-trivial subclass of instances for 2DVP. In Section 4, we develop *fully polynomial time approximation schemes (FPTAS)* for the subclasses of (i) data placement instances, and (ii) instances with constant number of compartment requirements.

Multiple knapsacks We show (in Section 4) that an iterative greedy algorithm achieves the ratio of $(2 + \varepsilon)$ for instances with arbitrary bin sizes, and $(\frac{e}{e-1} + \varepsilon)$ when the bins are identical. A PTAS is developed (in Section 5) for data placement instances in which the disks are identical (but may have arbitrary storage and load capacities), and the number of distinct file sizes is fixed. Finally, for instances in which M, the distinct number of colors, is fixed, we show (in section 6) that VPSD admits a PTAS, even if the items have arbitrary sizes and profits, and the bins are arbitrary.

In our PTAS for a single knapsack (in Section 2), we combine the guessing technique of [2] with a novel application of the approximation scheme of [3] to the *multidimensional multiple choice knapsack* problem. In our algorithms for uniform ratio instances (in Section 3), we show that a simple greedy algorithm and an approximation scheme proposed for the 0/1 knapsack problem can be extended to VPSD. The idea is to partially reduce VPSD to the knapsack problem, by first considering all the items of each color as a *single* item. Later, we map the *grouped* items back to the original items. While these extensions do not apply for general instances, it may be possible to apply similar ideas for other subclasses of VPSD and 2DVP.

Due to space constraints we state some of the results without proofs.[3]

1.2   Related Work

Packing problems in single dimension have been extensively studied. Since these problems are NP-hard, most of the research work in this area focused on finding

---

[3] The detailed proofs are given in [14].

approximation algorithms. The classic *0-1 knapsack problem* admits an FPTAS; that is, for any $\varepsilon > 0$, a $(1 - \varepsilon)$-approximation for the optimal solution can be found in $O(n/\varepsilon^2)$ steps [8,5]. In contrast, the *multiple knapsack (MK)* problem is known to be strongly NP-hard [4]. Chekuri and Khanna developed in [2] a PTAS for MK and showed that with slight generalizations this problem becomes APX-hard.

Packing problems in higher dimensions (also known as *d-dimensional vector packing*) are known to be substantially harder to solve, exactly or approximately. The best known result for a single knapsack is a PTAS due to Frieze and Clarke [3], for the case where d is a fixed constant. As opposed to the combinatorial schemes for the single dimension case, the PTAS in [3] uses as a procedure a *linear program*. To the best of our knowledge, none of the later published work on the d-dimensional knapsack problem gives a *combinatorial* scheme, even for the case where $d = 2$. For the case of $N > 1$ bins, Woeginger showed in [16] that 2-dimensional vector packing is MAX-SNP hard (see also in [1]). Chekuri and Khanna presented in [1] a PTAS for the *vector scheduling* problem, in which our goal is to schedule a set of jobs, given by d-dimensional vectors, on a set of machines, so as to minimize the maximum completion time (or makespan) over all dimensions. The scheme in [1] yields a *dual* PTAS for d-dimensional vector packing in $N \geq 1$ bins, where the bins have d equal-sized dimensions, and d is a fixed constant. The *class constrained multiple knapsack (CCMK)* problem introduced in [13] is a special case of VPSD, where $c_k = 1$ for all $1 \leq k \leq M$. The paper [13] presents a PTAS for any instance of CCMK in which M, the number of distinct colors of items, is fixed.

The data placement problem was initially studied in [12]. The paper presents an algorithm for the case where all the files are of the *same* (unit) size, and for all $1 \leq j \leq N$, we have the same ratio V_j/C_j for disk j (*uniform ratio disks*). The paper shows that the algorithm achieves a ratio of $1 - 1/(1 + C_{min})$ to the optimal, where $C_{min} = \min_j C_j$. Golubchik et al. gave in [6] a tighter analysis of this algorithm and showed that it achieves the ratio $1 - 1/(1 + \sqrt{C}_{min})^2$, and that this ratio is optimal for *any* algorithm for this problem. The paper [6] also presents a PTAS for the data placement problem with unit sized files and uniform ratio disks. Recently, Kashyap and Khuller [9] studied the problem with files of Δ distinct sizes, where Δ is fixed. They presented an algorithm that achieves a ratio of $\frac{C-\Delta}{C+\Delta}\left(1 - 1/(1 + \sqrt{\frac{C}{2\Delta}})^2)\right)$, where file sizes are in $\{1, \ldots, \Delta\}$, and C is the storage capacity of the disks. They also showed that this algorithm can be combined with an algorithm that runs in polynomial time when C is fixed, to get a PTAS for the data placement problem with constant number of file sizes.

2   Approximation Scheme for a Single Bin

In this section, we discuss the single knapsack version of VPSD. Assume that the knapsack has the volume V and C compartments. We first note that by using a two-level dynamic programming algorithm, VPSD with a single knapsack can be solved optimally in pseudo-polynomial time. (The details are given in [14].) Let $\hat{P}$ be an upper bound on the total profit (indeed, $\hat{P} \leq \sum_i p_i$).

Theorem 1. *VPSD can be solved optimally in $O(n\hat{P} + M\hat{P}^2 C)$ steps.*

We now describe a PTAS for a single knapsack. Note that, by the result of [10], this is the best we can expect, since our problem is strongly NP-hard. Assume that we know the optimal profit, P, for our instance. We reduce our problem to the *binary 2-dimensional multiple choice knapsack* (B2D-MCK) problem. That is, for given values of P and ε, we define an instance for B2D-MCK, whose optimal solution induces a solution for VPSD with profit at least $(1 - \varepsilon)P$. We then develop a PTAS for the B2D-MCK problem. By combining the reduction and the PTAS for B2D-MCK, we get a PTAS for VPSD. Note that P can be 'guessed' in polynomial time within factor $(1 + \varepsilon)$, using binary search over the range $(\max_i p_i, \sum_i p_i)$.

Reduction to the B2D-MCK Problem   Recall that an instance of B2D-MCK consists of a single 2-dimensional knapsack and M sets of items. Each item has a 2-dimensional size and is associated with a profit. We need to pack a subset of items of maximal total profit. A packing is feasible if it does not exceed the volume in any dimension, and at most one item is packed from each set.

Given the value of P, the parameter ε and a VPSD instance with n items of M distinct colors, we construct a B2D-MCK instance which consists of a single 2-dimensional knapsack with capacities $b_1 = V$ and $b_2 = C$, and M sets of items; each set S_k has $R = M/\varepsilon$ items, $1 \leq k \leq M$. Each of the items in S_k represents a subset of the items in the VPSD instance, which are of color k, and whose total profit is rounded down to the next integral multiple of $\varepsilon P/M$. In particular, the jth item in S_k, denoted as (k, j), is given by the triple $(s_{kj}, c_k, p(k,j))$: s_{kj} is the minimal total size of a subset of items in color k, whose total profit is $p(k,j) = (j\varepsilon P)/M$. This total size can be computed using dynamic programming for the items of S_k with the rounded profits (as in the FPTAS for the classic knapsack problem [8]).

Lemma 1. *If there exists a solution with profit P for the VPSD instance, then there exists a solution with profit at least $(1 - \varepsilon)P$ for the binary 2D-MCK instance.*

Approximating the Optimal Solution for B2D-MCK   Given an instance of B2D-MCK, we 'guess' the set S of most profitable items in the optimal solution, where $|S| = h = \min(M, \lfloor \frac{4(1-\varepsilon)}{\varepsilon} \rfloor)$. Let $E(S)$ be the subset of items with profits that are larger than the minimal profit of any item in S, that is, $E(S) = \{(k,j) \notin S \mid p(k,j) > p_{min}(S)\}$, where $p_{min}(S) = \min_{(k,j) \in S} p(k,j)$.

We pack all the items $(k,j) \in S$ and eliminate from the instance all the items $(k,j) \in E(S)$, and the sets S_k from which an item has been selected. In the next step we find an optimal *basic solution* for the following linear program, $LP(S)$

$$\max \quad \sum_{k=1}^{M}\sum_{j=1}^{R} p(k,j)x_{kj}$$

$$s.t. \quad \sum_{j=1}^{R} x_{kj} \leq 1 \text{ for } k=1,\ldots,M$$

$$\sum_{k=1}^{M}\sum_{j=1}^{R} x_{kj}s_{kj} \leq V$$

$$\sum_{k=1}^{M} c_k \sum_{j=1}^{R} x_{kj} \leq C$$

$$x_{kj} = 1 \text{ for } (k,j) \in S, \text{ and } x_{kj} = 0 \text{ for } (k,j) \in E(S)$$

$$x_{kj} \in \{0,1\} \text{ for } k=1,\ldots,M; \quad j=1,\ldots,R; \quad (k,j) \notin S \cup E(S)$$

In the linear programming relaxation we allow $0 \leq x_{kj} \leq 1$. Given an optimal fractional solution, we get an integral solution by rounding down to 0 the fractional variables in the solution. The output for B2D-MCK consists of the items in S and the items (k,j) for which $x_{kj} = 1$.

Theorem 2. *The above scheme achieves a ratio of $(1-\varepsilon)$ to the optimal B2D-MCK profit.*

Proof. Let $\mathbf{x}^*$ be an optimal solution for the linear program $\mathrm{LP}(S)$, and let S^* be the corresponding subset of items, that is, $S^* = \{(k,j)|\, x_{kj}^* = 1\}$. If $|S^*| < h$ then we are done (the scheme outputs a $(1-\varepsilon)$-approximation to the optimal profit: this is due to the initial guess of P); otherwise, let $S^* = \{(k_1,j_1),\ldots,(k_r,j_r)\}$, such that $p(k_1,j_1) \geq \cdots \geq p(k_r,j_r)$. Let $S_h^* = \{(k_1,j_1),\ldots,(k_h,j_h)\}$, and $\sigma = \sum_{\ell=1}^{h} p(k_\ell,j_\ell)$. Then, for any item $(k,j) \notin (S_h^* \cup E(S_h^*))$, we have $p(k,j) \leq \sigma/h$. Let $z^*, \hat{z}$ denote the optimal solution and the solution output by the scheme, respectively. We denote by $\mathbf{x}^B(S_h^*), \mathbf{x}^I(S_h^*)$ the basic and integral solutions of $\mathrm{LP}(S)$ as computed by the scheme, for the initial guess S_h^*. Now, we have that

$$z^* \leq \sum_{k=1}^{M}\sum_{j=1}^{R} p(k,j)x_{kj}^B(S_h^*) \leq \sum_{k=1}^{M}\sum_{j=1}^{R} p(k,j)x_{kj}^I(S_h^*) + \delta,$$

where $\delta = \sum_{(k,j)\in F} p(k,j)$, and F is the set of items for which the basic variable was a fraction, that is, $F = \{(k,j)|\, x_{kj}^B(S_h^*) > x_{kj}^I(S_h^*)\}$.

Recall that in any *basic* solution for a linear program, the number of non-zero variables is bounded by the number of tight constraints in some optimal solution (since non-tight constraints can be omitted). Assume that in the optimal (fractional) solution of $LP(S_h^*)$ there are L tight constraints, where $0 \leq L \leq M+2$. Then in the basic solution $\mathbf{x}^B(S_h^*)$, at most L variables can be strictly positive. Thus, at least $L-4$ variables get an integral value (i.e. '1'), and $|F| \leq 4$. Note that $\delta < \sigma/h$, since $F \cap (S_h^* \cup E(S_h^*)) = \emptyset$. Hence, we get that $z^* \leq \hat{z} + \frac{4\sigma}{h} \leq \hat{z} + \frac{4\hat{z}}{h} \leq \frac{\hat{z}}{1-\varepsilon}$.

3   Instances with Uniform Profit/Size Ratio

In this section we present algorithms for instances with uniform profit/size ratio, that is, for some $\alpha > 0$, $\forall i, p_i = \alpha s_i$. Our goal is to pack in a single bin a subset of the items whose total size is as large as possible. W.l.o.g. we assume that $\forall k$, $c_k \leq C$, $\forall i, s_i \leq V$.

3.1   A Greedy 2-Approximation Algorithm

Let S_k be the total size of items with color k, $1 \leq k \leq M$. Consider the following greedy algorithm $\mathcal{A}_G$..

1. Sort the colors such that $S_1/c_1 \geq S_2/c_2 \geq \ldots \geq S_M/c_M$.
2. Find in the sorted list the first j colors satisfying $\sum_{k=1}^{j} c_k \leq C$, and $\sum_{k=1}^{j+1} c_k > C$. Let A be the set of all items in the selected colors.
3. Pack the items in A in the knapsack from largest to smallest, ignoring colors, biggest to smallest, while there is enough space. Let a_1 denote the total size of items packed this way.
4. Let k^* be the color with maximal total size. Let a_2 be the total size of items that are packed from color k^* when adding items greedily, from largest to smallest, as long as there is enough space.
5. Select (and pack accordingly) the maximum between a_1 and a_2.

Theorem 3. *$\mathcal{A}_G$ yields a 2-approximation for* uniform-ratio *instances.*

Proof. If the total size of items in color k^* is more than V, then $a_2 > V/2$, otherwise, $a_2 = S_{k^*}$. If $a_2 > V/2$, we are done (since $OPT \leq V$). Consider the case that $a_2 = S_{k^*}$. Since we sort the colors by profit/compartment ratio, $OPT < S_1 + \ldots + S_{j+1}$. If in step 3 we pack all the items of A, then $a_1 = S_1 + \ldots + S_j$. $Alg = max(a_1, a_2) = max(S_1 + \ldots + S_j, S_{k^*}) \geq \frac{1}{2}(S_1 + \ldots + S_j + S_{k^*}) \geq \frac{1}{2}(S_1 + \ldots + S_{j+1}) \geq \frac{1}{2}OPT$. If in step 3 we pack only part of the items, then since we pack from largest to smallest we fill at least half of the volume, which is at least $\frac{1}{2}OPT$.

3.2   Approximation Scheme

We now describe a PTAS for the uniform ratio case. Our scheme extends the PTAS of Sahni [11] for the classical knapsack problem. Let k_1, k_2 be constants (to be determined). Algorithm $\mathcal{A}$ proceeds as follows. For any possible selection of at most k_1 items from I, and for any possible selection of at most k_2 colors among those that do not appear in the k_1 items, we do the following.

1. Let V' be the remaining volume (V' equals V minus the total size of the k_1 items). Let C' be the remaining number of compartments (C' equals C minus the total compartment demand of the k_2 colors and the k_1 items).

2. If this selection of items and colors is infeasible (that is, $V' < 0$ or $C' < 0$) stop; otherwise,
3. Let T be the set of the k_2 selected colors and the colors of the k_1 items.
   (a) Pack the k_1 items.
   (b) Add the other items of the T colors in arbitrary order as long as there is enough space.
   (c) If there is no space while adding these items, terminate with the packed items; otherwise,
   (d) Sort the colors that do not belong to T such that $S_1/c_1 \geq S_2/c_2 \geq \ldots$
   (e) Add items of color c_1 in arbitrary order, then items of color c_2 and so on, as long as there are enough space and enough compartments.

Theorem 4. *For all k_1, k_2, $\mathcal{A}$ has approximation ratio $R_\mathcal{A} \leq 1 + \frac{1}{\min(k_1, k_2)}$ and running time $O(n^{k_1+1} \cdot M^{k_2})$.*

By selecting $k_1 = k_2 = 1/\varepsilon$ we obtain a PTAS. Consider the subclass of 2DVP instances in which the size of any item i in each dimension is arbitrary, and the profit p_i is proportional to the size in one dimension. For such instances, we have a combinatorial approximation scheme, as formalized in the next result.

Corollary 1. *Algorithm $\mathcal{A}$ is a PTAS for uniform profit/size ratio instances of 2DVP.*

4   Better Algorithms for Special Instances

We now show that better approximations or more efficient algorithms can be obtained for several subclasses of instances.

Theorem 5. *If the compartment requirement of any color class can take one of the values $\eta_1, \ldots, \eta_w$, where w is fixed, then an optimal solution can be computed in $O(M^{w+1}\hat{P})$ steps.*

By scaling the item profits, using the upper bound $\hat{P}$ on the total profit, we may lose only a factor of ε in the approximation ratio.

Theorem 6. *There is an FPTAS for VPSD instances with a single knapsack, in which the compartment requirement of any color class can take one of the values $\eta_1, \ldots, \eta_w$, and w is fixed. The running time of the scheme is $O(n \log n + M^{w+2}\frac{n}{\varepsilon})$.*

For instances where all items have the same (unit) size and profit, and each color can have an arbitrary compartment requirement, we get an exact polynomial time algorithm.

Theorem 7. *If $\forall \, 1 \leq i \leq n \; s_i = p_i = 1$, an optimal solution can be computed in $O(Mn^2)$ steps.*

Recall that a data placement instance is given as M files, each having a specified size and a broadcast requirement, which takes a value in $(0, V]$. Thus, for such instances the optimal algorithm runs in $O(MV^2)$ steps. By scaling and rounding the load capacity of the disk, as well as the broadcast requirements of the files,[4] we obtain an FPTAS.

4.1 Packing in Multiple Bins: A Greedy Algorithm

Given an instance of VPSD with N bins, consider an algorithm, $\mathcal{A}_G$, which packs the bins sequentially. In step j, $1 \leq j \leq N$, we use an (approximation or exact) polynomial time algorithm for packing a 'good' subset of the remaining items in bin j. By the analysis of this iterative packing algorithm, as given in [13], and by the above results for a single knapsack, we get

Theorem 8. $\mathcal{A}_G$ *is a* $2 + \varepsilon$-*approximation algorithm for VPSD, and 2-approximation for instances with unit sizes and profits.*

In the special case where all the bins are identical, we can use a result for the *generalized assignment problem (GAP)* in [2] to get better approximation ratios.

Theorem 9. $\mathcal{A}_G$ *achieves a ratio of* $e/(e-1) + \varepsilon$ *for VPSD with identical bins, and the ratio* $e/(e-1)$ *for instances with unit sizes and profits, where* e *is the base of the natural logarithm.*

5   Approximation Scheme for Data Placement

In this section, we develop a PTAS for data placement instances in which the disks are identical, but may have arbitrary storage and load capacities, and the number of distinct file sizes is fixed.

In terms of the VPSD problem, we consider instances I consisting of n items, of M distinct colors; for any item $i, 1 \leq i \leq n$, $p_i = s_i = 1$. The compartment requirement of color k can take one of w possible values, $\eta_1, \ldots, \eta_w$, where $w > 1$ is a fixed constant. There are n_k items of color k, $1 \leq k \leq M$. We need to pack a subset of the items in N identical bins, where each bin has volume V and C compartments.

Given a parameter $\varepsilon > 0$, the scheme proceeds as follows. (i) Guess the optimal profit from the packing, P, to within factor $1 + \varepsilon$. (Recall that $1 \leq P \leq n$.) (ii) Guess the subset of items that are packed in the bins. (iii) Pack the selected items, distinguishing between items with 'large' and 'small' compartment requirements. In the latter case, we further distinguish between packings of 'large' and 'small' blocks (We define a *block* below).

---

[4] Similar to the proof of Theorem 6. We omit the details.

5.1   Guessing the Packed Items

Given a correct guess of P, we omit from the input the items in any color k such that $n_k \leq \varepsilon P/M$. By that we lose at most a factor of ε in the approximation ratio. Dividing the value of n_k, for each of the remaining colors, by $\varepsilon P/M$, and rounding down to the nearest integral power of $(1 + \varepsilon)$, we get an instance in which there are $h = O(\ln(M/\varepsilon))$ distinct n_k values.

Now, we partition the item sets to w groups, $S_1, \ldots, S_w$, by their compartment requirements. The item sets having the compartment requirement η_ℓ form the ℓ-th *compartment category*, S_ℓ. We find the subset of packed items by guessing the contribution of each compartment category to the overall profit. Let $P(S_\ell)$ denote the contribution of S_ℓ. We may assume that $P(S_\ell) \geq \varepsilon P/w$. By that, we lose at most a factor of ε from the overall profit. Then, we look for a vector $\boldsymbol{k}$, of integers, $w/\varepsilon \leq k_\ell \leq w/\varepsilon^2$; k_ℓ reflects the contribution of S_ℓ to the overall profit in some optimal packing, in multiples of $\frac{\varepsilon^2 P}{w}$, i.e., $k_\ell \frac{\varepsilon P}{w} \leq P(S_\ell) \leq (k_\ell + 1)\frac{\varepsilon^2 P}{w}$. We seek a vector $\boldsymbol{k} = (k_1, \ldots, k_w)$ satisfying $\sum_{\ell=1}^{w} k_\ell \leq \frac{w}{\varepsilon^2}$. The number of such vectors is at most $\binom{w+w/\varepsilon^2 - 1}{w - 1}$, which is a constant. Given the contribution of S_ℓ, we select from S_ℓ the minimum number of items that provide this profit. That is, we order the sets of items in S_ℓ in non-increasing order by sizes, and select sets of items, starting from the largest set, until we get the desired profit. Note that (at most) one set in S_ℓ may be 'partially selected', i.e., we pack in the bins only some of the items in this set.

5.2   Packing the Items

In packing the selected items, we choose in each step a subset of the items (or, *block*) in some color. We distinguish between the sizes of the packed blocks, and the compartment requirements of the corresponding items. We say that a block is *large* if its size is at least εV; otherwise, it is *small*. Also, the compartment requirement of color k is *large* if $c_k \geq \varepsilon C$; otherwise it is *small*.

Items with Large Compartment Requirements We first pack blocks of items with large compartment requirements. Note that we can pack at most $1/\varepsilon$ such blocks in each bin. We increase the volume of the bins to $V(1 + \varepsilon)$, and pack in each bin blocks in the sizes $\varepsilon^2 V, \ldots, V$. Thus, we round up the sizes of small blocks to $\varepsilon^2 V$. After packing the items, we eliminate extra volume, until we get that the sum of packed items in each bin is at most V. By that we lose at most a factor of ε in the approximation ratio. To obtain a constant number of possible block sizes, we modify the input, so that no set of items is too large. This can harm the approximation ratio at most by factor of ε, as formalized in the next lemma.[5]

Lemma 2. *The input I can be transformed to I', which satisfies (i) the size of any set of items is at most V/ε; (ii) any packing of items in I' can be mapped*

---

[5] See also in [6]. We omit the proof.

to a packing of items in I of the same profit. (iii) $OPT(I') \geq (1 - \varepsilon)OPT(I)$, where $OPT(I)$ is the optimal profit from packing I.

We now describe how we pack the set of items with large compartment requirements. Partition the item sets to the groups $I^1, \ldots, I^h$; all of the item sets in I^r, $1 \leq r \leq h$, are of the same size. We call I^r the r-th *profit category*. Initially, we 'guess' the partition of item sets to blocks. Each partition gives the number of blocks in the sizes $\varepsilon^2 V, ..., V$, taken from the profit categories $\varepsilon^2, ..., V/\varepsilon$. Assuming that block sizes are in multiples of $\varepsilon^2 V$, and the sizes of the item sets are given as integral powers of $(1 + \varepsilon)$, we get that the number of coordinates in each partition vector is $O(\lg(1/\varepsilon^3)/\varepsilon^2)$, which is a constant.

Note that in the above partition we define only the sizes of the blocks taken from each profit category I^r, $1 \leq r \leq h$; however, as the item sets in I^r may be of different compartment requirements and different colors, for each of the blocks we need to decide also on the item set in I^r from which it is extracted. This can be done greedily, without harming the approximation ratio of the scheme. Specifically, given the block configuration for I^r, we sort the item sets in I^r in non-decreasing order by their compartment requirements. The blocks are sorted in non-decreasing order by sizes. We now extract the blocks in the list from the sets, starting from the first item set. Once all of the items in the first set are allocated to blocks, we proceed to allocate blocks from the second set, and so on. In this process we allocate 'small' blocks from sets that have 'small' compartment requirement. Clearly, this decreases the potential number of compartments required for packing the item blocks. Now, we have a set of item blocks of given colors that need to be packed in the bins.

Given a correct guess of the partition of the item sets to blocks, we may now assume that each block is packed in a distinct set of compartments in some bin. Hence, we now have an instance of the 2DVP with fixed number of distinct item sizes in each dimension. By defining a *bin configuration* to be the number of items in each size (in both dimensions) in a bin, and guessing the number of bins having each configuration, we can find the optimal packing of the blocks in the bins.

Large Blocks with Small Compartment Requirements In this step we pack large blocks of items whose compartment requirement is at most εC. Since at most $1/\varepsilon$ large blocks can be packed in a bin, we look at items in the profit categories $\varepsilon V, \ldots, V/\varepsilon$. We guess as before the blocks extracted from each of the item sets in the small compartment categories. Then, we can find in polynomial time an optimal packing of these blocks in the bins, using bin configurations.

Packing the Remaining Items Finally, we pack small blocks of items with small compartment requirements, by using linear programming and rounding the (fractional) solution. The program takes as input the set of small blocks generated from the remaining sets of items. We need to allocate for each block of items of color k a set of c_k compartments in some bin. This is done by reducing our problem to GAP and using a technique of [15] for solving the GAP problem (see the details in [14]).

We summarize our discussion in the following result.

Theorem 10. *The above scheme yields a $(1 + \varepsilon)$-approximation for VPSD instances with unit sizes and profits, and w compartment categories. The running time of the scheme is $O(N^{(2e)^{w/\varepsilon^2}} \cdot \frac{1}{\varepsilon^{2w}})$.*

Thus,, we have a PTAS for the data placement problem, with N identical disks and w distinct file sizes, with running time as given in Theorem 10.

Finally, our scheme can be extended to apply to VPSD instances in which the item sizes take at most t distinct sizes, where t is fixed.

6   Approximation Scheme for Fixed Number of Colors

For instances in which the distinct number of colors is fixed, we show that VPSD admits a PTAS, even if the items have arbitrary sizes and profits, and the bins are arbitrary. Our scheme builds on an approximation technique presented in [13] for CCMK. However, since the scheme in [13] uses heavily the fact that the number of compartments in each bin can be bounded by some constant $\leq M$ (since the compartment requirement of any color k is $c_k = 1$), we need to use a different approach. Our technique relies on a partition of the bins to $O(\log(N/\varepsilon))$ types, by rounding up the volumes, and by eliminating compartments, such that in the resulting instance all the bins of the same type have (almost) the same volume and the same number of compartments. This is done without harming the approximation ratio of the scheme.

Theorem 11. *There is a PTAS for VPSD instances in which M is a fixed constant.*

Acknowledgments. We thank Chandra Chekuri for valuable discussions and suggestions.

References

1. C. Chekuri and S. Khanna, On Multi-dimensional Packing Problems. In *Proc. of SODA*, 185–194, 1999.
2. C. Chekuri and S. Khanna. A PTAS for the multiple knapsack problem. In *Proc. of SODA*, 213–222, 2000.
3. A. M. Frieze and M.R.B. Clarke, Approximation Algorithms for the m-dimensional 0-1 knapsack problem: worst-case and probabilistic analyses. In *European J. of Operational Research*, 15(1):100–109, 1984.
4. M.R. Garey and D.S. Johnson. Strong NP-completeness results: Motivations, examples, and implications. *J. of the ACM*, 25:499–508, 1978.
5. G.V. Gens and E.V. Levner. Computational complexity of approximation algorithms for combinatorial problems. In *Proc. of the 8th Int. Symp. on Mathematical Foundations of Computer Science*, 292–300, 1979.
6. L. Golubchik, S. Khanna, S. Khuller, R. Thurimella, and A. Zhu. Approximation algorithms for data placement on parallel disks. In *Proc. of SODA*, 223–232, 2000.

7. S. Ghandeharizadeh and R.R. Muntz. Design and implementation of scalable continuous media servers. *Parallel Computing J.*, 24(1):91–122, 1998.
8. O.H. Ibarra and C.E. Kim, Fast Approximation for the Knapsack and the Sum of Subset Problems, *J. of the ACM*, 22(4), pages 463–468, 1975.
9. S. Kashyap and S. Khuller, Algorithms for Non-Uniform Size Data Placement on Parallel Disks. Submitted 2003
10. B. Korte, R. Schrader. On the existence of fast approximation schemes. In *O. Magasarian, R. Meyer, S. Robinson (eds.): Nonlinear Programming 4*, 415–437, Academic Press, 1981
11. S. Sahni. Approximate Algorithms for the 0/1 knapsack problem, *J. of the ACM*, 22, 115–124, 1975.
12. H. Shachnai and T. Tamir, On Two Class-Constrained Versions of the Multiple Knapsack Problem. *Algorithmica*, vol. 29, 442–467, 2001.
13. H. Shachnai and T. Tamir, Polynomial Time Approximation Schemes for Class-Constrained Packing Problems. *J. of Scheduling*, vol. 4(6), pp. 313–338, 2001.
14. H. Shachnai and T. Tamir, Approximation Schemes for Generalized 2-dimensional Vector Packing with Application to Data Placement.
    http://www.cs.technion.ac.il/~hadas/PUB/vpsd.ps.
15. D. S. Shmoys and E. Tardos, Scheduling unrelated machines with Costs. In *Proc. of SODA*, 1993.
16. G.J. Woeginger. There is no asymptotic PTAS for two-dimensional vector packing. *Information Processing Letters* 64(6): 293-297, 1997.
17. J.L. Wolf, P.S. Yu, and H. Shachnai. Disk load balancing for video-on-demand systems. *ACM Multimedia Systems J.*, 5:358–370, 1997.

An Improved Algorithm for Approximating the Radii of Point Sets[*]

Yinyu Ye[1] and Jiawei Zhang[2]

[1] Management Science and Engineering and, by courtesy, Electrical Engineering,
Stanford University, Stanford, CA 94305, USA.
yinyu-ye@stanford.edu
[2] Department of Management Science and Engineering, School of Engineering,
Stanford University, Stanford, CA 94305, USA.
jiazhang@stanford.edu

Abstract. We consider the problem of computing the outer-radii of point sets. In this problem, we are given integers n, d, k where $k \leq d$, and a set P of n points in R^d. The goal is to compute the *outer k-radius* of P, denoted by $R_k(P)$, which is the minimum, over all $(d-k)$-dimensional flats F, of $\max_{p \in P} d(p, F)$, where $d(p, F)$ is the Euclidean distance between the point p and flat F. Computing the radii of point sets is a fundamental problem in computational convexity with significantly many applications. The problem admits a polynomial time algorithm when the dimension d is constant [9]. Here we are interested in the general case when the dimension d is not fixed and can be as large as n, where the problem becomes NP-hard even for $k = 1$.
It has been known that $R_k(P)$ can be approximated in polynomial time by a factor of $(1+\varepsilon)$, for any $\varepsilon > 0$, when $d - k$ is a fixed constant [15,2]. A factor of $O(\sqrt{\log n})$ approximation for $R_1(P)$, the width of the point set P, is implied from the results of Nemirovskii et al. [19] and Nesterov [18]. The first approximation algorithm for general k has been proposed by Varadarajan, Venkatesh and Zhang [20]. Their algorithm is based on semidefinite programming relaxation and the Johnson-Lindenstrauss lemma, and it has a performance guarantee of $O(\sqrt{\log n \cdot \log d})$.
In this paper, we show that $R_k(P)$ can be approximated by a ratio of $O(\sqrt{\log n})$ for any $1 \leq k \leq d$ and thereby improve the ratio of [20] by a factor of $O(\sqrt{\log d})$ that could be as large as $O(\sqrt{\log n})$. This ratio also matches the previously best known ratio for approximating the special case $R_1(P)$, the width of point set P. Our algorithm is based on semidefinite programming relaxation with a new mixed deterministic and randomized rounding procedure.

1   Introduction

Computing the outer k-radius of a point set is a fundamental problem in computational convexity with applications in global optimization, data mining, statistics and clustering, and has received considerable attention in the computational geometry literature [12,13,15]. In this problem, we are given integers

---

[*] Research supported in part by NSF grant DMI-0231600.

S. Arora et al. (Eds.): APPROX 2003+RANDOM 2003, LNCS 2764, pp. 178–187, 2003.

n, d, k where $k \leq d$, and a set P of n points in R^d. The goal is to compute the *outer k-radius* of P, denoted by $R_k(P)$, which is the minimum, over all $(d-k)$-dimensional flats F, of $\max_{p \in P} d(p, F)$, where $d(p, F)$ is the Euclidean distance between the point p and flat F. A $(d-k)$-flat is simply an affine subspace of R^d of dimension k. (See Section 2 for a more precise definition of $R_k(P)$). Roughly speaking, the outer k-radius $R_k(P)$ measures how well the point set P can be approximated by an affine subspace of dimension $d - k$. A few special cases of $R_k(P)$ which received particular attentions includes: $R_1(P)$, the *width* of P; $R_d(P)$, the radius of the minimum enclosing ball of P; and $R_{d-1}(P)$, the radius of the minimum enclosing cylinder of P.

When the dimension d is a fixed constant, $R_k(P)$ can be computed exactly in polynomial time [9]. It is also known that $R_k(P)$ can be approximated by a factor of $(1 + \varepsilon)$, for any $\varepsilon > 0$, in $O(n + (\frac{1}{\varepsilon})^{O(dk)})$ time [1,14]. In this paper, we are interested in the general scenario when the dimensions k and d are not fixed and d can be as large as n.

When the dimensions k and d are part of the input, the complexity of computing $R_k(P)$ depends on whether $d - k$ is constant or not. It is well-known that the problem is polynomial time solvable when $d - k = 0$, i.e., the minimum enclosing ball of a set of points can be computed in polynomial time (Gritzmann and Klee [12]). To the best of our knowledge, whether the problem is NP-hard or not is still open when $d - k = 1$. However, Bădoiu et al. [2] show that $R_{d-1}(P)$ can be approximated in polynomial time by a factor of $(1 + \varepsilon)$, for any $\varepsilon > 0$. Har-Peled and Varadarajan [15,16] generalize the result and show that $R_k(P)$ can be approximated by a factor of $(1 + \varepsilon)$ for any $\varepsilon > 0$ when $d - k$ is constant.

More hardness results are known when $d - k$ becomes large or k becomes small. Bodlaender et al. [4] show that the problem is NP-hard when $k = 1$. This is true even for the case $n = 2d$ ([12]). Gritzmann and Klee [12] also show that it is NP-hard to compute $R_k(P)$ if $k \leq c \cdot d$, for any fixed $0 < c < 1$. These negative results are further improved by Brieden et al. [5] and Brieden [8], the latter of which has shown that it is NP-hard to approximate $R_1(P)$, the width of a point set, to within *any* constant factor. Furthermore, Varadarajan, Venkatesh and Zhang [20] show that there exists some constant $\delta > 0$ such that for any $0 < \varepsilon < 1$, there is no quasi-polynomial time algorithm that approximates $R_k(P)$ within $(\log n)^\delta$ for $k \leq d - d^\varepsilon$ unless NP $\subseteq$ DTIME $[2^{(\log m)^{O(1)}}]$.

On the positive side, the algorithms of Nemirovskii et al. [18] and Nesterov [19] imply that $R_1(P)$, the width of point set P, can be approximated within a factor of $O(\sqrt{\log n})$. Another algorithm for approximating the width of a point set is given by Brieden et al. [6,7] and their algorithm has a performance guarantee $\sqrt{d/\log d}$ that is measured in the dimension d. The first approximation algorithm for general k have been proposed by Varadarajan et al [20]. Their algorithm is based on semidefinite programming relaxation and the Johnson-Lindenstrauss lemma, and has a performance factor of $O(\sqrt{\log n \cdot \log d})$.

Above mentioned results on computing $R_k(P)$ would give us a projection that the problem is harder when $d - k$ becomes larger or k becomes smaller. However, the result of Varadarajan et al [20] does not confirm this trend, since we have already known that $R_1(P)$ can be approximated by a factor of $O(\sqrt{\log n})$ while, for general k, the ratio proved in [20] is $O(\sqrt{\log n \cdot \log d})$, which is greater than $O(\sqrt{\log n})$. Therefore, they have conjectured that the factor of $O(\sqrt{\log n})$ applies to general k as well.

The main result of the present paper is to show that $R_k(P)$ can indeed be approximated by the factor of $O(\sqrt{\log n})$ for all $1 \leq k \leq d$, and thereby proves their conjecture. Note that the new approximation ratio is reduced by a factor of $O(\sqrt{\log d})$, which could be as large as $O(\sqrt{\log n})$. Our algorithm is based on semidefinite programming relaxation with a mixed deterministic and randomized rounding procedure, in contrast to all other purely randomized rounding procedures used for semidefinite programming approximation.

2   Preliminaries and SDP Relaxation

Generally speaking, the problem of computing $R_k(P)$ can be formulated as a quadratic minimization problem. Semidfinite programming (SDP) problems, where the unknowns are represented by positive semidefinite matrices, have recently been developed for approximating such problems; see, for example, Goemans and Williamson [10]. In the case of $k = 1$, computing $R_1(P)$ corresponds to a SDP problem plus an additional requirement that the rank of the unknown matrix equals 1. Removing the rank requirement, the SDP problem becomes a relaxation of the original problem and polynomially solvable for any given accuracy. Once obtaining an optimal solution, say X, of the SDP relaxation, one would like to generate a rank-1 matrix, say $\hat{X} = yy^T$, from X, where y is a column vector and serves as a solution to the original problem. Such rank reduction is called "rounding", and many rounding procedures are proposed and almost all of them are randomized, see, for example, [3].

One particular procedure has been proposed by Nemirovskii et al. [18] which can be used for approximating $R_1(P)$. Their procedure is a simple randomized rounding that can described as follows: an optimal solution X of the SDP relaxation, whose rank could be as large as d, can be represented as (e.g., by eigenvector decomposition)

$$X = \lambda_1 v_1 v_1^T + \lambda_2 v_2 v_2^T + \cdots + \lambda_d v_d v_d^T.$$

Then one can generate a single vector y by taking a random linear combination of the vectors $\sqrt{\lambda_1} v_1, \sqrt{\lambda_2} v_2, \cdots, \sqrt{\lambda_d} v_d$ where the coefficients of the combination takes values of -1 or 1 uniformly and independently.

When $k \geq 2$, one needs to generate k rank-1 matrices from X, the optimal solution of the corresponding SDP relaxation, such that

$$\hat{X} = \sum_{i=1}^{k} y_i y_i^T$$

where y_is are orthogonal to each other. The method of Varadarajan et al [20] first applies the Johnson-Lindenstrauss randomized dimension reduction technique [17] to reduce the rank of solution X to $k + O(\log n \cdot \log d)$ without losing much in the quality of the solution in terms of the objective value. Then they show that among the $k + O(\log n \cdot \log d)$ vectors, which are orthogonal to each other, k vectors can be randomly chosen as the solution with an approximate ratio $O(\sqrt{\log n \cdot \log d})$.

Our algorithm is based on the same SDP relaxation developed in [20]. However, our rounding procedure is different. Our procedure works as follows. Once obtaining an optimal solution for the SDP relaxation with

$$X = \lambda_1 v_1 v_1^T + \lambda_2 v_2 v_2^T + \cdots + \lambda_d v_d v_d^T,$$

we deterministically partition the vectors $v_1, v_2, \cdots, v_d$ into k groups where group j may contain n_j vectors and each group can be seen as a single semidefinite matrix with rank n_j. We can then generate one vector from each group using the randomized rounding procedure similar to that of Nemirovskii et al. [18]. The k vectors generated by this rounding procedure will automatically satisfy the condition that any pair of them must be orthogonal to each other, and the quality of these vectors have an approximation ratio no more than $O(\sqrt{\log n})$.

We now present the quadratic program formulation of the k-radii problem and its semidefinite programming relaxation. It will be helpful to first introduce some notations that will be used later. The trace of a given matrix A, denoted by $\text{Tr}(A)$, is the sum of the entries on the main diagonal of A. We use I to denote the identity matrix whose dimension will be clear in the context. The inner product of two vector p and q is denoted by $\langle p, q \rangle$. The 2-norm of a vector x, denoted by $\|x\|$, is defined by $\sqrt{\langle x, x \rangle}$. A positive semidefinite matrix X are represented by $X \succeq 0$. For simplicity, we assume that the P is symmetric in the sense that if $p \in P$ then $-p \in P$. Denote the set $\{-p | p \in P\}$ by $-P$. Let $Q = P \cup -P$. It is clear that $R_k(P) \leq R_k(Q) \leq 2R_k(P)$. Therefore, if we found a good approximation for $R_k(Q)$ then it must also be a good approximation for $R_k(P)$. Furthermore, since Q is a symmetric point set, the best $(d-k)$-flat for Q contains the origin so that it is a subspace.

Thus, the square of $R_k(P)$ can be defined by the optimal value of the following quadratic minimization problem:

$$
\begin{aligned}
R_k(P)^2 := \text{Minimize} \quad & \alpha \\
\text{Subject to} \quad & \sum_{i=1}^{k} \langle p, x_i \rangle^2 \leq \alpha, \ \forall p \in P, \\
& \|x_i\|^2 = 1, \ i = 1, ..., k, \\
& \langle x_i, x_j \rangle = 0, \ \forall i \neq j.
\end{aligned}
\tag{1}
$$

Assume that $x_1, x_2, \cdots, x_k \in R^d$ is the optimal solution of (1). Then one can easily verify that the matrix $X = x_1 x_1^T + x_2 x_2^T + \cdots + x_k x_k^T$ is a feasible solution for the following semidefinite program:

$$
\begin{aligned}
\alpha_k^* := \text{Minimize} \quad & \alpha \\
\text{Subject to} \quad & \text{Tr}(pp^T X) \ (= p^T X p) \leq \alpha, \ \forall p \in P, \\
& \text{Tr}(X) = k, \\
& I - X \succeq 0, \ X \succeq 0.
\end{aligned}
\tag{2}
$$

It follows that $\alpha_k^* \leq R_k(P)^2$. The following Lemma, which is proved in Varadarajan *et al* [20], follows from the above observations. We reproduce the proof below for completeness.

Lemma 1 *There exists an integer $r \geq k$ such that we can compute, in polynomial time, r nonnegative reals $\lambda_1, \lambda_2, \cdots, \lambda_r$ and r orthogonal unit vectors $v_1, v_2, \cdots, v_r$ such that*

(i). $\sum_{i=1}^r \lambda_i = k$.
(ii). $\max_{1 \leq i \leq r} \lambda_i \leq 1$.
(iii). $\sum_{i=1}^r \lambda_i \langle p, v_i \rangle^2 \leq R_k(P)^2$, *for any $p \in P$*.

Proof. We solve the semidefinte program (2), and let X^* be an optimal solution of (2). We claim that the rank of X^*, say r, is at least k. This follows from the fact that $\text{Tr}(X^*) = k$ and $I - X^* \succeq 0$. In other words, $\text{Tr}(X^*) = k$ implies that the sum of the eigenvalues of X^* is equal to k, and $I - X^* \succeq 0$ implies that the all eigenvalues are less than or equal to 1. Therefore, X^* has at least k non-zero eigenvalues, which implies that the rank of X^* is at least k. Let $\lambda_1, \lambda_2, \cdots, \lambda_r$ be the r nonnegative eigenvalues and $v_1, v_2, \cdots, v_r$ be the corresponding eigenvectors. Then we have $\sum_{i=1}^r \lambda_i = k$ and $\max_{1 \leq i \leq r} \lambda_i \leq 1$. Furthermore, for any $p \in P$,

$$\sum_{i=1}^r \lambda_i \langle p, v_i \rangle^2 = \text{Tr}(pp^T \sum_{i=1}^r \lambda_i v_i v_i^T) = \text{Tr}(pp^T X^*) \leq \alpha_k^* \leq R_k(P)^2.$$

$\square$

3   Deterministic First Rounding

In this section, we prove a lemma concerning how to deterministically group the eigenvalues and their eigenvectors. The proof of the lemma is elementary but it plays an important role for proving our main result.

Lemma 2 *The index set $\{1, 2, \cdots, r\}$ can be partitioned into k sets $I_1, I_2, \cdots, I_k$ such that*

(i). $\cup_{i=1}^k I_i = \{1, 2, \cdots, r\}$, *and for any $i \neq j$, $I_i \cap I_j = \emptyset$*.
(ii). *For any $i : 1 \leq i \leq k$, $\sum_{j \in I_i} \lambda_j \geq \frac{1}{2}$*.

Proof. Recall that $\sum_{j=1}^r \lambda_j = k$ and $0 \leq \lambda_j \leq 1$ for all j. Without loss of generality, we can assume that $\lambda_1 \geq \lambda_2 \geq \cdots \geq \lambda_r$. Our partitioning algorithm is the same as the Longest-Processing-Time heuristic algorithm for parallel machine scheduling problem. The algorithm works as follows:

STEP 1. For $i = 1, 2, \cdots, k$, set $I_i = \emptyset$ and let $L_i = 0$. Let $I = \{1, 2, \cdots, r\}$.
STPE 2. While $I \neq \emptyset$
        choose j from I with the smallest index;
        choose set i with the smallest value L_i;
        Let $I_i := I_i \cup \{j\}$, $L_i := L_i + \lambda_j$ and $I := I - \{j\}$.

It is clear that when the algorithm stops, the sets $I_1, I_2, \cdots, I_k$ satisfy condition (i). Now we prove condition (2) by contradiction. Assume that there exists some t such that $\sum_{j \in I_t} \lambda_j < \frac{1}{2}$.

We now claim that, for all i, $\sum_{j \in I_i} \lambda_j \leq 1$. Otherwise, suppose $\sum_{j \in I_{t'}} \lambda_j > 1$ for some t'. Note that $\lambda_j \leq 1$ for every j and thus there are at least two eigenvalues are assigned to $I_{t'}$. Denote the last eigenvalue by $\lambda_{s'}$. It follows that $\sum_{j \in I_{t'}} \lambda_j - \lambda_{s'} = \sum_{j \in I_{t'} \setminus \{s'\}} \lambda_j \leq \sum_{j \in I_t} \lambda_j$ since, otherwise, we would have not assigned $\lambda_{s'}$ to $I_{t'}$ in the algorithm. However, since $\sum_{j \in I_t} \lambda_j < \frac{1}{2}$, we must have $\sum_{j \in I_{t'}} \lambda_j - \lambda_{s'} = \sum_{j \in I_{t'} \setminus \{s'\}} \lambda_j < \frac{1}{2}$. Thus, $\lambda_{s'} > \sum_{j \in I_{t'}} \lambda_j - \frac{1}{2} > \frac{1}{2}$. This is impossible since $\lambda_{s'}$ is the last eigenvalue assigned to $I_{t'}$, which implies $\lambda_{s'} \leq \lambda_j$ for every $j \in I_{t'}$, and we have already proved that there must exist an l such that $s' \neq l \in I_{t'}$ and $\lambda_l \leq \sum_{j \in I_{t'} \setminus \{s'\}} \lambda_j < \frac{1}{2}$. Therefore, $\sum_{j \in I_i} \lambda_j \leq 1$ for all i, and in particular $\sum_{j \in I_t} \lambda_j < \frac{1}{2}$. It follows that $\sum_{i=1}^{k} \sum_{j \in I_i} \lambda_j < k$. However, we know that, by condition (i), $\sum_{i=1}^{k} \sum_{j \in I_i} \lambda_j = \sum_{j=1}^{r} \lambda_j = k$. This results a contradiction. Therefore, such t does not exists and we have proved condition (ii). □

Notice that the running time of the partitioning algorithm is bounded by $O(r \cdot k)$. An alternative way of partitioning the eigenvalues is the following: First, put the eigenvalues that are greater than or equal to $1/2$ into distinct subsets. If the number of such eigenvalues, say l, is not less than k, then we are done. Otherwise, arbitrarily put the remaining eigenvalues into $k - l$ subsets such that the sum of eigenvalues in each subset is greater than or equal to $1/2$. This method is suggested by an anonymous referee.

4   Randomized Second Rounding

Assume now that we have found $I_1, I_2, \cdots, I_k$. Then our next randomized rounding procedure works as follows.

STEP 1. Generate a r dimensional random vector ϕ such that each entry of ϕ takes value, independently, -1 or 1 with probability $\frac{1}{2}$ each way.

STEP 2. For $i = 1, 2, \cdots, k$, let

$$x_i = \frac{\sum_{j \in I_i} \phi_j \sqrt{\lambda_j} \cdot v_j}{\sqrt{\sum_{j \in I_i} \lambda_j}}.$$

The following Lemmas show that $x_1, x_2, \cdots, x_k$ form a feasible solution for the original problem. In other words, they are k orthogonal unit vectors.

Lemma 3 *For $i = 1, 2, \cdots, k$, $\|x_i\| = 1$.*

Proof. Recall that $\langle v_l, v_j \rangle = 0$ for any $l \neq j$ and $\|v_j\| = 1$. By definition,

$$
\begin{aligned}
\|x_i\|^2 &= \langle x_i, x_i \rangle \\
&= \left\langle \frac{\sum_{j \in I_i} \phi_j \sqrt{\lambda_j} v_j}{\sqrt{\sum_{j \in I_i} \lambda_j}}, \frac{\sum_{j \in I_i} \phi_j \sqrt{\lambda_j} v_j}{\sqrt{\sum_{j \in I_i} \lambda_j}} \right\rangle \\
&= \frac{1}{\sum_{j \in I_i} \lambda_j} \left\langle \sum_{j \in I_i} \phi_j \sqrt{\lambda_j} v_j, \sum_{j \in I_i} \phi_j \sqrt{\lambda_j} v_j \right\rangle \\
&= \frac{1}{\sum_{j \in I_i} \lambda_j} \sum_{j \in I_i} \langle \phi_j \sqrt{\lambda_j} v_j, \phi_j \sqrt{\lambda_j} v_j \rangle \\
&= \frac{1}{\sum_{j \in I_i} \lambda_j} \sum_{j \in I_i} \|\phi_j \sqrt{\lambda_j} v_j\|^2 \\
&= \frac{1}{\sum_{j \in I_i} \lambda_j} \sum_{j \in I_i} (\phi_j)^2 \lambda_j \|v_j\|^2 \\
&= \frac{1}{\sum_{j \in I_i} \lambda_j} \sum_{j \in I_i} \lambda_j \\
&= 1
\end{aligned}
$$

$\square$

Lemma 4 *If $s \neq t$ then $\langle x_s, x_t \rangle = 0$.*

Proof. The proof is similar as that of Lemma 3.

$$
\begin{aligned}
&\langle x_s, x_t \rangle \\
&= \left\langle \frac{\sum_{j \in I_s} \phi_j \sqrt{\lambda_j} v_j}{\sqrt{\sum_{j \in I_s} \lambda_j}}, \frac{\sum_{j \in I_t} \phi_j \sqrt{\lambda_j} v_j}{\sqrt{\sum_{j \in I_t} \lambda_j}} \right\rangle \\
&= \frac{1}{\sqrt{\sum_{j \in I_s} \lambda_j \cdot \sum_{j \in I_t} \lambda_j}} \left\langle \sum_{j \in I_s} \phi_j \sqrt{\lambda_j} v_j, \sum_{j \in I_t} \phi_j \sqrt{\lambda_j} v_j \right\rangle \\
&= 0.
\end{aligned}
$$

The last equality holds since for any $j \in I_s$ and $l \in I_t$, $\langle v_j, v_l \rangle = 0$. $\square$

Now we establish a bound on the performance of our algorithm. First, let us introduce Bernstein's Theorem (see, e.g., [18]), which is a form of the Chernoff Bound.

Lemma 5 *Let ϕ be a random vector whose entries are independent and either 1 or -1 with probability .5 each way. Then, for any vector e and $\beta > 0$,*

$$\text{prob}\{\langle \phi, e \rangle^2 > \beta \|e\|^2\} < 2 \cdot \exp(-\frac{\beta}{2}).$$

Let $C_{ip} = \sum_{j \in I_i} \lambda_j \langle p, v_j \rangle^2$. Then we have

Lemma 6 *For each $i = 1, 2, \cdots, k$ and each $p \in P$, we have*

$$\text{prob}\{\langle p, x_i \rangle^2 > 12 \log(n) \cdot C_{ip}\} < \frac{2}{n^3}.$$

Proof. Given i and p, define a $|I_i|$ dimensional vector e such that its entries are $\sqrt{\lambda_j} \langle p, v_j \rangle$, $j \in I_i$, respectively. Furthermore, we define the vector $\phi|_{I_i}$ whose entries are those of ϕ with indices in I_i. First notice that

$$\|e\|^2 = \sum_{j \in I_i} (\sqrt{\lambda_j} \langle p, v_j \rangle)^2 = \sum_{j \in I_i} \lambda_j \cdot \langle p, v_j \rangle^2 = C_{ip}.$$

On the other hand,

$$
\langle p, x_i \rangle^2
$$

$$
= \left\langle p, \frac{\sum_{j \in I_i} \sqrt{\lambda_j} v_j \phi_j}{\sqrt{\sum_{j \in I_i} \lambda_j}} \right\rangle^2
$$

$$
= \frac{1}{\sum_{j \in I_i} \lambda_j} \left\langle p, \sum_{j \in I_i} \sqrt{\lambda_j} v_j \phi_j \right\rangle^2
$$

$$
\leq 2 \left\langle p, \sum_{j \in I_i} \sqrt{\lambda_j} v_j \phi_j \right\rangle^2 \qquad (\text{since } \sum_{j \in I_i} \lambda_j \geq \frac{1}{2})
$$

$$
= 2 \left(\sum_{j \in I_i} \sqrt{\lambda_j} \phi_j \langle p, v_j \rangle \right)^2
$$

$$
= 2 \langle \phi|_{I_i}, e \rangle^2
$$

Thus

$$\text{prob}\{\langle p, x_i \rangle^2 > 12 \log(n) C_{ip}\} \leq \text{prob}\{\langle \phi|_{I_i}, e \rangle^2 > 6 \log(n) \|e\|^2\}.$$

Therefore, the conclusion of the lemma follows by using Lemma 5 and by letting $\beta = 6 \log(n)$. $\qquad \square$

Theorem 1 *We can computed in polynomial time, a $(d-k)$-flat such that, with probability at least $1 - \frac{2}{n}$, the distance between any point $p \in P$ and F is at most $\sqrt{12 \log(n)} \cdot R_k(P)$.*

Proof. For given $i = 1, 2, \cdots, k$ and $p \in P$, consider the event

$$B_{ip} = \{\phi | \langle p, x_i \rangle^2 > 12 \log(n) \cdot C_{ip}\}$$

and $B = \cup_{i,p} B_{ip}$. The probability that the event B happens is bounded by

$$\sum_{i,p} \text{prob}\{\langle p, x_i \rangle^2 > 12 \log(n) \cdot C_{ip}\} < \frac{2kn}{n^3} \leq \frac{2}{n}.$$

If B does not happen, then for any i and p,

$$\langle p, x_i \rangle^2 \leq 12 \log(n) \cdot C_{ip}.$$

Therefore, for each $p \in P$,

$$\sum_{i=1}^{k} \langle p, x_i \rangle^2 \leq 12 \log(n) \sum_{i=1}^{k} C_{ip} \leq 12 \log(n) \cdot R_k(P)^2.$$

The last inequality follows from Lemma 1. This completes the proof by taking F as the flat which is orthogonal to the vectors $x_1, x_2, \cdots, x_k$.   □

5   Final Remark

Finding efficient rounding methods for semidefinite programming relaxation plays a key role in constructing better approximation algorithms for various hard optimization problems. All of them developed to date are randomized in nature. Therefore, the mixed deterministic and randomized rounding procedure developed in this paper may have its own independent value. We expect to see more applications of the procedure in approximating various computational geometry and space embedding problems.

References

1. G. Barequet and S. Har-Peled, "Efficiently Approximating the Minimum-Volume Bounding Box of a Point Set in Three Dimensions," *J. Algorithms*, 38:91-109, 2001.
2. M. Bădoiu, S. Har-Peled and P. Indyk, "Approximate Clustering via Core-sets," In *Proc. ACM Symp. Theory of Computing*, 2002.
3. D. Bertsimas and Y. Ye, "Semidefinite relaxations, multivariate normal distributions, and order statistics," *Handbook of Combinatorial Optimization (Vol. 3)*, D.-Z. Du and P.M. Pardalos (Eds.) pp. 1-19, (1998 Kluwer Academic Publishers).
4. H.L. Bodlaender, P. Gritzmann, V. Klee and J. Van Leeuwen, "The Computational Complexity of Norm Maximization", *Combinatorica*, 10: 203-225, 1990.
5. A. Brieden, P. Gritzmann, and V. Klee, "Inapproximability of Some Geometric and Quadratic Optimization Problems," In P.M. Pardalos, editor, *Approximation and Complexity in Numerical Optimization: Continuous and Discrete Problems*, 96-115, Kluwer, 2000.

6. A. Brieden, P. Gritzmann, R. Kannan, V. Klee, L. Lovasz and M. Simonovits, "Deterministic and Randomized Polynomial-time Approximation of Radii," To appear in *Mathematika*.

7. A. Brieden, P. Gritzmann, R. Kannan, V. Klee, L. Lovasz and M. Simonovits, "Approximation of Diameters: Randomization Doesn't Help," In *Proc. IEEE Symp. Foundations of Comp. Sci.*, 244-251, 1998.

8. A. Brieden, "Geometric Optimization Problems Likely Not Contained in APX," *Discrete Comput. Geom.*, 28:201-209, 2002.

9. U. Faigle, W. Kern, and M. Streng, "Note on the Computaional Complexity of j-Radii of Polytopes in R^n," *Mathematical Programming*, 73:1-5, 1996.

10. M. X. Goemans and D.P. Williamson, "Improved Approximation Algorithms for Maximum Cut and Satisfiability Problems using Semi-definite Programming," *Journal of the ACM*, 42:1115-1145, 1995.

11. P. Gritzmann and V. Klee, "Inner and Outer j-Radii of Convex Bodies in Finite-Dimensional Normed Spaces," *Discrete Comput. Geom.*, 7:255-280, 1992.

12. P. Gritzmann and V. Klee, "Computational Complexity of Inner and Outer j-Radii of Polytopes in Finite-Dimenional Normed Spaces," *Math. Program.*, 59:162-213, 1993.

13. P. Gritzmann and V. Klee, "On the Complexity of Some basic Problems in Computational Convexity: I. Containment Problems," *Discrete Math.*, 136:129-174, 1994.

14. S. Har-Peled and K.R. Varadarajan, "Approximate Shape Fitting via Linearization," In *Proc. 42nd Annu. IEEE Sympos. Found. Comput. Sci.*, 66-73, 2001.

15. S. Har-Peled and K. Varadarajan, "Projective Clustering in High Dimensions Using Core-sets," In *Proc. ACM Symp. Comput. Geom.*, 2002.

16. S. Har-Peled and K. Varadarajan, "High-Dimensional Shap Fitting in Linear Time," In *Proc. ACM Symp. Comput. Geom.*, 2003.

17. W.B. Johnson and J. Lindenstrauss, "Extensions of Lipshitz Mapping into Hilbert Space," *Comtemporary Mathematics*, 26:189-206, 1984.

18. A. Nemirovskii, C. Roos and T. Terlaky, "On Maximization of Quadratic Forms Over Intersection of Ellipsoids with Common Center," *Math. Program.*, 86:463-473, 1999.

19. Yu. Nesterov, "Global Quadratic Optimization via Conic Relaxation," In eds H. Wolkowicz, R. Saigal and L. Vandenberghe, "Handbook of Semidefinite Programming Theory, Algorithms, and Applications", Kluwer Academic Publishers, Norwell, MA 02061 USA, 2000.

20. K. Varadarajan, S. Venkatesh and J. Zhang, "On Approximating the Radii of Point Sets in High Dimensions," In *Proc. 43rd Annu. IEEE Sympos. Found. Comput. Sci.*, 2002.

Testing Low-Degree Polynomials over $GF(2)$

Noga Alon[1][*], Tali Kaufman[2][**], Michael Krivelevich[3][* * *], Simon Litsyn[4][†], and
Dana Ron[4][‡]

[1] Department of Mathematics, Tel Aviv University, Tel Aviv 69978, Israel
and Institute for Advanced Study, Princeton, NJ 08540, USA.
nogaa@post.tau.ac.il
[2] School of Computer Science, Tel Aviv University,Tel Aviv 69978 Israel.
kaufmant@post.tau.ac.il
[3] Department of Mathematics, Tel Aviv University, Tel Aviv 69978, Israel.
krivelev@post.tau.ac.il
[4] Department of Electrical Engineering-Systems, Tel Aviv University, Tel Aviv 69978, Israel.
litsyn@eng.tau.ac.il,danar@eng.tau.ac.il

Abstract. We describe an efficient randomized algorithm to test if a given *binary* function $f : \{0,1\}^n \to \{0,1\}$ is a low-degree polynomial (that is, a sum of low-degree monomials). For a given integer $k \geq 1$ and a given real $\epsilon > 0$, the algorithm queries f at $O(\frac{1}{\epsilon} + k4^k)$ points. If f is a polynomial of degree at most k, the algorithm always accepts, and if the value of f has to be modified on at least an ϵ fraction of all inputs in order to transform it to such a polynomial, then the algorithm rejects with probability at least $2/3$. Our result is essentially tight: Any algorithm for testing degree-k polynomials over $GF(2)$ must perform $\Omega(\frac{1}{\epsilon} + 2^k)$ queries.

1 Introduction

In this work we consider the problem of testing whether a binary function $f : \{0,1\}^n \to \{0,1\}$ is a polynomial of degree at most k satisfying $f(0,\ldots,0) = 0$, for a given integer parameter k. Such a polynomial is simply a sum (modulo 2) of monomials each being a product of at most k variables, with the free term equal to zero. (The restriction $f(0,\ldots,0) = 0$ is imposed mainly for historical reasons, to make our definition and result consistent with the previously treated case of linear functions $k = 1$. With minor changes our algorithm can be adapted to test the class of all polynomials of degree at

---

[*] Research supported in part by a USA Israeli BSF grant and by a grant from the Israel Science
Foundation
[**] This work is part of the author's Ph.D. thesis prepared at Tel Aviv University under the supervision of Prof. Noga Alon, and Prof. Michael Krivelevich.
[* * *] Research supported in part by a USA Israeli BSF grant and by a grant from the Israel Science
Foundation.
[†] Research supported in part by a USA Israeli BSF grant and by a grant from the Israel Science
Foundation.
[‡] Research supported by the Israel Science Foundation (grant number 32/00-1).

S. Arora et al. (Eds.): APPROX 2003+RANDOM 2003, LNCS 2764, pp. 188–199, 2003.

most k in n variables, without the restriction on the free term.) The algorithm is required to accept functions that are polynomials of degree at most k (vanishing at zero), and to reject, with probability at least $2/3$, functions that are *far* from any such polynomial. More precisely, the algorithm is given a distance parameter ϵ, and is required to reject (with probability at least $2/3$) any function whose value should be modified on more that an ϵ-fraction of the domain to become a degree-k polynomial f satisfying $f(0,\ldots,0) = 0$. To this end the algorithm can query the function f on inputs of its choice, where our goal is to minimize the query complexity of the algorithm (as a function of k, $1/\epsilon$, and n).

The problem of testing multivariate low-degree polynomials has been studied quite extensively [4, 3, 13, 11, 17, 12, 2], and has important applications in the context of Probabilistically Checkable Proofs (PCP). However, with the exception of the case $k = 1$, that is, linear functions (which we discuss below), all results apply only to testing polynomials over fields that *are larger than k* (the degree bound). When the field F is sufficiently large, it is possible to reduce the problem of testing whether a function $f : F^n \rightarrow F$ is a multivariate degree-k polynomial to testing whether a function is a degree-k *univariate* polynomial, where the latter task is simply based on interpolation. Namely, the test for f selects random *lines* in F^n (more precisely, in the finite projective geometry $\mathrm{PG}(n-1, |F|)$), and verifies that the restriction of f to each of these lines is a (univariate) polynomial of degree at most k. This reduction does not hold for small fields, and in particular for $GF(2)$, which is our focus.

As noted above, in the case of $k = 1$ (linear functions), the linearity test of Blum, Luby and Rubinfeld [10] works also when the underlying field is $GF(2)$. In fact, our test can be viewed as an extension of the [10] algorithm, as we explain in more detail below. Linearity testing has also been studied in the following papers [4, 11, 6, 7, 5].

Our Results

We describe and analyze an algorithm that tests whether a function $f : \{0,1\}^n \rightarrow \{0,1\}$ is a degree-k polynomial satisfying $f(0,\ldots,0) = 0$, or is ϵ-far from any such polynomial, using $O(1/\epsilon + k \cdot 2^{2k})$ queries. As we show, the exponential dependency on k is unavoidable. This is in contrast to the case of testing degree-k polynomials over larger fields, where the sample complexity is polynomial in k. Our testing algorithm is simple. It repeats the following check $\Theta(\frac{1}{2^k \epsilon} + k2^k)$ times: It selects, uniformly and at random, $k + 1$ vectors $y_1, \ldots, y_{k+1} \in \{0,1\}^n$. It then evaluates f on the sum of every non-empty subset of the selected vectors, and checks that the sum of these evaluations is 0. If all checks succeed then it accepts, otherwise it rejects. Note that for the special case of $k = 1$, we obtain the linearity test of [10] which uniformly selects $O(1/\epsilon)$ pairs $y_1, y_2 \in \{0,1\}^n$, and verifies for each pair that $f(y_1) + f(y_2) = f(y_1 + y_2)$.

Our choice of the sets corresponds to a random selection of a $(k + 1)$-dimensional subspace in the affine geometry $\mathrm{AG}(n, 2)$ (see for example [14, Chap. 12]). In case $k = 1$ we deal with lines of the affine geometry $\mathrm{PG}(n, 2)$.

As a by-product of our analysis we obtain a *self-corrector* (as defined in [10]) for f, in case f is sufficiently close to a degree-k polynomial g. Specifically, for any given

$x \in \{0, 1\}^n$, it is possible to obtain the value $g(x)$ with high probability by querying f on additional, randomly selected, points.

Relation to Coding

Our setting and results have a very natural interpretation in terms of coding theory. The set of (evaluations of) all polynomials in n variables of degree at most k over $GF(2)$ is called the *Reed-Muller code* $\mathcal{R}(k, n)$ with parameters k and n. (See, e.g., [16] for relevant background). So our algorithm can be considered as (locally) testing Reed-Muller codes. To be more accurate, as we consider only polynomials f vanishing at zero, we in fact test the so-called *shortened Reed-Muller code* $\mathcal{R}(k, n)^*$, obtained from $\mathcal{R}(k, n)$ by choosing all codewords with the first bit (i.e. that corresponding to the zero vector) equal to zero, and deleting this bit. The Reed-Muller code $\mathcal{R}(k, n)$ is a linear code in $\{0, 1\}^{2^n}$ of minimum distance 2^{n-k}. The dual code of $\mathcal{R}(k, n)$ is the Reed-Muller code $\mathcal{R}(n - k - 1, n)$. The dual code of the shortened Reed-Muller code $\mathcal{R}(k, n)^*$ is the so called *punctured* Reed-Muller code with parameters $n - k - 1$ and n, obtained from $\mathcal{R}(n-k-1, n)$ by deleting the first bit of every codeword. The minimum distance of the punctured Reed-Muller code with parameters $n-k-1$ and n is $2^{k+1} - 1$, and its minimum weight codewords are obtained from the minimum weight codewords of $\mathcal{R}(n - k - 1, n)$, having the first bit equal to 1, by deleting this bit; the number of minimum weight vectors is proportional to $2^{(k+1)n}$.

For an arbitrary vector from $\{0, 1\}^{2^n}$ we want to distinguish between two cases: the vector belongs to the code, or, alternatively, it is at (Hamming) distance at least $\epsilon \cdot 2^n$ from the closest codeword of $\mathcal{R}(k, n)^*$. Our strategy is then to pick a random minimum weight vector from the punctured $\mathcal{R}(n - k - 1, n)$, and to check if it is orthogonal to the tested vector. Clearly, this will always confirm orthogonality if the considered vector is from the code. However, we prove that if the tested vector is far enough from the code, with positive probability the test will detect it, and give an estimate for this probability.

2   Preliminaries

For any integer ℓ, we denote by $[\ell]$ the set $\{1, \ldots, \ell\}$. For any $k \in [n]$, let $\mathcal{P}_k$ denote the family of all Boolean functions over $\{0, 1\}^n$ which are polynomials of degree at most k without a free term. That is, $f \in \mathcal{P}_k$ if and only if there exist coefficients $a_S \in \{0, 1\}$, for every $S \subseteq [n], 1 \leq |S| \leq k$, such that

$$f = \sum_{S \subseteq [n], |S| \in [k]} a_S \cdot \prod_{i \in S} x_i , \tag{1}$$

where the addition is in $GF(2)$. In particular, $\mathcal{P}_1$ is the family of all linear functions over $\{0, 1\}^n$, that is, all functions of the form $\sum_{i \in S} x_i$, where $S \subseteq [n]$.

For any two functions $f, g : \{0, 1\}^n \to \{0, 1\}$, the symmetric difference between f and g is $\Delta(f, g) \stackrel{\text{def}}{=} \{y \in \{0, 1\}^n : f(y) \neq g(y)\}$. The relative distance $\text{dist}(f, g) \in [0, 1]$ between f and g is: $\text{dist}(f, g) \stackrel{\text{def}}{=} |\Delta(f, g)|/2^n$. For a function g and a family of functions F, we say that g is ϵ-far from F, for some $0 < \epsilon < 1$, if, for every $f \in F$, $\text{dist}(g, f) > \epsilon$. Otherwise it is ϵ-close to F.

A testing algorithm (tester) for $\mathcal{P}_k$ is a probabilistic algorithm, that is given query access to a function f, and a distance parameter ϵ, $0 < \epsilon < 1$. If f belongs to $\mathcal{P}_k$ then with probability at least $\frac{2}{3}$, the tester should accept f, and if f is ϵ-far from $\mathcal{P}_k$, then with probability at least $\frac{2}{3}$ the tester should reject it. If the tester accepts every f in $\mathcal{P}_k$ with probability 1, then it is a one-sided tester.

The following notation will be used extensively in this paper. Given a function $f :$ $\{0,1\}^n \rightarrow \{0,1\}$, for $y_1, ..., y_\ell \in \{0,1\}^n$ let

$$T_f(y_1, \ldots, y_\ell) \overset{\text{def}}{=} \sum_{\emptyset \neq S \subseteq [\ell]} f\left(\sum_{i \in S} y_i\right), \tag{2}$$

where the first sum is over $GF(2)$ and the second one is over $(GF(2))^n$, and let

$$T_f^{y_1}(y_2, \ldots, y_\ell) \overset{\text{def}}{=} T_f(y_1, \ldots, y_\ell) + f(y_1). \tag{3}$$

3   Characterization of Low Degree Polynomials over $\{0,1\}^n$

Claim 1 *A function f belongs to $\mathcal{P}_k$ (i.e., it is a polynomial of total degree at most k satisfying $f(0,0,\ldots,0) = 0$), if and only if for every $y_1, \ldots, y_{k+1} \in \{0,1\}^n$ we have*

$$T_f(y_1, \ldots, y_{k+1}) = 0. \tag{4}$$

Proof. A polynomial from $\mathcal{P}_k$ can be viewed as a code word in the appropriate Reed-Muller code, see, e.g., [16]. Thus, the above characterization can be proved using known facts about its dual. For completeness we provide a direct, simple proof.

We first prove that if a function f belongs to $\mathcal{P}_k$ then $T_f(y_1, \ldots, y_{k+1}) = 0$ for every $y_1, \ldots, y_{k+1} \in \{0,1\}^n$.

As f is a sum of monomials of total degree at most k it suffices to show that for every monomial $m = \prod_{i \in I} x_i$, where $1 \leq |I| \leq k$, $T_m(y_1, \ldots, y_{k+1}) = 0$ for every $y_1, \ldots, y_{k+1} \in \{0,1\}^n$. The number of linear combinations $\sum_{j=1}^{k+1} b_j y_j$, where $b_j \in \{0,1\}$, for which $m(\sum_{j=1}^{k+1} b_j y_j) = 1$ is clearly the number of solutions of a linear system of $|I|$ equations in the $k+1$ variables b_j, and the trivial combination $b_j = 0$ for all j is not one of the solutions. Therefore, this number of solutions (which is possibly zero) is divisible by $2^{k+1-|I|}$, showing that there is an even number of sets S satisfying $\emptyset \neq S \subset [k+1]$ such that $m(\sum_{i \in S} y_i) = 1$. This implies that $T_m(y_1, \ldots, y_{k+1}) = 0$, as needed.

We next show that if $f = f(x_1, x_2, \ldots, x_n) : \{0,1\}^n \mapsto \{0,1\}$ satisfies Equation (4) for every $y_1, y_2, \ldots, y_{k+1} \in \{0,1\}^n$, then $f \in \mathcal{P}_k$. Every function from $\{0,1\}^n$ to $\{0,1\}$ can be written uniquely as a polynomial over $GF(2)$:

$$f = \sum_{I \subset [n]} a_I \prod_{i \in I} x_i.$$

Our objective is to show that $a_{\emptyset} = 0$ and that $a_I = 0$ for all $|I| > k$. Taking $y_j = (0, 0, \ldots, 0)$ for every j we conclude, by (4), that $a_{\emptyset} = 0$. Suppose, now, that there is a nonzero a_I with $|I| > k$. Take such an I of minimum cardinality, and assume, without loss of generality, that $I = [s]$ with $s \geq k + 1$.

Let e_i denote the i-th unit vector in $\{0, 1\}^n$, and define $y_1 = e_1, y_2 = e_2, \ldots, y_k = e_k$ and $y_{k+1} = e_{k+1} + \ldots + e_s$. Then the monomial $m = a_I \prod_{i \in I} x_i$ does not vanish on $\sum_{i=1}^{k+1} y_i$ and does vanish on $\sum_{i \in S} y_i$ for every $\emptyset \neq S \neq [k + 1]$. Thus $T_m(y_1, \ldots, y_{k+1}) \neq 0$. On the other hand, for any other monomial, say, $m' = \prod_{i \in I'} x_i$ with a nonzero coefficient in the representation of f, $T_{m'}(y_1, \ldots, y_{k+1}) = 0$. Indeed, if $|I'| \leq k$ this holds by the first part of the proof. Otherwise, by the minimality of I, $m'(\sum_{i \in S} y_i) = 0$ for all $S \subset [k + 1]$. Altogether this implies that $T_f(y_1, y_2, \ldots, y_{k+1}) = 1$, contradicting the assumption. This completes the proof of Claim 1.

4   A One-Sided Tester for Low Degree Polynomials over $\{0, 1\}^n$

In this section we present and analyze a one-sided tester for $\mathcal{P}_k$. This tester generalizes the linearity tester of Blum, Luby and Rubinfeld [10].

Algorithm Test-$\mathcal{P}_k$

1. Uniformly and independently select $\Theta(\frac{1}{2^k \epsilon} + k 2^k)$ groups of vectors. Each group contains $k + 1$ uniformly selected random vectors $y_1, \ldots, y_{k+1} \in \{0, 1\}^n$.
2. If for some group of vectors $y_1, \ldots, y_{k+1}$ it holds that $T_f(y_1, \ldots, y_{k+1}) \neq 0$, then reject, otherwise, accept.

Theorem 1 *The algorithm* **Test-$\mathcal{P}_k$** *is a one-sided tester for* $\mathcal{P}_k$ *with query complexity* $\Theta(\frac{1}{\epsilon} + k 2^{2k})$.

From the test definition and from Claim 1 it is obvious that if $f \in \mathcal{P}_k$, then the tester accepts. Thus, the crux of the proof is to show that if f is ϵ-far from $\mathcal{P}_k$, then the tester rejects with probability at least $2/3$. Our proof has a similar general structure to Sudan's analysis [18] of the linearity test in [10], but requires some additional ideas. In particular, if f is the function tested, we can define a function g as follows. For any $y \in \{0, 1\}^n$:

$$g(y) = 1 \text{ if } \Pr_{y_2, \ldots, y_{k+1} \in \{0,1\}^n} [T_f^y(y_2, \ldots, y_{k+1}) = 1] \geq 1/2 \text{ and } g(y) = 0 \text{ otherwise.} \tag{5}$$

Thus g is a kind of *majority* function. That is, for every vector $y \in \{0, 1\}^n$, $g(y)$ is chosen to satisfy most of the equations $T_f^y(y_2, \ldots, y_{k+1}) = g(y)$. We also define

$$\eta \overset{\text{def}}{=} \Pr_{y_1, \ldots, y_{k+1} \in \{0,1\}^n} [T_f(y_1, \ldots, y_{k+1}) \neq 0]$$
$$= \Pr_{y_1, \ldots, y_{k+1} \in \{0,1\}^n} [T_f^{y_1}(y_2, \ldots, y_{k+1}) \neq f(y_1)]. \tag{6}$$

Note that η is simply the probability that a single group of vectors $y_1, \ldots, y_{k+1}$ selected by the algorithm provides evidence that $f \notin \mathcal{P}_k$. We shall prove two claims. The first, and simpler claim (in Lemma 2), is that if η is small, then g is close to f. The second and more involved claim (in Lemma 5) is that if η is small, then g must belong to $\mathcal{P}_k$. This would suffice for proving the correctness of a slight variation on our algorithm that uses a larger sample size. In order to attain the sample complexity claimed in Theorem 1, we shall need to prove one more claim that deals with the case in which η is very small (see Lemma 6).

Lemma 2 *For a fixed function f, let g and η be as defined in Equations (5) and (6), respectively. Then,* $\mathrm{dist}(f,g) \leq 2\eta$.

Proof. Recall that for every $y \in \{0,1\}^n$, $\mathrm{Pr}_{y_2,\dots,y_{k+1}\in\{0,1\}^n}[T_f^y(y_2,\dots,y_{k+1}) = g(y)] \geq 1/2$. Hence

$$\eta = \mathrm{Pr}_{y,\,y_2,\dots,y_{k+1}\in\{0,1\}^n}[T_f^y(y_2,\dots,y_{k+1}) \neq f(y)]$$

$$= \frac{1}{2^n} \sum_{y\in\{0,1\}^n} \mathrm{Pr}_{y_2,\dots,y_{k+1}\in\{0,1\}^n}[T_f^y(y_2,\dots,y_{k+1}) \neq f(y)]$$

$$\geq \frac{1}{2^n} \sum_{y\in\Delta(f,g)} \mathrm{Pr}_{y_2,\dots,y_{k+1}\in\{0,1\}^n}[T_f^y(y_2,\dots,y_{k+1}) = g(y)]$$

$$\geq \frac{1}{2^n} \cdot |\Delta(f,g)| \cdot \frac{1}{2}$$

Thus, $\mathrm{dist}(f,g) = \frac{|\Delta(f,g)|}{2^n} \leq 2\eta$.

Recall that by the definition of g as a majority function, for every y, we have that for at least one half of the k-tuples of vectors $y_2,\dots,y_{k+1}$, $T_f^y(y_2,\dots,y_{k+1}) = g(y)$. In the next lemma we show that this equality actually holds for a vast majority of the k-tuples $y_2,\dots,y_{k+1}$ (assuming η is sufficiently small).

Lemma 3 *For every $y \in \{0,1\}^n$:* $\mathrm{Pr}_{y_2,\dots,y_{k+1}\in\{0,1\}^n}[g(y) = T_f^y(y_2,\dots,y_{k+1})] \geq 1 - 2k\eta$.

In order to prove Lemma 3 we shall first establish the following claim.

Claim 4 *For every $y, z, w, y_2, \dots, y_k \in \{0,1\}^n$,*

$$T_f(y, y_2, \dots, y_k, w) + T_f(y, y_2, \dots, y_k, z)$$
$$= T_f(y + w, y_2, \dots, y_k, y + w + z) + T_f(y + z, y_2, \dots, y_k, y + w + z) \quad (7)$$

Proof. Let $Y = \{y_2, \dots, y_k\}$, and consider any set $I \subseteq \{2, \dots, k\}$, which may be the empty set. For a vector $x \in \{0,1\}^n$ denote $f_{Y,I}(x) \stackrel{\mathrm{def}}{=} f(\sum_{i\in I} y_i + x)$.
For every set $I \subseteq \{2, \dots, k\}$, each element of type $f(\sum_{i\in I} y_i)$ appears twice in both sides of Equation (7) and thus cancels out. Now for every set $I \subset \{2, \dots, k\}$ (including the empty set), we get in the left hand side of Equation (7):

$$f_{Y,I}(y) + f_{Y,I}(w) + f_{Y,I}(y + w) + f_{Y,I}(y) + f_{Y_I}(z) + f_{Y,I}(y + z) \,.$$

In the right hand side of Equation (7) we get:

$$f_{Y,I}(y+w) + f_{Y,I}(y+z+w) + f_{Y,I}(z) + f_{Y,I}(y+z) + f_{Y,I}(y+w+z) + f_{Y,I}(w) .$$

This implies equality over $GF(2)$.

We now turn to prove Lemma 3.

Proof of Lemma 3:   We fix $y \in \{0,1\}^n$ and let $\gamma \overset{\text{def}}{=} \Pr_{y_2,\ldots,y_{k+1}\in\{0,1\}^n}[g(y) = T_f^y(y_2,\ldots,y_{k+1})]$. Recall that we are interested in proving that $\gamma \geq 1 - 2k\eta$. To this end, we shall bound a slightly different, but related probability. Let

$$\delta \overset{\text{def}}{=} \Pr_{y_2,\ldots,y_{k+1},z_2,\ldots,z_{k+1}\in\{0,1\}^n}[T_f^y(y_2,\ldots,y_{k+1}) = T_f^y(z_2,\ldots,z_{k+1})] . \quad (8)$$

Then, by the definitions of γ and δ,

$$\begin{aligned}
\delta &= \Pr[T_f^y(y_2,\ldots,y_{k+1}) = g(y) \text{ and } T_f^y(z_2,\ldots,z_{k+1}) = g(y)] \\
&\quad + \Pr[T_f^y(y_2,\ldots,y_{k+1}) \neq g(y) \text{ and } T_f^y(z_2,\ldots,z_{k+1}) \neq g(y)] \\
&= \gamma^2 + (1-\gamma)^2
\end{aligned} \quad (9)$$

where the probabilitites are over the choice of $y_2,\ldots,y_{k+1}, z_2,\ldots,z_{k+1} \in \{0,1\}^n$. Since we are working over $GF(2)$,

$$\delta = \Pr_{y_2,\ldots,y_{k+1},z_2,\ldots,z_{k+1}\in\{0,1\}^n}[T_f(y,y_2,\ldots,y_{k+1}) + T_f(y,z_2,\ldots,z_{k+1}) = 0] .$$

Now, for any choice of $y_2,\ldots,y_{k+1}$ and $z_2,\ldots,z_{k+1}$:

$$\begin{array}{lll}
T_f(y,y_2,\ldots,y_{k+1}) & + T_f(y,z_2,\ldots,z_{k+1}) & = \\
T_f(y,y_2,\ldots,y_{k+1}) & + T_f(y,y_2,\ldots,y_k,z_{k+1}) & + \\
T_f(y,y_2,\ldots,y_k,z_{k+1}) & + T_f(y,y_2,\ldots,y_{k-1},z_k,z_{k+1}) & + \\
T_f(y,y_2,\ldots,y_{k-1},z_k,z_{k+1}) & + T_f(y,y_2,\ldots,y_{k-2},z_{k-1},z_k,z_{k+1}) & + \\
\vdots & & \\
& + & \\
T_f(y,y_2,z_3,\ldots,z_{k+1}) & + T_f(y,z_2,\ldots,z_{k+1}). &
\end{array}$$

Consider any pair $T_f(y,y_2,\ldots,y_\ell,z_{\ell+1},\ldots,z_{k+1}) + T_f(y,y_2,\ldots,y_{\ell-1},z_\ell,\ldots,z_{k+1})$ that appears in the above sum. Note that $T_f(y,y_2,\ldots,y_\ell,z_{\ell+1},\ldots,z_{k+1})$ and $T_f(y,y_2,\ldots,y_{\ell-1},z_\ell,\ldots,z_{k+1})$ differ only in a single parameter. Since $T_f(\cdot)$ is a symmetric function we can apply Claim 4 and obtain that

$$\begin{aligned}
& T_f(y,y_2,\ldots,y_\ell,z_{\ell+1},\ldots,z_{k+1}) + T_f(y,y_2,\ldots,y_{\ell-1},z_\ell,\ldots,z_{k+1}) \\
&= T_f(y+y_\ell,y_2,\ldots,y_{\ell-1},z_{\ell+1},\ldots,z_{k+1},y+y_\ell+z_\ell) \\
&\quad + T_f(y+z_\ell,y_2,\ldots,y_{\ell-1},z_{\ell+1},\ldots,z_{k+1},y+y_\ell+z_\ell) \quad (10)
\end{aligned}$$

Recall that y is fixed and $y_2,\ldots,y_{k+1}, z_2,\ldots,z_{k+1} \in \{0,1\}^n$ are uniformly selected, and so all parameters on the right hand side in the above equation are uniformly distributed. Also recall that by the definition of η, for $T_f(r_1,\ldots,r_{k+1})$, where r_i are uniformly selected at random, $\Pr_{r_1,\ldots,r_{k+1}\in\{0,1\}^n}[T_f(r_1,\ldots,r_{k+1}) \neq 0] = \eta$. Hence, by the union bound:

$$\delta = \Pr_{y_2,\ldots,y_{k+1},z_2,\ldots,z_{k+1}\in\{0,1\}^n}[T_f(y,y_2,\ldots,y_{k+1}) + T_f(y,z_2,\ldots,z_{k+1}) = 0]$$
$$\geq 1 - 2k\eta. \tag{11}$$

By combining Equations (9) and (11) we get that $\gamma^2 + (1 - \gamma)^2 \geq 1 - 2k\eta$. Since $\gamma \geq 1/2$ it follows that $\gamma = \gamma^2 + \gamma(1 - \gamma) \geq \gamma^2 + (1 - \gamma)^2 \geq 1 - 2k\eta$. $\square$

Lemma 5 *If* $\eta < \frac{1}{(4k+2)2^k}$, *then the function g belongs to* $\mathcal{P}_k$.

Proof. By Claim 1 it suffices to prove that if $\eta < \frac{1}{(4k+2)2^k}$, then $T_g(y_1,\ldots,y_{k+1}) = 0$, for every $y_1,\ldots,y_{k+1} \in \{0,1\}^n$. Let us fix the choice of $y_1,\ldots,y_{k+1}$, and recall that as defined in Equation (2), $T_g(y_1,\ldots,y_{k+1}) = \sum_{\emptyset \neq I \subseteq [k+1]} g(\sum_{i \in I} y_i)$. Suppose we uniformly select $k \cdot (k + 1)$ random vectors $z_{i,j} \in \{0,1\}^n$, $1 \leq i \leq k + 1, 1 \leq j \leq k$. Then by Lemma 3, for every $I, \emptyset \neq I \subseteq [k + 1]$, with probability at least $1 - 2k\eta$ over the choice of the $z_{i,j}$'s,

$$g\left(\sum_{i\in I} y_i\right) = T_f\left(\sum_{i\in I} y_i, \sum_{i\in I} z_{i,1}, \sum_{i\in I} z_{i,2}, \ldots, \sum_{i\in I} z_{i,k}\right) + f\left(\sum_{i\in I} y_i\right). \tag{12}$$

Let E_1 be the event that Equation (12) holds for all $\emptyset \neq I \subseteq [k + 1]$. By the union bound:

$$\Pr[E_1] \geq 1 - (2^{k+1} - 1) \cdot 2k\eta \tag{13}$$

Assume that E_1 holds. Then

$$T_g(y_1,\ldots,y_{k+1})$$
$$= \sum_{\emptyset\neq I\subseteq[k+1]} \left[T_f\left(\sum_{i\in I} y_i, \sum_{i\in I} z_{i,1}, \sum_{i\in I} z_{i,2}, \ldots, \sum_{i\in I} z_{i,k}\right) + f\left(\sum_{i\in I} y_i\right)\right]$$
$$= \sum_{\emptyset\neq I\subseteq[k+1]} \sum_{\emptyset\neq J\subseteq[k]} \left[f\left(\sum_{i\in I}\sum_{j\in J} z_{i,j}\right) + f\left(\sum_{i\in I} y_i + \sum_{i\in I}\sum_{j\in J} z_{i,j}\right)\right]$$
$$= \sum_{\emptyset\neq J\subseteq[k]} \sum_{\emptyset\neq I\subseteq[k+1]} f\left(\sum_{i\in I}\sum_{j\in J} z_{i,j}\right)$$
$$+ \sum_{\emptyset\neq J\subseteq[k]} \sum_{\emptyset\neq I\subseteq[k+1]} f\left(\sum_{i\in I} y_i + \sum_{i\in I}\sum_{j\in J} z_{i,j}\right)$$
$$= \sum_{\emptyset\neq J\subseteq[k]} T_f\left(\sum_{j\in J} z_{1,j}, \ldots, \sum_{j\in J} z_{k+1,j}\right)$$
$$+ \sum_{\emptyset\neq J\subseteq[k]} T_f\left(y_1 + \sum_{j\in J} z_{1,j}, \ldots, y_{k+1} + \sum_{j\in J} z_{k+1,j}\right). \tag{14}$$

Let E_2 be the event that for every $\emptyset \neq J \subseteq [k]$, $T_f \left(\sum_{j \in J} z_{1,j}, \ldots, \sum_{j \in J} z_{k+1,j} \right) = 0$ and $T_f \left(y_1 + \sum_{j \in J} z_{1,j}, \ldots, y_{k+1} + \sum_{j \in J} z_{k+1,j} \right) = 0$. By the definition of η:

$$\Pr[E_2] \geq 1 - 2(2^k - 1)\eta \tag{15}$$

Suppose that $\eta < \frac{1}{(4k+2)2^k}$. Then, by Equations (13) and (15), the probability that both E_1 and E_2 hold, is strictly positive. In other words, there exists a choice of the $z_{i,j}$'s for which all summands in Equation (14) are 0. But this implies that $T_g(y_1, \ldots, y_{k+1}) = 0$. We conclude that if $\eta < \frac{1}{(4k+2)2^k}$, then g belongs to $\mathcal{P}_k$, and this completes the lemma's proof.

By combining Lemmas 2 and 5 we obtain that if f is $\Omega(1/(k2^k))$-far from $\mathcal{P}_k$, then $\eta = \Omega(1/(k2^k))$, and so the algorithm rejects f with sufficiently high constant probability (since it selects $\Omega(k2^k)$ groups of vectors $y_1, \ldots, y_{k+1}$). We next deal with the case in which η is small. By Lemma 2, in this case the distance $d = \text{dist}(f, g)$ between f and g is small, and we show that the test rejects f with probability that is close to $(2^{k+1} - 1)d$. This follows from the fact that in this case, the probability over the selection of $y_1, \ldots, y_{k+1}$, that among the $(2^{k+1} - 1)$ points $\sum_{\emptyset \neq I \subseteq [k+1]} y_i$, the functions f and g differ in precisely one point, is close to $(2^{k+1} - 1)d$. This is formally proved in the following lemma.

Lemma 6 *Suppose* $0 < \eta < \frac{1}{(4k+2)2^k}$. *Let* $d = \text{dist}(f, g)$ *denote the distance between* f *and* g, *and let*

$$p \stackrel{\text{def}}{=} \frac{1 - (2^{k+1} - 1)d}{1 + (2^{k+1} - 1)d} \cdot (2^{k+1} - 1)d.$$

Then, when $y_1, y_2, \ldots, y_{k+1}$ *are chosen randomly, the probability that for exactly one point* v *among the* $(2^{k+1} - 1)$ *points* $\sum_{i \in S} y_i$, $(\emptyset \neq S \subseteq [k + 1])$, $f(v) \neq g(v)$, *is at least* p.

By definition of η and the above lemma, $\eta \geq p$ (under the premise of the lemma). In particular, since (by Lemma 2) $d \leq 2\eta \leq \frac{1}{(2k+1)2^k}$ and $k \geq 1$, $\eta \geq \frac{1}{3}(2^{k+1} - 1)d$, and, for fixed k, as d tends to zero, $\eta \geq (2^{k+1} - 1)d - O(d^2)$.

Proof. For each subset S, $\emptyset \neq S \subseteq [k + 1]$, let X_S be the indicator random variable whose value is 1 if and only if $f(\sum_{i \in S} y_i) \neq g(\sum_{i \in S} y_i)$. Obviously, $\Pr[X_S = 1] = d$ for every S. It is not difficult to check that the random variables X_S are pairwise independent, since for any two distinct nonempty S_1, S_2, the sums $\sum_{i \in S_1} y_i$ and $\sum_{i \in S_2} y_i$ attain each pair of distinct values in $\{0, 1\}^n$ with equal probability when the vectors y_i are chosen randomly and independently. It follows that the random variable $X = \sum_S X_S$ which counts the number of points v of the required form in which $f(v) \neq g(v)$ has expectation $E[X] = (2^{k+1} - 1)d$ and variance $\text{Var}[X] = (2^{k+1} - 1)d(1 - d) \leq E[X]$. Our objective is to lower bound the probability that $X = 1$. We need the well known, simple fact that for a random variable X that attains nonnegative, integer values,

$$\Pr[X > 0] \geq \frac{(E[X])^2}{E[X^2]}.$$

Indeed, if X attains the value i with probability p_i for $i > 0$, then, by Cauchy-Schwartz,

$$(E[X])^2 = \left(\sum_{i>0} i p_i\right)^2 = \left(\sum_{i>0} i\sqrt{p_i}\sqrt{p_i}\right)^2 \leq \left(\sum_{i>0} i^2 p_i\right)\left(\sum_{i>0} p_i\right) = E[X^2]\Pr[X > 0].$$

In our case, this implies

$$\Pr[X > 0] \geq \frac{(E[X])^2}{E[X^2]} \geq \frac{(E[X])^2}{E[X] + (E[X])^2} = \frac{E[X]}{1 + E[X]}.$$

Therefore

$$E[X] \geq \Pr[X = 1] + \left(\frac{E[X]}{1 + E[X]} - \Pr[X = 1]\right) \cdot 2 = \frac{2E[X]}{1 + E[X]} - \Pr[X = 1],$$

implying that

$$\Pr[X = 1] \geq \frac{E[X] - (E[X])^2}{1 + E[X]}.$$

Substituting the value of $E[X]$, the desired result follows.

We are now ready to wrap-up the proof of Theorem 1.

Proof of Theorem 1: As we have noted previously, if f is in $\mathcal{P}_k$, then by Claim 1 the tester accepts (with probability 1). We next show that if f is ϵ-far from $\mathcal{P}_k$, then the tester rejects with probability at least $\frac{2}{3}$.

Suppose that $dist(f, \mathcal{P}_k) > \epsilon$. Denote $d = dist(f, g)$. If $\eta < \frac{1}{(4k+2)2^k}$ then by Lemma 5 $g \in \mathcal{P}_k$ and, by Lemma 6, $\eta \geq \Omega(2^k d) \geq \Omega(2^k \epsilon)$. Hence, $\eta \geq min\left(\Omega(2^k \epsilon), \frac{1}{(4k+2)2^k}\right)$. Clearly it is enough to perform $O(\frac{1}{\eta})$ rounds of the algorithm in order to detect a violation with probability at least $\frac{2}{3}$. This completes the proof of the theorem. $\square$

4.1   Self-correcting and a Lower Bound

From Lemmas 2, 3, and 5 one can immediately conclude the following:

Corollary 7 *Consider a function $f : \{0,1\}^n \to \{0,1\}$ that is ϵ-close to a degree-k polynomial $g : \{0,1\}^n \to \{0,1\}$, where $\epsilon < \frac{2}{(4k+2)2^k}$. Then the function f can be self-corrected. That is, for any given $x \in \{0,1\}^n$, it is possible to obtain the value $g(x)$ with probability at least $1 - \epsilon k$ by querying f on $2^k - 1$ points in $\{0,1\}^n$.*

The following is a lower bound on families of functions that correspond to linear codes.

Theorem 2 *Let $\mathcal{F}$ be any family of functions $f : \{0,1\}^n \to \{0,1\}$ that corresponds to a linear code $\mathcal{C}$. Let d denote the minimum distance of the code $\mathcal{C}$ and let $\bar{d}$ denote the minimum distance of the dual code of $\mathcal{C}$.*

Every testing algorithm for the family $\mathcal{F}$ must perform $\Omega(\bar{d})$ queries, and if the distance parameter ϵ is at most $d/2^{n+1}$, then $\Omega(1/\epsilon)$ is also a lower bound for the necessary number of queries.

As noted in the introduction, the family $\mathcal{P}_k$ corresponds to the shortened Reed-Muller code $\mathcal{R}(k,n)^*$. It is well known (see [16, Chap. 13]) that the distance of $\mathcal{R}(k,n)^*$ is 2^{n-k} and the distance of the dual code (which is a punctured Reed-Muller code) is $2^{k+1} - 1$. Hence we obtain the following corollary.

Corollary 8 *Every algorithm for testing* $\mathcal{P}_k$ *with distance parameter* ϵ *must perform* $\Omega\left(\max(\frac{1}{\epsilon}, 2^{k+1})\right)$ *queries.*

Proof of Theorem 2: We start with showing that $\Omega(\bar{d})$ queries are necessary. A well known fact from coding theory (see [16, Chap. 5]) states the following: for every linear code $\mathcal{C}$ whose dual code has distance $\bar{d}$, if we examine a sub-word having length d', $d' < \bar{d}$, of a uniformly selected codeword in $\mathcal{C}$, then the resulting sub-word is uniformly distributed in $\{0,1\}^{d'}$. Hence it is not possible to distinguish between a random codeword in $\mathcal{C}$ and a random word in 2^n (which with high probability is far from any codeword) using less than $\bar{d}$ queries.

We now turn to the case $\epsilon < d/2^{n+1}$. To prove the lower bound here, we apply, as usual, the Yao principle by defining two distributions, one of positive instances, and the other of negative ones, and then by showing that in order to distinguish between those distributions any algorithm must perform $\Omega(1/\epsilon)$ queries. The positive distribution has all its mass at the zero vector $\bar{0} = (0, \ldots, 0)$. To define the negative distribution, partition the set of all coordinates into $t = 1/\epsilon$ nearly equal parts $I_1, \ldots, I_t$ and give weight $1/t$ to each of the characteristic vectors w_i of I_i, $i = 1, \ldots, t$. (Observe that indeed $\bar{0} \in \mathcal{C}$ due to linearity, and $dist(w_i, \mathcal{C}) = \epsilon$ due to the assumption on the minimum distance of $\mathcal{C}$). Finally, a random instance is generated by first choosing one of the distributions with probability $1/2$, and then generating a vector according to the chosen distribution. It is easy to check (see, e.g., [1] for details) that in order to give a correct answer with probability at least $2/3$, the algorithm has to query $\Omega(1/\epsilon)$ bits of the input. $\square$

5   Concluding Remarks

We first note that in view of the above lower bound, our upper bound is almost tight.

It will be interesting to study analogous questions for other linear binary codes. Several recent papers, including [8], [9], deal with related questions. As shown above, a code is not testable with a constant number of queries if its dual distance is not a constant, and it seems plausible to conjecture that if the dual distance is a constant, and there is a doubly transitive permutation group acting on the coordinates that maps the dual code to itself, then the code can be testable with a constant number of queries. The automorphism group of punctured Reed-Muller codes contains the general linear group $GL(n,2)$, and thus those codes supply an example with these properties. Another interesting example is duals of BCH codes (this class also contains linear functions as a particular case). Another possible extension of the results could be the study of testability of low-degree multivariate polynomials over small fields $GF(q)$. This situation corresponds to generalized Reed-Muller codes [15].

References

[1] N. Alon, M. Krivelevich, I. Newman, and M. Szegedy. Regular languages are testable with a constant number of queries. In *Proceedings of the Fortieth Annual Symposium on Foundations of Computer Science*, pages 645–655, 1999.

[2] S. Arora and S. Safra. Improved low-degree testing and its applications. In *Proceedings of the Twenty-Ninth Annual ACM Symposium on the Theory of Computing*, pages 485–495, 1997.

[3] L. Babai, L. Fortnow, L. Levin, and M. Szegedy. Checking computations in polylogarithmic time. In *Proceedings of the Twenty-Third Annual ACM Symposium on Theory of Computing*, pages 21–31, 1991.

[4] L. Babai, L. Fortnow, and C. Lund. Non-deterministic exponential time has two-prover interactive protocols. *Computational Complexity*, 1(1):3–40, 1991.

[5] M. Bellare, D. Coppersmith, J. Håstad, M. Kiwi, and M. Sudan. Linearity testing in characteristic two. In *Proceedings of the Thirty-Sixth Annual Symposium on Foundations of Computer Science*, pages 432–441, 1995.

[6] M. Bellare, S. Goldwasser, C. Lund, and A. Russell. Efficient probabilistically checkable proofs and applications to approximation. In *Proceedings of the Twenty-Fifth Annual ACM Symposium on the Theory of Computing*, pages 294–304, 1993.

[7] M. Bellare and M. Sudan. Improved non-approximability results. In *Proceedings of the Twenty-Sixth Annual ACM Symposium on the Theory of Computing*, pages 184–193, 1994.

[8] E. Ben-Sasson, P. Harsha, and S. Raskhodnikova. 3CNF properties are hard to test. In *Proceedings of the Thirty-Fifth Annual ACM Symposium on the Theory of Computing*, 2003. To appear.

[9] E. Ben-Sasson, M. Sudan, S. Vadhan, and A. Wigderson. Derandomizing low degree tests via epsilon-biased spaces. In *Proceedings of the Thirty-Fifth Annual ACM Symposium on the Theory of Computing*, 2003. To appear.

[10] M. Blum, M. Luby, and R. Rubinfeld. Self-testing/correcting with applications to numerical problems. *Journal of Computer and System Sciences*, 47:549–595, 1993.

[11] U. Feige, S. Goldwasser, L. Lovász, S. Safra, and M. Szegedy. Approximating clique is almost NP-complete. *Journal of the Association for Computing Machinery*, pages 268–292, 1996.

[12] K. Friedl and M. Sudan. Some improvements to total degree tests. In *Proceedings of the 3rd Annual Israel Symposium on Theory of Computing and Systems*, pages 190–198, 1995. Corrected version available online at http://theory.lcs.mit.edu/~madhu/papers/friedl.ps.

[13] P. Gemmell, R. Lipton, R. Rubinfeld, M. Sudan, and A. Wigderson. Self-testing/correcting for polynomials and for approximate functions. In *Proceedings of the Twenty-Third Annual ACM Symposium on Theory of Computing*, pages 32–42, 1991.

[14] M. Hall. *Combinatorial Theory*. John Wiley & Sons, 1967.

[15] T. Kasami, S. Lin, and W.W. Peterson. New generalizations of the reed-muller codes, part i: Primitive codes. *IEEE Transactions on Information Theory*, pages 189–199, 1968.

[16] F. J. MacWilliams and N. J. A. Sloane. *The Theory of Error Correcting Codes*. North Holland, 1977.

[17] R. Rubinfeld and M. Sudan. Robust characterization of polynomials with applications to program testing. *SIAM Journal on Computing*, 25(2):252–271, 1996.

[18] M. Sudan. Private communications, 1995.

Computational Analogues of Entropy

Boaz Barak[1], Ronen Shaltiel[1], and Avi Wigderson[2]

[1] Department of Computer Science and Applied Mathematics,
Weizmann Institute of Science, Rehovot, Israel.
{boaz,ronens}@wisdom.weizmann.ac.il
[2] School of Mathematics, Institute for Advanced Study, Princeton, NJ and
Hebrew University, Jerusalem, Israel.
avi@ias.edu

Abstract. Min-entropy is a statistical measure of the amount of randomness that a particular distribution contains. In this paper we investigate the notion of *computational min-entropy* which is the computational analog of statistical min-entropy. We consider three possible definitions for this notion, and show equivalence and separation results for these definitions in various computational models.

We also study whether or not certain properties of statistical min-entropy have a computational analog. In particular, we consider the following questions:

1. Let X be a distribution with high computational min-entropy. Does one get a pseudo-random distribution when applying a "randomness extractor" on X?
2. Let X and Y be (possibly dependent) random variables. Is the computational min-entropy of (X, Y) at least as large as the computational min-entropy of X?
3. Let X be a distribution over $\{0,1\}^n$ that is "weakly unpredictable" in the sense that it is hard to predict a constant fraction of the coordinates of X with a constant bias. Does X have computational min-entropy $\Omega(n)$?

We show that the answers to these questions depend on the computational model considered. In some natural models the answer is false and in others the answer is true. Our positive results for the third question exhibit models in which the "hybrid argument bottleneck" in "moving from a distinguisher to a predictor" can be avoided.

1 Introduction

One of the most fundamental notions in theoretical computer science is that of *computaional indistinuishability* [1,2]. Two probability distributions are deemed close if no *efficient*[3] test can tell them apart - this comes in stark contrast to

---

[3] What is meant by "efficient" can naturally vary by specifying machine models and resource bounds on them

S. Arora et al. (Eds.): APPROX 2003+RANDOM 2003, LNCS 2764, pp. 200–215, 2003.
© Springer-Verlag Berlin Heidelberg 2003

the information theoretic view which allows *any* test whatsoever. The discovery [3,2,4] that simple computational assumptions (namely the existance of one-way functions) make the computational and information theoretic notions completely different has been one of the most fruitful in CS history, with impact on cryptography, complexity theory and computational learning theory.

The most striking result of these studies has been the efficient construction of nontrivial *pseudorandom* distributions, namely ones which are information theoretically very far from the uniform distribution, but are nevertheless indistinguishable from it. Two of the founding papers [2,4] found it natural to extend information theory more generally to the computational setting, and attempt to define its most fundamental notion of entropy[4]. The basic question is the following: when should we say that a distribution has (or is close to having) computational entropy (or pseudoentropy) k?. Interestingly, these two papers give two very different definitions! This point may be overlooked, since for the most interesting special case, the case of pseudorandomness (i.e., when the distributions are over n-bit strings and $k = n$), the two definitions coincide. This paper is concerned with the other cases, namely $k < n$, attempting to continue the project of building a computational analog of information theory.

1.1   Definitions of Pseudoentropy

To start, let us consider the two original definitions. Let X be a probability distribution over a set S.

A definition using "compression". Yao's definition of pseudoentropy [2] is based on compression. He cites Shannon's definition [5], defining $H(X)$ to be the minimum number of bits needed to describe a typical element of X. More precisely, one imagines the situation of Alice having to send Bob (a large number of) samples from X, and is trying to save on communication. Then $H(X)$ is the smallest k for which there are a compression algorithm A (for Alice) from S into k-bit strings, and a decompression algorithm B (for Bob) from k-bit strings into S, such that $B(A(x)) = x$ (in the limit, for typical x from X). Yao take this definition verbatim, adding the crucial computational constraint that both compression and decompression algorithms must be efficient. This notion of efficient compression is further studied in [6].

A definition using indistinguishability. Hastad et al's definition of pseudoentropy [4] extends the definition of pseudorandomness syntactically. As a distribution is said to be pseudorandom if it is indistinguishable from a distribution of maximum entropy (which is unique), they define a distribution to have pseudoentropy k is

---

[4] While we will first mainly talk about Shannon's entropy, we later switch to min-entropy and stay with it throughout the paper. However the whole introduction may be read when regarding the term "entropy" with any other of its many formal variants, or just as well as the informal notion of "information content" or "uncertainty"

it is indistinguishable from a distribution of Sahnnon entropy k (for which there are many possibilities).

It turns out that the two definitions of pseudoentropy above can be very different in natural computational settings, despite the fact that in the information theoretic setting they are identical for any k. Which definition, then, is the "natural one" to choose from? This question is actually more complex, as another natural point of view lead to yet another definition.

A definition using a natural metric space. The computational viewpoint of randomness may be thought of as endowing the space of *all* probability distributions with new, interesting metrics.

For every event (=test) T in our probability space we define: $d_T(X, Y) = |\Pr_X[T] - \Pr_Y[T]|$. In words, the distance between X and Y is the difference (in absolute value) of the probabilities they assign to T.[5]

Note that given a family of metrics, their maximum is also a metric. An information theoretic metric on distributions, the *statistical distance*[6] (which is basically $\frac{1}{2}L_1$-distance) is obtained by taking the maximum over the T-metrics above for *all* possible tests T. A natural computational metric, is given by taking the maximum over any class $\mathcal{C}$ of efficient tests. When should we say that a distribution X is indistinguishable from having Shannon entropy k? Distance to a set is the distance to the closest point in it, so X has to be close in this metric to *some* Y with Shannon entropy k.

A different order of quantifiers. At first sight this may look identical to the "indistinguishability" definition in [4]. However let us parse them to see the difference. The [4] definition say that X has pseudoentropy k if *there exists* a distribution Y of Shannon entropy k, such that *for all* tests T in $\mathcal{C}$, T has roughly the same probability under both X and Y. The metric definition above reverses the quantifiers: X has pseudoentropy k if *for every* a distribution Y of Shannon entropy k, *there exists* a test T in $\mathcal{C}$, which has roughly the same probability under both X and Y. It is easy to see that the metric definition is more liberal - it allows for at least as many distributions to have pseudoentropy k. Are they really different?

Relations between the three definitions. As all these definitions are natural and well-motivated, it makes sense to study their relationship. In the information theoretic world (when ignoring the "efficiency" constraints) all definitions are equivalent. It is easy to verify that regardless of the choice of a class $\mathcal{C}$ of "efficient" tests, they are ordered in permisiveness (allowing more distributions to have pseudoentropy k). The "indistinguishability" definition of [4] is the most stringent, then the "metric definition", and then the "compression" definition of

---

[5] This isn't precisely a metric as there may be different X and Y such that $d_T(X, Y) = 0$. However it is symmetric and satisfies the triangle inequality.

[6] Another basic distance measure is the so called KL-divergence, but for our purposes, which concern very close distributions, is not much different than statistical distance

[2]. What is more interesting is that we can prove collapses and separations for different computational settings and assumptions. For example, we show that the first two definitions drastically differ for logspace observers, but coincide for polynomial time observers (both in the uniform and nonuniform settings). The proof of the latter statement uses the "min-max" Theorem of [7] to "switch" the order of quantifiers. We can show some weak form of equivalence between all three definitions for circuits. We show that the "metric" coincides with the "compression" definition if $\mathbf{NP} \subseteq \mathbf{BPP}$. More precisely, we give a *non-deterministic* reduction showing the equivalence of the two definitions. This reduction guarantees high min-entropy according to the "metric" definition if the distribution has high min-entropy according to the "compression" distribution with respect to an **NP** oracle. A clean way to state this is that all three definitions are equivalent for **PH**/poly. We refer to this class as the class of poly-size **PH**-circuits. Such circuits are poly-size circuits which are allowed to compute an arbitrary function in the polynomial-hierarchy (**PH**). We remark that similar circuits (for various levels of the **PH** hierarchy) arise in related contexts in the study of "computational randomness": They come up in conditional "derandomization" results of **AM** [8,9,10] and "extractors for samplable distributions" [11].

1.2   Pseudoentropy versus Information Theoretic Entropy

We now move to another important part of our project. As these definitions are supposed to help establish a computational version of information theory, we attempt to see which of them respect some natural properties of information-theoretic entropy.

Using randomness extractors. In the information theoretic setting, there are *randomness extractors* which convert a high entropy[7] distribution into one which is statistically close to uniform. The theory of extracting the randomness from such distributions is by now quite developed (see surveys [12,13,14]). It is natural to expect that applying these randomness extractors on high pseudoentropy distributions produces a pseudorandom distribution. In fact, this is the motivation for pseudoentropy in some previous works [15,4,16].

It is easy to see that the the "indistinguishability" definition of [4] has this property. This also holds for the "metric" definition by the equivalence above. Interestingly, we do not know whether this holds for the "compression" definition. Nevertheless, we show that some extractor constructions in the literature (the ones based on Trevisan's technique [17,18,19,20,10]) do produce a pseudorandom distribution when working with the "compression" definition.

---

[7] It turns out that a different variant of entropy called "min-entropy" is the correct measure for this application. The min-entropy of a distribution X is $\log_2(\min_x 1/\Pr[X = x])$. This should be compared with Shannon's entropy in which the minimum is replaced by averaging.

The information in two dependent distributions. One basic principle in information theory is that two (possibly dependent) random variables have at least as much entropy as any one individually, e.g. $H(X, Y) \geq H(X)$. A natural question is whether this holds when we replace information-theoretic entropy with pseudoentropy. We show that the answer depends on the model of computation. If there exist one-way functions, then the answer is *no* for the standard model of polynomial-time distinguishers. On the other hand, if $\mathbf{NP} \subseteq \mathbf{BPP}$, then the answer is *yes*. Very roughly speaking, the negative part follows from the existence of pseudorandom generators, while the positive part follows from giving a *nondeterministic* reduction which relies on nondeterminism to perform approximate counting. Once again, this result can be also stated as saying that the answer is positive for poly-size $\mathbf{PH}$-circuits. We remark that the positive result holds for (nonuniform) online space-bounded computation as well.

Entropy and unpredictability. A deeper and interesting connection is the one between entropy and unpredictability. In the information theoretic world, a distribution which is unpredictable has high entropy.[8] Does this relation between entropy and unpredictability holds in the computational world?

Let us restrict ourselves here for a while to the metric definition of pseudoentropy. Two main results we prove is that this connection indeed holds in two natural computational notions of efficient observers. One is for logspace observers. The second is for $\mathbf{PH}$-circuits. Both results use one mechanism - a different characterization of the metric definition, in which distinguishers accept very few inputs (less than 2^k when the pseudoentropy is k). We show that predictors for the accepted set are also good for any distribution "caught" by such a distinguisher. This direction is promising as it suggests a way to "bypass" the weakness of the "hybrid argument".

The weakness of the hybrid argument. Almost all pseudorandom generators (whether conditional such as the ones for small circuits or unconditional such as the ones for logspace) use the hybrid argument in their proof of correctness. The idea is that if the output distribution can be efficiently distinguished from random, some bit can be efficiently predicted with nontrivial advantage. Thus, pseudorandomness is established by showing unpredictability.

However, in standard form, if the distinguishability advantage is ϵ, the prediction advantage is only ϵ/n. In the results above, we manage (for these two computational models) to avoid this loss and make the prediction advantage $\Omega(\epsilon)$ (just as information theory suggests).

While we have no concrete applications, this seem to have potential to improve various constructions of pseudorandom generators. To see this, it suffices to observe the consequences of the hybrid argument loss. It requires every output bit of the generator to be very unpredictable, for which a direct cost is paid in the

---

[8] We consider two different forms of prediction tests: The first called "next bit predictor" attempts to predict a bit from previous bits, whereas the second called "complement predictor" has access to all the other bits, both previous and latter.

seed length (and complexity) of the generator. For generators against circuits, a long sequence of works [2,21,22,16] resolved it optimally using efficient *hardness amplification*. These results allow constructing distributions which are unpredictable even with advantage $1/\text{poly}(n)$. The above suggests that sometimes this amplification may not be needed. One may hope to construct a pseudorandom distribution by constructing an unpredictable distribution which is only unpredictable with constant advantage, and then use a randomness extractor to obtain a pseudorandom distribution.[9]

This problem is even more significant when constructing generators against logspace machines [24,25]. The high unpredictability required seems to be the bottleneck for reducing the seed length in Nisan's generator [24] and its refinements from $O((\log n)^2)$ bits to the optimal $O(\log n)$ bits (that will result in $BPL = L$). The argument above gives some hope that for fooling logspace machines (or even just constant-width oblivious branching programs) the suggested approach may yield substantial improvements. However, in this setup there is another hurdle: In [26] it was shown that randomness extraction cannot be done by one pass log-space machines. Thus, in this setup it is not clear how to move from pseudoentropy to pseudorandomness.

1.3 Organization of the Paper

In Section 2 we give some basic notation. Section 3 formally defines our three basic notions of pseudoentropy, and proves a useful characterization of the metric definition. In Sections 5 and 6 we prove equivalence and separations results between the various definitions in several natural computational models. Section 7 is devoted to our results about computational analogs of information theory for concatenation and unpredictability of random variables. Because of space limitations many of the proofs do not appear in this version.

2   Preliminaries

Let X be a random variable over some set S. We say that X has *(statistical) min-entropy* at least k, denoted $H^\infty(X) \geq k$, if for every $x \in S$, $\Pr[X = x] \leq 2^{-k}$. We use U_n to denote the uniform distribution on $\{0,1\}^n$.

Let X, Y be two random variables over a set S. Let $f : S \to \{0,1\}$ be some function. The *bias* of X and Y with respect to f, denoted $\text{bias}_f(X,Y)$, is defined by $\big|\mathbb{E}[f(X)] - \mathbb{E}[f(Y)]\big|$. Since it is sometimes convenient to omit the absolute value, we denote $\text{bias}_f^*(X,Y) = \mathbb{E}[f(X)] - \mathbb{E}[f(Y)]$.

The *statistical distance* of X and Y, denoted $\text{dist}(X,Y)$, is defined to be the maximum of $\text{bias}_f(X,Y)$ over all functions f. Let $\mathcal{C}$ be a class of functions from S to $\{0,1\}$ (e.g., the class of functions computed by circuits of size m

---

[9] This approach was used in [16]. They show that even "weak" hardness amplification suffices to construct a high pseudoentropy distribution using the pseudo-random generator construction of [23]. However, their technique relies on the properties of the specific generator and cannot be applied in general.

for some integer m). The *computational distance* of X and Y w.r.t. $\mathcal{C}$, denoted comp-dist$_\mathcal{C}(X, Y)$, is defined to be the maximum of bias$_f(X, Y)$ over all $f \in \mathcal{C}$. We will sometimes drop the subscript $\mathcal{C}$ when it can be inferred from the context.

Computational models. In addition to the standard model of uniform and non-uniform polynomial-time algorithms, we consider two additional computational models. The first is the model of **PH**-*circuits*. A **PH**-circuit is a boolean circuit that allows queries to a language in the polynomial hierarchy as a basic gate.[10] The second model is the model of *bounded-width read-once oblivious branching programs*. A width-S read once oblivious branching program P is a directed graph with Sn vertices, where the graph is divided into n layers, with S vertices in each layer. The edges of the graph are only between from one layer to the next one, and each edge is labelled by a bit $b \in \{0, 1\}$ which is thought of as a variable. Each vertex has two outgoing edges, one labelled 0 and the other labelled 1. One of the vertices in the first layer is called the *source* vertex, and some of the vertices in the last layer are called the **accepting vertices**. A computation of the program P on input $x \in \{0, 1\}^n$ consists of walking the graph for n steps, starting from the source vertex, and in step i taking the edge labelled by x_i. The output of $P(x)$ is 1 iff the end vertex is accepting. Note that variables are read in the natural order and thus width-S read once oblivious branching programs are the non-uniform analog of one-pass (or online) space-$\log S$ algorithms.

3   Defining Computational Min-entropy

In this section we give three definitions for the notion of computational (or "pseudo") min-entropy. In all these definitions, we fix $\mathcal{C}$ to be a class of functions which we consider to be efficiently computable. Our standard choice for this class will be the class of functions computed by a boolean circuit of size $p(n)$, where n is the circuit's input length and $p(\cdot)$ is some fixed polynomial. However, we will also be interested in instantiations of these definitions with respect to different classes $\mathcal{C}$. We will also sometimes treat $\mathcal{C}$ as a class of *sets* rather then functions, where we say that a set D is in $\mathcal{C}$ iff its characteristic function is in $\mathcal{C}$. We will assume that the class $\mathcal{C}$ is closed under complement.

3.1   HILL-type Pseudoentropy: Using Indistinguishability

We start with the standard definition of computational (or "pseudo") min-entropy, as given by [4]. We call this definition *HILL-type pseudoentropy*.

Definition 1. *Let X be a random variable over a set S. Let $\epsilon \geq 0$. We say that X has ϵ-**HILL-type pseudoentropy** at least k, denoted $H_\epsilon^{\text{HILL}}(X) \geq k$, if there exists a random variable Y with (statistical) min-entropy at least k such that the computational distance (w.r.t. $\mathcal{C}$) of X and Y is at most ϵ.*

---

[10] Equivalently, the class languages accepted by poly-size **PH**-circuits is **PH**/poly.

We will usually be interested in ϵ-pseudoentroy for ϵ that is a small constant. In these cases we will sometimes drop ϵ and simply say that X has (HILL-type) pseudoentropy at least k (denoted $H^{\text{HILL}}(X) \geq k$).

3.2   Metric-Type Pseudoentropy: Using a Metric Space

In Definition 1 the distribution X has high pseudoentropy if there *exists* a high min-entropy Y such that X and Y are indistinguishable. As explained in the introduction, it is also natural to reverse the order of quantifiers: Here we allow Y to be a function of the "distinguishing test" f.

Definition 2. *Let X be a random variable over a set S. Let $\epsilon \geq 0$. We say that X has ϵ-metric-type pseudoentropy at least k, denoted $H_\epsilon^{\text{Metric}}(X) \geq k$, if for every test f on S there exists a Y which has (statistical) min-entropy at least k and $\text{bias}_f(X, Y) < \epsilon$.*

It turns out that metric-pseudoentropy is equivalent to a different formulation. (Note that the condition below is only meaningful for D such that $|D| < 2^k$.) The proof of Lemma 1 appears in the full version.

Lemma 1. *For every class $\mathcal{C}$ which is closed under complement and for every $k \leq \log|S| - 1$ and ϵ, $H_\epsilon^{\text{Metric}}(X) \geq k$ if and only if for every set $D \in \mathcal{C}$, $\Pr[X \in D] \leq \frac{|D|}{2^k} + \epsilon$*

3.3   Yao-Type Pseudoentropy: Using Compression

Let $\mathcal{C}$ be a class of functions which we consider to efficiently computable. Recall that we said that a set D is a member of $\mathcal{C}$ if its characteristic function was in $\mathcal{C}$. That is, a set D is in $\mathcal{C}$ if it is *efficiently decidable*. We now define a family $\mathcal{C}_{\text{compress}}$ of sets that are **efficiently compressible**. That is, we say that a set $D \subseteq S$ is in $\mathcal{C}_{\text{compress}}(\ell)$ if there exist functions $c, d \in \mathcal{C}$ ($c : S \to \{0,1\}^\ell$ stands for *compress* and $d : \{0,1\}^\ell \to S$ for *decompress*) such that $D = \{x | d(c(x)) = x\}$. Note that every efficiently compressible set is also efficiently decidable (assuming the class $\mathcal{C}$ is closed under composition). Yao-type pseudoentropy is defined by replacing the quantification over $D \in \mathcal{C}$ in the alternative characterization of metric-type pseudoentropy (Lemma 1) by a quantification over $D \in \mathcal{C}_{\text{compress}}(\ell)$ for all $\ell < k$. The resulting definition is the following:

Definition 3. *Let X be a random variable over a set S. X has ϵ-Yao-type pseudoentropy at least k, denoted $H_\epsilon^{\text{Yao}}(X) \geq k$, if for every $\ell < k$ and every set $D \in \mathcal{C}_{\text{compress}}(\ell)$, $\Pr[X \in D] \leq 2^{l-k} + \epsilon$*

4 Using Randomness Extractors

An extractor uses a short seed of truly random bits to extract many bits which are (close to) uniform.

Definition 4 ([27]). *A function $E : \{0,1\}^n \times \{0,1\}^d \to \{0,1\}^m$ is a (k, ϵ)-extractor if for every distribution X on $\{0,1\}^n$ with $H^\infty(x) \geq k$, the distribution $Z = E(X, U_d)$ has $\mathrm{dist}(Z, U_m) < \epsilon$.*

We remark that there are explicit (polynomial time computable) extractors with seed length $\mathrm{polylog}(n/\epsilon)$ and $m = k$. The reader is referred to survey papers on extractors [12,13,14]. The following standard lemma says that if a distribution X has HILL-type pseudoentropy at least k with respect to circuits, then for every randomness extractor the distribution $E(X, U_d)$ is pseudorandom.

Lemma 2. *Let C be the class of polynomial size circuits. Let X be a distribution with $H_\epsilon^{\mathrm{HILL}}(X) \geq k$ and let E be a (k, ϵ)-extractor computable in time $\mathrm{poly}(n)$ then $\mathsf{comp\text{-}dist}_C(E(X, U_d), U_m) < 2\epsilon$.*

Note that by Theorem 1 the same holds for the metric definition. Interestingly, we do not know whether this holds for Yao-type pseudoentropy. We can however show that this holds for the extractor of Trevisan [17]. Trevisan's extractor $E^{\mathrm{Tre}} : \{0,1\}^n \times \{0,1\}^{O(\log^2 n / \log k)} \to \{0,1\}^{\sqrt{k}}$ is a $(k, 1/n)$-extractor

Lemma 3. *Let C be the class of polynomial size circuits. Let X be a distribution with $H_\epsilon^{\mathrm{Yao}}(X) \geq k$, then $\mathsf{comp\text{-}dist}_C(E^{\mathrm{Tre}}(X, U_d), U_m) < 2\epsilon$.*

The proof of Lemma 3 appears in the full version. Loosely speaking, the correctness proof of Trevisan's extractor (and some later constructions, c.f., [14]) shows that if the output of the extractor isn't close to uniform, then the distribution X can be compressed (which is impossible for a distribution of sufficiently high min-entropy). For the lemma, one only needs to observe that in this argument an *efficient* distinguisher gives rise to an *efficient* compressing algorithm. Thus, running the extractor on an "incompressible" distribution gives a pseudorandom distribution.

5 Relationships between Definitions

5.1 Equivalence between HILL-type and Metric-Type

The difference between HILL-type and metric-type pseudoentropy is in the order of quantifiers. HILL-type requires that there exist a unique "reference distribution" Y with $H^\infty(Y) \geq k$ such that for every D, $\mathrm{bias}_D(X, Y) < \epsilon$, whereas metric-type allows Y to depend on D, and only requires that for every D there exists such a Y. It immediately follows that for every class C and every X, $H^{\mathrm{Metric}}(X) \geq H^{\mathrm{HILL}}(X)$. In this section we show that the other direction also applies (with small losses in ϵ and time/size) for small circuits.

Theorem 1 (Equivalence of HILL-type and metric-type for circuits).
Let X be a distribution over $\{0,1\}^n$. For every $\epsilon, \delta > 0$ and k, if $H_{\epsilon-\delta}^{\text{Metric}}(X) \geq k$ (with respect to circuits of size $O(ns/\delta^2)$) then $H_{\epsilon}^{\text{HILL}}(X) \geq k$ (with respect to circuits of size s)

The proof of Theorem 1 appears only in the full version. We now provide a sketch of the argument. It is sufficient to show that if $H_{\epsilon}^{\text{HILL}}(X) < k$ then then $H_{\epsilon-\delta}^{\text{Metric}}(X) < k$. Suppose indeed that $H_{\epsilon}^{\text{HILL}}(X) < k$. This implies that for every Y with $H^{\infty}(Y) \geq k$ there is a small circuit $D \in \mathcal{C}$ such that $\text{bias}_D(X, Y) \geq \epsilon$.

We consider a game between two players. The "circuit player" Alice chooses a small circuit D and the "distribution player" Bob chooses a "flat" distribution Y with $H^{\infty}(Y) \geq k$.[11] (Note that both players have a finite number of strategies in the game.) After the choices are made, Bob pays $\text{dist}_D(X, Y)$ dollars to Alice. Our assumption says that if Alice plays after Bob then she can always win ϵ dollars. Loosely speaking, the "min-max" theorem of [7] allows to switch the order of quantifiers and assert that Alice can guarantee the same amount even when playing first.[12] More formally, we conclude that there exists a *distribution* $\hat{D}$ over circuits for Alice such that she expects to get ϵ dollars for every reply Y of Bob. Note that we were able to switch the order of quantifiers to that of the "metric" definition. We are left with the task of converting $\hat{D}$ into a single circuit. This is done by sampling sufficiently many circuits $D_1, \cdots, D_t$ from $\hat{D}$ and taking their average. By a union bound there exists a choice of $D_1, \cdots, D_t$ which is good for every distribution Y.[13]

In the full version we also prove equivalence for *uniform* polynomial time machines.[14]

5.2 Equivalence between All Types for PH-circuits

We do not know whether the assumption that $H_{\epsilon}^{\text{Yao}}(X) \geq k$ for circuits implies that $H_{\epsilon}^{\text{Metric}}(X) \geq k'$ for slightly smaller k' and circuit size (and in fact, we conjecture that it's false). However, we can prove it assuming the circuits for the Yao-type definition have access to an NP-oracle.

---

[11] A "flat" distribution is a distribution which is uniformly distributed over a subset of S.

[12] There is a subtlety here. In order to apply the theorem, Alice must be able to win ϵ dollars even when Bob plays a *mixed strategy* (i.e., a convex combination of his choices). However, a convex combination of flat distributions with min-entropy k also has min-entropy k.

[13] It is crucial that this union bound is not performed over the $\binom{2^n}{2^k}$ choices for Y but rather on the 2^n inputs. More precisely, we show that there exist $D_1, \cdots, D_t$ such that for all inputs x, $\frac{1}{t}\sum D_i(x) \approx \mathbb{E}[\hat{D}(x)]$.

[14] We find this surprising because the argument above seems to exploit the non-uniformity of circuits: The "min-max theorem" works only for *finite* games and is non-constructive - it only shows existence of a distribution $\hat{D}$ and gives no clue to its complexity. The key idea is the observation that pseudoentropy with respect to uniform Turing machines implies also pseudoentropy for "slightly non-uniform" Turing machines. Exact details appear in the full version.

Theorem 2. *Let $k' = k+1$ There is a constant c so that if $H_\epsilon^{\mathbf{Yao}}(X) \geq k'$ (with respect to circuits of size $max(s, n^c)$ that use an NP-oracle) then $H_\epsilon^{\mathbf{Metric}}(X) \geq k$ (with respect to circuits of size s).*

The proof of Theorem 2 appears in the full version. The reduction in the proof of Theorem 2 uses an NP-oracle. The class of polynomial size **PH**-circuits are closed under the use of NP-oracles ($\mathbf{PH}^{NP}/poly = \mathbf{PH}/poly$). Applying the argument of Theorem 2 give the following corollary.

Corollary 1. *Let $\mathcal{C}$ be the class of polynomial size **PH**-circuits. If $H_\epsilon^{\mathbf{Yao}}(X) \geq 2k$ then $H_\epsilon^{\mathbf{Metric}}(X) \geq k$.*

6   Separation between Types

Given the results of the previous section it is natural to ask if HILL-type and metric-type pseudoentropy are equivalent in **all** natural computational models? We give a negative answer and prove that there's large gap between HILL-type and metric-type pseudoentropy in the model of bounded-width read-once oblivious branching programs.

Theorem 3. *For every constant $\epsilon > 0$ and sufficiently large $n \in \mathbb{N}$, and , there exists a random X variable over $\{0,1\}^n$ such that $H_\epsilon^{\mathbf{Metric}} X \geq (1 - \epsilon)n$ with respect to width-S read once oblivious branching programs, but $H_{1-\epsilon}^{\mathbf{HILL}}(X) \leq$ polylog(n, S) with respect to width-4 oblivious branching programs.*

Theorem 3 follows from the following two lemmas, whose proofs appear in the full version:

Lemma 4 (Based on [28]). *Let $\epsilon > 0$ be some constant and $S \in \mathbb{N}$ such that $S > \frac{1}{\epsilon}$. Let $l = \frac{10}{\epsilon} \log S$ and consider the distribution $X = (U_l, U_l, \dots, U_l)$ over $\{0,1\}^n$ for some $n < S$ which is a multiple of l. Then, $H_\epsilon^{\mathbf{Metric}}(X) \geq (1 - \epsilon)n$ with respect to width-S oblivious branching programs.*

Lemma 5. *Let $\epsilon > 0$ be some constant, and X be the random variable $(U_l, U_l, \dots, U_l)$ over $\{0,1\}^n$ (where $l > \log n$). Then, $H_{(1-\epsilon)}^{\mathbf{HILL}}(X) \leq \frac{100}{\log(1/\epsilon)} l^3$ with respect to width-4 oblivious branching programs.*

7   Analogs of Information-Theoretic Inequalities

7.1   Concatenation Lemma

A basic fact in information theory is that for every (possibly correlated) random variables X and Y, the entropy of (X, Y) is at least as large as the entropy of X. We show that if one-way-functions exist then this does not hold for all types of pseudoentropy with respect to polynomial time circuits. On the other hand, we show that the fact above does hold for polynomial-sized **PH**-circuits and for bounded-width oblivious branching programs.[15]

---

[15] With respect to the latter, we only prove that concatenation holds for metric-type pseudoentropy.

Negative result for standard model. Our negative result is the following easy lemma, whose proof is omitted:

Lemma 6. *Let $G : \{0,1\}^l \to \{0,1\}^n$ be a (poly-time computable) pseudorandom generator.[16] Let (X,Y) be the random variables $(G(U_l), U_l)$. Then $H_\epsilon^{\text{HILL}}(X) = n$ (for a negligible ϵ) but $H_{1/2}^{\text{Yao}}(X,Y) \leq l + 1$.*

Positive result for **PH**-*circuits.* Our positive result for **PH**-circuits is stated in the following lemma, whose proof appears in the full version:

Lemma 7. *Let X be a random variable over $\{0,1\}^n$ and Y be a random variable over $\{0,1\}^m$. Suppose that $H_\epsilon^{\text{Yao}}(X) \geq k$ with respect to s-sized* **PH**-*circuits. Then $H_\epsilon^{\text{Yao}}(X,Y) \geq k$ with respect to $O(s)$-sized* **PH**-*circuits.*

Applying the results of Section 5.2, we obtain that with respect to **PH**-circuit, the concatenation property is satisfied also for HILL-type and Metric-type pseudoentropy.

Positive result for bounded-width oblivious branching programs. We also show that the concatenation property holds also for metric-type pseudoentropy with respect to bounded-width read-once oblivious branching programs. This is stated in Lemma 8, whose proof appears in the full version. Note that the quality of this statement depends on the order of the concatenation (i.e., whether we consider (X,Y) or (Y,X)).

Lemma 8. *Let X be a random variable over $\{0,1\}^n$ and Y be a random variable over $\{0,1\}^m$. Suppose that $H_\epsilon^{\text{Metric}}(X) \geq k$ with respect to width-S read-once oblivious branching programs. Then, $H_\epsilon^{\text{Metric}}(X,Y) \geq k$ and $H_{2\epsilon S}^{\text{Metric}}(Y,X) \geq k - \log(1/\epsilon)$ with respect to such algorithms.*

7.2   Unpredictability and Entropy

Loosely speaking, a random variable X over $\{0,1\}^n$ is δ-**unpredictable** is for every index i, it is hard to predict X_i from $X_{[1,i-1]}$ (which denotes $X_1, \ldots, X_{i-1}$) with probability better than $\frac{1}{2} + \delta$.

Definition 5. *Let X be a random variable over $\{0,1\}^n$. We say that X is δ-unpredictable in index i with respect to a class of algorithms $\mathcal{C}$ if for every $P \in \mathcal{C}$, $\Pr[P(X_{[1,i-1]}) = X_i] < \frac{1}{2} + \delta$. X is δ-unpredictable if for every $P \in \mathcal{C}$ $\Pr[P(i, X_{[1,i-1]}) = X_i] < \frac{1}{2} + \delta$ where this probability is over the choice of X and over the choice of $i \leftarrow_R [n]$. We also define* complement *unpredictability by changing $X_{[1,i-1]}$ to $X_{[n]\setminus\{i\}}$ in the definition above.*

---

[16] We mean here a pseudorandom generator in the "cryptographic" sense of Blum, Micali and Yao [3,2]. That is, we require that G is polynomial time computable.

Yao's Theorem [2] says that if X is δ-unpredictable in all indices by polynomial-time (uniform or non-uniform) algorithms, then it is $n\delta$-indistinguishable from the uniform distribution. Note that this theorem can't be used for a constant $\delta > 0$. This loss of a factor of n comes from the use of the "hybrid argument" [1,2]. In contrast, in the context of information theory it is known that if a random variable X is δ-unpredictable (w.r.t. to all possible algorithms) for a small constant δ and for a constant fraction of the indices, then $H^\infty(X) \geq \Omega(n)$. Thus, in this context it is possible to extract $\Omega(n)$ bits of randomness even from δ-unpredictable distributions where δ is a *constant* [20].

In this section we consider the question of whether or not there exists a computational analog to this information-theoretic statement.

Negative result in standard model. We observe that if one-way functions exist, then the distribution $(G(U_l), U_l)$ where $|G(U_l)| = \omega(l))$ used in Lemma 6 is also a counterexample (when considering polynomial-time distinguishers). That is, this is a distribution that is δ-unpredictable for a negligible δ in almost all the indices, but has low pseudoentropy. We do not know whether or not there exists a distribution that is δ-unpredictable for a *constant* δ for *all* the indices, and has sublinear pseudoentropy.

Positive results. We also show some computational settings in which the information theoretic intuition *does* holds. We show this for **PH**-circuits, and for bounded-width oblivious branching programs using the metric definition of pseudoentropy. We start by considering a special case in which the distinguisher has distinguishing probability 1 (or very close to 1).[17]

Theorem 4. *Let X be a random variable over $\{0,1\}^n$. Suppose there exists a size-s **PH**-circuit (width-S oblivious branching program) D such that $|D^{-1}(1)| \leq 2^k$ and $\Pr[D(X) = 1] = 1$. Then there exists a size-$O(s)$ **PH**-circuit (width-S oblivious branching program) P such that $\Pr_{i\in[n], x\leftarrow_R X}[P(x_{[1,i]}) = x_i] \geq 1 - O(\frac{k}{n})$*

The main step in the proof of Theorem 4 is the following lemma:

Lemma 9. *Let $D \subseteq \{0,1\}^n$ be a set such that $|D| < 2^k$. Let $x = x_1 \ldots x_{i-1} \in \{0,1\}^{i-1}$, we define N_x to be the number of continuations of x in D (i.e., $N_x = |\{x' \in \{0,1\}^{n-i} \mid xx' \in D\}|$). We define $P(x)$ as follows: $P(x) = 1$ if $\frac{N_{x1}}{N_x} > \frac{2}{3}$ and $P(x) = 1$ if $\frac{N_{x1}}{N_x} < \frac{1}{3}$, where $P(x)$ is undefined otherwise. Then, for every random variable X such that $X \subseteq D$,*

$$\Pr_{i\in[n], x\leftarrow_R X}\left[P(x_{[1,i-1]}) \text{ is defined and equal to } x_i\right] \geq 1 - O\left(\frac{k}{n}\right)$$

Proof. For $x \in \{0,1\}^n$, we let $bad(x) \subseteq [n]$ denote the set of indices $i \in [n]$ such that $P(x_{[1,i-1]})$ is either undefined or different from x_i. We will prove the

---

[17] Intuitively, this corresponds to applications that use the high entorpy distribution for hitting a set (like a disperser) rather than for approximation of a set (like an extractor).

lemma by showing that $|bad(x)| \leq O(k)$ for *every* string $x \in D$. Note that an equivalent condition is that $|D| \geq 2^{-\Omega(|bad(x)|)}$. Indeed, we will prove that $|D| \geq (1 + \frac{1}{2})^{|bad(x)|}$. Let N_i denote the number of continuations of $x_{[1,i]}$ in D (i.e., $N_i = N_{x_{[1,i]}}$). We define $N_n = 1$. We claim that for every $i \in bad(x)$, $N_{i-1} \geq (1 + \frac{1}{2})N_i$. (Note that this is sufficient to prove the lemma). Indeed, $N_{i-1} = N_{x_{[1,i-1]}0} + N_{x_{[1,i-1]}1}$, or in other words, $N_{i-1} = N_i + N_{x_{[1,i-1]}\overline{x_i}}$ (where $\overline{x_i} \overset{def}{=} 1 - x_i$). Yet, if $i \in bad(x)$ then $N_{x_{[1,i-1]}\overline{x_i}} \geq \frac{1}{3}(N_i + N_{x_{[1,i-1]}\overline{x_i}}) \geq \frac{1}{2}N_i$. □

We obtain Theorem 4 from Lemma 9 for the case of **PH**-circuits by observing that deciding whether $P(x)$ is equal to 1 or 0 (in the cases that it is defined) can be done in the polynomial-hierarchy (using approximate counting [29]). The case of bounded-width oblivious branching programs is obtained by observing that the state of the width-S oblivious branching program D after seeing $x_1, \ldots, x_{i-1}$ completely determines the value $P(x_1, \ldots, x_{i-1})$ and so $P(x_1, \ldots, x_{i-1})$ can be computed (non-uniformly) from this state.[18]

We now consider the case that $\Pr_{x \leftarrow_R X}[x \in D] = \epsilon$ for an arbitrary constant ϵ (that may be smaller than $\frac{1}{2}$). In this case we are not able to use standard unpredictability and use *complement unpredictability*.

Theorem 5. *Suppose that X is δ-complement-unpredictable for a random index with respect to s-sized **PH**-circuits, where $\frac{1}{2} > \delta > 0$ is some constant. Let $\epsilon > \delta$ be some constant, then $H_\epsilon^{\text{Metric}}(X) \geq \Omega(n)$ with respect to $O(s)$-sized **PH**-circuits.*

Proof. We prove the theorem by the contrapositive. Let $\epsilon > \delta$ and suppose that $H_\epsilon^{\text{Metric}}(X) < k$ where $k = \epsilon' n$ (for a constant $\epsilon' > 0$ that will be chosen later). This means that there exists a set $D \in C$ such that $\Pr_{x \leftarrow_R X}[x \in D] \geq \frac{|D|}{2^k} + \epsilon$. In particular, this means that $|D| < 2^k$ and $\Pr_{x \leftarrow_R X}[x \in D] \geq \epsilon$. We consider the following predictor P': On input $i \in [n]$ and $x = x_1, \ldots, x_{i-1}, x_{i+1}, \ldots, x_n \in \{0,1\}^{n-1}$, P' considers the strings x^0, x^1 where $x^b = x_1, \ldots, x_{i-1}, b, x_{i+1}, \ldots, x_n$. If both x^0 and x^1 are not in D, then P' outputs a random bit. If $x^b \in D$ and $x^{\overline{b}} \notin D$ then P' outputs b. Otherwise (if $x^0, x^1 \in D$), P' outputs $P(x_1, \ldots, x_{i-1})$, where P is the predictor constructed from D in the proof of Lemma 9. Let $\Gamma(D)$ denote the set of all strings x such that $x \notin D$ but x is of Hamming distance 1 from D (i.e., there is $i \in [n]$ such that $x_1, \ldots, x_{i-1}, \overline{x_i}, x_{i+1}, \ldots, x_n \in D$). If $S \subseteq \{0,1\}^n$, then let $X_{\restriction S}$ denote the random variable $X|X \in S$. By Lemma 9 $\Pr_{i \in [i], x \leftarrow_R X_{\restriction D}}[P'(x_{[n] \setminus \{i\}}) = x_i] \geq 1 - O(\frac{k}{n})$ while it is clear that $\Pr_{i \in [i], x \leftarrow_R X_{\restriction \{0,1\}^n \setminus (D \cup \Gamma(D))}}[P'(x_{[n] \setminus \{i\}}) = x_i] = \frac{1}{2}$. Thus if it holds that $\Pr[X \in \Gamma(D)] < \epsilon'$ and $k < \epsilon' n$, where ϵ' is some small constant (depending on ϵ and δ) then $\Pr_{i \in [i], x \leftarrow_R X}[P'(x_{[n] \setminus \{i\}}) = x_i] \geq \frac{1}{2} + \delta$ and the proof is finished.

However, it may be the case that $\Pr[X \in \Gamma(D)] \geq \epsilon'$. In this case, we will consider the distinguisher $D^{(1)} = D \cup \Gamma(D)$, and use $D^{(1)}$ to obtain a

---

[18] Lemma 9 only gives a predictor given a distinguisher D such that $\Pr_{x \leftarrow_R X}[x \in D] = 1$. However, the proof of Lemma 9 will still yield a predictor with constant bias even if 1 is replaced by $\frac{9}{10}$ (or any constant greater than $\frac{1}{2}$).

predictor $P^{(1)'}$ in the same way we obtained P' from D. Note that $|D^{(1)}| \leq n|D|$ and that, using non-determinism, the circuit size of $D^{(1)}$ is larger than the circuit size of D by at most a $O(\log n)$ additive factor.[19] We will need to repeat this process for at most $\frac{1}{\epsilon'}$ steps,[20] to obtain a distinguisher $D^{(c)}$ (where $c \leq \frac{1}{\epsilon'}$) such that $|D^{(c)}| \leq n^{O(1/\epsilon')}|D| \leq 2^{k+O(\log n(1/\epsilon'))}$, $\Pr[X \in D^{(c)}] \geq \epsilon$ and $\Pr[X \in \Gamma(D^{(c)})] < \epsilon'$. The corresponding predictor $P^{(c)'}$ will satisfy that $\Pr_{i \in [i], x \leftarrow_R X}[P^{(c)'}(x_{[n] \setminus \{i\}}) = x_i] \geq \frac{1}{2} + \delta$ thus proving the theorem.   □

Acknowledgements We thank Oded Goldreich and the RANDOM 2003 referees for helpful comments.

References

1. Goldwasser, S., Micali, S.: Probabilistic encryption. Journal of Computer and System Sciences **28** (1984) 270–299 Preliminary version in STOC' 82.
2. Yao, A.C.: Theory and applications of trapdoor functions. In: 23rd FOCS. (1982) 80–91
3. Blum, M., Micali, S.: How to generate cryptographically strong sequences of pseudo-random bits. SIAM Journal on Computing **13** (1984) 850–864
4. Håstad, J., Impagliazzo, R., Levin, L.A., Luby, M.: A pseudorandom generator from any one-way function. SIAM Journal on Computing **28** (1999) 1364–1396 (electronic)
5. Shannon, C.E.: A mathematical theory of communication. Bell System Technical Journal **27** (1948) 379–423, 623–656
6. Goldberg, A.V., Sipser, M.: Compression and ranking. SIAM Journal on Computing **20** (1991) 524–536
7. von Neumann, J.: Zur theorie der gesellschaftsspiele. Math. Ann. **100** (1928) 295–320
8. Klivans, A.R., van Melkebeek, D.: Graph nonisomorphism has subexponential size proofs unless the polynomial-time hierarchy collapses. SIAM J. Comput. **31** (2002) 1501–1526 (electronic)
9. Miltersen, P.B., Vinodchandran, N.V.: Derandomizing Arthur-Merlin games using hitting sets. In: 40th FOCS. (1999) 71–80
10. Shaltiel, R., Umans, C.: Simple extractors for all min-entropies and a new pseudo-random generator. In: 42nd FOCS. (2001) 648–657
11. Trevisan, L., Vadhan, S.: Extracting randomness from samplable distributions. In: 41st FOCS. (2000) 32–42
12. Nisan, N.: Extracting randomness: How and why: A survey. In: Conference on Computational Complexity. (1996) 44–58
13. Nisan, Ta-Shma: Extracting randomness: A survey and new constructions. JCSS: Journal of Computer and System Sciences **58** (1999)
14. Shaltiel, R.: Recent developments in explicit constructions of extractors. Bulletin of the European Association for Theoretical Computer Science **77** (2002) 67– Also available on http://www.wisdom.weizmann.ac.il/~ronens.

---

[19] To compute $D^{(1)}(x)$, guess $i \in [n]$, $b \in \{0, 1\}$ and compute $D(x')$ where x' is obtained from x by changing x_i to b.

[20] Actually, a tighter analysis will show that we only need $O(\log \frac{1}{\epsilon'})$ steps.

15. Impagliazzo, R., Levin, L.A., Luby, M.: Pseudo-random generation from one-way functions. In: 21st STOC. (1989) 12–24
16. Sudan, M., Trevisan, L., Vadhan, S.: Pseudorandom generators without the XOR lemma. JCSS: Journal of Computer and System Sciences **62** (2001) Preliminary version in STOC' 99. Also published as ECCC Report TR98-074.
17. Trevisan, L.: Construction of extractors using pseudo-random generators. In: 31st STOC. (1999) 141–148
18. Raz, R., Reingold, O., Vadhan, S.: Extracting all the randomness and reducing the error in trevisan's extractors. JCSS: Journal of Computer and System Sciences **65** (2002) Preliminary version in STOC' 99.
19. Impagliazzo, R., Shaltiel, R., Wigderson, A.: Extractors and pseudo-random generators with optimal seed length. In ACM, ed.: 32nd STOC. (2000) 1–10
20. Ta-Shma, A., Zuckerman, D., Safra, S.: Extractors from Reed-Muller codes. In IEEE, ed.: 42nd FOCS. (2001) 638–647
21. Babai, L., Fortnow, L., Nisan, N., Wigderson, A.: BPP has subexponential time simulations unless EXPTIME has publishable proofs. Computational Complexity **3** (1993) 307–318
22. Impagliazzo, R., Wigderson, A.: P = BPP if E requires exponential circuits: Derandomizing the XOR lemma. In: 29th STOC. (1997) 220–229
23. Nisan, N., Wigderson, A.: Hardness vs. randomness. J. Comput. System Sci. **49** (1994) 149–167
24. Nisan, N.: Pseudorandom generators for space-bounded computations. In ACM, ed.: 22nd STOC. (1990) 204–212
25. Impagliazzo, R., Nisan, N., Wigderson, A.: Pseudorandomness for network algorithms. In ACM, ed.: 26th STOC. (1994) 356–364
26. Bar-Yossef, Z., Reingold, O., Shaltiel, R., Trevisan, L.: Streaming computation of combinatorial objects. In: Conference on Computational Complexity (CCC). Volume 17. (2002)
27. Nisan, N., Zuckerman, D.: Randomness is linear in space. Journal of Computer and System Sciences **52** (1996) 43–52 Preliminary version in STOC' 93.
28. Saks, M.: Randomization and derandomization in space-bounded computation. In: Conference on Computational Complexity (CCC). (1996) 128–149
29. Jerrum, M.R., Valiant, L.G., Vazirani, V.V.: Random generation of combinatorial structures from a uniform distribution. Theoretical Computer Science **43** (1986) 169–188

Bounds on 2-Query Codeword Testing

Eli Ben-Sasson[1], Oded Goldreich[2], and Madhu Sudan[3]

[1] Division of Engineering and Applied Sciences, Harvard University and Laboratory
for Computer Science, Massachusetts Institute of Technology, Cambridge, MA.
eli@eecs.harvard.edu
[2] Department of Computer Science, Weizmann Institute of Science, Rehovot, Israel.
oded@wisdom.weizmann.ac.il
[3] Laboratory for Computer Science, Massachusetts Institute of Technology,
200 Technology Square, Cambridge, MA 02139.
madhu@mit.edu

Abstract. We present upper bounds on the size of codes that are locally testable by querying only two input symbols. For linear codes, we show that any 2-locally testable code with minimal distance δn over any finite field $\mathbb{F}$ cannot have more than $|\mathbb{F}|^{3/\delta}$ codewords. This result holds even for testers with two-sided error. For general (non-linear) codes we obtain the exact same bounds on the code size as a function of the minimal distance, but our bounds apply only for binary alphabets and one-sided error testers (i.e. with perfect completeness). Our bounds are obtained by examining the graph induced by the set of possible pairs of queries made by a codeword tester on a given code. We also demonstrate the tightness of our upper bounds and the essential role of certain parameters.

1 Introduction

Locally testable codes are error-correcting codes that admit very efficient codeword testers. Specifically, using a constant number of (random) queries, non-codewords are rejected with probability proportional to their distance from the code.

Locally testable codes arise naturally from the study of probabilistically checkable proofs, and were explicitly defined in [5] and systematically studied in [7]. The task of testing a code locally may also be viewed as a special case of the general task of property testing initiated by [9,6], where the property being tested here is that of being a codeword. In this paper we explore codes that can be tested with constant number of queries.

We focus on codes $\mathcal{C} \subset \Sigma^n$ that have large distance (i.e., each pair of codewords differ in at least $\Omega(n)$ coordinates) and large size (i.e., at the very least, $|\mathcal{C}|$ should grow with n and $|\Sigma|$). Such codes are known to exist. Specifically, in [7] locally testable codes are shown such that $|\mathcal{C}| = |\Sigma|^k$ for $k = n^{1-o(1)}$. We highlight two of these results:

S. Arora et al. (Eds.): APPROX 2003+RANDOM 2003, LNCS 2764, pp. 216–227, 2003.

1. For $\Sigma = \{0, 1\}$, *three* queries are shown to suffice. Furthermore, these codes are *linear*.
2. For $|\Sigma| > 2$, *two* queries are shown to suffice.[4]

This raises the question of whether binary codes and/or linear codes can have codeword tests that make only *two queries*. In this paper, we show that the answer is essentially negative; that is, *for codes of linear distance, such codes can contain only a constant number of codewords*. More general statements are provided by Theorems 3.1 and 4.1, which address linear codes over arbitrary fields and non-linear binary codes, respectively. We also address the tightness of our upper-bounds and the essential role of certain parameters (i.e., our upper-bounds apply either to linear codes or to binary codes that have a tester of perfect completeness).

Organization: In Section 2 we present the main definitions used in this paper, and state our main results. In Section 3 we study *linear* codes that admit two-query codeword testers. In Section 4 we study general *binary* codes that admit two-query codeword testers of perfect completeness. Due to space considerations, the rests of our results appear only in our technical report [3]: In [3, Sec. 5] we show that our upper-bounds cease to hold for *ternary non-linear* codes (rather than for non-linear codes over much larger alphabets as considered in [7] and mentioned in Item 2 above). In [3, Sec. 6] we show that perfect completeness is essential for the results regarding *non-linear binary* codes (presented in Section 4).

2   Formal Setting

We consider words over an alphabet Σ. For $w \in \Sigma^n$ and $i \in [n]$, we denote by w_i the i-th symbol of w; that is, $w = w_1 \cdots w_n$.

2.1   Codes

We consider codes $\mathcal{C} \subseteq \Sigma^n$ over a finite size alphabet Σ. The blocklength of $\mathcal{C}$ is n, and the size of $\mathcal{C}$ is its cardinality $|\mathcal{C}|$. We use normalized Hamming distance as our distance measure; that is, for $u, v \in \Sigma^n$ the distance $\Delta(u, v)$ is defined as the number of locations on which u and v differ, divided by n (i.e., $\Delta(u, v) = |\{i : u_i \neq v_i\}|/n$). The relative minimal distance of a code, denoted $\delta(\mathcal{C})$, is the minimal normalized Hamming distance between two distinct codewords. Formally

$$\delta(\mathcal{C}) = \min_{u \neq v \in \mathcal{C}} \{\Delta(u, v)\}$$

The distance of a word w from the code, denoted $\Delta(w, \mathcal{C})$, is $\min_{v \in \mathcal{C}} \{\Delta(w, v)\}$.

---

[4] We comment that these codes are "linear" in a certain sense. Specifically, Σ is a vector space over a field F, and the code is a linear subspace over F (rather than over Σ). That is, if $\Sigma = F^\ell$ then $\mathcal{C} \subset \Sigma^n$ is a linear subspace of $F^{n \cdot \ell}$ (but not of Σ^n, no matter what finite field we associate with Σ). In the coding literature such codes are called F-linear.

A code is called redundant if its projection on some coordinate is constant (i.e., there exists $i \in \{1, \ldots, n\}$ such that for any two codewords w, w' it holds that $w_i = w'_i$). A redundant code can be projected on all non-redundant coordinates, yielding a code with the same size and distance, but smaller blocklength. Thus, w.l.o.g., *we assume all codes to be non-redundant*.

Typically (in this paper) Σ is a finite field $\mathbb{F}$ and we view $\mathbb{F}^n$ as a vector space over $\mathbb{F}$. In particular, for $u, v \in \mathbb{F}^n$ the inner product of the two is $\langle v, u \rangle = \sum_{i=1}^{n} v_i \cdot u_i$ (all arithmetic operations are in $\mathbb{F}$). The weight of $v \in \mathbb{F}^n$, denoted $\mathrm{wt}(v)$, is the number of non-zero elements in v. In this case $\Delta(u, v) = \mathrm{wt}(u - v)/n$.

2.2   Testers and Tests

By a codeword tester (or simply tester) with query complexity q, completeness c and soundness s (for the code $\mathcal{C} \subseteq \Sigma^n$) we mean a *randomized* oracle machine that given oracle access to $w \in \Sigma^n$ (viewed as a function $w : \{1, \ldots, n\} \to \Sigma$) satisfies the following three conditions:

- Query Complexity q: The tester makes at most q queries to w.
- Completeness: For any $w \in \mathcal{C}$, given oracle access to w the tester accepts with probability at least c.
- Soundness: For any w that is at relative distance at least $\delta(\mathcal{C})/3$ from $\mathcal{C}$, given oracle access to w, the tester accepts with probability at most s.[5]

If $\mathcal{C}$ has a codeword tester with query complexity q, completeness c and soundness s we say $\mathcal{C}$ is $[q, c, s]$-locally testable.

A deterministic test (or simply test) with query complexity q is a *deterministic* oracle machine that given oracle access to $w \in \Sigma^n$ makes at most q queries to w, and outputs 1 (= accept) or 0 (= reject). Any (randomized) tester can be described as a distribution over deterministic tests, and we adopt this view throughout the text.

A (deterministic) test is called adaptive if its queries depend on previous answers of the oracle, and otherwise it is called non-adaptive. A test has perfect completeness if it accepts all codewords. Both notions extend to (randomized) testers. Alternatively, we say that a tester is non-adaptive (resp., has perfect completeness) if all the deterministic tests that it uses are non-adaptive (resp., have perfect completeness resp.), and otherwise it is adaptive (resp., has non-perfect completeness).

2.3   Our Results

We study 2-query codeword testers. Our main results are upper-bounds on the sizes of linear (resp., binary) codes admitting such testers (resp., testers of perfect completeness):

---

[5] We have set the *detection radius* of the tester at third its distance (i.e., for any w whose distance from $\mathcal{C}$ is at least $\frac{1}{3} \cdot \delta(\mathcal{C})$ the test rejects with probability at least s). As will be evident from the proofs, our results hold for any radius less than half the distance.

Theorem 2.1 *For any constants $c > s$, any $[2, c, s]$-locally testable* linear *code over Σ has at most $|\Sigma|^{3/\delta}$ codewords, where δ is its relative distance.*

Theorem 2.2 *For any constant $s < 1$, any $[2, 1, s]$-locally testable* binary *code has at most $2^{3/\delta}$ codewords, where δ is its relative distance.*

In contrast to the above, we state the following facts:

1. The upper-bounds stated in Theorems 2.1 and 2.2 are reasonablly tight: For some constants $s < 1$ and $\delta > 0$, and every finite field $\mathbb{F}$, there exists a linear $[2, 1, s]$-locally testable code of size $|\mathbb{F}|^{1/\delta}$ and minimal relative distance δ over $\mathbb{F}$ (see, Proposition 3.6).
2. Non-linearity of the code is essential to Theorem 2.1 and binary alphabet is essesntial to Theorem 2.2: there exists good *non-linear codes over ternary alphabets* that have 2-query codeword testers (of perfect completeness). That is, for some constants $s < 1$ and $\delta > 0$, there exists a $[2, 1, s]$-locally testable *ternary* code of relative distance δ that has size that grows almost linearly with the blocklength (see [3, Thm. 5.6]).
3. Perfect completeness is essesntial to Theorem 2.2: there exists good *non-linear codes over binary alphabets* that have 2-query codeword testers of *non-perfect completeness*. That is, for some constants $c > s > 0$ and $\delta > 0$, there exists a $[2, c, s]$-locally testable *binary* code of relative distance δ that has size that grows almost linearly with the blocklength (see [3, Thm. 6.1]).
4. Regarding the difference between linearity and "semi-linearity" (as in Footnote 1), we note that there exists good $GF(2)$-*linear codes over* $\{0, 1\}^2$ that have 2-query codeword testers (of perfect completeness): (see [3, Thm. 5.7]).

We mention that some of our results are analogous to results regarding probabilistic checkable proof (PCP) systems. In particular, let $\mathcal{PCP}^{\Sigma}_{c,s}[\log, q]$ denote the class of languages having PCP systems with logarithmic randomness, making q queries to oracles over the alphabet Σ, and having completeness and soundness bounds c and s respectively. Then, it is known that $\mathcal{PCP}^{\{0,1\}}_{1,s}[\log, 2] = \mathcal{P}$ for every $s < 1$, whereas $\mathcal{PCP}^{\{0,1\}}_{c,s}[\log, 2] = \mathcal{NP}$ for some $c > s > 0$ and $\mathcal{PCP}^{\{0,1,2\}}_{1,s}[\log, 2] = \mathcal{NP}$ for some $s < 1$.[6] Folllowing [7], we warn that the translation between PCPs and locally-checkable codes is not obvious. In particular, we do not know whether it is possible to obtain our coding results from the known PCP results or the other way around.

3   Linear Codes

In this section we show that $[2, c, s]$-locally testable *linear* codes with constant minimal relative distance must have very small size. Throughout this section $\mathbb{F}$ is a finite field of size $|\mathbb{F}|$. A code $\mathcal{C} \subseteq \mathbb{F}^n$ is called linear if it is a linear subspace of $\mathbb{F}^n$. The main result of this section is the following.

---

[6] The first two results are proven in [2], whereas the third result is lolklore that is based on the NP-Hardness of approximating Max3SAT as established in [1].

Theorem 3.1 (Theorem 2.1, restated): *Let $\mathcal{C} \subset \mathbb{F}^n$ be a $[2, c, s]$-locally testable* linear *code with minimal relative distance δ. If $c > s$ then*

$$|\mathcal{C}| \leq |\mathbb{F}|^{3/\delta}$$

We start by pointing out that, when considering testers for linear codes, the tester can be assumed to be non-adaptive and with perfect completeness. This holds by the following result of [4].

Theorem 3.2 [4]: *If a linear code (over any finite field) is $[q, c, s]$-locally testable using an adaptive tester, then it is $[q, 1, 1 - (c - s)]$-locally testable using a non-adaptive tester.*

Notice that if we start off with a tester having completeness greater than soundness ($c > s$), then the resulting non-adaptive, perfect-completeness tester (guaranteed by Theorem 3.2) will have soundness strictly less than 1. Thus, in order to prove Theorem 3.1 it suffices to show the following.

Theorem 3.3 *Let $\mathcal{C} \subseteq \mathbb{F}^n$ be a $[2, 1, s]$-locally* (non-adaptively) *testable linear code, with $s < 1$, and let the minimal relative distance be δ. Then:*

$$|\mathcal{C}| \leq |\mathbb{F}|^{3/\delta}$$

In the rest of the section we prove Theorem 3.3. The proof idea is as follows. Each possible test of query complexity 2 and perfect completeness imposes a constraint on the code, because all codewords must pass the test. Thus, we view the n codeword coordinates as *variables* and the set of tests as inducing constraints on these variables (i.e., codewords correspond to assignments (to the variables) that satisfy all these constraints). Since the code is linear, each test imposes a *linear* constraint on the pair of variables queried by it. (A linear constraint on the variables x, y has the form $ax + by = 0$ for some fixed $a, b \in \mathbb{F}$). We will show that in a code of large distance, these constraints induce very few satisfying assignments. Specifically, we look at the graph in which the vertices are the (n) codeword-coordinates (or variables) and edges connect two vertices that share a test. The main observation is that in any codeword, the values of all variables in a connected component are determined by the value of any one variable in the component; that is, the assignment to a single variable determines the assignment to the whole component. By perfect completeness, any word that satisfies all constraints in all connected components will pass *all* tests. Hence there cannot be many variables in small connected components, for then we could find a word that is far from the code and yet is accepted with probability 1. But this means that the code is essentially determined by the (small number of) large connected components, and hence the size of the code is small. We now give the details, starting with a brief discussion of *dual codes* which is followed by the proof.

3.1   Linearity Tests and Dual Codes

Recall that $\mathcal{C} \subseteq \mathbb{F}^n$ is linear iff for all $u, v \in \mathcal{C}$ we have $u + v \in \mathcal{C}$. In this case $\delta(\mathcal{C}) = \min_{w \in \mathcal{C}}\{\text{wt}(w)/n\}$. As pointed out in [8], codeword tests for linear codes are intimately related to the "dual" of the code. For a linear code $\mathcal{C}$, the *dual code* $\mathcal{C}^\perp$ is defined as the subspace of $\mathbb{F}^n$ orthogonal to $\mathcal{C}$, i.e.

$$\mathcal{C}^\perp = \{v : v \perp \mathcal{C}\}$$

where $v \perp \mathcal{C}$ iff for all $u \in \mathcal{C}$, $v \perp u$ (recall $v \perp u$ iff $\langle v, u \rangle = 0$).

The support of a vector v, denoted $\textbf{Supp}(v)$, is the set of indices of non-zero entries. Similarly, the support of a test T is the set of indices it queries. Notice that a non-adaptive test with query complexity q has support size q. For $v, u \in \mathbb{F}^n$ we say that v covers u if $\textbf{Supp}(v) \supseteq \textbf{Supp}(u)$. A test is called trivial if it always accepts. Elementary linear algebra gives the following claim.

Proposition 3.4 *The support of any non-trivial perfect-completeness test for* $\mathcal{C}$ *covers an element of* $\mathcal{C}^\perp \setminus \{0^n\}$.

Proof: Let T be a test and $\mathcal{C}_T$ be the projection of (the linear space) $\mathcal{C}$ onto $\textbf{Supp}(T)$. The projection is a linear operator, so $\mathcal{C}_T$ is a linear space over $\mathbb{F}$. The linear space $\mathcal{C}_T$ must be a strict subspace of $\mathbb{F}^{\textbf{Supp}(T)}$, because $|\mathcal{C}_T| = |\mathbb{F}^{\textbf{Supp}(T)}|$ (i.e. $\mathcal{C}_T$ includes all vectors in $\mathbb{F}^{\textbf{Supp}(T)}$) implies that either T reject some valid codeword in $\mathcal{C}$ (in violation of perfect completeness) or T always accepts (in violation of non-triviality). It follows that $(\mathcal{C}_T)^\perp$ has a non-zero element, denoted w. However, $\textbf{Supp}(w) \subseteq \textbf{Supp}(T)$ and $w \in \mathcal{C}^\perp$, completing the proof. □

Clearly one can assume that all tests used by a tester are non-trivial. We also assume $\mathcal{C}^\perp$ has no element of weight 1, because otherwise $\mathcal{C}$ is redundant. Since we consider only testers that make two queries, it follows that all tests they use have support size exactly two. Furthermore, without loss of generality, all the tests are linear.[7]

3.2   Upper Bounds on Code Size

By the above discussion (i.e., end of Section 3.1), we may assume (w.l.o.g.) that the $[2, 1, s]$-tester for $\mathcal{C}$ is described by a distribution over

$$\mathcal{C}_2^\perp \overset{\text{def}}{=} \{v \in \mathcal{C}^\perp : \text{wt}(v) = 2\}$$

The test corresponding to $v \in \mathcal{C}_2^\perp$ refers to the orthogonality of v and the oracle w; that is, the test accepts w if $v \perp w$ and rejects otherwise.[8] We now look at $\mathcal{C}_2^\perp$ and bound the size of $(\mathcal{C}_2^\perp)^\perp$. Our theorem will follow because $\mathcal{C} \subseteq (\mathcal{C}_2^\perp)^\perp$.

---

[7] In genenral, without loss of generality, a one-sided tester for a property P accepts y if and only if its view of y is consistent with its view of some $x \in P$. In our case P is a linear space, so consistecy means satisfying a linear system. For further details see Appendix.

[8] Notice that since $\text{wt}(v) = 2$ such a test amounts to two queries into w.

The set $\mathcal{C}_2^\perp$ gives rise to a natural graph, denoted $G_\mathcal{C}$. The vertex set of $G_\mathcal{C}$ is $V(G_\mathcal{C}) = \{1,\ldots,n\}$ and $(i,j) \in E(G_\mathcal{C})$ iff there exists $v_{ij} \in \mathcal{C}_2^\perp$ with $\mathbf{Supp}(v_{ij}) = \{i,j\}$.

The key observation is that, for any edge $(i,j) \in E(G_\mathcal{C})$ there is some $c_{ij} \in \mathbb{F} \setminus \{0\}$ such that for any $w \in \mathcal{C}$ it holds that $w_i = c_{ij} \cdot w_j$. To see this, notice the constraint corresponding to (i,j) can be written as $a_{ij}w_i + b_{ij}w_j = 0$, where $a_{ij}, b_{ij} \in \mathbb{F}\setminus\{0\}$ (if either a_{ij} or b_{ij} are 0 then v_{ij} has support size one, meaning $\mathcal{C}$ is redundant). So, by transitivity, the value of w on all variables in the connected component of i, is determined by w_i. (Moreover, all these values are non-zero iff $w_i \neq 0$.) Assuming that the number of connected components is k, this implies that there can be at most $|\mathbb{F}|^k$ different codewords (because there are only k degrees of freedom corresponding to the settings (of all variables) in each of the k components). To derive the desired bound we partition the components into big and small ones, and bound the number of codewords as a function of the number of big components (while showing that the small components do not matter).

Let $C_1, \ldots, C_k$ be the connected components of $G_\mathcal{C}$. We call a component small if its cardinality is less than $\delta n/3$. Without loss of generality, let $C_1, \ldots C_s$ be all the small components, and let $S = \bigcup_{i=1}^s C_i$ denote their union.

Claim 3.5 $|S| \leq 2\delta n/3$.

Proof: Otherwise there exists $I \subset \{1,\ldots,s\}$ such that

$$\delta n/3 \leq \sum_{i \in I} |C_i| < 2\delta n/3$$

For every $i \in I$, we consider a vector $w^i \in (\mathcal{C}_2^\perp)^\perp$ with $\mathbf{Supp}(w^i) = C_i$. To see that such a vector exists, set an arbitrary coordinate of C_i to 1 (which is possible because the code is not redundant) and force non-zero values to all other coordinates in C_i (by virtue of the above discussion). Furthermore, note that this leaves all coordinates out of C_i unset, and that the resulting w^i satisfy all tests in $\mathcal{C}_2^\perp$ (where the tests that correspond to the edges in C_i are satified by our setting of the non-zero values, whereas all other tests refer to vertices out of C_i and are satisfied by zero values). Now, define $w = \sum_{i \in I} w^i$. By definition, we have $\mathbf{Supp}(w) = \cup_{i \in I} C^i$, and $\delta n/3 \leq \mathrm{wt}(w) < 2\delta n/3$ follows by the hypothesis. Hence, $\Delta(w,\mathcal{C}) \geq \delta/3$.

On the other hand, w is orthogonal to $\mathcal{C}_2^\perp$. To see this, consider any $v \in \mathcal{C}^\perp$. If $\mathbf{Supp}(v) \subseteq C_i$, for some $i \in I$, then the "view v has of w" (i.e. the values of the coordinates v queries) is identical to the view v has of the codeword w^i, and so $\langle v, w \rangle = \langle v, w^i \rangle = 0$. Otherwise (i.e., $\mathbf{Supp}(v)$ has empty intersection with S), by definition v "sees" only zeros, and so $\langle v, w \rangle = 0$.

We conclude w is $\frac{\delta}{3}$-far from $\mathcal{C}$, yet it passes all possible tests of query complexity two. This contradicts the soundness condition, and the claim follows. $\square$

Proof (of Theorem 3.3):   *Assume for the sake of contradiction that*

$$|\mathcal{C}| > |\mathbb{F}|^{3/\delta}$$

Recall that (by the "key observation") the values of all variables in a connected component are determined by the value of a single variable in this component. Since there are at most $3/\delta$ large connected components in G_C (because each has cardinality at least $\delta n/3$), the contradiction hypothesis implies that there exist two codewords $x \neq y$ that agree on all variables that reside in the large connected components. Indeed, these two codewords $x \neq y$, may differ on variables that reside in the small connected components (i.e., variables in S), but Claim 3.5 says that there are few such variables (i.e.. $|S| \leq 2\delta n/3$). By linearity $x - y \in C$ (but $x - y \neq 0^n$), and so $0 < \mathrm{wt}(x - y) \leq |S| < \delta n$. We have reached a contradiction (because C has distance δ), and Theorem 3.3 follows. ∎

3.3   Tightness of the Upper Bound

We remark that our upper bound is quite tight. For any $\delta < 1$, consider the following code $C_n \subset \mathbb{F}^n$ formed by taking $1/\delta$ elements of $\mathbb{F}$ and repeating each one of them δn times. Thus, a codeword in C_n is formed of $1/\delta$ blocks, each block of the form $e^{\delta n}$ for some $e \in \mathbb{F}$ (here e^k means k repetitions of e).

Proposition 3.6 C_n *is a linear* $[2, 1, 1 - \frac{2\delta}{3|\mathbb{F}|}]$-*locally testable code with minimal relative distance δ and size* $|\mathbb{F}|^{1/\delta}$.

For instance, taking $\mathbb{F} = GF(2)$, the soundness parameter in the proposition is $1 - \delta/3$.

Proof: The linearity, distance and size of C_n are self-evident. Consider the following natural tester for C_n: Select a random block, read two random elements in it, and accept iff the two are equal. This tester has perfect completeness and query complexity 2. As to the soundness, let $k = 1/\delta$ and write $v \in \mathbb{F}^n$ as $(v^{(1)}, \ldots, v^{(k)})$, where $v^{(i)}$ is the i-th block of v (i.e., $|v^{(i)}| = \delta n$). The Hamming distance of v from C_n is the sum of the Hamming distances of the individual blocks $v^{(i)}$ from the code $B = \{e^{\delta n} : e \in \mathbb{F}\}$.

Suppose v has relative distance at least $\delta/3$ from C_n. Let δ_i denote the relative distance of $v^{(i)}$ from B. Then, $\frac{1}{k}\sum_{i=1}^{k} \delta_i \geq \delta/3$ (and $\delta_i \leq 1 - \frac{1}{|\mathbb{F}|}$). The acceptance probability of the tester equals

$$\frac{1}{k}\sum_{i=1}^{k}\left(\delta_i^2 + (1 - \delta_i)^2\right) = 1 - \frac{2}{k}\sum_{i=1}^{k}(1 - \delta_i) \cdot \delta_i$$

$$\leq 1 - \frac{2}{k|\mathbb{F}|}\sum_{i=1}^{k}\delta_i$$

$$\leq 1 - \frac{2\delta}{3|\mathbb{F}|}$$

where the first inequality is due to $\delta_i \leq 1 - \frac{1}{|\mathbb{F}|}$. Thus, the soundness parameter is as claimed. □

4   Non-linear Codes

In this section we provide upper bounds on the code size of arbitrary (i.e., possibly non-linear) 2-locally testable codes. Our bounds apply only to binary codes and testers with perfect completeness, and with good reason: There exist good 2-testable binary codes with non-perfect completeness (see [3, Sec. 6]) and there exist good 2-testable codes with perfect completeness over ternary alphabets (see [3, Sec. 5]). Our main result is:

Theorem 4.1 (Theorem 2.2, restated): *If* $\mathcal{C} \subseteq \{0, 1\}^n$ *is a* $[2, 1, s]$-*locally testable code with minimal relative distance* δ *and* $s < 1$, *then*

$$|\mathcal{C}| \leq 2^{3/\delta}$$

The proof (presented below) generalizes that of the *binary* linear case (binary means $\mathbb{F} = GF(2)$), with some necessary modifications, which we briefly outline now. In the binary linear case a test querying x_i and x_j forces $x_i = x_j$ for all codewords (this is the only possible linear constraint of size two over $GF(2)$). In that case, the set of all tests corresponds to an undirected graph in which each connected component forces all variables to have the same value. In the non-linear case a test (adaptive or non-adaptive) corresponds to a 2-CNF. (Recall that in both cases we deal with perfect completeness testers.) The set of all tests (which is itself a 2-CNF) corresponds to a *directed* graph of constraints on codewords, where the constraint $x_i \vee x_j$ translates to the pair of directed edges $\bar{x}_i \to x_j$ and $\bar{x}_j \to x_i$. In the resulting *directed* graph, a *strongly* connected component takes the role played by the connected component in the linear case. Namely, for any codeword, all variables in a strongly connected component are fixed by the value of a single variable in the component. As in the linear case, we use the properties of the code and its tester (i.e., the code's large distance and the fact that the tester rejects any word that is far from the code with non-zero probability) to show that the weight of the *small* strongly connected components is small. Hence, the code is determined by a small number of *large* connected components.

Proof of Theorem 4.1

Again, we view the n codeword coordinates as *variables* and the set of tests (which are 2-CNFs) as inducing constraints on these variables. We stress that each test (even an adaptive one) can be represented by a 2-CNF.[9] Let $\mathcal{F}$ be the conjunction of all non-trivial deterministic tests that are used by a 2-query tester that has perfect completeness with respect to $\mathcal{C}$. We look at the satisfying assignments of $\mathcal{F}$, and use this to bound the size of $\mathcal{C}$. If $\mathcal{F}$ includes a clause of

---

[9] In general, an adaptive test querying k variables is a decision tree of depth k. It is easy to verify that (the function computed by) such a tree can be represented both as a k-CNF and as a k-DNF.

size 1 then C is redundant. Thus, assuming non-redundancy of C implies that $\mathcal{F}$ can be represented by a 2-CNF in which each clause has *exactly* two literals.

We examine the following directed graph $G_{\mathcal{F}}$. The vertex set of $G_{\mathcal{F}}$ is the set of literals $\{x_1, \bar{x}_1 \ldots, x_n, \bar{x}_n\}$. For each clause $(\ell \vee \ell') \in \mathcal{F}$ we introduce in $G_{\mathcal{F}}$ one directed edge from $\bar{\ell}$ to ℓ', and one from $\bar{\ell'}$ to ℓ. We use the notation $\ell \leadsto \ell'$ to indicate the existence of a directed path from ℓ to ℓ' in $G_{\mathcal{F}}$. We use the notation $w(\ell)$ to denote the value of literal ℓ under assignment w to the underlying variables. Identifying *True* with 1 and *False* with 0, we have

Claim 4.2 (folklore): *The following two conditions are equivalent*

1. *The assignment w satisfies $\mathcal{F}$.*
2. *For every directed edge $\ell \leadsto \ell'$ it holds that $w(\ell) \leq w(\ell')$.*

A strongly connected component in a directed graph G is a maximal set of vertices $C \subseteq V(G)$ such that for any $v, v' \in C$ it holds that $v \leadsto v'$. For two strongly connected components C and C' in G, we say $C \leadsto C'$ iff there exist $v \in C$ and $v' \in C'$ such that $v \leadsto v'$. (Indeed, this happens iff for all $v \in C, v' \in C'$ it holds that $v \leadsto v'$.)

By Claim 4.2, w satisfies all constraints corresponding to edges of a strongly connected component C iff $w(\ell) = w(\ell')$ for all $\ell, \ell' \in C$. So, any satisfying assignment w either sets to 1 *all* literals in C, or sets them all to 0. In the first case we say that $w(C) = 1$ and in the latter we say $w(C) = 0$.

Let L be the set of literals belonging to large strongly-connected components, where a component is called large iff its cardinality is at least $\delta n/3$. Consider an arbitrary assignment ρ' to the variables of L that can be extended to a satisfying assignment (to $\mathcal{F}$). In particular, ρ' does not falsify any clause of $\mathcal{F}$ (i.e., no clause of $\mathcal{F}$ is set to 0 by ρ'). A literal $\ell \notin L$ is said to be forced by ρ' if there exists $\ell' \in L$ such that $\ell' \leadsto \ell$ and $\rho'(\ell') = 1$. This is because any satisfying assignment to $\mathcal{F}$ that extends ρ' must set ℓ to 1 (since for such an assignment ρ it must holds that $\rho(\ell) \geq \rho(\ell') = 1$. Indeed, the complementary literal (i.e., $\bar{\ell}$) is forced to 0. Let ρ be the *closure* of ρ' obtained by (iteratively) fixing all forced literals to the value 1 (and their complementary literals to 0). By definition, ρ does not falsify $\mathcal{F}$. Let S_ρ be the set of unfixed variables under ρ.

Claim 4.3 *For any closure ρ of an assignment that satisfies L, it holds that $|S_\rho| \leq 2\delta n/3$.*

Proof: Otherwise, let $C_1, \ldots, C_k$ be a topological ordering of the unfixed strongly connected components comprising S_ρ, where the ordering is according to $\leadsto$ (as defined above). (Indeed, the digraph defined on the C_i's by $\leadsto$ is acyclic.) For $j = 0, \ldots, k$, let $v^{(j)}$ be the assignment extending ρ defined by:

$$v^{(j)}(C_i) = \begin{cases} 0 & i \leq j \\ 1 & i > j \end{cases}$$

By Claim 4.2, each assignment $v^{(j)}$ satisfies $\mathcal{F}$. Since C is 2-locally testable with soundness $s < 1$, each word that is at distance at least $\delta/3$ from C must falsify some clause in $\mathcal{F}$. But since $v^{(j)}$ satisfies $\mathcal{F}$, it must be that $v^{(j)}$ is within distance $\delta/3$ from some codeword, denoted $w^{(j)}$. By the contradiction hypothesis, we have $\Delta(v^{(0)}, v^{(k)}) = |S_\rho|/n > 2\delta/3$, which implies $w^{(0)} \neq w^{(k)}$ (because $\Delta(v^{(0)}, v^{(k)}) \leq \Delta(v^{(0)}, w^{(0)}) + \Delta(w^{(0)}, w^{(k)}) + \Delta(w^{(k)}, v^{(k)})$, which is upper-bounded by $2 \cdot (\delta/3) + \Delta(w^{(0)}, w^{(k)}))$. It follows that

$$\Delta(v^{(k)}, w^{(0)}) \geq \Delta(w^{(k)}, w^{(0)}) - \Delta(w^{(k)}, v^{(k)}) \geq \delta - (\delta/3) = 2\delta/3$$

On the other hand, recall that $\Delta(v^{(0)}, w^{(0)}) \leq \delta/3$. Since, for each j, it holds that $\Delta(v^{(j)}, v^{(j+1)}) < \delta/3$ (because $|C_j| < \delta n/3$), there must be $j \in \{0, 1, \ldots, k\}$ such that $\delta/3 \leq \Delta(v^{(j)}, w^{(0)}) \leq 2\delta/3$. For this j, it holds that $\Delta(v^{(j)}, \mathcal{C}) \geq \delta/3$. But $v^{(j)}$ satisfies $\mathcal{F}$ and so it is accepted by the tester with probability 1, in contradiction to the soundness condition. $\square$

Our proof is nearly complete. As in the proof of Theorem 3.3, assume for the sake of contradiction that

$$|\mathcal{C}| > 2^{\delta/3}$$

In this case, there must be two distinct codewords $w \neq u$ that agree on all large connected components. Let ρ' be the restriction of w to the variables of the large connected components. That is, ρ' agrees with w and with u on the assignment to all variables in L and is unfixed otherwise. Let ρ be the closure of ρ' (obtained by forcing as above). Note that w and u are satisfying assignments to $\mathcal{F}$ that agree on ρ', so they also must agree on ρ (which is forced by ρ'). Thus, by Claim 4.3

$$0 < \Delta(u, w) \leq |S_\rho|/n < \delta$$

This contradicts the hypothesis that the minimal distance of $\mathcal{C}$ is δ, and the theorem follows. ∎

Acknowledgments

Eli Ben-Sasson was supported by NSF grants CCR-0133096, CCR-9877049, CCR 0205390, and NTT Award MIT 2001-04. Oded Goldreich was supported by the MINERVA Foundation, Germany. Madhu Sudan was supported in part by NSF Awards CCR 9912342, CCR 0205390, and NTT Award MIT 2001-04.

References

1. S. Arora, C. Lund, R. Motwani, M. Sudan and M. Szegedy. Proof Verification and Intractability of Approximation Problems. *Journal of the ACM*, Vol. 45, pages 501–555, 1998.
2. M. Bellare, O. Goldreich and M. Sudan. Free Bits, PCPs and Non-Approximability – Towards Tight Results. *SIAM Journal on Computing*, Vol. 27, No. 3, pages 804–915, 1998.
3. E. Ben-Sasson, O. Goldreich and M. Sudan. Bounds on 2-Query Codeword Testing. ECCC, TR03-019, 2003.

4. E. Ben-Sasson, P. Harsha, S. Raskhodnikova. Some 3-CNF Properties are Hard to Test. In *35th STOC*, 2003.
5. K. Friedl and M. Sudan. Some Improvements to Total Degree Tests. In *Proc. of ISTCS*, pages 190-198, 1995.
6. O. Goldreich, S. Goldwasser, D. Ron. Property Testing and its connection to Learning and Approximation. *Journal of the ACM*, 45(4):653–750, July 1998.
7. O. Goldreich and M. Sudan. Locally Testable Codes and PCPs of Almost-Linear Length. In *43rd FOCS*, pages 13–22, 2002.
8. M. Kiwi. *Probabilistically Checkable Proofs and the Testing of Hadamard-like Codes.* Ph.D. Thesis, MIT, 1996.
9. R. Rubinfeld and M. Sudan. Robust characterization of polynomials with applications to program testing. *SIAM Journal on Computing*, Vol. 25 (2), pages 252–271, 1996.

Appendix: A General Proposition Regarding Property Testing

In Section 3.1, we used the fact that, without loss of generality, a perfect-completeness codeword-tester for a linear code makes only linear tests. This fact is a special case of the following general (folklore) proposition:

Proposition A.1 *Let M be an oracle machine for the promise problem (Π_{YES}, Π_{NO}) such that for every $x \in \Pi_{\text{YES}}$ it holds that $\mathbf{Pr}[M^x = 1] = 1$ (i.e., M has perfect completeness). Then, modifying M such that it outputs 1 if and only if its view is consistent with some $x' \in \Pi_{\text{YES}}$ may only improve its performance. That is, denoting the modified machine by $\widetilde{M}$, we have $\mathbf{Pr}[\widetilde{M}^x = 1] = 1$ for every $x \in \Pi_{\text{YES}}$ and $\mathbf{Pr}[\widetilde{M}^x = 1] \leq \mathbf{Pr}[M^x = 1]$ for every x.*

In our case, the property being tested is belonging to a certain linear subspace, and thus in our case consistecy (among two answers) means satisfying a linear condition.

Proof: Let us fix a contents r to the random-tape of M, and denote by $\text{view}^x_M(r)$ the view of machine M on random-tape r and access to oracle x. Then, machine $\widetilde{M}$ accepts on random-tape r and access to oracle x if and only if $\text{view}^x_M(r)$ equals $\text{view}^{x'}_M(r)$ for some $x' \in \Pi_{\text{YES}}$ (where the condition may be determined by scanning all $x' \in \Pi_{\text{YES}}$ and computing the corresponding $\text{view}^{x'}_M(r)$'s). Clearly, $\mathbf{Pr}[\widetilde{M}^x = 1] = 1$ for every $x \in \Pi_{\text{YES}}$ (by considering $x' = x$). On the other hand, for every x and r, if $M^x(r) \neq 1$ then by the one-sided feature of M it must be that $\text{view}^x_M(r)$ differs from $\text{view}^{x'}_M(r)$ for all $x' \in \Pi_{\text{YES}}$. It follows that $\widetilde{M}^x(r) \neq 1$ too. Thus, $\mathbf{Pr}[\widetilde{M}^x \neq 1] \geq \mathbf{Pr}[M^x \neq 1]$, and the proposition follows. $\square$

The Lovász Number of Random Graphs

Amin Coja-Oghlan[*]

Humboldt-Universität zu Berlin, Institut für Informatik,
Unter den Linden 6, 10099 Berlin, Germany
coja@informatik.hu-berlin.de

Abstract. We study the Lovász number ϑ along with two further SDP relaxations $\vartheta_{1/2}$, ϑ_2 of the independence number and the corresponding relaxations $\bar{\vartheta}$, $\bar{\vartheta}_{1/2}$, $\bar{\vartheta}_2$ of the chromatic number on random graphs $G_{n,p}$. We prove that $\bar{\vartheta}, \bar{\vartheta}_{1/2}, \bar{\vartheta}_2(G_{n,p})$ in the case $p < n^{-1/2-\varepsilon}$ are concentrated in intervals of constant length. Moreover, we estimate the probable value of $\vartheta, \bar{\vartheta}(G_{n,p})$ etc. for essentially the entire range of edge probabilities p. As applications, we give improved algorithms for approximating $\alpha(G_{n,p})$ and for deciding k-colorability in polynomial expected time.

1 Introduction and Results

Given a graph $G = (V, E)$, let $\alpha(G)$ be the independence number, let $\omega(G)$ be the clique number, and let $\chi(G)$ be the chromatic number of G. Further, let $\bar{G}$ signify the complement of G. Since it is NP-hard to compute any of $\alpha(G)$, $\omega(G)$ or $\chi(G)$, it is remarkable that there exists an efficiently computable function $\vartheta(G)$ that is "sandwiched" between $\alpha(G)$ and $\chi(\bar{G})$, i.e. $\alpha(G) \leq \vartheta(G) \leq \chi(\bar{G})$. Passing to complements, and letting $\bar{\vartheta}(G) = \vartheta(\bar{G})$, we have $\omega(G) \leq \bar{\vartheta}(G) \leq \chi(G)$. The function ϑ was introduced by Lovász, and is called the *Lovász number* of G (cf. [16,21]).

Though $\vartheta(G)$ is sandwiched between $\alpha(G)$ and $\chi(\bar{G})$, Feige [7] proved that the gap between $\alpha(G)$ and $\vartheta(G)$ or between $\chi(\bar{G})$ and $\vartheta(G)$ can be as large as $n^{1-\varepsilon}$, $\varepsilon > 0$. Indeed, unless NP=coRP, none of $\alpha(G)$, $\omega(G)$, $\chi(G)$ can be approximated within a factor of $n^{1-\varepsilon}$, $\varepsilon > 0$, in polynomial time [17,9]. However, though there exist graphs G such that $\vartheta(G)$ is not a good approximation of $\alpha(G)$ (or $\bar{\vartheta}(G)$ of $\chi(G)$), it might be the case that the Lovász number performs well on "average" instances. In fact, several algorithms for random and semirandom graph problems are based on computing ϑ [4,5,8]. Therefore, the aim of this paper is to study the Lovász number of random graphs more thoroughly.

The standard model of a random graph is the binomial model $G_{n,p}$, pioneered by Erdős and Renyi. We let $0 < p = p(n) < 1$ be a number that may depend on n. Let $V = \{1, \ldots, n\}$. Then the random graph $G_{n,p}$ is obtained by including each of the $\binom{n}{2}$ possible edges $\{v, w\}$, $v, w \in V$, with probability p independently. Though $G_{n,p}$ may fail to model some types of input instances appropriately, both the combinatorial structure and the algorithmic theory of

[*] supported by the Deutsche Forschungsgemeinschaft (grant DFG FOR 413/1-1).

S. Arora et al. (Eds.): APPROX 2003+RANDOM 2003, LNCS 2764, pp. 228–239, 2003.

$G_{n,p}$ are of fundamental interest [18,12]. We say that $G_{n,p}$ has some property A *with high probability* (whp.), if $\lim_{n \to \infty} P(G_{n,p}$ has property $A) = 1$.

We also address two further SDP relaxations $\vartheta_{1/2}, \vartheta_2$ of α (cf. [27]) on random graphs. These relaxations satisfy $\alpha(G) \leq \vartheta_{1/2}(G) \leq \vartheta(G) \leq \vartheta_2(G) \leq \chi(\bar{G})$, for all G. Passing to complements, and setting $\bar{\vartheta}_i(G) = \vartheta_i(\bar{G})$ ($i = 1/2, 2$), one gets $\omega(G) \leq \bar{\vartheta}_{1/2}(G) \leq \bar{\vartheta}(G) \leq \bar{\vartheta}_2(G) \leq \chi(G)$. The relaxation $\bar{\vartheta}_{1/2}(G)$ coincides with the well-known *vector chromatic number* $\chi(G)$ of Karger, Motwani, and Sudan [20].

The Concentration of $\bar{\vartheta}$, $\bar{\vartheta}_{1/2}$, $\bar{\vartheta}_2$.

A remarkable fact concerning the chromatic number of sparse random graphs $G_{n,p}$, $p \leq n^{-\varepsilon-1/2}$, is that $\chi(G_{n,p})$ is concentrated in an interval of constant length. Indeed, Shamir and Spencer [26] proved that there is a function $u = u(n, p)$ such that in the case $p = n^{-\beta}$, $1/2 < \beta < 1$, we have $P(u \leq \chi(G_{n,p}) \leq u + \lceil(2\beta+1)/(2\beta-1)\rceil) = 1 - o(1)$. Furthermore, Łuczak [25] showed that in the case $5/6 < \beta < 1$, the chromatic number is concentrated in width one. In fact, Alon and Krivelevich [2] could prove that two point concentration holds for the entire range $p = n^{-\beta}$, $1/2 < \beta < 1$. The two following theorems state similar results as given by Shamir and Spencer and by Łuczak for the relaxations $\bar{\vartheta}_{1/2}(G_{n,p})$, $\bar{\vartheta}(G_{n,p})$, and $\bar{\vartheta}_2(G_{n,p})$ of the chromatic number.

Theorem 1. *Suppose that $c_0/n \leq p \leq n^{-\beta}$ for some large constant $c_0 > 0$ and some number $1/2 < \beta < 1$. Then $\bar{\vartheta}_{1/2}(G_{n,p})$, $\bar{\vartheta}(G_{n,p})$, $\bar{\vartheta}_2(G_{n,p})$ are concentrated in width $s = \frac{2}{2\beta-1} + o(1)$, i.e. there exist numbers u, u', u'' depending on n and p such that whp. $u \leq \bar{\vartheta}_{1/2}(G_{n,p}) \leq u + s$, $u' \leq \bar{\vartheta}(G_{n,p}) \leq u' + s$, and $u'' \leq \bar{\vartheta}_{1/2}(G_{n,p}) \leq u'' + s$.*

Theorem 2. *Suppose that $c_0/n < p \leq n^{-5/6-\delta}$ for some large constant c_0 and some $\delta > 0$. Then $\bar{\vartheta}_{1/2}(G_{n,p})$, $\bar{\vartheta}(G_{n,p})$, and $\bar{\vartheta}_2(G_{n,p})$ are concentrated in width 1.*

In contrast to the chromatic number, $\bar{\vartheta}_{1/2}$, $\bar{\vartheta}$, and $\bar{\vartheta}_2$ need not be integral. Therefore, the above results do *not* imply that $\bar{\vartheta}_{1/2}(G_{n,p})$, $\bar{\vartheta}(G_{n,p})$, $\bar{\vartheta}_2(G_{n,p})$ are concentrated on a constant number of points.

The Probable Value of $\vartheta(G_{n,p})$, $\bar{\vartheta}(G_{n,p})$, etc.

Concerning the probable value of $\vartheta(G_{n,p})$ and $\bar{\vartheta}(G_{n,p})$, Juhász [19] gave the following partial answer: If $\ln(n)^6/n \ll p \leq 1/2$, then with high probability we have $\vartheta(G_{n,p}) = \Theta(\sqrt{n/p})$ and $\bar{\vartheta}(G_{n,p}) = \Theta(\sqrt{np})$. However, we shall indicate in Sec. 4 that Juhász's proof fails in the case of sparse random graphs (e.g. $np = O(1)$). Making use of concentration results on ϑ, $\bar{\vartheta}$ etc., we can compute the probable value not only of $\vartheta(G_{n,p})$ and $\bar{\vartheta}(G_{n,p})$, but also of $\vartheta_i(G_{n,p})$ and $\bar{\vartheta}_i(G_{n,p})$, $i = 1/2, 2$, for essentially the entire range of edge probabilities p.

Theorem 3. *Suppose that $c_0/n \leq p \leq 1/2$ for some large constant $c_0 > 0$. Then there exist constants $c_1, c_2, c_3, c_4 > 0$ such that*

$$c_1\sqrt{n/p} \leq \vartheta_{1/2}(G_{n,p}) \leq \vartheta(G_{n,p}) \leq \vartheta_2(G_{n,p}) \leq c_2\sqrt{n/p} \tag{1}$$
$$\text{and} \quad c_3\sqrt{np} \leq \bar{\vartheta}_{1/2}(G_{n,p}) \leq \bar{\vartheta}(G_{n,p}) \leq \bar{\vartheta}_2(G_{n,p}) \leq c_4\sqrt{np}$$

with high probability. More precisely,

$$P(c_3\sqrt{np} \leq \bar{\vartheta}_{1/2}(G_{n,p}) \leq \bar{\vartheta}(G_{n,p}) \leq \bar{\vartheta}_2(G_{n,p})) \geq 1 - \exp(-n). \tag{2}$$

Assume that $c_0/n \leq p = o(1)$. Then $\alpha(G_{n,p}) \sim 2\ln(np)/p$ and $\chi(G_{n,p}) \sim np/(2\ln(np))$ whp. (cf. [18]). Hence, Thm. 3 shows that $\vartheta_2(G_{n,p})$ $(\bar{\vartheta}_{1/2}(G_{n,p}))$ approximates $\alpha(G_{n,p})$ $(\chi(G_{n,p}))$ within a factor of $O(\sqrt{np})$. In fact, if $np = O(1)$, then we get a constant factor approximation. Our estimate on the vector chromatic number $\bar{\vartheta}_{1/2}(G_{n,p})$ answers a question of Krivelevich [22].

Finally, consider the *random regular graph* $G_{n,r}$. The proof of the following theorem is somewhat technical, and is omitted.

Theorem 4. *Let c_0 be a sufficiently large constant, and let $c_0 \leq r = o(n^{1/4})$. There are constants $c_1, c_2 > 0$ such that whp. the random regular graph $G_{n,r}$ satisfies $c_1 n/\sqrt{r} \leq \vartheta_{1/2}(G_{n,r}) \leq \vartheta(G_{n,r}) \leq \vartheta_2(G_{n,r}) \leq c_2 n/\sqrt{r}$. Moreover, there is a constant $c_3 > 0$ such that in the case $c_0 \leq r = o(n^{1/2})$ we have $P(c_3\sqrt{r} \leq \bar{\vartheta}_{1/2}(G_{n,r}) \leq \bar{\vartheta}(G_{n,r}) \leq \bar{\vartheta}_2(G_{n,r})) \geq 1 - \exp(-n)$.*

Algorithmic Applications. There are two types of algorithms for NP-hard random graph problems. First, there are *heuristics* that *always* run in polynomial time, and *almost always* output a good solution. On the other hand, there are algorithms that guarantee some approximation ratio on *any* input instance, and which have a polynomial *expected* running time when applied to $G_{n,p}$. In this paper, we deal with algorithms with a polynomial expected running time.

First, we consider the maximum independent set problem in random graphs. Krivelevich and Vu [23] gave an algorithm that in the case $p \gg n^{-1/2}$ approximates the independence number of $G_{n,p}$ in polynomial expected time within a factor of $O(\sqrt{np}/\ln(np))$. Moreover, they ask whether a similar algorithm exists for smaller values of p. As a first answer, Coja-Oghlan and Taraz [4], gave an $O(\sqrt{np}/\ln(np))$-approximative algorithm for the case $p \gg \ln(n)^6/n$.

Theorem 5. *Suppose that $c_0/n \leq p \leq 1/2$. There is an algorithm* ApproxMIS *that for any input graph G outputs an independent set of size at least $\frac{\alpha(G)\ln(np)}{c_1\sqrt{np}}$, and which applied to $G_{n,p}$ runs in polynomial expected time. Here $c_0, c_1 > 0$ denote constants.*

As a second application, we give an algorithm for deciding within polynomial expected time whether the input graph is k-colorable. Instead of $G_{n,p}$, we shall even consider the *semirandom model* $G_{n,p}^+$ that allows for an adversary to add edges to the random graph. We say that *the expected running time of an algorithm $\mathcal{A}$ is polynomial over $G_{n,p}^+$*, if there is some constant l such that the expected running time of $\mathcal{A}$ is $O(n^l)$ regardless of the behavior of the adversary.

Theorem 6. *Suppose that $k = o(\sqrt{n})$, and that $p \geq c_0 k^2/n$, for some constant $c_0 > 0$. There exists an algorithm $\mathtt{Decide}_k$ that for any input graph G decides whether G is k-colorable, and that applied to $G_{n,p}^+$ has a polynomial expected running time.*

The algorithm $\mathtt{Decide}_k$ is essentially identical with Krivelevich's algorithm for deciding k-colorability in polynomial expected time [22]. However, the analysis given in [22] requires that $np \geq \exp(\Omega(k))$. The improvement results from the fact that the analysis given in this paper relies on the asymptotics for $\bar\vartheta_{1/2}(G_{n,p})$ derived in Thm. 3 (instead of the concept of semi-colorings). Finally, we mention that our algorithm $\mathtt{Decide}_k$ also applies to random regular graphs $G_{n,r}$.

Theorem 7. *Suppose that $c_0 k^2 \leq r = o(n^{1/2})$ for some constant $c_0 > 0$. Then, applied to $G_{n,r}$, the algorithm $\mathtt{Decide}_k$ has polynomial expected running time.*

Notation. Throughout we let $V = \{1, \ldots, n\}$. If $G = (V, E)$ is a graph, then $A(G)$ is the adjacency matrix of G. By $\mathbf{1}$ we denote the vector with all entries $= 1$, and J denotes a square matrix with all entries $= 1$. If M is a real symmetric $n \times n$-matrix, then $\lambda_1(M) \geq \cdots \lambda_n(M)$ signify the eigenvalues of M.

2 Preliminaries

Let $G = (V, E)$ be a graph, let $(v_1, \ldots, v_n)$ be an n-tuple of unit vectors in $\mathbf{R}^n$, and let $k > 1$. Then $(v_1, \ldots, v_n)$ is a *vector k-coloring* of G if $\langle v_i, v_j \rangle \leq -1/(k-1)$ for all edges $\{i, j\} \in E$. Furthermore, $(v_1, \ldots, v_n)$ is a *strict* vector k-coloring if $\langle v_i, v_j \rangle = -1/(k-1)$ for all $\{i, j\} \in E$. Finally, we say that $(v_1, \ldots, v_n)$ is a *rigid* vector k-coloring if $\langle v_i, v_j \rangle = -1/(k-1)$ for all $\{i, j\} \in E$ and $\langle v_i, v_j \rangle \geq -1/(k-1)$ for all $\{i, j\} \notin E$. Following [20,14,3], we define

$$\bar\vartheta_{1/2}(G) = \inf\{k > 1 |\ G \text{ admits a vector } k\text{-coloring}\},$$
$$\bar\vartheta(G) = \bar\vartheta_1(G) = \inf\{k > 1 |\ G \text{ admits a strict vector } k\text{-coloring}\}, \qquad (3)$$
$$\bar\vartheta_2(G) = \inf\{k > 1 |\ G \text{ admits a rigid vector } k\text{-coloring}\}.$$

Observe that $\bar\vartheta_{1/2}(G)$ is precisely the *vector chromatic number* introduced by Karger, Motwani, and Sudan [20]; $\bar\vartheta_2$ occurs in [14,27]. Further, we let $\vartheta_{1/2}(G) = \bar\vartheta_{1/2}(\bar G)$, $\vartheta(G) = \vartheta_1(G) = \bar\vartheta(\bar G)$, and $\vartheta_2(G) = \bar\vartheta_2(\bar G)$. It is shown in [20] that the above definition of ϑ is equivalent with Lovász's original definition (cf. [16]).

Proposition 8. *Let $G = (V, E)$ be a graph of order n, and let $S \subset V$. Let $G[S]$ denote the subgraph of G induced on S. Then $\vartheta_i(G) \leq \vartheta_i(G[S]) + \vartheta_i(G[V \setminus S])$.*

It is obvious from the definitions that for any weak subgraph H of G we have $\bar\vartheta_i(H) \leq \bar\vartheta_i(G)$, $i \in \{1/2, 1, 2\}$. In addition to ϑ, $\vartheta_{1/2}$, and ϑ_2, we consider the semidefinite relaxation of MAX CUT invented by Goemans and Williamson [15]: $\mathrm{SMC}(G) = \max \sum_{i<j} \frac{a_{ij}}{2} (1 - \langle v_i, v_j \rangle)$ s.t. $\|v_i\| = 1$, where the max is taken over $v_1, \ldots, v_n \in \mathbf{R}^n$.

Finally, we need the following concentration result on $\vartheta_{1/2}, \vartheta, \vartheta_2$. For ϑ, the proof can be found in [5]. Using suitable characterizations of $\vartheta_{1/2}, \vartheta_2$, the argument given in [5] can be adapted to cover these cases as well.

Theorem 9. *Suppose that $p \leq 0.99$, and that $n \geq n_0$ for a certain constant $n_0 > 0$. Let m be a median of $\vartheta(G_{n,p})$.*

   i. Let $\xi \geq \max\{10, m^{1/2}\}$. Then $P(\vartheta(G_{n,p}) \geq m+\xi) \leq 30 \exp(-\xi^2/(5m+10\xi))$.
   ii. Let $\xi > 10$. Then $P(\vartheta(G_{n,p}) \leq m - \xi) \leq 3 \exp(-\xi^2/10m)$.

The same holds with ϑ replaced by $\vartheta_{1/2}$ or by ϑ_2.

3 The Concentration Results

Proof of Thm. 1. Let p and β be as in Thm. 1. The proof is based on the following large deviation result, which is a consequence of Azuma's inequality.

Lemma 10. *Suppose that $X : G_{n,p} \to \mathbf{R}$ is a random variable that satisfies the following conditions for all graphs $G = (V, E)$.*

   – *For all $v \in V$ the following holds. Let $G^* = G + \{\{v, w\}| \; w \in V, \; w < v\}$, and let $G_* = G - \{\{v, w\}| \; w \in V, \; w < v\}$. Then $|X(G^*) - X(G_*)| \leq 1$.*
   – *If H is a weak subgraph of G, then $X(H) \leq X(G)$.*

Then $P(|X - E(X)| > t\sqrt{n}) \leq 2 \exp(-t^2/2)$.

Let $\omega = \omega(n)$ be a sequence tending to infinity slowly, e.g. $\omega(n) = \ln \ln(n)$. Furthermore, let $k = k(n, p) = \inf\{x > 0| \; P(\bar{\vartheta}_2(G_{n,p}) \leq x) \geq \omega^{-1}\}$. For any graph $G = (V, E)$ let $Y(G) = \min\{\#U| \; U \subset V, \; \bar{\vartheta}_2(G-U) \leq k\}$. Then $\bar{\vartheta}_2(G) \leq k$ if and only if $Y(G) = 0$. Hence, $P(Y = 0) \geq \omega^{-1}$. Moreover, by Prop. 8, the random variable Y satisfies the assumptions of L. 10. Let $\mu = E(Y)$. Then $\mu \leq \sqrt{n}\omega$. Thus, by L. 10, $Y \leq 2\sqrt{n}\omega$ with high probability. The following lemma is implicit in [26] (cf. the proof of L. 8 in [26]).

Lemma 11. *Let $\delta > 0$. Whp. the random graph $G = G_{n,p}$ enjoys the following property. If $U \subset V$, $\#U \leq 2\sqrt{n}\omega$ then $\chi(G[U]) \leq s$, where $s > \frac{2}{2\beta-1} + \delta$.*

To conclude the proof of Thm. 1, let $G = G_{n,p}$, and suppose that there is some $U \subset V$, $\#U \leq 2\sqrt{n}\omega$, such that $\bar{\vartheta}_2(G - U) \leq k \leq \bar{\vartheta}_2(G)$. Since by L. 11 $\bar{\vartheta}_2(G[U]) \leq \chi(G[U]) \leq s$ whp., Prop. 8 entails that $k \leq \bar{\vartheta}_2(G) \leq k + s$ whp.

Proof of Thm. 2. Let ω be a sequence tending to infinity slowly. The random graph $G = G_{n,p}$ admits no $U \subset V$, $\#U \leq \omega^3 \sqrt{n}$, spanning more than $3(\#U - \varepsilon)/2$ edges whp., where $\varepsilon > 0$ is a small constant. Let k be as in the proof of Thm. 1. Then whp. there is a set $U \subset V$, $\#U \leq \omega\sqrt{n}$, such that $\bar{\vartheta}_2(G-U) \leq k$. Following Łuczak [25], we let $U = U_0$, and construct a sequence $U_0, \ldots, U_m$ as follows. If there is no edge $\{v, w\} \in E$ with $v, w \in N(U_i) \setminus U_i$, then we let $m = i$ and finish. Otherwise, we let $U_{i+1} = U_i \cup \{v, w\}$ and continue. Then $m \leq$

$m_0 = \omega^2 \sqrt{n}$, because otherwise $\#U_{m_0} = (2 + o(1))\omega^2 \sqrt{n}$ and $\#E(G[U_{m_0}]) \geq 3(1 - o(1))\#U_{m_0}/2$. Let $R = U_m$.

By L. 11, $\bar{\vartheta}_2(G[R]) \leq \chi(G[R]) \leq 3$. Furthermore, $I = N(R) \setminus R$ is an independent set. Let $G_1 = G[R \cup I]$, $S = V \setminus (R \cup I)$, and $G_2 = G[S \cup I]$. Then $\bar{\vartheta}_2(G_2) \leq k$, and $\bar{\vartheta}_2(G_1) \leq 4$. In order to prove that $\bar{\vartheta}_2(G) \leq k+1$, we shall first construct a rigid vector $k + 1$-coloring of G_2 that assigns the same vector to all vertices in I. Thus, let $(x_v)_{v \in S \cup I}$ be a rigid vector k-coloring of G_2. Let x be a unit vector perpendicular to x_v for all $v \in S$. Moreover, let $\alpha = (k^2 - 1)^{-1/2}$, and set $y_v = (\alpha^2 + 1)^{-1/2}(x_v - \alpha x)$ for $v \in S$, and $y_v = x$ for $v \in I$. Then $(y_v)_{v \in S \cup I}$ is a rigid vector $(k + 1)$-coloring of G_2. In a similar manner, we can construct a rigid vector 4-coloring $(y'_v)_{v \in R \cup I}$ of G_1 that assigns the same vector x' to all vertices in I.

Applying a suitable orthogonal transformation if necessary, we may assume that $x = x'$. Let $l = \max\{4, k+1\}$. Since $N(R) \subset R \cup I$, we obtain a rigid vector l-coloring $(z_v)_{v \in V}$ of G, where $z_v = y_v$ if $v \in S \cup I$, and $z_v = y'_v$ if $v \in R$. By the lower bound on $\bar{\vartheta}_2(G_{n,p})$ in Thm. 3 (which does not rely on Thm. 2 of course), choosing c_0 large enough we may assume that $k \geq 4$, whence $k \leq \bar{\vartheta}_2(G) \leq k+1$.

4   The Probable Value of $\vartheta(G_{n,p})$, $\bar{\vartheta}(G_{n,p})$, etc.

4.1   The Lower Bound on $\bar{\vartheta}_{1/2}(G_{n,p})$

To bound $\bar{\vartheta}_{1/2}(G_{n,p})$ from below, we make use of an estimate on the probable value of the SDP relaxation SMC of MAX CUT (cf. Sec. 2). Suppose that $c_0/n \leq p \leq 1 - c_0/n$ for some large constant $c_0 > 0$. Combining Thms. 4 and 5 of [6] instantly yields that there is a constant $\lambda > 0$ such that

$$P\left(\text{SMC}(G_{n,p}) > \frac{1}{2}\binom{n}{2}p + \lambda n^{3/2}p^{1/2}(1 - p)^{1/2}\right) \leq \exp(-2n). \quad (4)$$

Let $G = (V, E)$ be a graph with adjacency matrix $A = (a_{ij})_{i,j=1,\dots,n}$. Let $v_1, \dots, v_n$ be a vector k-coloring of G, where $k = \bar{\vartheta}_{1/2}(G) \geq 2$. Then $\|v_i\| = 1$ for all i, and $\langle v_i, v_j \rangle \leq -1/(k - 1)$ whenever $\{i, j\} \in E$. Therefore,

$$\text{SMC}(G) \geq \sum_{i<j} \frac{a_{ij}}{2}(1 - \langle v_i, v_j \rangle) \geq \#E\left(\frac{1}{2} + \frac{1}{k - 1}\right). \quad (5)$$

Let $c_0/n \leq p \leq 1 - c_0/n$ for some large constant $c_0 > 0$. By Chernoff bounds (cf. [18, p. 26]),

$$P\left(\#E(G_{n,p}) < \binom{n}{2}p - 8n^{3/2}p^{1/2}(1 - p)^{1/2}\right) \leq \exp(-2n). \quad (6)$$

Combining (4), (5), and (6), we conclude that

$$\bar{\vartheta}_{1/2}(G_{n,p}) \geq \bar{\vartheta}_{1/2}(G_{n,p}) - 1 \geq \frac{\binom{n}{2}p - 8n^{3/2}p^{1/2}(1 - p)^{1/2}}{(\lambda + 4)n^{3/2}p^{1/2}(1 - p)^{1/2}} \geq \frac{1}{2(\lambda + 4)}\sqrt{\frac{np}{1 - p}}$$

holds with probability at least $1 - \exp(-n)$. As $\bar{G}_{n,p} = G_{n,1-p}$, this proves (2) and the lower bounds in Thm. 3.

4.2   Spectral Considerations

Let us briefly recall Juhász's proof that $\vartheta(G_{n,p}) \leq (2 + o(1))\sqrt{n(1-p)/p}$ for constant values of p, say. Given a graph $G = (V, E)$, we consider the matrix $M = M(G) = (m_{ij})_{i,j=1,\ldots,n}$, where

$$m_{ij} = \begin{cases} 1 & \text{if } \{i, j\} \notin E \\ (p-1)/p & \text{otherwise,} \end{cases} \quad (i \neq j), \tag{7}$$

and $m_{ii} = 1$ for all i. Then $\lambda_1(M) \geq \vartheta(G)$. Moreover, as p is constant, the result of Füredi and Komlos [13] on the eigenvalues of random matrices applies and yields that $\vartheta(G_{n,p}) \leq \lambda_1(M) \leq (2 + o(1))\sqrt{n(1-p)/p}$ whp. This argument carries over to the case $\ln(n)^7/n \leq p \leq 1/2$ (cf. [4]):

Lemma 12. *Let* $\ln(n)^7/n \leq p \leq 1/2$. *Then* $\|M(G_{n,p})\| \leq 3\sqrt{n/p}$ *whp.*

However, it is easily seen that in the sparse case, e.g. if $np = O(1)$, we have $\lambda_1(M) \gg n$ whp. The reason is that in the case $np \geq \ln(n)^7$ the random graph $G_{n,p}$ is "almost regular", which is not true if $np = O(1)$. We will get around this problem by chopping off all vertices of degree considerably larger than np, as first proposed in [1]. Thus, let $\varepsilon > 0$ be a small constant, and consider the graph $G' = (V', E')$ obtained from $G = G_{n,p}$ by deleting all vertices of degree greater than $(1 + \varepsilon)np$.

Lemma 13. *Suppose that* $c_0/n \leq p \leq \ln(n)^7/n$ *for some large constant* c_0. *Let* $G = G_{n,p}$, *and let* $M' = M(G')$. *Then* $P(\|M'\| \leq c_1\sqrt{n/p}) \geq 9/10$, *where* $c_1 > 0$ *denotes some constant.*

To prove L. 13, we make use of the following lemma, which is implicit in [10, Sections 2 and 3]; the proof is based on the method of Kahn and Szemerédi [11].

Lemma 14. *Let* $G = G_{n,p}$ *be a random graph, where* $c_0/n \leq p \leq \ln(n)^7/n$ *for some large constant* $c_0 > 0$. *Let* $n' = \#V(G')$, $e = n'^{-1/2}\mathbf{1} \in \mathbf{R}^{n'}$, *and* $A' = A(G')$. *For each* $\delta > 0$ *there is a constant* $C(\delta) > 0$ *such that in the case* $np \geq C(\delta)$ *with probability* $\geq 1 - \delta$ *we have*

$$\max\{|\langle A'v, e\rangle|, |\langle A'v, w\rangle|\} \leq c_1\sqrt{np} \text{ for all } v, w \perp \mathbf{1}, \|v\| = \|w\| = 1. \tag{8}$$

Here $c_1 > 0$ *denotes a certain constant.*

In addition, the proof of Lemma 13 needs the following observation.

Lemma 15. *Let* c_1 *be a large constant. The probability that in* $G = G_{n,p}$ *there exists a set* $U \subset V$, $\#U \geq n/2$, *such that* $|\#E(G[U]) - \#U^2 p/2| \geq c_1(\#U)^{3/2}p^{1/2}$ *is less than* $\exp(-n)$.

Proof. There are at most 2^n sets U. By Chernoff bounds (cf. [18, p. 26]), for a fixed U the probability that $|\#E(G[U]) - \#U^2 p/2| \geq c_1(\#U)^{3/2}p^{1/2}$ is at most $\exp(-2n)$, provided that c_0, c_1 are large enough. $\qquad\square$

Proof of Lemma 13. Let $G = G_{n,p}$, let $n' = \#V(G')$, and let A', e be as in L. 14. Without loss of generality, we may assume that $V' = V(G') = \{1, \ldots, n'\}$. Let $c_1 > 0$ be a sufficiently large constant. Let J signify the $n' \times n'$ matrix with all entries equal to 1. Letting $\delta > 0$ be sufficiently small and $c_0 \geq C(\delta)$, we assume in the sequel that (8) holds, and that G has the property stated in L. 15. Let $z \in \mathbf{R}^{n'}$, $\|z\| = 1$. Then we have a decomposition $z = \alpha e + \beta v$, $\|v\| = 1$, $v \perp \mathbf{1}$, $\alpha^2 + \beta^2 = 1$. Since $\|M'z\| \leq \|M'e\| + \|M'v\|$, if suffices bound $\max_{v \perp e, \|v\|=1} \|M'v\|$ and $\|M'e\|$.

Let $\rho : \mathbf{R}^{n'} \to \mathbf{R}^{n'}$ be the projection on the space $\mathbf{1}^\perp$. Then $A'v = \rho A'v + \langle A'v, e \rangle e$, whence $\|A'v\| \leq \|\rho A'v\| + c_1\sqrt{np}$, for all unit vectors $v \perp \mathbf{1}$. In order to bound $\|\rho A'v\|$, we estimate $\|\rho A'\rho\|$ via (8):

$$\|\rho A'\rho\| = \sup_{\|y\|=1} |\langle \rho A'\rho y, y \rangle| = \sup_{\|y\|=1} |\langle A'\rho y, \rho y \rangle| = \sup_{\|y\|=1,\ \mathbf{1}\perp y} |\langle A'y, y \rangle| \leq c_1\sqrt{np}.$$

Consequently, $\|M'v\| = \|(J - \frac{1}{p}A')v\| = \frac{1}{p}\|A'v\| \leq 2c_1\sqrt{n/p}$ ($v \perp \mathbf{1}$, $\|v\| = 1$).

To bound $\|M'e\|$, note that $-pM' = A' - pJ$. Let $\bar{d} = 2\#E(G')/n'$, and $x = A'e - (\bar{d}/n')Je$. Then $x \perp \mathbf{1}$, and by (8) we have $\|x\|^2 = \langle A'e, x \rangle - \langle (\bar{d}/n')Je, x \rangle = \langle A'e, x \rangle \leq c_1\sqrt{np}\|x\|$, whence $\|x\| \leq c_1\sqrt{np}$. By L. 15, $|\bar{d} - n'p| \leq c_1\sqrt{np}$. As a consequence, $\|(\bar{d}/n')Je - pJe\| \leq c_1\sqrt{np}$. Therefore, $\|pM'e\| \leq \|x\| + \|(\bar{d}/n')Je - pJe\| \leq 2c_1\sqrt{np}$, i.e. $\|M'e\| \leq 2c_1\sqrt{n/p}$. $\quad\square$

4.3   Bounding $\vartheta_2(G_{n,p})$ from Above

Let $c_0/n \leq p \leq 1/2$ for some large constant $c_0 > 0$. The following lemma is a consequence of the characterization of $\bar{\vartheta}_2$ as an eigenvalue minimization problem given in [27].

Lemma 16. *Let G be any graph. Let $M = M(G)$. Then $\lambda_1(M) \geq \bar{\vartheta}_2(G)$.*

In the case $\ln(n)^7/n \leq p \leq 1/2$, combining L. 12 and L. 16 yields that $\vartheta_2(G_{n,p}) \leq c_2\sqrt{n/p}$ whp. for some constant $c_2 > 0$, as desired. Thus, let us assume that $c_0/n \leq p \leq \ln(n)^7/n$ in the sequel. Let $\varepsilon > 0$ be a small constant.

Lemma 17. *With probability at least $9/10$ the random $G_{n,p}$ has at most $1/p$ vertices of degree greater than $(1 + \varepsilon)np$.*

Proof. For each vertex v of $G_{n,p}$, the degree $d(v)$ is binomially distributed with mean $(n - 1)p$. By Chernoff bounds (cf. [18, p. 26]), the probability that $d(v) > (1 + \varepsilon)np$ is at most $\exp(-\varepsilon^2 np/100)$. Hence, the expected number of vertices v such that $d(v) > (1 + \varepsilon)np$ is at most $n\exp(-\varepsilon^2 np/100) < 1/(10p)$, provided $np \geq c_0$ for some large constant $c_0 > 0$. Therefore, the assertion follows from Markov's inequality. $\quad\square$

Let $G = G_{n,p}$, and let $G' = (V', E')$ be the graph obtained from G by deleting all vertices of degree greater than $(1 + \varepsilon)np$. Let $V'' = V \setminus V'$, and $G'' = G[V'']$. Combining L. 17 and L. 13, we obtain that

$$\mathrm{P}\left(\vartheta_2(G') \leq c_2\sqrt{n/p} \text{ and } \vartheta_2(G'') \leq \#V(G'') \leq 1/p \leq \sqrt{n/p}\right) > 1/2,$$

where c_2 denotes a suitable constant. Consequently, Prop. 8 yields that

$$P(\vartheta_2(G_{n,p}) \leq (c_2 + 1)\sqrt{n/p}) > 1/2.$$

Let $\mu = (c_2 + 1)\sqrt{n/p}$, $t = \ln(n)\sqrt{n}$, and note that $t = o(\sqrt{n/p})$. Then, by Thm. 9, $P(\vartheta_2(G_{n,p}) > \mu + t) \leq 30\exp\left(-\Omega(\ln(n)^2)\right) = o(1)$. Since $t < \sqrt{n/p}$, we get that $\vartheta_2(G_{n,p}) \leq (c_2 + 2)\sqrt{n/p}$ with high probability.

4.4   Bounding $\bar{\vartheta}_2(G_{n,p})$ from Above

Let us first assume that $\ln(n)^7/n \leq p \leq 1/2$. Let $G = (V, E) = G_{n,p}$ be a random graph, and consider the matrix $\bar{M} = \frac{1}{1-p}E_n - \frac{p}{1-p}M(G)$, where E_n is the $n \times n$-unit matrix, and $M(G)$ is the matrix defined in (7). Combining L. 12 and L. 16, we have $\bar{\vartheta}_2(G) \leq \lambda_1(\bar{M}) \leq \|\frac{1}{1-p}E - \frac{p}{1-p}M'\| \leq \frac{p}{1-p}\|M\| + 2 \leq c_4\sqrt{np}$ whp., where $c_4 > 0$ is a certain constant.

Now let $c_0/n \leq p \leq \ln(n)^7/n$ for some large constant $c_0 > 0$. In this case, the proof of our upper bound on $\bar{\vartheta}_2(G_{n,p})$ relies on the concentration result Thm. 2.

Lemma 18. *Whp. the random graph $G = G_{n,p}$ admits no set $U \subset V$, $\#U \leq 1/p$, such that $\chi(G[U]) > \sqrt{np}$.*

Proof. We shall prove that for all $U \subset V$, $\#U = \nu \leq 1/p$, we have $\#E(G[U]) < \nu\sqrt{np}/2$. Then each subgraph $G[U]$ has a vertex of degree $< \sqrt{np}$, a fact which immediately implies our assertion. Thus, let $\nu \leq 1/p$. The probability that there exists some $U \subset V$, $\#U = \nu$, $\#E(G[U]) \geq \nu\sqrt{np}/2$, is at most

$$\binom{n}{\nu}\binom{\binom{\nu}{2}}{\nu\sqrt{np}/2}p^{\nu\sqrt{np}/2} \leq \left(\frac{en}{\nu}\left(\frac{e\nu\sqrt{p}}{\sqrt{n}}\right)^{\sqrt{np}/2}\right)^{\nu}$$

Let $b_\nu = (en/\nu)(e\nu\sqrt{p}/\sqrt{n})^{\sqrt{np}/2}$. Observe that the sequence $(b_\nu)_{\nu=1,\dots,n}$ is monotone increasing, and that $b_{1/p} = enp(e/\sqrt{np})^{\sqrt{np}/2} \leq \exp(-2)$. Therefore, $\sum_{\nu=\ln(n)}^{1/p} b_\nu \leq b_{1/p}^{\ln(n)}/p \leq n^{-2}p^{-1} = o(1)$. Moreover, if $\nu \leq \ln(n)$, then $b_\nu \leq en\nu^{-1}(e\nu\sqrt{p}/\sqrt{n})^{\sqrt{np}/2} \leq 1/n$, whence $\sum_{\nu=1}^{\ln n} b_\nu = o(1)$. Thus, $\sum_{\nu=1}^{1/p} b_\nu = o(1)$, thereby proving the lemma.   □

Let $G = (V, E) = G_{n,p}$ be a random graph, and let $G' = (V', E')$ be the graph obtained from G by removing all vertices of degree greater than $(1+\varepsilon)np$, where $\varepsilon > 0$ is small but constant. Let $V'' = V \setminus V'$, and let $G'' = G[V'']$. By L. 17, with probability at least $9/10$ we have $\#V'' \leq 1/p$. Therefore, by L. 18, $P(\bar{\vartheta}_2(G'') \leq \sqrt{np}) \geq P(\chi(G'') \leq \sqrt{np}) \geq 9/11$. To bound $\bar{\vartheta}_2(G')$, we consider the matrix $\bar{M} = \frac{1}{1-p}E_{n'} - \frac{p}{1-p}M(G')$. By L. 16, $\bar{\vartheta}_2(G') \leq \lambda_1(\bar{M})$. Moreover, by L. 13, with probability $\geq 9/10$ we have $\bar{\vartheta}_2(G') \leq \lambda_1(\bar{M}) \leq \frac{p}{1-p}\|M'\| + 2 \leq c_4\sqrt{np}$, for some constant $c_4 > 0$. Prop. 8 implies that $\bar{\vartheta}_2(G) \leq \bar{\vartheta}_2(G') + \bar{\vartheta}_2(G'')$, whence we conclude that $P(\bar{\vartheta}_2(G_{n,p}) \leq (c_4 + 1)\sqrt{np}) > 1/2$. Since Thm. 2 shows that $\vartheta_2(G_{n,p})$ is concentrated in width one, we have

$$P\left(\bar{\vartheta}_{1/2}(G_{n,p}) \leq \bar{\vartheta}(G_{n,p}) \leq \bar{\vartheta}_2(G_{n,p}) \leq (c_4 + 1)\sqrt{np} + 1\right) = 1 - o(1),$$

thereby completing the proof of Thm. 3.

Remark 19. One could prove slightly weaker results on the probable value of $\vartheta(G_{n,p})$ and $\bar{\vartheta}(G_{n,p})$ than provided by Thm. 3 without applying any concentration results, or bounds on the SDP relaxation SMC of MAX CUT. Indeed, using only L. 17, 18, 13 (thus implicitly [10]) and the estimates proposed in [19], one could show that for each $\delta > 0$ there is $C(\delta) > 0$ such that $P(c_1\sqrt{n/p} \leq \vartheta(G_{n,p}) \leq c_2\sqrt{n/p}) \geq 1-\delta$ and $P(c_3\sqrt{np} \leq \bar{\vartheta}(G_{n,p}) \leq c_4\sqrt{np}) \geq 1-\delta$, provided $np \geq C(\delta)$. Such an approach is mentioned without proof independently in the latest version of [10].

5 Approximating the Independence Number and Deciding k-Colorability

Approximating the Independence Number. The algorithm ApproxMIS for approximating the independence number consists of two parts. First, we employ a certain greedy procedure that on input $G = G_{n,p}$ finds a large independent set whp. Secondly, we compute $\vartheta(G)$ to bound $\alpha(G)$ from above. Following [23], to find a large independent set of $G = G_{n,p}$, we run the greedy algorithm for graph coloring and pick the largest color class it produces.

Lemma 20. *The probability that the largest color class produced by the greedy coloring algorithm contains $< \ln(np)/(2p)$ vertices is at most $\exp(-n)$.*

Proof. The proof given in [23] for the case that $p \geq n^{\varepsilon-1/2}$ carries over. □

The following algorithm is essentially identical with the one given in [4].

Algorithm 21. ApproxMIS(G)
Input: A graph $G = (V, E)$. *Output:* An independent set of G.

1. Run the greedy algorithm for graph coloring on input G. Let I be the largest resulting color class. If $\#I < \ln(np)/(2p)$, then go to 5.
2. Compute $\vartheta(G)$. If $\vartheta(G) \leq C\sqrt{n/p}$, then output I and terminate. Here C denotes some sufficiently large constant (cf. the analysis below).
3. Check whether there exists a subset S of V, $\#S = 25\ln(np)/p$, such that $\#V \setminus (S \cup N(S)) > 12(n/p)^{1/2}$. If no such set exists, then output I and terminate.
4. Check whether in G there is an independent set of size $12(n/p)^{1/2}$. If this is not the case, then output I and terminate.
5. Enumerate all subsets of V and output a maximum independent set.

Lemma 22. *The expected running time of ApproxMIS($G_{n,p}$) is polynomial.*

Proof. The first two steps can be implemented in polynomial time. By Thm. 3, the median μ of $\vartheta(G_{n,p})$ is at most $c\sqrt{n/p}$, for some constant c. Therefore, Thm. 9 entails that the probability that ApproxMIS runs step 3 is less than $\exp(-(n/p)^{1/2})$, provided C is large enough. Furthermore, up to polynomial factors, step 3 consumes time $\leq \exp(25\ln(np)^2/p) < \exp(\sqrt{n/p})$. Hence, the expected time spent executing step 3 is polynomial. Taking into account L. 20, the expected running time of the remaining steps can be estimated as in the proof of Thm. 4 in [4]. $\qquad\square$

Finally, it is not hard to show that `ApproxMIS` guarantees the desired approximation ratio.

Deciding k-Colorability. Following [22], we decide k-colorability by computing the vector chromatic number of the input graph. Let $k = k(n)$ be a sequence of positive integers such that $k(n) = o(\sqrt{n})$. Since the vector chromatic number is always a lower bound on the chromatic number, the answer of the following algorithm is correct for all input graphs G.

Algorithm 23. $\texttt{Decide}_k(G)$
Input: A graph $G = (V, E)$. *Output:* Either "$\chi(G) \leq k$" or "$\chi(G) > k$".

1. If $\bar{\vartheta}_{1/2}(G) > k$ then terminate with output "$\chi(G) > k$".
2. Otherwise, compute $\chi(G)$ in time $o(\exp(n))$ using Lawler's algorithm [24], and answer correctly.

Lemma 24. *Suppose that $p \geq Ck^2/n$ for some large constant C. Then the expected running time of $\texttt{Decide}_k(G_{n,p}^+)$ is polynomial.*

Proof. In [20] it is shown that $\bar{\vartheta}_{1/2}$ can be computed in polynomial time. Since the second step consumes time $o(\exp(n))$, (2) shows that the expected running time of $\texttt{Decide}_k$ on input $G_{n,p}^+$ is polynomial. $\qquad\square$

The analysis of $\texttt{Decide}_k$ on input $G_{n,r}$, $r \geq Ck^2$, is based on Thm. 4 and yields the proof of Thm. 7.

Acknowledgment. I am grateful to M. Krivelevich and C. Helmberg for helpful discussions, and to U. Feige and E. Ofek for providing me with their technical report [10].

References

1. Alon, N., Kahale, N.: A spectral technique for coloring random 3-colorable graphs. SIAM J. Comput. **26** (1997) 1733–1748
2. Alon, N., Krivelevich, M.: The concentration of the chromatic number of random graphs. Combinatorica **17** 303–313
3. Charikar, M.: On semidefinite programming relaxations for graph coloring and vertex cover. Proc. 13th SODA (2002) 616–620
4. Coja-Oghlan, A., Taraz, A.: Exact and approximative algorithms for colouring $G(n, p)$. preprint (available from http://www.informatik.hu-berlin.de/~coja/). A preliminary version has appeard in Proc. 20th STACS (2003) 487–498

5. Coja-Oghlan, A.: Finding large independent sets in polynomial expected time. Proc. 20th STACS (2003) 511–522
6. Coja-Oghlan, A., Moore, C., Sanwalani, V.: MAX k-CUT and approximating the chromatic number of random graphs. available from http://www.informatik.hu-berlin.de/~coja/. An extended abstract version has appeared in Proc. ICALP 2003.
7. Feige, U.: Randomized graph products, chromatic numbers, and the Lovász theta function. Combinatorica **17**(1) 79–90
8. Feige, U., Kilian, J.: Heuristics for semirandom graph problems. J. Comput. and System Sci. **63** (2001) 639–671
9. Feige, U. and Kilian, J.: Zero knowledge and the chromatic number. Proc. 11th IEEE Conf. Comput. Complexity (1996) 278–287.
10. Feige, U., Ofek, E.: Spectral techniques applied to sparse random graphs, report MCS03-01, Weizmann Institute (2003)
    (available from http://www.wisdom.weizmann.ac.il/math/research.shtml)
11. Friedman, J., Kahn, J., Szemeredi, E.: On the second eigenvalue in random regular graphs. Proc. 21st STOC (1989) 587–598.
12. Frieze, A., McDiarmid, C.: Algorithmic theory of random graphs. Random Structures & Algorithms **10** (1997) 5–42
13. Füredi, Z., Komlós, J.: The eigenvalues of random symmetric matrices, Combinatorica **1** (1981) 233–241
14. Goemans, M.X., Kleinberg, J.: The Lovasz theta function and a semidefinite programming relaxation of vertex cover. SIAM J. on Discrete Math. **11** (1998) 1–48
15. Goemans, M.X., Williamson, D.P.: Improved approximation algorithms for maximum cut and satisfiability problems using semidefinite programming. J. ACM **42** (1995) 1115–1145.
16. Grötschel, M., Lovász, L., Schrijver, A.: Geometric algorithms and combinatorial optimization. Springer (1988)
17. Håstad, J.: Clique is hard to approximate within $n^{1-\varepsilon}$. Proc. 37th FOCS (1996) 627–636
18. Janson, S., Łuczak, T., Ruciński, A.: Random Graphs. Wiley (2000)
19. Juhász, F.: The asymptotic behaviour of Lovász ϑ function for random graphs. Combinatorica **2** (1982) 269–270
20. Karger, D., Motwani, R., Sudan, M.: Approximate graph coloring by semidefinite programming. J. ACM **45** (1998) 246–265
21. Knuth, D.: The sandwich theorem, Electron. J. Combin. **1** (1994)
22. Krivelevich, M.: Deciding k-colorability in expected polynomial time, Information Processing Letters **81** (2002), 1–6
23. Krivelevich, M., Vu, V.H.: Approximating the independence number and the chromatic number in expected polynomial time. J. of Combinatorial Optimization **6** (2002) 143–155
24. Lawler, E.L.: A note on the complexity of the chromatic number problem, Information Processing Letters **5** (1976) 66–67
25. Łuczak, T.: A note on the sharp concentration of the chromatic number of random graphs. Combinatorica **11** (1991) 45–54
26. Shamir, E., Spencer, J.: Sharp concentration of the chromatic number of random graphs $G_{n,p}$. Combinatorica **7** (1987) 121–129
27. Szegedy, M.: A note on the θ number of Lovász and the generalized Delsarte bound. Proc. 35th FOCS (1994) 36–39

Perfectly Balanced Allocation[*]

Artur Czumaj[1], Chris Riley[2], and Christian Scheideler[2]

[1] Department of Computer Science, New Jersey Institute of Technology, University Heights, Newark, NJ 07102-1982, USA,
czumaj@cis.njit.edu

[2] Department of Computer Science, Johns Hopkins University, 3400 N. Charles Street, Baltimore, MD 21218, USA,
{chrisr,scheideler}@cs.jhu.edu

Abstract. We investigate randomized processes underlying load balancing based on the multiple-choice paradigm: m balls have to be placed in n bins, and each ball can be placed into one out of 2 randomly selected bins. The aim is to distribute the balls as evenly as possible among the bins. Previously, it was known that a simple process that places the balls one by one in the least loaded bin can achieve a maximum load of $m/n + \Theta(\log \log n)$ with high probability. Furthermore, it was known that it is possible to achieve (with high probability) a maximum load of at most $\lceil m/n \rceil + 1$ using maximum flow computations.

In this paper, we extend these results in several aspects. First of all, we show that if $m \geq cn \log n$ for some sufficiently large c, then a *perfect* distribution of balls among the bins can be achieved (i.e., the maximum load is $\lceil m/n \rceil$) with high probability. The bound for m is essentially optimal, because it is known that if $m \leq c' n \log n$ for some sufficiently small constant c', the best possible maximum load that can be achieved is $\lceil m/n \rceil + 1$ with high probability. Next, we analyze a simple, randomized load balancing process based on a local search paradigm. Our first result here is that this process always converges to a best possible load distribution. Then, we study the convergence speed of the process. We show that if m is sufficiently large compared to n, then *no matter* with which ball distribution the system starts, if the imbalance is Δ, then the process needs only $\Delta \cdot n^{O(1)}$ steps to reach a perfect distribution, with high probability. We also prove a similar result for $m \approx n$, and show that if $m = O(n \log n / \log \log n)$, then an optimal load distribution (which has the maximum load of $\lceil m/n \rceil + 1$) is reached by the random process after a polynomial number of steps, with high probability.

Keywords: load balancing, local search algorithms, stochastic processes.

1 Introduction

The study of balls-into-bins games or occupancy problems has a long history (see e.g. [1,2,3,4,5,8,10,11,12,18]). These problems have numerous applications, e.g., in graph theory, queueing theory, hashing, and randomized rounding. In general, the goal of a

[*] Research supported in part by NSF grant CCR-0105701.

S. Arora et al. (Eds.): APPROX 2003+RANDOM 2003, LNCS 2764, pp. 240–251, 2003.

balls-and-bins algorithm is to assign a set of independent objects (tasks, jobs, memory blocks) to a set of resources (servers, disks) so that the load is distributed among the bins as evenly as possible.

In the classical single-choice game, each ball is placed into a bin chosen *independently and uniformly at random (i.u.r.)*. For the case of n bins and $m \geq n \log n$ balls it is well known that there exists a bin receiving $m/n + \Theta(\sqrt{m \log n/n})$ balls. This result holds not only in expectation but also with high probability. (We say that an event A occurs *with high probability (w.h.p.)* if $\mathbf{Pr}[A] \geq 1 - n^{-\alpha}$ for an arbitrarily chosen constant $\alpha \geq 1$.) On the other hand, it was shown by Azar *et al.* [1] and Berenbrink *et al.* [2] that if the balls are placed in a sequential (on-line) fashion and each ball is assigned to the currently least loaded of the two locations (ties broken arbitrarily), then the maximum load of any bin is $m/n + \Theta(\log \log n)$ with high probability. It can also be proven [1,2] that any protocol that assigns the balls to the bins in an on-line fashion (that is, the decision where the ball is placed is performed only on the base of the placement of the previously placed balls) cannot be stochastically better than the scheme above. In particular, this implies that in any on-line scheme, with high probability, there is a bin with load $m/n + \Theta(\log \log n)$.

On the other side, some authors have been studying *off-line assignments*. In off-line assignments, after first selecting the two locations for all the balls, one seeks an optimal placement of the balls assuming each ball can choose only among its two locations and the locations of all balls are known to the algorithm (off-line case). This problem arises naturally in numerous applications, for example, in hashing, scheduling, load balancing, and video on demand (see, e.g., [1,7,9,14,15,16]). (For example, Sanders *et al.* [16] discussed in depth applications to support fast parallel access to external memory systems with parallel disks and Karp [7] discussed applications in video on demand; Karp called our problem k-orientability.)

Let the *minmax load* be the minimum, over all possible placements of the balls into bins, of the maximum load in the system. Azar *et al.* [1] showed that for $n = \Theta(m)$, the minmax load is $\Theta(1)$, with high probability. Later, Frieze (personal communication in [1]) and, independently, Czumaj and Stemann [5], tightened this bound and, in particular, showed that for $n = m$, the minmax load is exactly 2, with high probability. Sanders *et al.* [16] extended the result from [1,5] to arbitrary m and proved the following result.

Theorem 1. [16]   *The minmax load is at most $\lceil m/n \rceil + 1$, with high probability.*   □

Notice that since the minmax load cannot be smaller than $\lceil m/n \rceil$, this bound is optimal up to an additive constant 1. Furthermore, it is easy to see that there exists a positive constant λ, such that if $m \leq \lambda n \ln n$, then the bound in Theorem 1 is tight[3]. Our first contribution is that this bound for m is asymptotically tight in the following sense: there is a constant c such that if $m \geq cn \ln n$, then a *perfect balance* is possible:

Theorem 2. *There exists a positive constant c such that for every $m \geq cn \ln n$, the minmax load is exactly $\lceil m/n \rceil$, with high probability.*

---

[3] Indeed, if we choose at random two locations for each of the $\lambda n \ln n$ balls, then there will be a bin that has not been chosen by any ball. Therefore, there is a bin whose load is 0 w.h.p. and hence it is impossible that all bins have identical load of m/n, w.h.p.

Stochastic load balancing. Next, we present a novel approach to off-line assignments and discuss a new stochastic process (algorithm) that achieves *optimal maximum* load. Sanders *et al.* [16] described a polynomial time algorithm that finds an optimal assignment of the balls into bins minimizing the maximum load (which in this optimal allocation is equal to the minmax load). Their algorithm uses maximum flow computations.

A drawback of the approach by Sanders *et al.* is that it requires global (centralized) knowledge about locations of all balls, which is far too space consuming if m is large. This makes also the algorithm difficult (if suitable at all) for implementations in distributed or decentralized systems (like, for example, systems of parallel disks as discussed in [9,16]). Therefore, as our second contribution, we present a simple, memoryless, *local search* algorithm that can balance the load of the bins in the system as much as this is possible. The idea behind our algorithm is to begin with an arbitrary assignment of the balls to the bins, and then to use a stochastic replacement process that gradually improves the balance of the bins' load.

Suppose that initially all the balls have chosen their locations in $\{1, \ldots, n\}$ and each ball is (arbitrarily) placed in one of its two locations. The *Self-Balancing Algorithm* repeats the following *Self-Balancing Step*:

Self-Balancing Step:

Pick independently and uniformly at random a pair of bins (b_1, b_2).
If there is a ball placed in b_1 with alternative location in bin b_2, then
    Pick any ball x that is placed in b_1 with alternative location in bin b_2;
    Place x into the least loaded bin (among b_1 and b_2);
        If tie, that is, bin b_1 has (without x) the same load as bin b_2, then
            place x into a randomly chosen of the two bins.

We prove two theorems about the Self-Balancing Algorithm (throughout our analysis, unless stated otherwise, terms *"with high probability"* are with respect to the random choices of the two locations of each ball, as well as the random choices of balls in the Self-Balancing Algorithm).

The first theorem shows that the Self-Balancing Algorithm will gradually converge to states in which the maximum load is best possible.

Theorem 3. *If the Self-Balancing Algorithm is run sufficiently long (i.e., the Self-Balancing Step is repeated sufficiently many times), then the maximum load of any bin in the system is equal to the minmax load with probability 1. (The probability 1 is with respect to the random choices of balls in the Self-Balancing Algorithm only.)*

In particular, if the Self-Balancing Algorithm is run sufficiently long then the maximum load of any bin in the system is smaller than or equal to $\lceil m/n \rceil + 1$ with high probability. If, additionally, $m > cn \ln n$ for a sufficiently large constant c, then the maximum load is exactly $\lceil m/n \rceil$ with high probability.

The Self-Balancing Algorithm is a simple example of a local search algorithm, similar to load balancing algorithms existing in the literature before, see, e.g., [6,13]. Theorem 3 shows the non-trivial property that no matter with which state (i.e., assignment of balls to bins) the Self-Balancing Algorithm starts, it will always converge to a state in which the maximum load is optimally small. Notice that in many local search approaches one frequently arrives at a "dead-lock" situation, in which the balancing may be far away from optimal and no re-balancing progress is possible (that is, a locally optimal solution is not in a global optimum). Theorem 3 shows that this is not the case for the Self-Balancing Algorithm. (Observe, however, that if we removed the randomized rule for tie breaking, then — as one can easily show — the algorithm would not necessarily converge to an optimal state.)

The next theorem considers the heavily loaded case and deals with the speed of the "convergence" of the Self-Balancing Algorithm to a state in which the maximum load is upper bounded by $\lceil m/n \rceil$. Let the *imbalance* of the system be its distance from a best possible distribution, or more precisely, $\sum_{i=1}^{n} \max\{0, \text{load of bin } i - \lceil m/n \rceil\}$.

Theorem 4. *If $m \gg n$, then after a polynomial number (with respect to n only) of Self-Balancing Steps the maximum load in the system is equal to $\lceil m/n \rceil$, with high probability. Furthermore, if the system imbalance is Δ, then the number of steps is $\Delta \cdot n^{O(1)}$, with high probability.*

Notice that if the balls are allocated to the bins in the on-line fashion using the least loaded bin approach, as in [1,2], the system imbalance is $\Delta = O(n \log \log n)$, with high probability [2]. Therefore, Theorem 4 implies the following corollary.

Corollary 1. *If $m \gg n$, then in time $O(m) + n^{O(1)}$ one can find a perfect load distribution with the maximum load of the system equal to $\lceil m/n \rceil$, with high probability.* □

As we argued before, one cannot extend the result from Theorem 4 to the case $m \approx n$, because then the minmax load is expected to be equal to $\lceil m/n \rceil + 1$ (instead of $\lceil m/n \rceil$). Our next theorem shows however that if m is close to n, then the Self-Balancing Algorithm still rapidly converges to the optimal distribution.

Theorem 5. *If $m = O(n \log n / \log \log n)$, then after a polynomial number (with respect to n) of Self-Balancing Steps the maximum load in the system is smaller than or equal to $\lceil m/n \rceil + 1$, with high probability.*

Notational conventions. To simplify the presentation of the paper, we will use a shorthand μ to denote m/n and $\widehat{\mu}$ to denote $\lceil m/n \rceil = \lceil \mu \rceil$. We shall identify the balls with the integers in $\{1, \ldots, m\} = [m]$ and the bins with the integers in $\{1, \ldots, n\} = [n]$. Let the *load* of a bin $b \in [n]$ be equal to the number of balls placed in b. Notice that the average load among all the bins is μ.

2   Perfect Balancing for $\Omega(n \log n)$ Balls

In this section we prove Theorem 2, that is, we show that if $m \geq cn \log n$ for certain suitable constant c, then the minmax load is $\lceil m/n \rceil = \widehat{\mu}$, with high probability. It is easy to see that it is sufficient to prove this bound in the case $\mu = \widehat{\mu}$, and therefore from now on we assume that μ is an integer.

Let $\mathfrak{B}$ denote the set of n bins in the system. Let us fix an allocation of m balls to n bins in $\mathfrak{B}$ such that each ball has two locations in $\mathfrak{B}$ (we allow a ball to have both locations in the same bin). For any $U \subseteq \mathfrak{B}$, let $\Psi[U]$ denote the number of balls having all locations in the bins in U. Then, one can show the following result (see [16,17]).

Lemma 1. *[17, Theorem 1] The minmax load is equal to* $\max_{U \subseteq \mathfrak{B}, U \neq \emptyset} \left\lceil \frac{\Psi[U]}{|U|} \right\rceil$. $\quad\square$

Consider the stochastic process of assigning two locations of the m balls to the n bins in $\mathfrak{B}$ i.u.r. For any set $U \subseteq \mathfrak{B}$, let C_U be the random variable denoting the value of $\Psi[U]$. Furthermore, let $\mathcal{E}_U$ be the random indicator of the event that $C_U > \mu \cdot |U|$ and let $\mathcal{E} = \bigvee_{U \subseteq \mathfrak{B}, U \neq \emptyset} \mathcal{E}_U$. Our goal is to show that

$$\mathbf{Pr}[\mathcal{E}] \leq n^{-\gamma} . \tag{1}$$

for a constant γ depending on c.

Let $\mathfrak{B}_k = \{U \subseteq \mathfrak{B} : |U| = k\}$. Then, by the union bound, to prove (1) it is enough to prove the following bound for every[4] k, $1 \leq k \leq n-1$, and for every set $U \in \mathfrak{B}_k$:

$$\mathbf{Pr}[\mathcal{E}_U] \leq \frac{1}{n^{\gamma+1} \cdot \binom{n}{k}} . \tag{2}$$

From now on, we concentrate on proving inequality (2). Let us observe that for any set $U \in \mathfrak{B}_k$, the value of C_U is a binomial random variable with the parameters m and $(k/n)^2$, which we denote by $\mathbb{B}(m, (k/n)^2)$. Therefore, $\mathbf{Pr}[\mathcal{E}_U] = \mathbf{Pr}[\mathbb{B}(m, (k/n)^2) > m \cdot k/n] \leq \mathbf{Pr}[\mathbb{B}(m, (k/n)^2) \geq m \cdot k/n]$ and our goal now is to investigate bounds for $\mathbf{Pr}[\mathbb{B}(m, (k/n)^2) \geq m \cdot k/n]$.

We begin with three simple results about concentration of binomial random variables.

Lemma 2.

1. *For any* $t \geq 6\, m\, q^2$, $\mathbf{Pr}[\mathbb{B}(m, q^2) \geq t] \leq 2^{-t}$.
2. *For any* $0 < q < 1$, $\mathbf{Pr}[\mathbb{B}(m, q^2) \geq q \cdot m] \leq \exp(-2\, q^2\, (1-q)^2\, m)$.
3. *For any* $0 < q < 1$, *if* $\frac{1}{u-1} \leq q$ *for certain* $u > 1$, *then* $\mathbf{Pr}[\mathbb{B}(m, q^2) \geq q \cdot m] \leq (u/e)^{m\,(1-q)}$. $\quad\square$

Let $m \geq c\, n \ln n$ for a large constant c. Let $U \in \mathfrak{B}_k$ and $q = k/n$. Let us first consider the case $k/n = q \leq 0.1$. Then, if we set $t = m\, k/n$, then we have $t \geq 6 \cdot \mathbf{E}[\mathbb{B}(m, q^2)]$, and hence by Lemma 2 (1) and by the inequality $\binom{n}{k} \leq n^k$, we get (provided c is a large enough constant):

$$\mathbf{Pr}[\mathcal{E}_U] \leq 2^{-t} = 2^{-m\,k/n} \leq e^{-(\gamma+1)\ln n - k \ln n} = \frac{1}{n^{\gamma+1} \cdot n^k} \leq \frac{1}{n^{\gamma+1} \cdot \binom{n}{k}} . \tag{3}$$

Next, we consider $0.1\, n \leq k < 2/3\, n$. Then, by Lemma 2 (2) and by observing that $\binom{n}{k} \leq 2^n$, we have (again, if we set $m = c\, n \ln n$ for a large enough constant c)

$$\mathbf{Pr}[\mathcal{E}_U] \leq \exp(-2(\tfrac{k\,(n-k)}{n^2})^2 m) \leq e^{-0.001m} \leq e^{-(\gamma+1)\ln n - k \ln n} \leq \frac{1}{n^{\gamma+1} \cdot \binom{n}{k}} . \tag{4}$$

---

[4] We do not have to consider the case $U = \mathfrak{B}_n$ because in that case $\mathcal{E}_U$ trivially never holds.

The remaining case is when $k/n = q \geq 2/3$. Then, we can apply Lemma 2 (3) with $u = 2.5$ to obtain

$$\mathbf{Pr}[\mathcal{E}_U] \leq \mathbf{Pr}[\mathbb{B}(m, q^2) \geq qm] \leq (2.5/e)^{m/3} \leq e^{-(\gamma+1)\ln n - k \ln n} \leq \frac{1}{n^{\gamma+1} \cdot \binom{n}{k}}. \tag{5}$$

Therefore, from inequalities (3 – 5), we have that for every integer k, $1 \leq k \leq n-1$, and for every $U \in \mathfrak{B}_k$, we have $\mathbf{Pr}[\mathcal{E}_U] \leq \frac{1}{n^{\gamma+1} \cdot \binom{n}{k}}$. This implies that $\mathbf{Pr}[\mathcal{E}] \leq n^{-\gamma}$, which in turn yields Theorem 2. □

3 Convergence to Optimal Assignment

In this section we sketch the proof of Theorem 3. We begin with basic definitions and notation. A placement of the balls after performing t repetitions of the Self-Balancing Step, $t \geq 0$, is called the *tth assignment*, and is denoted by $\mathcal{A}_t$. To each assignment $\mathcal{A}_t$ we assign a *load vector*, which is vector $\mathbb{L}_t = \langle \mathbb{L}_t(1), \ldots, \mathbb{L}_t(n) \rangle$ such that $\mathbb{L}_t(j)$ denotes the load of the jth fullest bin in $\mathcal{A}_t$. For any two load vectors $\mathbb{L} = \langle \mathbb{L}(1), \ldots, \mathbb{L}(n) \rangle$ and $\mathbb{L}^* = \langle \mathbb{L}^*(1), \ldots, \mathbb{L}^*(n) \rangle$, we say $\mathbb{L}$ *majorizes* $\mathbb{L}^*$, denoted by $\mathbb{L} \succeq \mathbb{L}^*$, if for every j, $1 \leq j \leq n$, we have $\sum_{r=1}^{j} \mathbb{L}(r) \geq \sum_{r=1}^{j} \mathbb{L}^*(r)$. Furthermore, we write $\mathbb{L} \succ \mathbb{L}^*$ if $\mathbb{L} \succeq \mathbb{L}^*$ and there is at least one j with $\sum_{r=1}^{j} \mathbb{L}(r) > \sum_{r=1}^{j} \mathbb{L}^*(r)$.

Our first lemma describes the way the load vector can change in the course of the algorithm. Informally, it says that after any repetition of Self-Balancing Step the load vector will never worsen.

Lemma 3. *For any $t \geq 0$, independently of the random choices performed by the Self-Balancing Algorithm, we always have $\mathbb{L}_t \succeq \mathbb{L}_{t+1}$.* □

Let us observe two important consequences of Lemma 3. Firstly, this lemma implies that the maximum load never increases. Secondly, Lemma 3 yields the following claim:

Lemma 4. *The number of changes in the load vector is upper bounded by $m \cdot n$.* □

Now, since we know the algorithm gradually converges to a more balanced distribution of the bins' loads, we formally describe the states to which it converges. We say, a system is *stable in step τ*, if independently of the random choices performed in the iterations $T > \tau$ of the Self-Balancing Algorithm we will have $\mathbb{L}_\tau = \mathbb{L}_T$ for every $T > \tau$. In order to characterize stable states formally, we define a *directed multigraph* representing the state of the system (see also, e.g., [5,16], for similar representations).

Definition 1. *A directed multigraph $G = (V, E)$ representing the system is a directed multigraph with the vertex set $V = \{1, \ldots, n\}$ corresponding to the bins in the system and the edge multiset E (loops are allowed) corresponding to the assignment of the balls in the system. Each edge is associated with a ball, has as the endpoints the two locations of the associated ball, and it is directed from (outwards) the bin containing the associated ball.*

We denote by $G_t = (V, E_t)$ the directed multigraph representing $\mathcal{A}_t$. For any vertex v of G we denote by out-deg(v) the *out-degree* of v in G; if G is not clear from the context, then we also use the notation out-deg$_G(v)$. The *in-degree* is defined analogously. Notice that since the choices of the locations of each bin are performed at random, the *undirected* version of any G_t is a *random multigraph* with n vertices and m edges (where each endpoint of each edge is selected independently and uniformly at random).

The following lemma follows directly from Definition 1.

Lemma 5. *If $G_t = (V, E_t)$ is a directed multigraph representing $\mathcal{A}_t$, then for any j, $1 \le j \le n$, the out-degree of vertex j is equal to the load of bin j in $\mathcal{A}_t$.*   □

Let $G_\tau = (V, E_\tau)$ be the directed multigraph representing $\mathcal{A}_\tau$. A directed path $(v_1, v_2, \ldots, v_\ell)$ in G_τ is called a *slope* if out-deg(v_1) $\ge$ out-deg(v_ℓ)+2 and out-deg(v_i) $\ge$ out-deg(v_{i+1}) for every i, $1 \le i < \ell$. If $(v_1, v_2, \ldots, v_\ell)$ is a slope in G_τ, then we can *straighten* $(v_1, v_2, \ldots, v_\ell)$ by modifying the directions of the edges in G_τ (following the rules in the Self-Balancing Algorithm) so that the load vector will change (see also a scheme presented in Figure 1). Indeed, let us consider the case that $\ell \ge 3$ (the case $\ell = 2$ can be handled similarly), and assume (actually, without loss of generality) that out-deg(v_1) $=$ out-deg(v_2) $+ 1$, out-deg(v_j) $=$ out-deg(v_{j+1}) for $2 \le j < \ell - 1$, and that out-deg($v_{\ell-1}$) $=$ out-deg(v_ℓ) $+ 1$. Then, we reverse directions of the edges (v_j, v_{j+1}) for all $1 \le j < \ell - 1$ (this can be easily done according to the rules in the Self-Balancing Algorithm). After applying these changes, the bin corresponding to the vertex v_1 decreased its load by 1, the bin corresponding to the vertex v_ℓ increased its load by 1, and the load of all other bins remains the same. This implies that the load vectors $\mathbb{L}$ of $\mathcal{A}_\tau$ and $\mathbb{L}'$ of the new system state fulfill $\mathbb{L} \succ \mathbb{L}'$.

The following key lemma provides a necessary and sufficient condition for a system to be stable at step t. (Notice that the *only if* part follows from our arguments above.)

Lemma 6. *A system is stable at step τ if and only if the directed multigraph $G_\tau = (V, E_\tau)$ representing $\mathcal{A}_\tau$ has no slope.*   □

The next lemma describes a relationship between stable states and the maximum load in the system.

Lemma 7. *Consider a system of m balls and n bins with the minmax load κ. Then, if the system is stable in step τ then the maximum load of $\mathcal{A}_\tau$ is κ.*

Proof. The proof is by contradiction. Let us consider a system of m balls and n bins with the minmax load κ. Let us suppose the system is in a stable state $\mathcal{A}_\tau$ represented by the directed multigraph $G_\tau = (V, E_\tau)$, and, for the purposes of contradiction, let us assume that the maximum out-degree in G_τ is greater than κ.

Since $\mathcal{A}_\tau$ is a stable state, we know by Lemma 6 that G_τ has no slope. Let us pick any vertex $v \in V$ with out-deg$_{G_\tau}(v) > \kappa$. Let U be the set of all vertices in G_τ (not including v) that are reachable from v by a directed path in G_τ. Since G_τ has no slope, all vertices in U must have the out-degree at least out-deg$_{G_\tau}(v) - 1 \ge \kappa$. Therefore, if we define $U^* = U \cup \{v\}$, then there are at least $|U| \cdot \kappa + (\kappa + 1)$ balls having both locations in the bins corresponding to the vertices in U^*. This, however, by Lemma 1, means that minmax load is at least $\frac{1}{|U|+1} \cdot (|U| \cdot \kappa + (\kappa+1)) > \kappa$, which is a contradiction to our initial assumption that the minmax load of the system is κ.   □

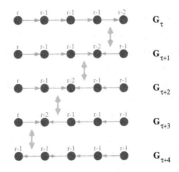

Fig. 1. Illustration describing the *straightening* procedure that changes the out-degrees of the vertices $v_1, v_2, \ldots, v_s$ (with $s = 5$) on the slope performed in the proof of Lemma 6. In this case, initially we have out-deg$(v_1) = r$, out-deg$(v_2) =$ out-deg$(v_3) =$ out-deg$(v_4) = r - 1$, and out-deg$(v_5) = r - 2$.

Now we are ready to complete the proof of Theorem 3. By Lemma 7, the system is not stable if and only if the directed multigraph $G_\tau = (V, E_\tau)$ representing $\mathcal{A}_\tau$ has a slope $(v_1, \ldots, v_\ell)$ for certain positive ℓ. Thus, if the system is not stable, then let us consider any shortest slope. Then, with a positive probability, in the next $\ell - 1$ iterations in the Self-Balancing Algorithm we will perform slope straightening of $(v_1, \ldots, v_\ell)$, which will decrease the load of v_1 by 1, increase the load of v_ℓ by 1, and leave the remaining loads the same. Hence, if $\mathcal{A}_t$ is not stable, then after sufficiently many iterations of Self-Balancing Step, with probability 1 the load vector will be modified. Since the load vector may change at most $n\,m$ times, if we combine the arguments above with Lemma 7, after sufficiently many iterations in Self-Balancing Step, with probability 1 the system will be in a stable state in which the maximum load equals the minmax load. □

4   Convergence to Optimal Assignment for $m \gg n$

In this section we briefly sketch the proof of Theorem 4, which estimates the convergence speed of the Self-Balancing Algorithm for $m \gg n$. First of all, let us recall that by Lemma 4, the load vector may change at most $n\,m$ times. Therefore, we only have to show that if the system is not stable, then after a polynomial number of steps of the Self-Balancing Algorithm the system will change its load vector with high probability. The following is the key theorem of our analysis (the proof is deferred to the full version of the paper).

Theorem 6. *Let $n^5 \log m = o(\mu)$. Let ξ be an arbitrary constant. Let b be a bin with any load greater than or equal to $\mu + \xi$. Then, with probability at least $1 - m^{-O(1)}$,*

- *either every bin has load greater than or equal to $\mu + \xi$,*
- *or the directed multigraph representing the current state of the system has a directed path of length at most 2 from the vertex corresponding to b to some other vertex u whose out-degree is strictly smaller than $\mu + \xi$.*

In view of this theorem, with high probability, as long as the maximum load in the system is strictly larger than $\widehat{\mu}$, the directed multigraph representing the state of the system has *always* a slope $(v_0, \dots, v_\tau)$ with $\tau \leq 2$, *no matter* how the directions of the edges are set. (Indeed, in that case there is a bin b with the load larger than $\widehat{\mu}$, and if we set $\xi = 0$, then it is impossible that every bin in the system has load greater than or equal to μ. Therefore, by Theorem 6, there must exist a directed path of length at most 2 from the vertex corresponding to b to some other vertex u, such that the out-degree of u is strictly smaller than μ. Therefore, either this path or its sub-path must be a slope.) Therefore, with probability at least $O(1/n^4)$, the Self-Balancing Algorithm will, in at most two steps, perform slope straightening of $(v_0, \dots, v_\tau)$ such that the out-degree of v_0 decreases from some ℓ to $\ell-1$ and no other vertex on the path increases its out-degree to more than $\ell - 1$. Therefore, the system will change its load vector with probability at least $O(1/n^4)$. Hence, with high probability the system will change its load after $O(n^4)$ Self-Balancing Steps, and thus, after $O(mn^5)$ steps the Self-Balancing Algorithm will reach a state in which the maximum load equals to the minmax load.

Actually, it is easy to see that our arguments above can be used to show that if the imbalance of the system is Δ (where $\Delta = \sum_{i=1}^n \max\{\mathbb{L}(i) - \widehat{\mu}, 0\}$), then the process needs only $\Delta \cdot n^{O(1)}$ steps to reach a perfect distribution, with high probability. This yields the proof in the heavily loaded case. □

5   Convergence to Optimal Assignment for $m = O(n)$

In this section we deal with the proof of Theorem 5 and consider the convergence speed of the Self-Balancing Algorithm in the lightly loaded case. We focus only on the case $m = O(n)$; we believe that this is the most challenging case and therefore we will elaborate on its proof. The analysis of the case $m = O(n \log n / \log \log n)$, $m = \omega(n)$, is deferred to the full version of the paper.

The main idea behind the proof is to use similar arguments as in the previous section, but this time we cannot assume that we have a slope of a constant length. The analysis requires the following three key properties. The first property, proven in [16], is that if the pairs of locations for all the balls are chosen i.u.r., then (with high probability, depending only on the random choices of the locations) in any state of the system, if there is a bin with load greater than $\widehat{\mu} + 1$ then there is a slope of length $O(\log n)$. The second property is that the sum of the degrees (in- and out-degrees) of all vertices on this slope path is at most $O(\log n)$. The third property is that the probability that a given slope path will be straightened is inversely proportional to the sum of the degrees of the vertices on this path. With these properties, we can show that the probability that in the next $O(mn \log n)$ Self-Balancing Steps a slope of length $O(\log n)$ is chosen and then straightened by the algorithm (without interfering with the other bins (vertices)) is at least $O(1/n^{O(1)})$. This implies that (with high probability) in the next $O(n^{O(1)})$ steps the Self-Balancing Algorithm will change the load vector. Therefore, (with high probability) after $n^{O(1)}$ steps the Self-Balancing Algorithm will reach a state, in which, by Theorem 1, the maximum load is at most $\widehat{\mu} + 1$, with high probability.

We describe now our analysis in more detail. We first develop some properties of the directed multigraphs discussed in Section 3. We begin with a lemma proven implicitly in [16, Lemma 14].

Lemma 8. [16] *Let* $G_t = (V, E_t)$ *be a directed multigraph representing certain* $\mathcal{A}_t$. *Let* $m = O(n)$. *Then, with high probability (depending only on the random locations of the balls), either* $\mathcal{A}_t$ *has the maximum load of at most* $\widehat{\mu} + 1$ *or* G_t *has a slope of length* $O(\log n)$.   □

Our approach is to explore Lemma 8. First of all, from now on, we shall condition on the fact that there is an assignment of the balls among the bins with maximum load $\widehat{\mu} + 1$. (By Theorem 1, this fact holds with high probability.) Then, by Lemma 8, we know that the system is either in the state when the maximum load is $\widehat{\mu} + 1$, in which case we do not have to prove anything, or there is slope in G_t of length $O(\log n)$. We consider only the latter case.

We work in rounds, each round corresponding to $O(n^3 \log^2 n)$ repetitions of Self-Balancing Step. All rounds are independent. At the beginning of each round we take any slope π in G of length $O(\log n)$ that is promised by Lemma 8 (if no such a path exists, then we know that we are already in a state with maximum load smaller than or equal to $\widehat{\mu} + 1$). We prove in Lemma 10 that with probability greater than or equal to $\frac{1}{poly(n)}$ we will successfully straighten the slope in this round. From this and Theorem 3 it follows easily that after a polynomial number of rounds of the Self-Balancing Algorithm we reach a stable state having the maximum load at most $\widehat{\mu} + 1$, with high probability.

Now, our ultimate goal is to analyze the probability that a slope of length $O(\log n)$ will be straightened in $O(n^3 \log^2 n)$ iterations of the Self-Balancing Algorithm. We begin with an auxiliary lemma about random (undirected) multigraphs (the proof is deferred to the full version of the paper).

Lemma 9. *Let* b *and* c *be arbitrary positive constants. If* G *is a random* undirected *multigraph with* n *vertices and* $m \leq bn$ *edges, then, with high probability* G *does not have any simple path of length less than or equal to* $c \log n$ *for which the sum of the degrees of the vertices on the path is greater than* $d \cdot \log n$, *where* d *is a constant.*   □

Our next and key result shows that the probability that the Self-Balancing Algorithm will straighten a given slope path is inversely proportional to the sum of the degrees of the vertices on this path.

Lemma 10. *Let* b *and* c *be arbitrary positive constants. Let* G *be an arbitrary directed multigraph with* n *vertices and* $m \leq bn$ *edges. Suppose there is a slope path* $\pi = (v_1, \ldots, v_\ell)$ *in* G. *Then, with probability greater than*

$$\left(1 - \frac{1}{n^{10}}\right) \cdot \frac{1}{n^2} \cdot \left(\prod_{i=2}^{\ell} \frac{1}{1 + out\text{-}deg(v_1) + in\text{-}deg(v_i)}\right),$$

the load vector will change after less than or equal to $2\,\ell\,m\,\log n$ *iterations.*

Proof. We only sketch the proof and defer more details to the full version of the paper.

Consider any slope $\pi = (v_1, v_2, \ldots, v_\ell)$ of shortest length in the system. Recall that $out\text{-}deg(v_1) - 1 = out\text{-}deg(v_2) = out\text{-}deg(v_3) = \cdots = out\text{-}deg(v_{\ell-1}) = out\text{-}deg(v_\ell) + 1$. If $\ell = 2$, then the probability that the load vector will change in the next step is at least as large as the probability that we will choose the edge (v_1, v_2), which is equal to

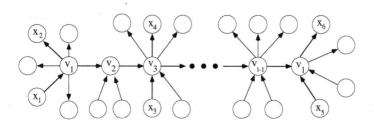

Fig. 2. A slope $\pi = (v_1, v_2, \ldots, v_\ell)$ with incident edges. We only include those edges (y, z) with out-deg$(y) =$ out-deg$(z) + 1$

$1/n^2$. Hence, in this case the lemma easily follows. Therefore, from now on we shall assume that $\ell \geq 3$, i.e. there are no edges (y, z) in G with out-deg$(y) >$ out-deg$(z) + 1$.

We use the terminology from Figure 2. Initially, we have a slope $\pi = (v_1, \ldots, v_\ell)$ of length $\ell - 1$. In each iteration of the Self-Balancing Algorithm we will hit a certain edge chosen at random and in this way we may modify the graph and the load vector. We observe that if we hit an edge that does not belong to π nor is incident to π, then any eventual modification of that edge will not influence path π. Therefore, we only have to consider the following eight cases, when an edge of the following form is chosen: (i) (v_1, v_2), (ii) $(v_{\ell-1}, v_\ell)$, (iii) (x_1, v_1), (iv) (v_1, x_2), (v) (x_3, v_3), (vi) (v_3, x_4), (vii) (x_5, v_ℓ), and (viii) (v_ℓ, x_6). We say a *very good edge is hit* if we hit an edge from cases (iv) or (ix); a *good edge is hit* if we hit an edge from cases (i), (iii), (vi), or (vii); a *bad edge is hit* if we hit an edge from cases (v), or (viii). Very good edges create an edge (y, z) with out-deg$(y) >$ out-deg$(z) + 1$, good edges make the slope shorter, and bad edges make it longer.

Now, we consider a round lasting $2 \ell n^2 \log n$ iterations and observe only very good edge hits, good edge hits, and bad edge hits. A round is called *successful* if no bad edge is hit until we either have a very good edge hit and then straighten the obtained path or we modify the slope path (we straighten it) by only good edges. One can show that with probability greater than or equal to $1 - 1/n^{10}$ a round is either successful or we made a bad edge hit. Notice that there are at most out-deg$(v_1) +$ in-deg(v_ℓ) bad edges at the beginning, and there is at least one good edge at any time. Certainly, under the assumption that either a bad edge or $(v_{\ell-1}, v_\ell)$ is picked, the probability that $(v_{\ell-1}, v_\ell)$ is picked is at least $1/(1 +$ out-deg$(v_1) +$ in-deg$(v_\ell))$. Once $(v_{\ell-1}, v_\ell)$ is picked, we concentrate on the edge $(v_{\ell-2}, v_{\ell-1})$, and so on. Using this approach, we get that the probability that a round is successful is lower bounded by $\frac{1}{n^2} \cdot \left(\prod_{i=2}^{\ell} \frac{1}{1+\text{out-deg}(v_1)+\text{in-deg}(v_i)} \right)$. This completes the proof. $\qquad\square$

We can reduce our analysis to the case when for the slope $\pi = (v_1, \ldots, v_\ell)$ we have out-deg$(v_1) = \widehat{\mu} + 2$ and $\ell = O(\log n)$. Therefore, by Lemma 9 we know that $\sum_{i=1}^{\ell}$ in-deg$(v_i) = O(\log n)$, with high probability. Hence, by Lemma 10, the probability that in a round lasting $2nm \log n$ iterations we change the load vector is greater than or equal to $\frac{1}{poly(n)}$. Hence, after $poly(n)$ rounds (iterations) of the Self-Balancing Algorithm we shall modify the load vector with high probability. Now, since the load vector can be modified at most $m \cdot n$ times before we reach the stable state, the theorem follows. $\square$

References

1. Y. Azar, A. Z. Broder, A. R. Karlin, and E. Upfal. Balanced allocations. *SIAM J. Comput.*, 29(1):180–200, 1999.
2. P. Berenbrink, A. Czumaj, A. Steger, and B. Vöcking. Balanced allocations: The heavily loaded case. *STOC*, pp. 745–754, 2000.
3. R. Cole, A. Frieze, B. M. Maggs, M. Mitzenmacher, A. W. Richa, R. K. Sitaraman, and E. Upfal. On balls and bins with deletions. *RANDOM*, pp. 145–158, 1998.
4. R. Cole, B. M. Maggs, F. Meyer auf der Heide, M. Mitzenmacher, A. W. Richa, K. Schröder, R. K. Sitaraman, and B. Vöcking. Randomized protocols for low-congestion circuit routing in multistage interconnection networks. *STOC*, pp. 378–388, 1998.
5. A. Czumaj and V. Stemann. Randomized allocation processes. *Random Structures and Algorithms*, 18(4):297–331, 2001. A preliminary version appeared in *FOCS*, pp. 194–203, 1997.
6. B. Ghosh, F. T. Leighton, B. M. Maggs, S. Muthukrishnan, C. G. Plaxton, R. Rajaraman, A. W. Richa, R. E. Tarjan, and D. Zuckerman. Tight analyses of two local load balancing algorithms. *SIAM J. Comput.*, 29(1):29–64, September 1999.
7. R. M. Karp. Random graphs, random walks, differential equations and the probabilistic analysis of algorithms. *STACS*, pp. 1–2, 1998.
8. R. M. Karp, M. Luby, and F. Meyer auf der Heide. Efficient PRAM simulation on a distributed memory machine. *Algorithmica*, 16(4/5):517–542, 1996.
9. J. Korst. Random duplicated assignment: An alternative to striping in video servers. *ACM MULITIMEDIA*, pp. 219–226, 1997.
10. M. J. Luczak and E. Upfal. Reducing network congestion and blocking probability through balanced allocation. *FOCS*, pp. 587–595, 1999.
11. M. Mitzenmacher. Load balancing and density dependent jump Markov processes. *FOCS*, pp. 213–222, 1996.
12. M. Mitzenmacher, A. W. Richa, and R. Sitaraman. The power of two random choices: A survey of techniques and results. In *Handbook of Randomized Computing*, Rajasekaran et al., eds., Volume I, pp. 255-312, Kluwer Academic Press, 2001.
13. Y. Rabani, A. Sinclair, and R. Wanka. Local divergence of Markov chains and the analysis of iterative load-balancing schemes. *FOCS*, pp. 694–703, 1998.
14. P. Sanders. Asynchronous scheduling of redundant disk arrays. *SPAA*, pp. 89–98, 2000.
15. P. Sanders. Reconciling simplicity and realism in parallel disk models. *SODA*, pp. 67–76, 2001.
16. P. Sanders, S. Egner, and J. Korst. Fast concurrent access to parallel disks. *Algorithmica*, 35(1):21–55, 2003. A preliminary version appeared in *SODA*, pp. 849–858, 2000.
17. L. A. M. Schoenmakers. A new algorithm for the recognition of series parallel graphs. Technical Report CS-R9504, CWI — Centrum voor Wiskunde en Informatica, January 1995.
18. B. Vöcking. How asymetry helps load balancing. *FOCS*, pp. 131–141, 1999.

On Extracting Private Randomness over a Public Channel

Yevgeniy Dodis[1]* and Roberto Oliveira[2]**

[1] Department of Computer Science
New York University
251 Mercer Street
New York, NY 10012, USA.
dodis@cs.nyu.edu

[2] Department of Mathematics
New York University
251 Mercer Street
New York, NY 10012, USA.
oliveira@cims.nyu.edu

Abstract. We introduce *strong blender*s. A strong blender $\text{BLE}(\cdot, \cdot)$ uses weak sources X, Y to produce $\text{BLE}(X, Y)$ that is statistically random even if one is given Y. Strong blenders generalize strong extractors [15] and extractors from two weak random sources [25,6]. We show that non-constructive strong blenders can extract all the randomness from X, as long as Y has logarithmic min-entropy. We also give explicit strong blenders which work provided the sum of the min-entropies of X and Y is at least their block length. Finally, we show that strong blenders have applications to cryptographic systems for parties that have independent weak sources of randomness. In particular, we extend the results of Maurer and Wolf [12] and show that parties that are not able to sample even a single truly random bit can still perform privacy amplification over an adversarially controlled channel.

1 Introduction

IMPERFECT RANDOMNESS. Randomization has proved to be extremely useful and fundamental in many areas of computer science. Unfortunately, in many situations one does not have ideal sources of randomness, and has to base a given application on *imperfect sources of randomness*. Among many imperfect sources considered so far, perhaps the most general and realistic source is the *weak* source [28,6]. The only thing guaranteed about a *weak* source is that no string (of some given length ℓ) occurs with probability more than 2^{-b}, where b is the so-called *min-entropy* of the source. We will call this source (ℓ, b)-weak. Handling such weak sources is often necessary in many applications, as it is typically hard to assume much structure on the source beside the fact that

---

* Partially supported by the NSF CAREER Award.
** Supported by a doctoral fellowship from CNPq, Brazil.

S. Arora et al. (Eds.): APPROX 2003+RANDOM 2003, LNCS 2764, pp. 252–263, 2003.
© Springer-Verlag Berlin Heidelberg 2003

it contains some randomness. Thus, by now a universal goal in basing some application on imperfect sources is to make it work with the weak source.

The most direct way of utilizing weak sources would be to extract nearly perfect randomness from such a source. Unfortunately, it is trivial to see [6] that no *deterministic* function can extract even one random bit from a weak source, as long as $b < \ell$ (i.e., the source is not random to begin with). This observation leaves two possible options. First, one can try to use weak sources for a given application *without* an intermediate step of extracting randomness from it. Second, one can try designing *probabilistic* extractors, and later justify where and how one can obtain the additional randomness needed for extraction.

USING A SINGLE WEAK SOURCE. A big successful line of research [26,24,6,7,28,3] following the first approach showed that a single weak source is sufficient to simulate any probabilistic computation of decision or optimization problems (i.e., problems with a unique "correct" output which are potentially solved more efficiently using randomization; this class is called BPP). Unfortunately, most of the methods in this area are not applicable in situations where randomness is needed by the application *itself*, and not mainly for the purposes of efficiency. One prime example of this is cryptography. For example, secret keys have to be random, and many cryptographic primitives (such as public-key encryption) *must* be probabilistic. Thus, new methods are needed to base cryptographic protocols on weak sources. So far, this question has only been studied in the setting of information-theoretic symmetric-key cryptography. In this scenario, the shared secret key between the sender and the recipient is no longer random, but comes from a weak source. As a very negative result, McInnes and Pinkas [13] proved that one cannot securely encrypt even a single bit, even when using an "almost random" $(\ell, \ell - 1)$-weak source. Thus, one cannot base symmetric-key encryption on weak sources. Dodis and Spencer [9] also consider the question of message authentication and show that one cannot (non-interactively) authenticate even one bit using $(\ell, \ell/2)$-weak source (this bound is tight as Maurer and Wolf [12] showed how to authenticate up to $\ell/2$ bits when $b > \ell/2$).

Basing more advanced cryptographic primitives on a single weak random sources also promises to be challenging. For example, it is not clear how to meaningfully model access to a single weak source by many users participating in a given cryptographic protocol. Additionally, moving to the *computational* setting will likely require making very non-standard cryptographic assumptions.

USING SEVERAL WEAK SOURCES. Instead, we will assume that each party will have its own weak source, which is *independent* from all the other weak random sources. In other words, while each individual party cannot assume that his source is truly random, the parties are located "far apart" so that their imperfect sources are independent from each other. For simplicity, we will restrict the number of independent sources to two for the remainder of this paper. One of the questions we will consider if it is possible to construct cryptographic protocols, like secret-key encryption or key exchange, in this new setting. In fact, rather than construct these primitives from scratch, we will try to extract nearly ideal randomness from two weak sources, and then simply use whatever standard methods exist for the cryptographic task at hand!

This brings us to the question of randomness extraction from two or more independent random sources, a question originated in the works of Sántha and Vazirani [17,25]. Chor and Goldreich [6] were the first to consider general weak sources of equal block length; let us say that the sources X and Y are (ℓ_1, b_1)-weak and (ℓ_2, b_2)-weak, for concreteness, while here we also assume $\ell_1 = \ell_2 = \ell$. They showed that a random function can extract almost $(b_1 + b_2 - \ell)$ nearly random bits in this setting.[3] They also gave an explicit number-theoretic construction that can essentially match this (non-optimal) bound. Moreover, they showed that the simple inner product function is also a good bit-extractor under the same condition that $b_1 + b_2 > \ell$. Recently, Trevisan and Vadhan [23] broke this the "barrier" $b_2 + b_2 > \ell$, but only for the very "imbalanced" case when $b_1 = \varepsilon^2 \ell$, $b_2 = (1 - O(\varepsilon))\ell$ (for any $\varepsilon > 0$). To summarize, while non-trivial randomness extraction is possible, the known constructions and parameters seem far from optimal. Unfortunately, improving this situation seems to be extremely challenging. Indeed, it is easy to see that the question of extracting randomness from two independent sources beyond what is currently known is even harder than a notoriously hard problem of explicitly constructing certain bipartite Ramsey graphs (see [27,16]).

STRONG EXTRACTORS. A special case of the above question has received a huge amount of attention recently. It involved the case when one of the two sources, say Y, is perfect: $b_2 = \ell_2$. In this case, one invests b_2 bits of true randomness Y (called the *seed*) and hopes to extract nearly $b_1 + b_2$ random bits from Y and a given (b_1, ℓ_1)-weak source X. A deterministic function EXT achieving this task has simply been called an *extractor* [15]. A *strong extractor* additionally requires Y itself to be part of the extracted randomness. In this case, Y is usually excluded from the output of EXT, so that the goal becomes to extract up to b_1 random bits from X. By now, it is well known that one can indeed achieve this goal provided $b_1 \gg \log \ell_2$. Moreover, many explicit constructions of strong extractors which come very close to this bound are known by now (see [14,22,11,19,18] and the references therein). Not surprisingly, strong extractors have found many applications (e.g., see [18]).

OUR QUESTION. The general question of extracting randomness from two weak sources [17,25,6] concentrated on regular, non-strong extractors. If the extracted randomness is to be used as the secret key of the conventional cryptographic systems, this means that one should sample X and Y from two independent weak sources, and securely "transport" X to Y. Consider, for example, the following application. Alice and Bob stay together and wish to securely communicate when Alice goes away. They can agree on an auxiliary secret key X sampled from their common weak source. When Bob leaves far away, he gets access to an independent source Y. Assuming the parameters are right, Bob can now extract a nearly random secret key $S = \text{EXT}(X, Y)$. However, Alice only knows X, so

---

[3] A trivial strengthening of their technique can push this number to $\min(b_1, b_2)$; we will later *non-trivially* push this to $b_1 + b_2$.

Bob has to send Y to Alice. In cryptography, it is conventional to assume that the communication channel between Alice and Bob is public. Thus, Bob has to send Y "in the clear". With regular extractors, even from two weak sources, there is no guarantee that $\textsc{Ext}(X, Y)$ will look random to the eavesdropper who learns Y. On the other hand, conventional strong extractors resolve this problem, but rely on a strong assumption that Alice can sample a truly random Y and send it over the channel. In the world with no "true randomness" and only weak sources, this assumption is not realizable, unless eventually two independent sources are secretly brought together.

The above example motivates our common generalization of previous work. We wish to consider *strong* extractors with *weak* seeds. We will call such functions *strong blenders*.[4] Namely, we want to design a function BLE such that $\textsc{Ble}(X, Y)$ looks random even for an observer who knows Y, for any X and Y sampled from their corresponding (ℓ_1, b_1) and (ℓ_2, b_2) weak sources.

OUR RESULTS. As we demonstrate, such remarkable strong blenders exist. In particular, we show that a random function can be used to extract essentially all the randomness from X (i.e., nearly b_1 bits), provided only that $b_2 \geq \log \ell_1$ (and also $b_1 \geq \log \ell_2$). The latter condition says that as long as the public seed Y has barely enough randomness, we can extract almost all the randomness from our target source X. Clearly, this bound generalizes the standard setting of (strong) extractors, where one needs $\ell_2 = b_2 \geq \log \ell_1$ to extract all the randomness from X. We also remark that our analysis *non-trivially* extends the previous work.[5] It involves a martingale construction, and then applying Azuma's inequality to bound its deviation from the mean, thus strengthening what was known for regular (non-strong) extraction from two weak sources [6]. As mentioned, their result gave only $b_1 + b_2 - \ell$ bits. It is easy to improve it to $\min(b_1, b_2)$ bits, but getting $b_1 + b_2$ bits — which follows from our *more general* bound — does not seem possible when using standard Chernoff type bounds used by [6].

Next, we address explicit constructions of strong blenders. Unfortunately, the large body of work on strong extractors does not seem to be applicable to strong blenders. Intuitively, standard extractors use the seed to pick a bit from a codeword of a list-decodable code [22], or to select a hash function from a small family of functions. These arguments seem to fall apart completely once the seed comes from a weak random source. On the other hand, any explicit constructions of strong blenders will in particular imply extraction from two independent weak sources, for which any improvement seems very hard, as we mentioned earlier. Thus, the best we can hope for is to extend the best known constructions in this latter setting to yield strong blenders. And, indeed, this is exactly what we achieve. First, we show that the inner product function is a one-bit strong blender for the case $\ell_1 = \ell_2 = \ell$, provided $b_1 + b_2 > \ell$. This argument

----

[4] We propose that functions that extract randomness from two weak sources be called *blenders*. With this choice of terminology, strong blenders are related to (regular) blenders in the same way strong extractors are related to their regular counterparts.

[5] We are not aware of any written proof for the existence of *strong* extractors, as all the references we found point to [20,21].

involves extending the combinatorial lemma of Lindsey (see Section 4). Second, we show that Vazirani's multi-bit extraction for SV-sources can be applied to weak sources as well. This allows to extract $\Omega(\ell)$ bits provided $b_1 + b_2 \gg 3\ell/2$. Finally, we show that the explicit extractor of [6] based on discrete logarithms can also be also be extended to our setting, which gives a way to extract nearly $(b_1 + b_2 - \ell)/2$ random bits. Again, we remark than all these extensions actually involve non-trivial modifications to the existing arguments.

PRIVACY AMPLIFICATION. Finally, we return to applications of strong blenders to the setting where different parties have independent weak sources, but all the communication between them is public. The most natural such application is that of key agreement (aka *privacy amplification* [5,4]) by public discussion: sending Y over the channel allows Alice and Bob to agree on a (nearly) random key $S = \text{BLE}(X, Y)$, provided the communication channel is *authentic*. Therefore, the remaining interesting case to consider is what happens when the channel is not only public, but *adversarially controlled* [12]. In particular, the question is whether we can build any kind of message authentication with a shared key coming from a (ℓ_1, b_1)-weak block source, and without any local randomness. Specifically, assume Alice and Bob share a key A, B, X coming from 3 (possibly correlated) samples from the (ℓ_1, b_1)-weak block source. When Bob gets his hands on an independent source Y, he would like to authenticate Y using A, B. Then, they both can agree on the key $S = \text{BLE}(X, Y)$, where BLE is our strong blender. As we mentioned, [9] showed that non-interactive one-time authentication from Alice to Bob is impossible when $b_1 \le \ell_1/2$. On the other hand, Maurer and Wolf [12] gave a way to non-interactively authenticate up to ℓ bits using two blocks of min-entropy $\gg \ell/2$.[6] Assuming (as we shall) that $\ell_1 = \ell_2$, Bob can indeed authentically transmit Y over the channel, so that both can apply a strong blender to agree on a random $S = \text{BLE}(X, Y)$. Combining this observation with our explicit constructions of strong blender for $\ell_1 = \ell_2 = \ell$, we get that the first efficient privacy amplification *without ideal local randomness*, provided $b_1 \gg \ell/2$ and $b_2 \gg \ell - b_1$.

ONLINE VERSION. We refer to the online manuscript [8] for the proofs of results that are only stated in this work.

2   Preliminaries

2.1   Basic Notation

We mostly employ standard notation. The symbol log is reserved for the base 2 logarithm. For a positive integer t, U_t denotes a random variable that is uniform over $\{0, 1\}^t$ and independent of all other random variables under consideration. We also write $[t] \equiv \{1, 2, \ldots t\}$. For two random variables A, B taking values in

---

[6] We remark that unlike our setting, Alice and Bob had *ideal* local randomness in the setting of [12], and used it at later stages of their application. Luckily, the authentication step was deterministic, which makes it "coincidentally applicable" to our situation.

the finite set $\mathcal{A}$, their *statistical distance* is $\|A - B\| \equiv \frac{1}{2}\sum_{a\in\mathcal{A}} |\Pr(A = a) - \Pr(B = a)|$, and the min-entropy of A is $H_\infty(A) \equiv \min_{a\in\mathcal{A}} -\log(\Pr(A = a))$. Finally, if C is another random variable, $C|_{A=a}$ represents the distribution of C conditioned on $A = a \in \mathcal{A}$.

2.2   Strong Extractors Vs. Strong Blenders

Min-entropy quantifies the amount of hidden randomness in a source X. The objective of extractors is to purify this randomness with the aid of a (small amount of) truly random bits.

Definition 1. *Let $k \geq 0$, $\varepsilon > 0$. A (k,ε)-extractor $\mathrm{EXT} : \{0,1\}^n \times \{0,1\}^d \to \{0,1\}^m$ is a function such that for all n-bit random variables X with min-entropy $H_\infty(X) \geq k$ $\|\mathrm{EXT}(X, U_d) - U_m\| \leq \varepsilon$. EXT is a (k,ε)-strong extractor if the function $\mathrm{EXT}' : (x, y) \mapsto y \circ \mathrm{EXT}(x, y)$ is an extractor.*

In this work, however, we are interested in strong randomness extraction from 2 weak sources, as defined below.

Definition 2. *[6] The set $\mathsf{CG}(\ell_1, \ell_2, b_1, b_2)$ of pairs of independent (Chor-Goldreich) weak sources is the set of all pairs of independent random variables (X, Y) where X (respectively Y) is ℓ_1 (resp. ℓ_2) bits long and $H_\infty(X) \geq b_1$ (resp. $H_\infty(Y) \geq b_2$).*

Definition 3. *A (b_1, b_2, ε)-strong blender (SB) is a function*

$$\mathrm{BLE} : \{0,1\}^{\ell_1} \times \{0,1\}^{\ell_2} \to \{0,1\}^m$$

such that for all pairs (X, Y) in $\mathsf{CG}(\ell_1, \ell_2, b_1, b_2)$, we have

$$\|\langle Y, \mathrm{BLE}(X,Y)\rangle - \langle Y, U_m\rangle\| \leq \varepsilon$$

We state for later convenience the following proposition, which can be deduced from the linear programming argument in [6] (i.e. the fact that general sources of a given min-entropy are convex combinations of flat distributions with the same min-entropy).

Proposition 1. *If b_1 and b_2 are integers, then for any function $f : \{0,1\}^{\ell_1} \times \{0,1\}^{\ell_2} \to \{0,1\}^m$ the maximum of $\|\langle Y, f(X,Y)\rangle - \langle Y, U_m\rangle\|$ over all (X, Y) contained in the set $\mathsf{CG}(\ell_1, \ell_2, b_1, b_2)$ is achieved by flat random variables, that is, by a pair (X, Y) for which X is uniform over a subset $S_X \subset \{0,1\}^{\ell_1}$ with $|S_X| = 2^{b_1}$, and Y is uniform over $S_Y \subset \{0,1\}^{\ell_2}$, $|S_Y| = 2^{b_2}$.*

3   Existence of Strong Blenders

From now on m, $\ell_1 \geq b_1 \geq 2$ and $\ell_2 \geq b_2 \geq 2$ are positive integers and $\varepsilon > 0$ is a real number. The aim of this section is to prove non-constructively that strong blenders exist for certain choices of parameters, and to provide lower bounds that almost match the existence result. The theorems below are proven in [8].

Theorem 1. *There exists a* (b_1, b_2, ε)-*SB* BLE $: \{0,1\}^{\ell_1} \times \{0,1\}^{\ell_2} \to \{0,1\}^m$ *for any choice of parameters satisfying* $m \leq b_1 - 2\log\frac{1}{\varepsilon}$, $b_1 \geq \log_2(\ell_2 - b_2) + 2\log\frac{1}{\varepsilon} + O(1)$ *and* $b_2 \geq \log_2(\ell_1 - b_1) + 2\log\frac{1}{\varepsilon} + O(1)$.

By noticing that strong extractors are special cases of strong blenders (e.g., using $Y = U_{b_2}0^{\ell_2 - b_2}$), and applying the lower bounds of Ta-Shma and Radhakrishnan [21], we obtain (see [8] for details)

Theorem 2. *For some constant c, if* $b_1 \leq \ell_1 - c$ *and* $b_2 \leq \ell_2 - c$, *then conditions* $m \leq b_1 - 2\log\frac{1}{\varepsilon}$ *and* $b_2 \geq \log_2(\ell_1 - b_1) + 2\log\frac{1}{\varepsilon} + O(1)$ *of Theorem 1 are in fact necessary for the existence of a* (b_1, b_2, ε)-*SB* BLE $: \{0,1\}^{\ell_1} \times \{0,1\}^{\ell_2} \to \{0,1\}^m$.

4   Efficient Constructions

4.1   Hadamard Matrices and Extraction of One Bit

A class of 1-bit strong blenders which includes the inner product function is now considered, thus providing a strengthening of a result of Chor and Goldreich [6]. Identify $[L] \equiv [2^\ell] \approx \{0,1\}^\ell$ and let $H = \{H_{xy}\}^L_{x,y=1}$ be a $L \times L$ Hadamard matrix (i.e. a ± 1 matrix with pairwise orthogonal rows/columns). Define

$$\text{BLE}_H : \{0,1\}^\ell \times \{0,1\}^\ell \to \{0,1\} \qquad (1)$$
$$(x,y) \longmapsto \frac{1+H_{xy}}{2}$$

We shall prove the following two results.

Theorem 3. BLE$_H$ *as defined above is a* (b_1, b_2, ε)-*SB with* $\log\frac{1}{\varepsilon} = \frac{b_1 + b_2 - \ell}{2} + 1$.

Corollary 1. *The inner product function on* ℓ-*bit strings is a* (b_1, b_2, ε)-*SB with* ε *as above.*

Proof. (*of Corollary 1*) Inner product is of the form BLE$_H$ for some Hadamard matrix H (as one can easily show). □

Proof. (*of Theorem 3*) The proof parallels that of the corresponding theorem in [6]. In particular, we also employ Lindsay's Lemma.

Lemma 1. (Lindsay's Lemma cf. [6]) *Let* $G = (G_{ij})^T_{i,j=1}S$ *be a* $T \times T$ *Hadamard matrix, and* R *and* C *be subsets of* $[T]$ *corresponding to choices of rows and columns of* G *(respectively). Then* $|\sum_{i \in R}\sum_{j \in C} G_{ij}| \leq \sqrt{|R||C|T}$.

For any choice of $(q_1, \ldots, q_L) \in \{-1, +1\}^L$, the matrix $\tilde{H} = (\tilde{H}_{ij})$ whose ith row is q_i times the ith row of H is Hadamard. Hence Lindsay's Lemma applies and for all sets $R, C \subset [L]$ the sum $\sum_{i \in R}\sum_{j \in C} \tilde{H}_{ij}$, which is just $\sum_{i \in R}\left(q_i \sum_{j \in C} H_{ij}\right)$, is bounded by $\sqrt{|R||C|L}$. From this fact it is easy to deduce a stronger form of Lemma 1.

$$\forall R, C \subset [N] \quad \sum_{i \in R}\left|\sum_{j \in C} H_{ij}\right| \leq \sqrt{|R||C|L} \qquad (2)$$

Now let $(X, Y) \in \mathsf{CG}(\ell, \ell, b_1, b_2)$ be flat random variables and assume that X is uniform on S_X, $|S_X| = 2^{b_1}$ and Y is uniform on S_Y, $|S_Y| = 2^{b_2}$. Applying (2), one obtains precisely the desired inequality

$$\| \langle Y, \mathrm{BLE}_H(X, Y) \rangle - \langle Y, U_1 \rangle \| = \frac{1}{|S_Y|} \sum_{y \in S_Y} \| \mathrm{BLE}_H(X, y) - U_1 \|$$

$$= \frac{1}{2|S_Y|} \sum_{y \in S_Y} \left| \sum_{x \in S_X} \frac{H_{xy}}{|S_X|} \right| \leq \frac{1}{2} \sqrt{\frac{L}{|S_X||S_Y|}} = 2^{-\frac{b_1 + b_2 - \ell}{2} - 1} \quad (3)$$

$\square$

4.2   Extracting Many Bits Using Error-Correcting Codes

We now adapt a construction from [25] based on error-correcting codes to obtain many bits from weak sources of same length and sufficiently high min-entropy. In what follows $\mathrm{ECC} : \{0,1\}^m \to \{0,1\}^\ell$ is a linear error correcting code with distance d, $\{e_i : 1 \leq i \leq m\}$ is the canonical basis of $\{0,1\}^m$ as a vector space over $\mathbb{Z}_2$, and for $(x, y) = \left((x_1, \ldots, x_\ell), (y_1, \ldots, y_\ell) \right) \in \{0,1\}^\ell \times \{0,1\}^\ell$ we let $v(x, y) \in \{0,1\}^\ell$ be the vector whose ith coordinate is $x_i y_i$. The proposed SB is

$$\begin{array}{rcl} \mathrm{BLE} : \{0,1\}^\ell \times \{0,1\}^\ell & \to & \{0,1\}^m \\ (x, y) & \longmapsto & \left(\mathrm{ECC}(e_1) \cdot v(x, y) \right) \circ \cdots \circ \left(\mathrm{ECC}(e_m) \cdot v(x, y) \right) \end{array} \quad (4)$$

Note that each bit that BLE outputs corresponds to the inner product of matching segments of the input strings x and y. We show in below that

Theorem 4. *The function* BLE *constructed above is a* (b_1, b_2, ε)-*SB with* $\log \frac{1}{\varepsilon} = 1 + \frac{b_1 + b_2 + d}{2} - (\ell + m)$.

There exist efficiently encodable linear codes of codeword length ℓ, dimension $m = \delta^3 \ell$ and distance $d = (\frac{1}{2} - \delta)\ell$, for all fixed $0 < \delta < \frac{1}{2}$. Plugging one such code into Theorem 4 yields an efficiently computable (b_1, b_2, ε)-SB with $\varepsilon = \ell^{-\omega(1)}$ for all min-entropies satisfying $\frac{b_1 + b_2}{2} \geq (3/4 + \delta)\ell + \omega(\log \ell)$, and the number of extracted bits is $m = \delta^3 n$. We prove Theorem 4 below, with the aid of the following two lemmas (the second being fairly standard).

Lemma 2. *(Parity Lemma, [25]) For any* t-*bit random variable* T, $\|T - U_t\|$ *is upper-bounded by* $\sum_{a \in \{0,1\}^t \setminus \{0\}} \| (T \cdot a) - U_1 \|$.

Lemma 3. *If* $Z = Z_1 Z_2 \ldots Z_t$ *is a* t-*bit random variable and* $W \subset [t]$, *let* $Z|_W$ *denote the concatenation of all* Z_i *with* $i \in W$. *Then* $H_\infty(Z|_W) \geq H_\infty(Z) - t + |W|$.

Proof. (of Theorem 4) By Lemma 2 and some simple calculations, it suffices to show that for any $a \in \{0,1\}^m \setminus \{0\}$

$$\| \langle Y, (\mathrm{BLE}(X, Y) \cdot a) \rangle - \langle Y, U_1 \rangle \| \leq \frac{\varepsilon}{2m} = 2^{\ell - \frac{b_1 + b_2 + d}{2} - 1}. \quad (5)$$

Fix some non-zero $a = \sum_{i=1}^{m} a_i e_i$ and note that by the linearity of ECC

$$\text{BLE}(X,Y) \cdot a = \sum_{i=1}^{m} a_i\big(\text{ECC}(e_i) \cdot v(X,Y)\big) = \text{ECC}(a) \cdot v(X,Y) = (X|_S) \cdot (Y|_S) \tag{6}$$

where S is the set of all non-zero coordinates of $\text{ECC}(a)$, and $X|_S$ and $Y|_S$ are defined as in Lemma 3. Applying that Lemma, we conclude that $X|_S$ ($Y|_S$) has min-entropy at least $b_1 - \ell + |S|$ (respectively $b_2 - \ell + |S|$). It now follows from (6), Corollary 1, Lemma 3 and the fact that $X|_S$ and $Y|_S$ have length $|S|$ that

$$\big\| \langle Y, (\text{BLE}(X,Y) \cdot a)\rangle - \langle Y, U_1 \rangle \big\| \leq 2^{\frac{|S|-b_1-b_2+2\ell-2|S|}{2}} - 1 = 2^{\ell-1-\frac{b_1+b_2+|S|}{2}} \tag{7}$$

Since $|S| = (\text{weight of } \text{ECC}(a)) \geq d$ by definition of S, equation (7) proves (5) and finishes the proof.   □

4.3   A Number-Theoretic Construction

A third efficient SB construction is now presented. Its minimal min-entropy requirement is basically $b_1 + b_2 > \ell$, which roughly matches the Hadamard matrix construction for 1-bit extraction. However, this SB has the drawback of requiring a pre-processing stage for efficiency to be achieved. The construction dates back to [6], in which it was shown that $\text{BLE}(X,Y)$ is close to random. We claim that the same is true even if Y is given to the adversary, thus establishing that this construction satisfies our definition of SB. In what follows, $p > 2$ is a prime and we take $\ell = \lfloor \log p \rfloor$ so that we can assume $\{0,1\}^\ell \subseteq \mathbb{Z}_p$. Let k be a divisor of $p-1$; our SB will output elements of $\mathbb{Z}_k$ (the definition of a SB easily generalizes to this case). Finally, let g be a generator of the multiplicative group $\mathbb{Z}_p^*$ and denote by $\log_g$ the base-g discrete logarithm in $\mathbb{Z}_p^*$. We define

$$\begin{aligned} \text{BLE} : \{0,1\}^\ell \times \{0,1\}^\ell &\to \mathbb{Z}_k \\ (x,y) &\longmapsto \log_g(x-y) \mod k \end{aligned} \tag{8}$$

We prove below that approximately $m = \log k \approx \frac{b_1+b_2-\ell}{2} - \log\frac{1}{\varepsilon}$ bits can be extracted by this construction.

Theorem 5. *The function* BLE *defined above is a* (b_1, b_2, ε)-SB *with* $\log\frac{1}{\varepsilon} = \frac{b_1+b_2-\ell}{2} + 1 - \log k$.

We refer to [6] for details on the efficient implementation of BLE and the pre-computation of p, k and g.

Proof. (of Theorem 5) The following inequality (proven in [8]) holds for all sub-sets $A, B, C \subseteq \mathbb{Z}_p$: setting $\Phi_C \equiv \max_{1 \leq j \leq p-1} |\sum_{c \in C} e^{\frac{2\pi i c j}{p}}|$,

$$\sum_{a \in A} \left| \#\{b \in B : a - b \in C\} - \frac{|B| \, |C|}{p} \right| \leq \Phi_C \sqrt{|A| \, |B|} \tag{9}$$

Assuming (9), choose $\alpha \in \mathbb{Z}_k$ and set $\mathcal{C} \equiv \{c \in C \mid \log_g(c) = \alpha \mod k\}$. Following [6, Section 3.2], we note that $|C| = p/k$ and $\Phi_C < \sqrt{p}$. Hence for all $A, B \subseteq \{0,1\}^\ell \subseteq \mathbb{Z}_p$

$$\sum_{a \in A} \left| \#\{b \in B : \log_g(a-b) = \alpha\} - \frac{|B|}{k} \right| \leq \sqrt{p|A|\,|B|} \tag{10}$$

We deduce from (10) that for any choice of flat random variables (X, Y) in the subset $\mathsf{CG}(\ell_1, \ell_2, b_1, b_2)$ with respective supports S_X, S_Y of sizes 2^{b_1}, 2^{b_2}

$$\|\langle Y, \mathrm{BLE}(X, Y)\rangle - \langle Y, U\rangle\|$$
$$= \frac{1}{2} \sum_{\alpha \in \mathbb{Z}_k} \sum_{y \in S_Y} \left| \frac{\#\{x \in S_X : \log_g(x-y) = \alpha\}}{2^{b_1+b_2}} - \frac{1}{2^{b_1}k} \right| \leq \frac{k}{2}\sqrt{\frac{p}{2^{b_1+b_2}}} = \varepsilon \tag{11}$$

and this implies the theorem by Proposition 1. □

5   Simple Authentication with Weak Sources

5.1   Motivation

As noted in the Introduction, strong blenders trivially solve the problem of privacy amplification over passive public channels with weak random sources are used. In this Section we provide a simple protocol PA for privacy amplification over an *adversarially controlled channel*, when only weak sources of randomness are available. Following [12], we show that weak sources can be used in conjunction with the simple "$ay + b$" message authentication code (MAC) to transmit the non-secret input Y over the adversarial channel.

In our simplified model, Bob can either be close (to Alice) or far (from Alice), and each one of them has a weak source (specified below). If Bob is close, they can share secret information, but their sources could be arbitrarily correlated. On the other hand, if Bob is far, the sources can be assumed to be independent, but only active adversarial communication channels are available. Bob's source outputs a ℓ-bit long string Y with min-entropy $H_\infty(Y) \geq b_2$, and Alice's source outputs three ℓ-bit strings A, B, X, which are assumed that they form a b_1-*block source* [6]. That is, for any $a, b \in \{0,1\}^\ell$, $A, B|_{A=a}$ and $X|_{A=a, B=b}$ all have min entropy at least b_1.

Our scenario differs from that of previous work on privacy amplification (e.g. [12]) in that Alice and Bob are not capable of sampling perfectly random bits. Moreover, geographical distance between the sources is necessary for independence, which is a reasonable assumption for physical and adversarial sources. Whereas in [12] (for instance) it is not clear that it would not be possible for the parties to agree on a perfectly random secret key when they meet in the first place, this is impossible in our case. Therefore, the need for privacy amplification is arguably better motivated in the present work. We also note that, although our assumption on Alice's source is stronger than that on Bob's source, it is still much weaker than the capability to generate truly random bits.

5.2   The Protocol

Alice and Bob's aim is to agree on a secret key S that is very close to being random from Eve's point of view. This is achieved by the protocol PA which we now describe (see also Table 1 in [8]), in which we identify $\{0,1\}^\ell$ with the finite field $\mathbb{F}_{2^\ell}$ for the purpose of arithmetic operations, and BLE : $\{0,1\}^\ell \times \{0,1\}^\ell \rightarrow \{0,1\}^m$ is a function (we will later choose it to be a suitable SB). Briefly, Alice and Bob share (A, B, X) when Bob is close. Then Bob moves to far, samples, Y and sends $Y, Z = AY + B$ to Alice. Eve then intercepts (Y, Z) and retransmits a possibly different pair $(\tilde{Y}, \tilde{Z})$ to Alice. Alice checks if $A\tilde{Y} + B = \tilde{Z}$ and, if this is satisfied, she computes $\tilde{S} = \text{BLE}(X, \tilde{Y})$, rejecting otherwise. In the meantime, Bob has computed $S = \text{BLE}(X, Y)$. Theorem 6 (proven in [8]) shows that with high probability either $S = \tilde{S}$ and Alice and Bob share a secret key, or else Alice has rejected. This is true as long as $b_1 = \frac{\ell}{2} + \omega(\log \ell)$ and a $(b_1, b_2, \ell^{-\omega(1)})$-SB   exists. For instance, the number-theoretic SB (Theorem 5) permits agreement on a key of length $m \approx \frac{b_1 + b_2 - \ell}{2} - \omega(\log \ell)$.

Theorem 6. *If BLE is a (b_1, b_2, ε)-SB, the protocol PA has the following property. If Eve is passive, Alice never rejects, $\tilde{S} = S$ and $\|\langle Y, S\rangle - \langle Y, U_m\rangle\| \le \varepsilon$. If Eve is active, the probability of either Alice rejecting or $S = \tilde{S}$ and*

$$\|\langle Y, S\rangle - \langle Y, U_m\rangle\| \le \varepsilon$$

is at least $1 - 2^{\ell - 2b_1}$.

6   Acknowledgments

We thank Yan Zhong Ding, Amit Sahai, Joel Spencer and Salil Vadhan for useful discussions, and an anonymous referee for suggesting the name "blender".

References

1. M. Ajtai, L. Babai, P. Hajnal, J. Komlos, P. Pudlak. Two lower bounds for branching programs. In *Proceedings of the eighteenth annual ACM symposium on Theory of computing*, 30–38, 1986.
2. N. Alon and J. Spencer. The Probabilistic Method - 2nd ed. Wiley Interscience, New York, 2000.
3. A. Andreev, A. Clementi, J. Rolim, L. Trevisan. Dispersers, deterministic amplification, and weak random sources. In *SIAM J. on Comput.*, 28(6):2103–2116, 1999.
4. C. H. Bennett, G. Brassard, C. Crépeau, U. Maurer. Generalized Privacy Amplification. In *IEEE Transaction on Information Theory*, vol. 41, no. 6, pp. 1915–1923, 1995
5. C. H. Bennett, G. Brassard, and J.-M. Robert. How to reduce your enemy's information. In *Proc. of CRYPTO '85*, Lecture Notes in Computer Science, vol. 218, pp. 468–476, Springer-Verlag, 1986.

6. B. Chor, O. Goldreich. Unbiased bits from sources of weak randomness and probabilistic communication complexity. *SIAM J. Comput.*, 17(2):230–261, 1988.
7. A. Cohen, A. Wigderson. Dispersers, deterministic amplification, and weak random sources. In *Proc. of FOCS*, pp. 14–19, 1989.
8. Y. Dodis, R. Oliveira. On extracting private randomness over a public channel (extended version). Available from
    `http://www.math.nyu.edu/~oliveira/extracting.pdf` .
9. Y. Dodis, J. Spencer. On the (Non-)Universality of the One-Time Pad. In *Proc. of FOCS*, 2002.
10. S. Janson, T. Luksak and A. Ruciński. Random Graphs. Wiley Interscience, New York, 2000.
11. C. Lu, O. Reingold, S. Vadhan and A. Wigderson. Extractors: Optimal Up to Constant Factors. In *Proc. of STOC*, 2003.
12. U. Maurer and S. Wolf. Privacy Amplification Secure Against Active Adversaries. In *Proc. of CRYPTO*, Lecture Notes in Computer Science, Springer-Verlag, vol. 1294, pp. 307–321, 1997.
13. J. McInnes, B. Pinkas. On the Impossibility of Private Key Cryptography with Weakly Random Keys. In *Proc. of CRYPTO*, pp. 421–435, 1990.
14. N. Nisan, A. Ta-Shma. Extracting Randomness: a survey and new constructions. In *JCSS*, 58(1):148–173, 1999.
15. N. Nisan, D. Zuckerman. Randomness is Linear in Space. In *JCSS*, 52(1):43–52, 1996.
16. L. Rónyai, L. Babai, M. Ganapathy On the number of zero-patterns in a sequence of polynomials *Journal of the AMS*, 2002.
17. M. Sántha, U. Vazirani. Generating Quasi-Random Sequences from Semi-Random Sources. *Journal of Computer and System Sciences*, 33(1):75–87, 1986.
18. R. Shaltiel. Recent developments in Explicit Constructions of Extractors. *Bulletin of the EATCS*, 77:67–95, 2002.
19. R. Shaltiel and C. Umans. Simple extractors for all min-entropies and a new pseudo-random generator. In *Proceedings of FOCS 2001*, pp.648-657, IEEE Computer Society, 2001.
20. M. Sipser. Expanders, Randomness or Time versus Space. In *Journal of Computer and Systems Sciences* 36, pp. 379-383, 1988.
21. A. Ta-Shma and J. Radhakrishnan. Bounds for Dispersers, Extractors, and Depth-Two Superconcentrators. In *SIAM Journal on Discrete Mathematics*, 13(1):2–24, 2000.
22. L. Trevisan. Construction of Extractors Using PseudoRandom Generators. In Proc. of STOC, pp. 141–148, 1999.
23. L. Trevisan, S. Vadhan. Extracting Randomness from Samplable Distributions. In *Proc. of FOCS*, 2000.
24. U. Vazirani. Randomness, Adversaries and Computation. *PhD Thesis*, University of California, Berkeley, 1986.
25. U. Vazirani. Strong Communication Complexity or Generating Quasi-Random Sequences from Two Communicating Semi-Random Sources. *Combinatorica*, 7(4):375–392, 1987.
26. U. Vazirani, V. Vazirani. Random polynomial time is equal to slightly-random polynomial time. In *Proc. of 26th FOCS*, pp. 417–428, 1985.
27. A. Wigderson. Open problems. Notes from *DIMACS Workshop on Pseudorandomness and Explicit Combinatorial Constructions*, 1999
28. D. Zuckerman. Simulating BPP Using a General Weak Random Source. *Algorithmica*, 16(4/5):367-391, 1996.

High Degree Vertices and Eigenvalues in the Preferential Attachment Graph

Abraham Flaxman[*1], Alan Frieze[**1], and Trevor Fenner[2]

[1] Department of Mathematical Sciences
Carnegie Mellon University
Pittsburgh, PA, 15213, USA
abie@cmu.edu, alan@random.math.cmu.edu
[2] School of Computer Science
Birkbeck College, University of London
Malet Street, London WC1E 7HX
trevor@dcs.bbk.ac.uk

Abstract. The preferential attachment graph is a random graph formed by adding a new vertex at each time step, with a single edge which points to a vertex selected at random with probability proportional to its degree. Every m steps the most recently added m vertices are contracted into a single vertex, so at time t there are roughly t/m vertices and exactly t edges. This process yields a graph which has been proposed as a simple model of the world wide web [BA99]. For any constant k, let $\Delta_1 \geq \Delta_2 \geq \cdots \geq \Delta_k$ be the degrees of the k highest degree vertices. We show that at time t, for any function f with $f(t) \to \infty$ as $t \to \infty$, $\frac{t^{1/2}}{f(t)} \leq \Delta_1 \leq t^{1/2}f(t)$, and for $i = 2, \ldots, k$, $\frac{t^{1/2}}{f(t)} \leq \Delta_i \leq \Delta_{i-1} - \frac{t^{1/2}}{f(t)}$, with high probability (**whp**). We use this to show that at time t the largest k eigenvalues of the adjacency matrix of this graph have $\lambda_k = (1 \pm o(1))\Delta_k^{1/2}$ **whp**.

1 Introduction

Recently there has been much interest in understanding the properties of real-world large-scale networks such as the structure of the Internet and the World Wide Web. For a general introduction to this topic, see Bollobás and Riordan [BR02], Hayes [Hay00], or Watts [Wat99]. One approach is to model these networks by random graphs. Experimental studies by Albert, Barabási, and Jeong [ABJ99], Broder et al [BKM+00], and Faloutsos, Faloutsos, and Faloutsos [FFF99] have demonstrated that in the World Wide Web/Internet the proportion of vertices of a given degree follows an approximate inverse power law i.e. the proportion of vertices of degree k is approximately $Ck^{-\alpha}$ for some constants C, α. The classical models of random graphs introduced by Erdős and Renyi [ER59] do not have power law degree sequences, so they are not suitable for

---

[*] Supported in part by NSF VIGRE Grant DMS-9819950
[**] Supported in part by NSF grant CCR-0200945

S. Arora et al. (Eds.): APPROX 2003+RANDOM 2003, LNCS 2764, pp. 264–274, 2003.
© Springer-Verlag Berlin Heidelberg 2003

modeling these networks. This has driven the development of various alternative models for random graphs.

One approach to remedy this situation is to study graphs with a prescribed degree sequence (or prescribed expected degree sequence). This is proposed as a model for the web graph by Aiello, Chung, and Lu in [ACL00]. Mihail and Papadimitriou also use this model [MP02] in their study of large eigenvalues, as do Chung, Lu, and Vu in [CLV].

An alternative approach, which we will follow in this paper, is to sample graphs via some generative procedure which yields a power law distribution. There is a long history of such models, outlined in the survey by Mitzenmacher [Mit01]. We will use the preferential attachment model to generate our random graph. The preferential attachment random graph has been the subject of recently revived interest. It dates back to Yule [Yul25] and Simon [Sim55]. It was proposed as a model for the web by Barabási and Albert [BA99], and their description was elaborated by Bollobás, Riordan, Spencer, and Tusnády [BRST01] who proved that the degree sequence does follow a power law distribution. Bollobás and Riordan obtained several additional results regarding the diameter and connectivity of such graphs [BR]. We use the generative model of [BRST01] (see also [BR02]) and build a graph sequentially as follows:

- At each time step t, we add a vertex v_t, and we add an edge from v_t to some other vertex u, where u is chosen at random according to the distribution:

$$\Pr[u = v_i] = \begin{cases} \frac{d_t(v_i)}{2t-1}, & \text{if } v_i \neq v_t; \\ \frac{1}{2t-1}, & \text{if } v_i = v_t; \end{cases}$$

  where $d_t(v)$ denotes the degree of vertex v at time t. This means that each vertex receives an additional edge with probability proportional to its current degree. The probability of choosing v_t (and forming a loop) is consistent with this, since we've already committed "half" an edge to v_t and are deciding where to put the other half.
- For some constant m, every m steps we contract the most recently added m vertices to form a supervertex.

Let G_t^m denote the random graph at time step t with contractions of size m. Note that contracting each set of vertices $\{im + 1, im + 2, \ldots, (i + 1)m\}$ of G_t^1 yields a graph identically distributed with G_t^m.

It is worth mentioning that there are several alternative simple models for the World Wide Web and for general power law graphs. A generalization of the preferential attachment model is described by Drinea, Enachescu, and Mitzenmacher in [DEM01], and degree sequence results analogous to [BRST01] are proved for this model by Buckley and Osthus in [BO01]. A completely different generative model, based on the idea that new webpages are often consciously or unconsciously copies of existing pages, is developed by Kleinberg et al and Kumar et al in [KKR+99], [KRRT99], [KRR+00b], [KRR+00a]. Cooper and Frieze analyze a model combining these approaches in [CF01].

The results in previous papers on preferential attachment graphs concern low degree vertices. For example the results in [BRST01] concern degrees up to $t^{1/15}$. Our firt theorem deals with the highest degree vertices:

Theorem 1. *Let m and k be fixed positive integers, and let $f(t)$ be a function with $f(t) \to \infty$ as $t \to \infty$. Let $\Delta_1 \ge \Delta_2 \ge \cdots \ge \Delta_k$ denote the degrees of the k highest degree vertices of G_t^m. Then*

$$\frac{t^{1/2}}{f(t))} \le \Delta_1 \le t^{1/2} f(t)$$

and for $i = 2, \ldots, k$,

$$\frac{t^{1/2}}{f(t)} \le \Delta_i \le \Delta_{i-1} - \frac{t^{1/2}}{f(t)},$$

whp[3].

The next theorem relates maximum eigenvalues and maximum degrees. It mirrors results of Mihail and Papadimitriou [MP02] and Chung, Liu and Vu [CLV] for fixed degree expectation models and at a high level, the proof follows the same lines as these two papers. Experimentally, a power law distribution for eigenvalues was observed in "real-world" graphs in [FFF99].

Theorem 2. *Let m and k be fixed positive integers, and let $f(t)$ be a function with $f(t) \to \infty$ as $t \to \infty$. Let $\lambda_1 \ge \lambda_2 \ge \cdots \ge \lambda_k$ be the k largest eigenvalues of the adjacency matrix of G_t^m. Then for $i = 1, \ldots, k$ we have $\lambda_i = (1 \pm o(1)) \Delta_i^{1/2}$* **whp.**

Our proofs of these theorems require two lemmas.

Lemma 1. *Let $d_t^m(s)$ denote the degree of vertex s in G_t^m. Then for any positive integer k,*

$$E\left[(d_t^m(s))^k\right] \le 8m^k 2^{k^6} \left(\frac{t}{s}\right)^{k/2}.$$

To simplify the exposition, we speak of a *supernode*, which is simply a collection of vertices viewed as one vertex. So the degree of a supernode is the sum of the degrees of the vertices in the supernode, and an edge is incident to a supernode if it is incident to some vertex in the supernode.

Lemma 2. *Let $\mathbf{S} = (S_1, S_2, \ldots, S_\ell)$ be a collection of disjoint supernodes, and let $p_{\mathbf{S}}(\mathbf{r}; \mathbf{d}, t_0, t)$ denote the probability that each supernode S_i has degree $r_i + d_i$ at time t conditioned on $d_{t_0}(S_i) = d_i$. Let $d = \sum_{i=1}^\ell d_i$ and $r = \sum_{i=1}^\ell r_i$. If $d = o(t^{1/2})$ and $r = o(t^{2/3})$, then*

$$p_{\mathbf{S}}(\mathbf{r}; \mathbf{d}, t_0, t) \le \left(\prod_{i=1}^\ell \binom{r_i + d_i - 1}{d_i - 1}\right) \left(\frac{t_0 + 1}{t}\right)^{d/2} \exp\left\{2 + t_0 - \frac{d}{2} + \frac{2r}{t^{1/2}}\right\}.$$

In the next section we prove Theorems 1 and 2. The proofs of Lemmas 1 and 2 are too long to fit in here and we leave them for the final version.

---

[3] In this paper an event $\mathcal{E}$ is said to hold *with high probability* (**whp**) if $\Pr[\mathcal{E}] \to 1$ as $t \to \infty$.

2 Proof of Theorems

2.1 Proof of Theorem 1

We partition the vertices into those added before time t_0, before t_1, and after t_1 and argue about the maximum degree of vertices in each set. Here

$$t_0 = \log \log \log f(t) \text{ and } t_1 = \log \log f(t).$$

We break the proof of Theorem 1 into 5 Claims.

Claim. In G_t^m the degree of the supernode of vertices added before time t_0 is at least $t_0^{1/3} t^{1/2}$ **whp**.

Proof Let $\mathcal{A}_1$ denote the event that the supernode consisting of the first t_0 vertices has degree less than $t_0^{1/3} t^{1/2}$. We bound the probability of this event using Lemma 2 with $\ell = 1$. Since at time t_0 the supernode of all vertices added by this time has all of the edges, we take $\mathbf{d} = d_1 = 2t_0$. Then

$$\Pr[\mathcal{A}_1] \leq \sum_{r_1=0}^{t_0^{1/3} t^{1/2} - 2t_0} \binom{r_1 + 2t_0 - 1}{2t_0 - 1} \left(\frac{t_0+1}{t}\right)^{d/2} e^{2+t_0-d/2+2r/t^{1/2}}$$

$$\leq (t_0^{1/3} t^{1/2}) \frac{(t_0^{1/3} t^{1/2})^{2t_0-1}}{(2t_0-1)!} \left(\frac{t_0+1}{t}\right)^{t_0} e^{2+t_0+2t_0^{1/3}}$$

$$\leq t_0^{2t_0/3} \frac{e^{2t_0-1}}{(2t_0-1)^{2t_0-1}} (t_0+1)^{t_0} e^{2+t_0+2t_0^{1/3}}$$

$$\leq \frac{e^{3t_0+2t_0^{1/3}+2}}{(2t_0-1)^{t_0/3-1}}$$

$$= o(1).$$

Claim. In G_t^m no vertex added after time t_1 has degree exceeding $t_0^{-2} t^{1/2}$ **whp**.

Proof Let $\mathcal{A}_2$ denote the event that some vertex added after time t_1 has degree exceeding $t_0^{-2} t^{1/2}$. Then we have

$$\Pr[\mathcal{A}_2] \leq \sum_{s=t_1}^{t} \Pr[d_t(s) \geq t_0^{-2} t^{1/2}] = \sum_{s=t_1}^{t} \Pr\left[(d_t(s))^3 \geq \left(t_0^{-2} t^{1/2}\right)^3\right]$$

$$\leq \sum_{s=t_1}^{t} t_0^6 t^{-3/2} E[d_t(s)^3]$$

Using Lemma 1 this bound becomes

$$\Pr[\mathcal{A}_2] \leq \sum_{s=t_1}^{t} t_0^6 t^{-3/2} 8m^3 2^{729} \left(\frac{t}{s}\right)^{3/2} = m^3 2^{735} t_0^6 \sum_{s=t_1}^{t} s^{-3/2}$$

$$\leq m^3 2^{736} t_0^6 t_1^{-1/2} = o(1).$$

Claim. In G_t^m no vertex added before time t_1 has degree exceeding $t_0^{1/6}t^{1/2}$ **whp**.

Proof Let $\mathcal{A}_3$ denote the event that some vertex added before t_1 has degree exceeding $t_0^{1/6}t^{1/2}$. Then by using Lemma 1 for a third moment argument as above we have

$$\Pr[\mathcal{A}_3] \le \sum_{s=1}^{t_1} (t_0^{1/6}t^{1/2})^{-3} 8m^3 2^{729} \left(\frac{t}{s}\right)^{3/2}$$

$$= m^3 2^{732} t_0^{-1/2} \sum_{s=1}^{t_1} s^{-3/2} \le m^3 2^{734} t_0^{-1/2} = o(1).$$

$\square$

Claim. The k highest degree vertices of G_t^m are added before time t_1 and have degree Δ_i bounded by $t_0^{-1}t^{1/2} \le \Delta_i \le t_0^{1/6}t^{1/2}$ **whp**.

Proof

- **(Upper bound on Δ_i)** By Claim 2, all vertices added after time t_1 have degree at most $t_0^{-2}t^{1/2}$ **whp**. Combining this with Claim 3 we have $\Delta_1 \le t_0^{1/6}t^{1/2}$ **whp**.
- **(Lower bound on Δ_i)** The conditions from Claims 1,2, and 3 imply the lower bound. To see this, suppose the conditions of these claims are satisfied, but assume for contradiction that at most $k - 1$ vertices added before t_1 have degree exceeding $t_0^{-1}t^{1/2}$. Then the total degree of vertices added before t_0 is less than $k(t_0^{1/6}t^{1/2}) + t_0(t_0^{-1}t^{1/2}) \le 2kt_0^{1/6}t^{1/2}$. But this contradicts the condition of Claim 1, which says the total degree of vertices added before t_0 at least $t_0^{1/3}t^{1/2}$.
- **(Added before t_1)** By Claim 2 all vertices added after time t_1 have degree at most $t_0^{-2}t^{1/2}$ **whp**. So the lower bound on Δ_i shows the k highest degree vertices are added before time t_1 **whp**.

$\square$

Claim. The k highest degree vertices of G_t^m have $\Delta_i \le \Delta_{i-1} - t^{1/2}/f(t)$ **whp**.

Proof Let $\mathcal{A}_4$ denote the event that there are 2 vertices among the first t_1 with degrees exceeding $t_0^{-1}t^{1/2}$ and within $t^{1/2}/f(t)$ of each other.

Let $p_{\ell,s_1,s_2} = \Pr[d_t(s_1) - d_t(s_2) = \ell \mid \overline{\mathcal{A}_3}]$, for $|\ell| \le \sqrt{t}/f(t)$. Then

$$\Pr[\mathcal{A}_4 \mid \overline{\mathcal{A}_3}] \le \sum_{1 \le s_1 < s_2 \le t_1} \sum_{\ell=-t^{1/2}/f(t)}^{t^{1/2}/f(t)} p_{\ell,s_1,s_2}.$$

Since

$$p_{\ell,s_1,s_2} \le \sum_{r_1=t_0^{-1}t^{1/2}}^{t_0^{1/6}t^{1/2}} \sum_{d_1,d_2=1}^{2t_1} p_{(s_1,s_2)}((r_1,r_1-\ell);(d_1,d_2),t_1,t)$$

$$\le t_0^{1/6}t^{1/2} \sum_{d_1,d_2=1}^{2t_1} \binom{2t_0^{1/6}t^{1/2}}{d_1-1}\binom{2t_0^{1/6}t^{1/2}}{d_2-1}\left(\frac{t_1+1}{t}\right)^{(d_1+d_2)/2} e^{t_0+2+2t_0^{1/6}}$$

$$\le t_0^{1/6}t^{1/2} \sum_{d_1,d_2=1}^{2t_1} \left(2t_0^{1/6}t^{1/2}\right)^{d_1+d_2-2}(t_1+1)^{2t_1}t^{-(d_1+d_2)/2}e^{3t_0}$$

$$\le t_0^{1/6}(2t_1)^2 2^{4t_1}t_0^{2t_1/3}(t_1+1)^{2t_1}e^{3t_0}t^{-1/2}$$

$$= o(t_1^{-2}t^{-1/2}f(t)),$$

we have

$$\Pr[\mathcal{A}_4 \mid \overline{\mathcal{A}_3}] \le \sum_{1 \le s_1 < s_2 \le t_1} \sum_{\ell=-t^{1/2}/f(t)}^{t^{1/2}/f(t)} p_{\ell,s_1,s_2} = o(1).$$

So

$$\Pr[\mathcal{A}_4] = \Pr[\mathcal{A}_4 \mid \mathcal{A}_3]\Pr[\mathcal{A}_3]+\Pr[\mathcal{A}_4 \mid \overline{\mathcal{A}_3}]\Pr[\overline{\mathcal{A}_3}] \le \Pr[\mathcal{A}_3]+\Pr[\mathcal{A}_4 \mid \overline{\mathcal{A}_3}] = o(1).$$

$\square$

2.2   Proof of Theorem 2

We partition the vertices into 3 sets; let S_i be the vertices added after time t_{i-1} and at or before time t_i, for

$$t_0 = 0, \quad t_1 = t^{1/8}, \quad t_2 = t^{9/16}, \quad t_3 = t.$$

To reduce the number of subscripts necessary, we use G to denote the graph G_t.

For any graph H, we let M_H denote the adjacency matrix of H, and we let $\lambda_i(H)$ denote the i-th largest eigenvalue of M_H. We will use the identity (*Rayleigh's Principle*)

$$\lambda_i(H) = \min_L \max_{v \in L, v \ne 0} \frac{v^T M_H v}{v^T v} \qquad (1)$$

where L ranges over all $(n-i+1)$-dimensional subspaces of $\mathbb{R}^n$. (See, for example, [Str88]).

Our approach, as in [MP02], [CLV], is to show that **whp** G contains a star forest F with stars of degree asymptotic to the maximum degree vertices of G. Then we will show $G \setminus F$ has small eigenvalues, and conclude that the large eigenvalues of G cannot be too different from the large eigenvalues of F.

To do this, we need reasonable bounds on the degrees and codegrees in G. Recall that $d_s^m(r)$ is the degree at time s of the vertex added at time r with contractions of size m.

Claim. For any $\epsilon > 0$ and any $f(t)$ with $f(t) \to \infty$ as $t \to \infty$ the following holds **whp**: for all s with $f(t) \leq s \leq t$, for all vertices $v \in G_s^m$, if v was added at time r, then $d_s^m(v) \leq s^{1/2+\epsilon}r^{-1/2}$.

Proof   We use Lemma 1 and the union bound. Let $\ell = \lceil 3/\epsilon \rceil$.

$$\Pr\left[\bigcup_{s=f(t)}^{t} \bigcup_{r=1}^{s} \{d_s^m(r) \geq s^{1/2+\epsilon}r^{-1/2}\} \right]$$

$$\leq \sum_{s=f(t)}^{t} \sum_{r=1}^{s} \Pr[d_s^m(r) \geq s^{1/2+\epsilon}r^{-1/2}]$$

$$= \sum_{s=f(t)}^{t} \sum_{r=1}^{s} \Pr[(d_s^m(r))^\ell \geq \left(s^{1/2+\epsilon}r^{-1/2}\right)^\ell]$$

$$\leq \sum_{s=f(t)}^{t} \sum_{r=1}^{s} s^{-\ell(1/2+\epsilon)}r^{\ell/2} E[(d_s^m(r))^\ell]$$

$$\leq \sum_{s=f(t)}^{t} \sum_{r=1}^{s} s^{-\ell(1/2+\epsilon)}r^{\ell/2}8m^\ell 2^{\ell^6}(s/r)^{\ell/2}$$

$$= 8m^\ell 2^{\ell^6} \sum_{s=f(t)}^{t} s^{1-\epsilon\ell}.$$

Since $\ell \geq 3/\epsilon$,

$$\sum_{s=f(t)}^{t} s^{1-\epsilon\ell} \leq \int_{f(t)-1}^{\infty} x^{1-\epsilon\ell}dx = \frac{1}{\epsilon\ell - 2}(f(t) - 1)^{2-\epsilon\ell} = o(1).$$

$\square$

Claim. Let S_3' be the set of vertices in S_3 which are adjacent to more than 1 vertex of S_1 in G. Then $|S_3'| \leq t^{7/16}$ **whp**.

Proof   Let $\mathcal{B}_1$ be the event that the conditions of Claim 2.2 hold with $f(t) = t_2$ and $\epsilon = 1/16$. Then for a vertex $v \in S_3$ added at time s,

$$\Pr[|N(v) \cap S_1| \geq 2 \mid \mathcal{B}_1] \leq \binom{m}{2}\left(\frac{s^{1/2+\epsilon}t_1}{2s - 1}\right)^2 \leq m^2 s^{-7/8}t^{1/4}.$$

Let X denote the number of $v \in S_3$ adjacent to more than 1 vertex of S_1. Then

$$E[X \mid \mathcal{B}_1] \leq \sum_{s=t_2+1}^{t} m^2 s^{-7/8}t^{1/4} \leq m^2 t^{1/4} \int_{t_2}^{t} x^{-7/8}dx \leq 8m^2 t^{3/8}.$$

We finish the claim with Markov's inequality,

$$\Pr[X \geq t^{7/16} \mid \mathcal{B}_1] \leq E[X \mid \mathcal{B}_1]/t^{7/16} = o(1).$$

□

Now, let $F \subseteq G$ be the star forest consisting of edges between S_1 and $S_3 \setminus S_3'$.

Claim. Let $\Delta_1 \geq \Delta_2 \geq \cdots \geq \Delta_k$ denote the degrees of the k highest degree vertices of G. Then $\lambda_i(F) = (1 - o(1))\Delta_i^{1/2}$ **whp**.

Proof   Let H be the star forest $H = K_{1,d_1} \cup K_{1,d_2} \cup \cdots \cup K_{1,d_k}$, with $d_1 \geq d_2 \geq \cdots \geq d_k$. Then for $i = 1, \ldots, k$, $\lambda_i(H) = d_i^{1/2}$. So it is sufficient to show that $\Delta_i(F) = (1 - o(1))\Delta_i(G)$ for $i = 1, \ldots, k$.

Claim 2.1 shows that the k highest degree vertices of G are added before time t_1, so these vertices are all in F. The only edges to these vertices that are not in F are those added before time t_2 and those incident to S_3'. By Theorem 1 we have $\Delta_1(G_{t_2}^m) \leq t_2^{7/9} = t^{7/16}$ and, also by Theorem 1, $\Delta_i(G) \geq t^{1/2}/\log t$ for $i = 1, \ldots, k$, **whp**. Claim 7 says that **whp** $|S_3'| \leq t^{7/16}$, and so **whp**

$$\Delta_i(F) \geq \Delta_i(G) - t^{7/16} - mt^{7/16} = (1 - o(1))\Delta_i(G).$$

□

Let $H = G \setminus F$. We complete the proof of Theorem 2 by showing that $\lambda_1(H)$ is small.

Claim. $\lambda_1(H) \leq 6mt^{15/64}$ **whp**.

Proof   We bound the eigenvalues of H in 6 parts. Let

$$H_i = H[S_i], \qquad H_{ij} = H(S_i, S_j),$$

where $H[S]$ is the subgraph of H induced by the vertex set S, and $H(S,T)$ is the subgraph containing only edges with one vertex in S and the other in T.

To bound $\lambda_1(H_i)$ we use the fact that the maximum eigenvalue of a graph is at most the maximum degree of the graph. This is easily verified from (1).

We use Claim 6 with $f(t) = t_1$ and $\epsilon = 1/64$ to conclude that **whp**

$$\lambda_1(H_1) \quad \leq \Delta_1(H_1) \quad = \max_{v \leq t_1}\{d_{t_1}^m(v)\} \quad \leq t_1^{1/2+\epsilon} \quad = t^{33/512},$$

$$\lambda_1(H_2) \quad \leq \Delta_1(H_2) \quad \leq \max_{t_1 \leq v \leq t_2}\{d_{t_2}^m(v)\} \quad \leq t_2^{1/2+\epsilon}t_1^{-1/2} \quad = t^{233/1024},$$

$$\lambda_1(H_3) \quad \leq \Delta_1(H_3) \quad \leq \max_{t_2 \leq v \leq t_3}\{d_{t_3}^m(v)\} \quad \leq t_3^{1/2+\epsilon}t_2^{-1/2} \quad = t^{15/64}.$$

To bound $\lambda_1(H_{ij})$, we begin by considering the case $m = 1$. Then, for $i < j$, each vertex in S_j has at most 1 edge in H_{ij}, so H_{ij} is a star forest. As observed in Claim 8, the eigenvalues of a star forest are directly related to the degrees of the stars.

When $m > 1$, we let G' denote a preferential attachment graph with t edges and $m = 1$. Recall that by contracting vertices $\{(i-1)m+1, \ldots, im\}$ into a single vertex i, we obtain a graph identically distributed with G. There is a simple representation of this observation in terms of linear algebra: we can write the adjacency matrix of G in terms of the adjacency matrix of the graph G':

$$M_G = C_m^T M_{G'} C_m,$$

where C_m is the $t \times t/m$ matrix with i-th column

$$[\ \underbrace{0 \ \cdots \ 0}_{(i-1)m} \ \underbrace{1 \ \cdots \ 1}_{m} \ \underbrace{0 \ \cdots \ 0}_{(t/m-i)m} \ ]^T.$$

Similarly, we can write the adjacency matrix of H_{ij} in terms of the adjacency matrix of H'_{ij} using this "contraction matrix" C_m.

Note that for $w = C_m v$ we have $w^T w = m(v^T v)$. So

$$\lambda_1(H_{ij}) = \max_{v \neq 0} \frac{v^T M_{H_{ij}} v}{v^T v} = \max_{v \neq 0} \frac{v^T C_m^T M_{H'_{ij}} C_m v}{v^T v} = \max_{w:\ w = C_m v \neq 0} m \frac{w^T M_{H'_{ij}} w}{w^T w}$$

$$\leq m \max_{w \neq 0} \frac{w^T M_{H'_{ij}} w}{w^T w} = m \lambda_1(H'_{ij}).$$

We use Claim 6 with $f(t) = t_1$ and $\epsilon = 1/64$ as above to conclude that **whp**

$$\Delta_1(H'_{12}) = \max_{v \leq t_2} \{ d^1_{t_2}(v) \} \leq t_2^{1/2+\epsilon} = t^{297/1024}$$

$$\Delta_1(H'_{23}) = \max_{t_1 \leq v \leq t_3} \{ d^1_{t_3}(v) \} \leq t_3^{1/2+\epsilon} t_1^{-1/2} = t^{29/64}$$

Finally, all edges in H'_{13} are between S_1 and S'_3, so Claim 7 shows that $\Delta_1(H'_{13}) \leq t^{7/16}$ **whp**.

We now conclude that **whp**

$$\lambda_1(H_{ij}) \leq m \lambda_1(H'_{ij}) \leq m \Delta_1(H'_{ij})^{1/2} \leq m t^{15/64},$$

and so **whp**

$$\lambda_1(H) \leq \sum_{i=1}^{3} \lambda_1(H_i) + \sum_{i<j} \lambda_1(H_{ij}) \leq 6m t^{15/64}.$$

$\square$

References

[ABJ99]   R. Albert, A. Barabási, and H. Jeong. Diameter of the world wide web. *Nature*, 401:103–131, 1999.

[ACL00]   W. Aiello, F. R. K. Chung, and L. Lu. A random graph model for massive graphs. In *Proc. of the 32nd Annual ACM Symposium on the Theory of Computing*, pages 171–180, 2000.

[BA99]   A. Barabási and R. Albert. Emergence of scaling in random networks. *Science*, 286:509–512, 1999.

[BKM+00]   A. Broder, R. Kumar, F. Maghoul, P. Raghavan, S. Rajagopalan, R. Stata, A. Tomkins, and J. Wiener. Graph structure in the web. In *Proc. of the 9th Intl. World Wide Web Conference*, pages 309–320, 2000.

[BO01]   G. Buckley and D. Osthus. Popularity based random graph models leading to a scale-free degree distribution, 2001.

[BR]   B. Bollobás and O. Riordan. The diameter of a scale-free random graph. To appear.

[BR02]   B. Bollobás and O. Riordan. Mathematical results on scale-free random graphs. In *Handbook of Graphs and Networks*. Wiley-VCH, Berlin, 2002.

[BRST01]   B. Bollobás, O. Riordan, J. Spencer, and G. Tusanády. The degree sequence of a scale-free random graph process. *Random Structures and Algorithms*, 18:279–290, 2001.

[CF01]   C. Cooper and A. M. Frieze. A general model of undirected Web graphs. In *Proc. of ESA*, pages 500–511, 2001.

[CLV]   F.R.K. Chung, L. Lu, and V. Vu. Eigenvalues of random power law graphs. To appear.

[DEM01]   E. Drinea, M. Enachescu, and M. Mitzenmacher. Variations on random graph models for the web. Technical report, Harvard University, 2001.

[ER59]   P. Erdös and A. Rényi. On random graphs I. *Publicationes Mathematicae Debrecen*, 6:290–297, 1959.

[FFF99]   M. Faloutsos, P. Faloutsos, and C. Faloutsos. On power-law relationships of the internet topology. In *SIGCOMM*, pages 251–262, 1999.

[Hay00]   B. Hayes. Graph theory in practice: Part II. *American Scientist*, 88:104–109, 2000.

[KKR+99]   J. M. Kleinberg, R. Kumar, P. Raghavan, S. Rajagopalan, and A. S. Tomkins. The Web as a graph: Measurements, models and methods. *Lecture Notes in Computer Science*, 1627:1–??, 1999.

[KRR+00a]   R. Kumar, P. Raghavan, S. Rajagopalan, D. Sivakumar, A. Tomkins, and E. Upfal. Stochastic models for the web graph. In *FOCS: IEEE Symposium on Foundations of Computer Science (FOCS)*, 2000.

[KRR+00b]   R. Kumar, P. Raghavan, S. Rajagopalan, D. Sivakumar, A. Tomkins, and E. Upfal. The Web as a graph. In *Proc. 19th ACM SIGACT-SIGMOD-AIGART Symp. Principles of Database Systems, PODS*, pages 1–10. ACM Press, 15–17 2000.

[KRRT99]   R. Kumar, P. Raghavan, S. Rajagopalan, and A. Tomkins. Trawling the Web for emerging cyber-communities. *Computer Networks (Amsterdam, Netherlands: 1999)*, 31(11–16):1481–1493, 1999.

[Mit01]   M. Mitzenmacher. A brief history of generative models for power law and lognormal distributions. In *Proc. of the 39th Annual Allerton Conf. on Communication, Control, and Computing*, pages 182–191, 2001.

[MP02]   M. Mihail and C. H. Papadimitriou. On the eigenvalue power law. In *Proc. of 6th Intl. Workshop on Randomization and Approximation Techniques*, pages 254–262, 2002.

[Sim55]   H. A. Simon. On a class of skew distribution functions. *Biometrika*, 42(3/4):425–440, 1955.

[Str88]   G. Strang. *Linear algebra and its applications.* Hardcourt Brace Jo-
            vanovich Publishing, 1988.
[Wat99]   D. J. Watts. *Small Worlds: The Dynamics of Networks between Order
            and Randomness.* Princeton: Princeton University Press, 1999.
[Yul25]   G. Yule. A mathematical theory of evolution based on the conclusions
            of Dr. J.C. Willis. *Philosophical Transactions of the Royal Society of
            London (Series B)*, 213:21–87, 1925.

The Satisfiability Threshold for Randomly Generated Binary Constraint Satisfaction Problems

Alan Frieze[1]* and Michael Molloy[2]

[1] Department of Mathematical Sciences, Carnegie Mellon University,
Pittsburgh PA15213, USA.
[2] Department of Computer Science, University of Toronto, Toronto,
Ontario M5S 3G4 and Microsoft Research, Redmond WA.

Abstract. We study two natural models of randomly generated constraint satisfaction problems. We determine how quickly the domain size must grow with n to ensure that these models are robust in the sense that they exhibit a non-trivial threshold of satisfiability, and we determine the asymptotic order of that threshold. We also provide resolution complexity lower bounds for these models.

1 Introduction

The *Constraint Satisfaction Problem* (CSP) is a fundamental problem in Artificial Intelligence, with applications ranging from scene labeling to scheduling and knowledge representation. See for example Dechter [12], Mackworth [18] and Waltz [26]. An instance of the CSP comprises a set of n *variables*, each taking a value in some given domain, and a set of *constraint relations*, each of which determines the permitted joint values of a given subset of the variables. The problem is either to determine any set of values for the variables which respects all the constraint relations, or determine that none exists. In recent years, there has been a strong interest in studying the relationship between the input parameters that define an instance of CSP (e.g. number of variables, domain sizes, tightness of constraints) and certain solution characteristics, such as the likelihood that the instance has a solution or the difficulty with which a solution may be discovered. An extensive account of relevant results, both experimental and theoretical, can be found in Hogg, Hubermann and Williams [15].

One of the most commonly used practices for conducting experiments with CSP is to generate a large set of random instances, all with the same defining parameters, and then for each instance in the set to use heuristics for deciding if a solution exists. Note that, in general CSP is NP-complete. The proportion of random instances that have a solution is used as an indication of the likelihood that an instance will be soluble, and the average time taken per instance (by

---

* Supported in part by NSF grant CCR0200945. Research carried out during a visit to the Microsoft Research, Theory Group

S. Arora et al. (Eds.): APPROX 2003+RANDOM 2003, LNCS 2764, pp. 275–289, 2003.
© Springer-Verlag Berlin Heidelberg 2003

some standard algorithm) gives some measure of the hardness of such instances. A characteristic of many of these experiments is that the fraction of assignments of values that are permissible for each constraint is kept *constant* as the number of variables increases. The very active experimental study of random models of CSP has necessitated a rigorous analysis of such models. Various models of random CSP's for which m, the domain-size, is constant have been studied in several papers, for example [2,21,11,22,23]. One of the earliest such studies, [2] discovered that the most natural models suffer a fatal flaw (described below). The first study of the case where m grows with n was [13], where one of these most natural models was studied. Implicit in that study was the fact that for certain settings of the relevant parameters, the fatal flaw did not occur and we had a rich random model to study. One the main contributions of this paper is to determine which parameter settings avoid that fatal flaw, and thus provide random models that are both natural and robust.

In this paper we consider only *binary* CSPs (BCSPs). These can be succinctly described in the following way: A graph $G = (V, E)$ is given, where $V = \{x_1, x_2, \ldots, x_n\}$ denotes the set of variables of the problem, and E the set of binary relations of the instance. We assume, without loss of generality, that each variable can take values in the same set $[m] = \{1, 2, \ldots, m\}$. For each edge $e = \{x_i, x_j\} \in E$, the relation can then be represented by an $m \times m$ 0-1 matrix M_e, where 0 indicates that the pair of values is forbidden and 1 that it is allowed. A solution to the associated BCSP is an assignment $f : V \to [m]$ of values to the variables, such that $M_e(f(x_i), f(x_j)) = 1$ for all $e = \{x_i, x_j\} \in E$.

The aim of this paper is to conduct a probabilistic analysis of some aspects of the following simple random models of BCSP:

Model A: The underlying graph G is G_{n,p_1} for some $p_1 = p_1(n) < 1$ where $p_1 \neq o(1/n)$. (This means that, with $V = \{x_1, x_2, \ldots, x_n\}$, we let each of the $\binom{n}{2}$ possible edges occur independently in E with probability p_1.) We let $d = np_1$. For each edge e of G there is a random $m \times m$ constraint matrix M_e where $M_e(i, j) = 1$ or 0 independently with probability p_2 or $q_2 = 1 - p_2$ respectively, for some constant $0 < p_2 < 1$. (In the final paper we will consider $p_2 \to 0$ and $p_2 \to 1$ as well.)

For $p_1 = o(1/n)$, the graph G_{n,p_1} is very sparse, and consists of a collection of small vertex-disjoint trees in which all but $o(n)$ of the vertices have degree 0. This is why we restrict our attention to $p_1 \neq o(1/n)$.

Given m, p_2 we wish to know: for what values of p_1 is our random CSP almost surely satisfiable? This question has been asked for many similar models of CSP, SAT and other problems. Traditionally, one of the first steps is to determine some values of p_1 for which it is not satisfiable as follows:

Fact: For $p_1 \geq \frac{2\ln m}{q_2 n}$, the random CSP is unsatisfiable **whp**.

The proof follows easily from the fact that the expected number of satisfying solutions is $m^n (1 - p_1 q_2)^{\binom{n}{2}}$.

Inspired by a familiar pattern of similar random models, it is tempting to assume that $\frac{\ln m}{n}$ is the asymptotic order of a so-called "satisfiability threshold" and so hypothesize that:

Hypothesis A: There is some constant $c > 0$ so that for $p_1 \leq c\frac{\ln m}{n}$, the random CSP is satisfiable **whp**.

See [16] for a lengthy list of papers in which the authors fell to the temptation of assuming an equivalent hypothesis. In [2], it was observed that for most of those papers, and in fact whenever m, p_2 are both constants, the hypothesis is wrong. In fact, if $p_1 \geq \omega(n)/n^2$ for any $\omega(n)$ that tends to infinity with n, then almost surely the random CSP is trivially unsatisfiable in the sense that it has an edge whose constraint forbids every pair of values; we call such an edge a *blocked edge*

In this paper we asymptotically determine which values of m meet Hypothesis A.

Theorem 1. *(a) If $m \leq (1 - \epsilon)\sqrt{\ln nd/\ln(1/q_2)}$ for some constant $\epsilon > 0$, then provided $nd \to \infty$, the random CSP has a blocked edge **whp***
*(b) If $m \geq (1 + \epsilon)\sqrt{\ln nd/\ln(1/q_2)}$ then there is some constant $c > 0$ so that for $p_1 \leq c\frac{\ln m}{n}$, the random CSP is satisfiable **whp**. Furthermore, an assignment can be found in $O(mn)$ time **whp***.

For m, p_2 as in case (b), Hypothesis A holds, and so $\frac{\ln m}{n}$ is, indeed, the order of the satisfiability threshold. In case (a), **whp** the fact that the random CSP is unsatisfiable can be demonstrated easily by examining a single edge. We show that for $m \geq (\ln n)^{1+\epsilon}$ for any $\epsilon > 0$, this is far from the case. In particular, we show that **whp** there is. no short resolution proof of unsatisfiability when p_1 is of the same asymptotic order as the threshold of satisfiability.

Theorem 2. *If $m \geq (\ln n)^{1+\epsilon}$, $d = c\ln m$, for any constants $\epsilon, c > 0$, then **whp** the resolution complexity of the random CSP is $2^{\Omega(n/m)}$.*

The resolution complexity of various models of random boolean formula has been well-studied, starting with [10], and continuing through [4],[5],[3] and other papers. This line of inquiry was first extended to random models of CSP in [20,19] and was then continued in [23]. In both of those studies, the domain-size was constant. Our Theorem 2 is the first result on the resolution complexity for a model of random CSP where the domain-size grows with n.

We now consider another model.

Model B: Here we generate a random $m \times m$ *symmetric matrix M* with density p_2 and put $M_e = M$ for every edge of $G = G_{n,p_1}$.

Theorem 3. *Let ϵ be a small positive constant, and consider a random CSP from Model B.*

*(a) If $d \leq (4 - \epsilon)(\ln(1/q_2))^{-1}\ln m \ln\ln\ln m$ then **whp** the CSP is satisfiable **whp**.*
*(b) If $d \leq (1 - \epsilon)(\ln(1/q_2))^{-1}\ln m \ln\ln\ln m$ then an assignment can be found in polynomial time **whp**.*
*(c) If $0 < q_2 < 1$ is constant and if $d \geq K \ln m \ln\ln\ln m$ for sufficiently large K then **whp** the CSP is unsatisfiable.*

We can prove high resolution complexity in a restricted range of d, m, p_2.

Theorem 4. *If $m \to \infty$ and $d = c\ln m \ln\ln\ln m$ for some constant $c > 0$, then **whp** the resolution complexity of a random CSP from Model B is $2^{\Omega(n/(d^3 m))}$.*

2   Model A: Unsatisfiable Region

2.1   Blocked Edges and Vertices

Let an edge $e = (x, y)$ of G be *blocked* if $M_e = \mathbf{O}$ (the matrix with all zero entries). Of course, any CSP with a blocked edge is unsatisfiable, since there is no possible consistent assignment to x, y. We start with a simple lemma:

Lemma 1. *Let $\epsilon > 0$ be a small positive constant and assume that $nd \to \infty$ (so that **whp** G has edges). Let $m_0 = \sqrt{(\ln n + \ln d)/\ln(1/q_2)}$. Then*

(a) $m \geq (1 + \epsilon)m_0$ *implies that there are no blocked edges,* **whp.**
(b) $m \leq (1 - \epsilon)m_0$ *implies that there are blocked edges,* **whp.**

Proof   Let Z be the number of blocked edges in our instance. Given the graph G, the distribution of Z is $Bin(|E|, q_2^{m^2})$.

$$\mathbf{E}(Z) = \binom{n}{2} p_1 q_2^{m^2} \tag{1}$$

If $m \geq (1 + \epsilon)m_0$ then (1) implies that

$$\mathbf{E}(Z) \leq (nd)^{-\epsilon} \to 0$$

and then $Z = 0$ **whp** and (a) follows.
   If $m \leq (1 - \epsilon)m_0$ then (1) implies that

$$\mathbf{E}(Z) \geq \frac{1}{3}(nd)^\epsilon \to \infty.$$

Part (b) now follows from the Chernoff bounds.
   This proves Theorem 1(a).   □
   We now consider another simple cause of unsatisfiability that [2] also discovered to be prevalent amongst the models commonly used for experimentation. We say that a vertex (variable) x is *blocked* if for every possible assignment $i \in [m]$ there is some neighbour y which blocks the assignment of i to x, because the ith row of M_e, $e = (x, y)$ is all zero.

Lemma 2. *Let ϵ be a small positive constant, and suppose that $m - \sqrt{\ln n/\ln(1/q_2)} \to \infty$. Then*

(a) $m \geq (1+\epsilon)\sqrt{(\ln n + m \ln d)/\ln(1/q_2)}$ *implies that there are no blocked vertices,* **whp.**
(b) $m \leq (1-\epsilon)\sqrt{(\ln n + m \ln d)/\ln(1/q_2)}$ *implies that there are blocked vertices,* **whp.**

   Remark: Note that $m = \sqrt{(\ln n + m \ln d)/\ln(1/q_2)}$, for m slightly smaller than m_0 from Lemma 1.
   Proof   If the graph G is given and vertex v has degree d_v then

$$\mathbf{Pr}(v \text{ is blocked} \mid G) = (1 - (1 - q_2^m)^{d_v})^m.$$

This is because for $i \in [m]$, $(1 - q_2^m)^{d_v}$ is the probability that no neighbour w of v is such that row i of $M_{(v,w)}$ is all zero.

Part (a) now follows from an easy first moment calculation, which we omit. We turn our attention to proving part (b). Rearranging our assumption yields $\ln d \geq (1 - \epsilon)^{-1}(m \ln(1/q_2) - \frac{1}{m} \ln n)$. So we choose d such that $\ln d = (1 - \epsilon)^{-1}(m \ln(1/q_2) - \frac{1}{m} \ln n)$, i.e. $d = (q_2^{-m^2} n)^{1/m(1-\epsilon)}$ as proving the result for that value of d clearly implies that it holds for all larger values.

Our assumption implies that $d \to \infty$ and so **whp** $n - o(n)$ vertices v have $d_v \in I = [(1 - \epsilon)d, (1 + \epsilon)d]$. Thus if Z is the number of blocked vertices with $d_v \in I$ then

$$\mathbf{E}(Z) \geq (n - o(n))(1 - (1 - q_2^m)^{d(1-\epsilon)})^m \geq (n - o(n))(d(1 - \epsilon)q_2^m)^m$$

$$\geq (1 - o(1)) \left(q_2^{-m^2} n\right)^{\epsilon/(1-\epsilon)} (1 - \epsilon)^m$$

$$\geq (1 - o(1)) n^{\epsilon/(1-\epsilon)}(1 - \epsilon)^{m_0} \quad \text{(see the Remark preceding this proof)}$$

$$\geq n^{\epsilon/2} \to \infty.$$

To show that $Z \neq 0$ **whp** we use Talagrand's inequality [25]. We condition on G. Then we let each $\Omega_e, e \in E$ be an independent copy of $\{0, 1\}^{m^2}$ (the set of $m \times m$ 0-1 matrices). Now changing a single M_e can change z by at most 2 and so Assumption 1 holds with $a = 2$. Then to show that a vertex v is blocked we only have to expose M_e for e incident with v. Thus Assumption 2 holds with $c(\xi) = (1 + \epsilon)d\xi$. Thus if $M = Med(Z)$, the inequality gives

$$\mathbf{Pr}(|Z - M| \geq t(1 + \epsilon)dM^{1/2}) \leq 2e^{-t^2/16} \qquad (2)$$

for any $t > 0$.

Our assumptions imply that $d^2 = o(\mathbf{E}(Z))$ and so (2) implies the result. $\quad \square$

3 Model A: Satisfiable Region

We assume for this section that

$$m = (1 + \epsilon) \left(\frac{\ln n}{\ln q_2^{-1}}\right)^{1/2}, \; d = c \ln m \text{ and } p_2 \text{ is constant}$$

where c, ϵ are small. (Note that this also implies the result for larger m).

Now let a vertex v be *troublesome* if it has degree $\geq D = 10d$ or there are assignments to its neighbours which leave v without a consistent assignment. Let $\mathcal{T}$ denote the set of troublesome vertices. A subset of $\mathcal{T}$ is called a troublesome set.

Let $\mathcal{A}$ be the event that every set of k_0 vertices contains at most k_0 edges where

$$k_0 = \left\lceil \frac{2 \ln n}{d} \right\rceil.$$

Then

Lemma 3.
$$\mathbf{Pr}(\mathcal{A}) = 1 - o(1).$$

Proof

$$\mathbf{Pr}(\overline{\mathcal{A}}) \le \binom{n}{k_0}\binom{\binom{k_0}{2}}{k_0+1}\left(\frac{d}{n}\right)^{k_0+1} \le \left(\frac{ne}{k_0}\right)^{k_0} \cdot \left(\frac{d}{n}\right)^{k_0+1} \cdot \left(\frac{k_0 e}{2}\right)^{k_0+1}$$
$$= \frac{k_0 e^{2k_0+1} d^{k_0+1}}{2^{k_0+1}} \cdot \frac{d}{n} = o(1).$$

$\square$

We show next that **whp** the sub-graph induced by $\mathcal{T}$ has no large trees.

Lemma 4. Whp *there are no troublesome trees with* $\ge k_0$ *vertices.*

Proof   If $\mathcal{T}$ contains a tree of size greater than k_0 then it contains one of size k_0. Let Z be the number of troublesome trees with k_0 vertices. Let Ω be the set of trees/unicyclic graphs spanning $[k_0]$. Then for any subset J of $[k_0]$ we may write

$$\mathbf{E}(Z \cdot 1_{\mathcal{A}}) \le \binom{n}{k_0} \sum_{T \in \Omega} \left(\frac{d}{n}\right)^{k_0-1} \prod_{i \in J} \mathbf{Pr}(x_i \in \mathcal{T} \mid \mathcal{G}_T, x_j \in \mathcal{T}, \forall j \in J, j < i). \quad (3)$$

Here $\mathcal{G}_T$ is the event that the sub-graph of G induced by $[k_0]$ is T.

Fix $T \in \Omega$ and let I_1 be the set of vertices of T with degree at most 4 in T. Then $|I_1| \ge k_0/2$. Note next that I_1 contains an independent set I of size at least $k_0/10$.

Now if $i \in I$ then

$$\mathbf{Pr}(x_i \in \mathcal{T} \mid \mathcal{G}_T, x_1, x_2, \dots, x_{i-1} \in \mathcal{T}) \le \binom{n}{D-4}\left(\frac{d}{n}\right)^{D-4} + \sum_{t=0}^{D} m^t (1 - p_2^t)^m.$$

The first term bounds the probability that x_i has at least $D - 4$ neighbours outside the tree and assuming the degree of x_i is at most D, the second term bounds the probability that the $\le D$ neighbours have an assignment which can not be extended to x_i. We use the fact that I is an independent set to gain the stochastic independence we need.

Thus, applying (3) with $J = I$ we obtain

$$\mathbf{E}(Z \cdot 1_{\mathcal{A}})$$
$$\le \binom{n}{k_0} k_0^{k_0-2} k_0^2 \left(\frac{d}{n}\right)^{k_0-1} \left(\binom{n}{D-4}\left(\frac{d}{n}\right)^{D-4} + \sum_{t=0}^{D} m^t (1 - p_2^t)^m\right)^{k_0/10} \quad (4)$$
$$\le n(de)^{k_0} \left(\left(\frac{de}{D-4}\right)^{D-4} + Dm^D e^{-mp_2^D}\right)^{k_0/10} = o(1).$$

$\square$

Now we deal with troublesome cycles in a similar manner.

Lemma 5. Whp *there are no troublesome cycles.*

Proof   It follows from Lemma 4 that we need only consider cycles of length k_0 or less. If Z now denotes the number of troublesome cycles of length k_0 or less then arguing as in (3), (4) we see that

$$\mathbf{E}(Z) \leq$$

$$\sum_{k=3}^{k_0} \binom{n}{k} \frac{(k-1)!}{2} \left(\frac{d}{n}\right)^k \left(\binom{n}{D-2}\left(\frac{d}{n}\right)^{D-2} + \sum_{t=0}^{D} m^t (1-p_2^t)^m\right)^{\lfloor k/2 \rfloor}$$

$$= o(1).$$

$\square$

Let a tree be *small* if it contains at most k_0 vertices.

We have therefore shown that **whp** the troublesome vertices $\mathcal{T}$ induce a forest of small trees.

We show next that **whp** there at most $n^{1+o(1)}$ small trees.

Lemma 6. Whp *there are at most $n^{1+o(1)}$ small trees.*

Proof   Let σ_T denote the number of small trees. Then

$$\mathbf{E}(\sigma_T) = \sum_{k=1}^{k_0} \binom{n}{k} k^{k-2} \left(\frac{d}{n}\right)^{k-1} \leq \sum_{k=1}^{k_0} n(de)^k = n^{1+o(1)}.$$

The result now follows from the Markov inequality. $\square$

Our method of finding an assignment to our CSP is to (i) make a consistent assignment to the vertices of $\mathcal{T}$ first and then (ii) extend this assignment "greedily" to the non-troublesome vertices.

It is clear from the definition of troublesome that it is possible to carry out Step (ii). We wish to show that (i) can be carried out successfully **whp**. For this purpose we show that **whp** G does not contain a small tree which cannot be given a consistent assignment.

So we fix a small tree T and a vertex $v \in T$ and root T at v. Then let $X_i, 0 \leq i \leq k_0$ denote the vertices at distance i from v in T. Then let d_ℓ be the maximum number of descendants of a vertex in X_ℓ and let L denote the depth of T.

For $u \in X_\ell$ let $S_\ell(u)$ be the the set of values δ such that there is a consistent assignment to the sub-tree of T rooted at u in which u receives δ. We let $t = \lceil 10/\epsilon \rceil$ and define the events

$$\mathcal{B}_{u,i}^\ell = \left\{ \frac{(i-1)m}{t} \leq |S_\ell(u)| \leq \frac{im}{t} \right\}.$$

Then for $1 \leq i \leq t$ let

$$\pi_{i,\ell} = \max_{u \in X_\ell} \mathbf{Pr}\left(\bigcup_{j=1}^{i} \mathcal{B}_{u,j} \right).$$

Note that $\pi_{t,\ell} = 1$.

We claim that for $\ell > 1$,

$$\pi_{i,\ell} \leq \sum_{k_1+\cdots+k_t=d_\ell} \sum_{r=\frac{t-i}{t}m}^{\frac{t-i+1}{t}m} \binom{m}{r} \prod_{j=1}^{t}(1 - (1 - q_2^{\frac{j-1}{t}})^{k_j})^r \pi_{j,\ell-1}^{k_j} \tag{5}$$

$$\leq \sum_{j=1}^{t} t^{d_\ell} 2^m (1 - (1 - q_2^{\frac{j-1}{t}m})^{d_\ell})^{\frac{t-i}{t}m} \pi_{j,\ell-1}.$$

$$\leq \sum_{j=1}^{t} t^{d_\ell} 2^m (d_\ell q_2^{\frac{j-1}{t}m})^{\frac{t-i}{t}m} \pi_{j,\ell-1}. \tag{6}$$

Explanation of (5): Suppose that there are k_j descendants w of u for which $\mathcal{B}_{w,j}^{\ell-1}$ occurs. If $u \in \mathcal{B}_{u,i}^{\ell}$ then r assignment values will be forbidden to it, $\frac{t-i}{t}m \leq r \leq \frac{t-i+1}{t}m$. The product bounds the probability that these values are forbidden and that $\mathcal{B}_{w,j}^{\ell-1}$ occurs for the corresponding descendants.

Now let us prove by induction on ℓ that for $\eta = \epsilon/3$ and for $1 \leq j \leq t$ we have

$$\pi_{j,\ell} \leq t^{\ell} n^{-(1+\eta)\frac{t-j}{t}}. \tag{7}$$

This is clearly true for $\ell = 0$ since $\pi_{j,0} = 0$ for $j < t$ and $\pi_{t,0} = 1$. Then from (6) we obtain

$$\pi_{i,\ell} \leq t^{\ell-1} \sum_{j=1}^{t} t^{d_\ell} 2^m d_\ell^{\frac{t-i}{t}m} q_2^{\frac{(j-1)(t-i)}{t^2}m^2} n^{-(1+\eta)\frac{t-j}{t}}$$

$$\leq t^{\ell-1} \sum_{j=1}^{t} n^{-\frac{(j-1)(t-i)}{t^2}(1+\frac{\epsilon}{2})-\frac{t-j}{t}(1+\eta)}.$$

Notice that in going from the first to second inequality we use the fact that since $\ell, d_\ell \leq k_0$ we find that $2^m t^{d_\ell} d_\ell^{\frac{t-i}{t}m} = n^{o(1)}$. This term is then absorbed by using $1 + \epsilon/2$ in place of $1 + \epsilon$.

Now consider the expression

$$\Delta = \frac{(j-1)(t-i)}{t^2}(1+\frac{\epsilon}{2}) + \frac{t-j}{t}(1+\eta) - \frac{t-i}{t}(1+\eta)$$

$$= \frac{(j-1)(t-i)}{t^2}(1+\frac{\epsilon}{2}) + \frac{i-j}{t}(1+\eta).$$

To complete the inductive proof of (7) we have only to show that it is non-negative.

Now Δ is clearly non-negative if $i \geq j$ and so assume that $j > i$. Now for a fixed j, Δ can be thought of as a linear function of i and so we need only check non-negativity for $i = 1$ or $i = j - 1$.

For $i = 1$ we need

$$(j-1)(t-1)(1+\frac{\epsilon}{2}) \geq (j-1)t(1+\eta) \tag{8}$$

and this holds for $\epsilon \leq 1$.

For $i = j - 1$ we need

$$(j - 1)(t - j + 1)(1 + \frac{\epsilon}{2}) \geq t(1 + \eta).$$

But here $j \geq 2$ and the LHS is at least $(t - 1)(1 + \frac{\epsilon}{2})$ and the inequality reduces to (8) (after dividing through by $j - 1$). This competes the proof of (7). In particular

$$\pi_{1,k_0} \leq t^{k_0} n^{-(1+\eta)(t-1)/t}.$$

$\mathbf{Pr}(\exists$ a troublesome tree which cannot be consistently assigned)

$$\leq o(1) + n^{1+o(1)} t^{k_0} n^{-(1+\eta)(t-1)/t} = o(1)$$

which implies that Step (i) can be completed **whp**. This proves the satisfiability claim in Theorem 1(b).

It only remains to discuss the time to find an assignment. Once we have assigned values to $\mathcal{T}$ then we can fill in an assignment in $O(mn)$ time. So let us now fix a small tree T of troublesome vertices. Choose a root $v \in T$ arbitrarily. Starting at the lowest levels we compute the set of values $S_\ell(u)$ available to a vertex $u \in X_\ell$. For each descendant w of u we compute $T_\ell(w) = \{a \in S_{\ell+1}(w) : M_{(u,w)}(a) = 1\}$ and then we have $S_\ell(u) = \bigcap_w T_\ell(w)$. At the leaves, $S_L = [m]$ and so in this way we can assign a value to the root and then work back down the tree to the leaves giving an assignment to the whole of T. Thus the whole algorithm takes $O(mn)$ time as claimed. □

4 Model A: Resolution Complexity

For a boolean CNF-formula F, a *resolution refutation* of F with length r is a sequence of clauses $C_1, ..., C_r = \emptyset$ such that each C_i is either a clause of F, or is derived from two earlier clauses $C_j, C_{j'}$ for $j, j' < i$ by the following rule: $C_j = (A \vee x), C_{j'} = (B \vee \bar{x})$ and $C_i = (A \vee B)$, for some variable x. The *resolution complexity* of F, denoted $\mathbf{RES}(F)$, is the length of the shortest resolution refutation of F. (If F is satisfiable then $\mathbf{RES}(F) = \infty$.)

Mitchell[20] discusses two natural ways to extend the notion of resolution complexity to the setting of a CSP. These two measures of resolution complexity are denoted $\mathbf{C} - \mathbf{RES}$ and $\mathbf{NG} - \mathbf{RES}$. Here, our focus will be on the $\mathbf{C} - \mathbf{RES}$ measure, as it was in [19] and in [23].

Given an instance $\mathcal{I}$ of a CSP in which every variable has domain $\{1, ..., m\}$, we construct a boolean CNF-formula $\mathrm{CNF}(\mathcal{I})$ as follows. For each variable x of $\mathcal{I}$, there are m variables in $\mathrm{CNF}(\mathcal{I})$, denoted $x : 1, x : 2, ..., x : m$, and there is a *domain clause* $(x : 1 \vee ... \vee x : m)$. For each pair of variables x, y and each *restriction* (i, j) such that $M_{(x,y)}(i, j) = 0$, $\mathrm{CNF}(\mathcal{I})$ has a *conflict clause* $(\overline{x : i} \vee \overline{y : j})$. We also add $\binom{m}{2}$ 2-clauses for each x which specify that $x : i$ can be true for at most one value of i. It is easy to see that $\mathrm{CNF}(\mathcal{I})$ has a satisfying assignment iff $\mathcal{I}$ does. We define the resolution complexity of $\mathcal{I}$, denoted $\mathbf{C} - \mathbf{RES}(\mathcal{I})$ to be equal to $\mathbf{RES}(\mathrm{CNF}(\mathcal{I}))$.

A variable x is *free* if any assignment which satisfies $\mathcal{I} - x$ can be extended to a satisfying assignment of $\mathcal{I}$. The *boundary* $\mathcal{B}(\mathcal{I})$ is the set of *free* variables. We extend a key result from [20] to the case where m grows with n:

Lemma 7. *Suppose that there exist $s, \zeta > 0$ such that*

(a) Every subproblem on at most s variables is satisfiable, and
(b) Every subproblem $\mathcal{I}'$ on v variables where $\frac{1}{2}s \leq v \leq s$ has $|\mathcal{B}(\mathcal{I}')| \geq \zeta n$.

then $\mathbf{C} - \mathbf{RES}(\mathcal{I}) \geq 2^{\Omega(\zeta^2 n/m)}$.

The proof is a straightforward adaptation of the proof of the corresponding work in [20] and so we omit it.

We assume now that ϵ is a small positive constant and

$$m \geq (\ln n)^{1+\epsilon}, \; d = c\ln m \text{ and } p_2 \text{ is constant.} \tag{9}$$

Let γ be a sufficiently small constant. Let $\mathcal{T}_1$ denote the set of vertices v for which there are γd neighbours W and a set of assignments of values to W for which v has no consistent assignment.

Lemma 8.
$$\mathbf{Pr}(\mathcal{T}_1 \neq \emptyset) = o(1).$$

Proof

$$\mathbf{E}(|\mathcal{T}_1|) \leq n \sum_{t=\gamma d}^{n-1} \binom{n}{t} \left(\frac{d}{n}\right)^t \binom{t}{\gamma d} m^{\gamma d} (1 - p_2^{\gamma d})^m$$

$$\leq n \sum_{t=\gamma d}^{n-1} \left(\frac{de}{t}\right)^t \left(\frac{tem}{\gamma d}\right)^{\gamma d} e^{-mp_2^{\gamma d}}$$

$$\leq ne^{-m^{1-\epsilon/2}} \left(\sum_{t=\gamma d}^{10d} (de)^{10d} (10e\gamma^{-1}m)^{\gamma d} + \sum_{10d}^{n-1} (mn)^{\gamma d} \right) = o(1).$$

$\square$

Now we show that **whp** every set of $s \leq s_0 = \alpha n$ vertices, $\alpha = \gamma/3$ has less than $\gamma ds/2$ edges. Let $\mathcal{B}$ denote this event.

Lemma 9.
$$\mathbf{Pr}(\mathcal{B}) = 1 - o(1).$$

Proof

$$\mathbf{Pr}(\overline{\mathcal{B}}) \leq \sum_{s=\gamma d}^{\alpha n} \binom{n}{s} \binom{\binom{s_0}{2}}{\gamma ds/2} \left(\frac{d}{n}\right)^{\gamma ds/2} \leq \sum_{s=\gamma d}^{\alpha n} \left(\left(\frac{se}{\gamma n}\right)^{-1+\gamma d/2} \cdot \frac{e^2}{\gamma} \right)^s = o(1).$$

$\square$

Let us now check the conditions of Lemma 7. Condition (a) holds because Lemma 9 implies that if $s = |S| \leq \alpha n$ then we can order S as $v_1, v_2, \ldots, v_s$ so that v_j has less than αd neighbours among $v_1, v_2, \ldots, v_{j-1}$ for $1 \leq j \leq s$. Because we can assume that $T_1 = \emptyset$ (Lemma 8) we see that it will be possible to sequentially assign values to $v_1, v_2, \ldots, v_s$ in order. Lemma 9 implies that at least $\frac{1}{2}$ the vertices of S have degree $\leq \alpha d$ in S and now $T_1 = \emptyset$ implies that (b) holds with $\zeta = 1/2$.

We conclude that with the parameters as stated in (9), $\mathbf{C} - \mathbf{RES}(\mathcal{I})$ is **whp** as large as is claimed by Theorem 2.

5   Model B: Satisfiability

We have a blocked edge iff $M = \mathbf{O}$ and this happens with probability $q_2^{m(m-1)}$ and so there is not much more to say on this point.

Secondly, if $M \neq \mathbf{O}$ then there are two values x, y which can be assigned to adjacent vertices. This implies that for any bipartite subgraph H of G there is a satisfying assignment for H just using x, y. So, in particular there will be no blocked vertices.

Let us now consider Theorem 3. Let H be the graph defined by treating M as its adjacency matrix. Thus $H = G_{m, p_2}$. As such it has a clique I of size $(2 - o(1)) \ln m / (\ln 1/q_2)$.

If we can properly colour G with I (i.e. give adjacent vertices different values in I) then we will have a satisfying assignment for our CSP. Now the chromatic number of G is $(1 + o(1))d/(2 \ln d)$ **whp**. So the CSP is satisfiable **whp** if

$$(2 - o(1)) \ln m / (\ln 1/q_2) \geq (1 + o(1))d/(2 \ln d)$$

and this holds under assumption (a).

For (b) we observe that we can find a clique of size $(1 - o(1)) \ln m / (\ln 1/q_2)$ in polynomial time and we can colour G with $(1 + o(1))d/\ln d$ colours in polynomial time.

We now prove part (c) of Theorem 3. We first observe

Lemma 10. *There exists a constant ϵ_0 such that for $\epsilon \leq \epsilon_0$ there exist $R_0 = R_0(\epsilon), Q_0 = Q_0(\epsilon)$ such that if $Q \geq Q_0, R \geq R_0$ and $s_0 = R \ln m$ then*

(a) **whp** *every pair of disjoint sets $S_1, S_2 \subseteq [m], |S_1| = s_1 \geq s_0, |S_2| = s_2 \geq s_0$ contains at most $(1 - \epsilon)s_1 s_2$ $S_1 : S_2$ edges of H;*

(b) **whp** *every $S \subseteq [m], |S| = s \geq s_0$ contains at most $Q \ln m$ members with degree greater than $(1 - \epsilon)s$ in the subgraph of H induced by S.*

Proof
(a) We can bound the probability that there are sets S_1, S_2 with more than the stated number of $S_1 : S_2$ edges by

$$\sum_{s_1=s_0}^{m}\sum_{s_2=s_0}^{m}\binom{m}{s_1}\binom{m}{s_2}\binom{s_1 s_2}{\epsilon s_1 s_2}p_2^{(1-\epsilon)s_1 s_2}$$

$$\leq \sum_{s_1=s_0}^{m}\sum_{s_2=s_0}^{m}\left(\frac{me}{s_1}\right)^{s_1}\left(\frac{me}{s_2}\right)^{s_2}\left(\left(\frac{e}{\epsilon}\right)^{\epsilon}p_2^{1-\epsilon}\right)^{s_1 s_2} = o(1).$$

(b) We choose $\epsilon > 0$ so that $p_2 < 1 - 3\epsilon$. Given S, we consider a set $L \subset S$ of size $Q \ln m$. For $R > Q\epsilon^{-1}$ we have $|L| < \epsilon|S|$ and so if each $i \in L$ has at least $(1-\epsilon)s$ neighbours in S then it has at least $(1-2\epsilon)s$ neighbours in $S - L$. By the Chernoff bound, this occurs with probability at most $\left(e^{-\zeta s}\right)^{|L|}$, for some $\zeta > 0$ and this is less than m^{-2s} for Q sufficiently high. Therefore, the expected number of S, L violating part (b) is at most

$$\sum_{s=s_0}^{m}\binom{m}{s}\binom{s}{Q\ln m}m^{-2s} < \sum_{s=s_0}^{m}\left(\frac{em}{s}\right)^{s}2^s m^{-2s} < \sum_{s \geq s_0}m^{-s} = o(1).$$

$\square$

Now consider an assignment σ for our CSP and let N_i be the set of variables that are assigned the value i by σ. We observe that if σ is consistent then each N_i is an independent set in G and so **whp** G is such that we must have

$$|N_i| \leq \frac{3n \ln d}{d} < \frac{4n}{K \ln m} \qquad \text{for } i = 1, 2, \ldots, m. \tag{10}$$

Thus, we will restrict our attention to assignments which satisfy (10). We will prove that the expected number of such assignments that are consistent is $o(1)$, thus proving part (c) of Theorem 3.

We say that a pair of vertices is *forbidden* by σ if that pair cannot form an edge of G without violating σ. Note that every pair in the same set N_i is forbidden, and a pair in $N_i \times N_j$ is forbidden iff ij is not an edge of H. We will show that the number of forbidden pairs is at least $n^2/\ln\ln m$. It follows that

$$\mathbf{Pr}(\sigma \text{ is consistent}) \leq (1-p_1)^{n^2/\ln\ln m} \leq e^{-nd/\ln\ln m} = o(m^{-n}),$$

assuming that $d \geq K \ln m \ln\ln\ln m$ for sufficiently large K. Since this probability is $o(m^{-n})$ we can multiply by m^n, which is an overcount of the number of assignments satisfying (10), and so obtain the desired first moment bound.

Let $n_i = |N_i|$ and let $I = \{i : n_i \geq n/(2m)\}$. Now

$$\sum_{i \in I}n_i = n - \sum_{i \notin I}n_i \geq n - m \cdot \frac{n}{2m} = \frac{n}{2}. \tag{11}$$

For the following analysis we choose constants:

$$\epsilon, \qquad Q = \max\{Q_0, 100\epsilon^{-1}\}, \qquad K_1 = 100R_0, \qquad K = 100K_1 Q$$

where $\epsilon \leq \epsilon_0, Q_0, R_0$ are from Lemma 10.

We partition I into 3 parts:

- $I_1 = \{i : n/(K_1 \ln m \ln \ln \ln m) \le n_i < 4n/K \ln m\}$
- $I_2 = \{i : n/(K_1 \ln m)^2 \le n_i < n/(K_1 \ln m \ln \ln \ln m)\}$
- $I_3 = \{i : n/(2m) \le n_i < n/(K_1 \ln m)^2\}$

Case 1: $\sum_{i \in I_1} n_i \ge \frac{n}{6}$ Let H_1 be the subgraph of H induced by I_1, and for each $i \in I_1$, we let $\overline{d}(i)$ be the degree of i in $\overline{H_1}$. Note that the total number of forbidden pairs of vertices for G is at least

$$\frac{1}{2} \sum_{i \in I_1} \overline{d}(i) n_i \times \frac{n}{K_1 \ln m \ln \ln \ln m}, \tag{12}$$

since for all $i' \in I_1, n_{i'} \ge n/(K_1 \ln m \ln \ln \ln m)$.

By (10), we have $|I_1| \ge (K \ln m)/24$, so $(K \ln m)/Q < \epsilon |I_1|$. Thus, by Lemma 10(b) then there are at most $Q \ln m$ members $i \in I_1$ with $\overline{d}(i) < (K \ln m)/Q$. Again using (10), these members contribute at most $4Qn/K < n/12$ to $\sum_{i \in I_1} n_i$. Therefore, the sum in (12) is at least

$$\frac{1}{2} \times \frac{K \ln m}{Q} \times \frac{n}{12} \times \frac{n}{K_1 \ln m \ln \ln \ln m} \ge \frac{n^2}{\ln \ln m}.$$

Case 2: $\sum_{i \in I_2} n_i \ge \frac{n}{6}$ We let $I(j) = \{i \in I_2 : n/2^j \le n_i \le n/2^{j-1}\}$, for
$\log_2(K_1 \ln m \ln \ln \ln m) \le j \le 2 \log_2(K_1 \ln m)$. We set $t_j = \sum_{i \in I(j)} n_i$ and $s_j = |I(j)| \ge t_j \times (K_1 \ln m \ln \ln \ln m/n)$. We set $J = \{j : t_j \ge n/(100 \ln \ln m)\}$ and note that $s_j \ge s_0$ (from Lemma 10) for each $j \in J$. Note also that

$$\sum_{j \in J} t_j \ge \frac{n}{6} - 2 \log_2(K_1 \ln m) \times \frac{n}{100 \ln \ln m} \ge \frac{n}{8}.$$

Consider $I(j)$ for any $j \in J$. By Lemma 10, there are at least $\epsilon \binom{s_j}{2}$ pairs $i, i' \in I(j)$ such that every pair of vertices in $N_i \times N_{i'}$ is forbidden. Also, for any i, every pair in $N_i \times N_i$ is forbidden. Since the sizes of the sets $N_i, i \in I(j)$ differ by at most a factor of 2, this implies that the number of forbidden pairs in $\cup_{i \in I(j)} N_i$ is at least $\frac{\epsilon}{8} t_j^2$. Now consider any pair $I(j), I(j')$ with $j, j' \in J$. By Lemma 10(a), there are at least $\epsilon s_j s_{j'}$ pairs $i \in I(j), i' \in I(j')$ such that every pair of vertices in $N_i \times N_{i'}$ is forbidden, and this implies that the number of forbidden pairs in $\cup_{i \in I(j)} N_i \times \cup_{i \in I(j')} N_i$ is at least $\frac{\epsilon}{4} t_j t_{j'}$. Thus, the total number of forbidden pairs is at least

$$\frac{\epsilon}{8} \left(\sum_{j \in J} t_j^2 + \sum_{j,j' \in J; j < j'} 2 t_j t_{j'} \right) = \frac{\epsilon}{8} \left(\sum_{j \in J} t_j \right)^2 \ge \frac{\epsilon n^2}{8^3} > \frac{n^2}{\ln \ln m}.$$

Case 3: $\sum_{i \in I_3} n_i \ge \frac{n}{6}$.   Here we follow essentially the same argument as in Case 2. Again, let $I(j) = \{i \in I : n/2^j \le n_i \le n/2^{j-1}\}$, but this time we consider $2\log_2(K_1 \ln m) < j \le \log_2(2m)$. Again, $t_j = \sum_{i \in I(j)} n_i$ and $s_j = |I(j)|$, but note that this time we have

$$s_j \ge \frac{t_j}{n/(K_1 \ln m)^2}.$$

Here, we set $J = \{j : t_j \ge n/K_1 \ln m\}$ and so again we have $s_j \ge s_0$ for every $j \in J$.

$$\sum_{j \in J} t_j \ge \frac{n}{4} - \log_2(2m) \times \frac{n}{K_1 \ln m} \ge \frac{n}{8}.$$

The same argument as in Case 2 now goes through to imply that the total number of forbidden pairs is at least

$$\frac{\epsilon}{8} \left(\sum_{j \in J} t_j \right)^2 > \frac{n^2}{\ln \ln m}.$$

□

6   Model B: Resolution Complexity

First note that **whp** every set of 10 vertices in H has a common neighbour, since the probability of at least one such set not having a common neighbour is less than $\binom{m}{10} q_2^{m-10} = o(1)$. Assuming that H has this property, every vertex of degree at most 10 in G will be in the boundary.

A straightforward first moment argument shows that a.s. every subgraph G' of G with at most $n/d^{3/2}$ vertices has at most $5|G'|$ edges. (We omit the standard calculation.) Therefore, every such G' has at least $|G'|/11$ vertices of degree at most 10. This implies both conditions of Lemma 7 with $s = n/d^{3/2}$ and $\zeta = 1/(22d^{3/2})$ and thus implies Theorem 4.   □

We remark that the exponent "3" of d in the statement of Theorem 4 can be replaced by values arbitrarily close to 2 by replacing "10" with a larger value in this proof.

References

1. D. Achlioptas, P. Beame and M. Molloy. *A sharp threshold in proof complexity.* Proceedings of STOC 2001, 337 - 346.
2. D. Achlioptas, L. Kirousis, E. Kranakis, D. Krizanc, M. Molloy, and Y. Stamatiou. *Random constraint satisfaction: a more accurate picture.* Constraints **6**, 329 - 324 (2001). Conference version in Proceedings of CP 97, 107 - 120.
3. P. Beame, J. Culberson and D. Mitchell. *The resolution complexity of random graph k-colourability.* In preparation.
4. P. Beame and T. Pitassi. *Simplified and improved resolution lower bounds.* Proceedings of FOCS 1996, 274 - 282.

5. P. Beame, R. Karp, T. Pitassi and M. Saks. *The efficiency of resolution and Davis-Putnam procedures*. Proceedings of STOC 1998 and SIAM Journal on Computing, **31**, 1048 - 1075 (2002).
6. E. Ben-Sasson and A. Wigderson. *Short proofs are narrow - resolution made simple*. Proceedings of STOC 1999 and Journal of the ACM **48** (2001)
7. B. Bollobás, Random graphs, Second Edition, Cabridge University Press, 2001.
8. B. Bollobás, *A probabilistic proof of an asymptotic formula for the number of labelled regular graphs*, European Journal on Combinatorics **1** (1980) 311–316.
9. E. A. Bender and E. R. Canfield, *The asymptotic number of labelled graphs with given degree sequence*, Journal of Combinatorial Theory (A) **24** (1978) 296–307.
10. V. Chvatal and E. Szemeredi. *Many hard examples for resolution*. Journal of the ACM **35** (1988) 759 - 768.
11. N. Creignou and H. Daude. *Random generalized satisfiability problems*. Proceedings of SAT 2002.
12. R. Dechter, *Constraint networks*, in Encyclopedia of Artificial Intelligence, S. Shapiro (ed.), Wiley, New York, 2nd ed. (1992) 276–285.
13. M. Dyer, A. Frieze and M. Molloy, *A probabilistic analysis of randomly generated binary constraint satisfaction problems*. Theoretical Computer Scince **290**, 1815 - 1828 (2003).
14. E. C. Freuder, *A sufficient condition for backtrack-free search*, Journal of the ACM **29** (1982) 24–32.
15. D. G. Bobrow and M. Brady, eds., Special Volume on Frontiers in Problem Solving: Phase Transitions and Complexity, Guest Editors: T. Hogg, B. A. Hubermann, and C. P. Williams, *Artificial Intelligence* **81** (1996), nos. 1 and 2.
16. I. Gent, E. MacIntyre, P. Prosser, B. Smith and T. Walsh. *Random constraint satisfaction: flaws and structure*. Constraints **6**, 345 - 372 (2001).
17. S. Janson, T. Łuczak and A. Ruciński, *Random Graphs*, Wiley, 2000.
18. A. K. Mackworth, *Constraint satisfaction*, in Encyclopedia of Artificial Intelligence, S. Shapiro (ed.), Wiley, New York, 2nd ed. (1992) 285-293.
19. D. Mitchell, *The Resolution complexity of random constraints*. Proceedings of Principles and Practices of Constraint Programming - CP 2002.
20. D. Mitchell, *The Resolution Complexity of Constraint Satisfaction*. Ph.D. Thesis, University of Toronto, 2002.
21. M. Molloy, *Models for Random Constraint Satisfaction Problems*. Proceedings of STOC 2002, 209 - 217. Longer version to appear in SIAM J. Computing.
22. M. Molloy, *When does the giant component bring unsatisfiability?* Submitted.
23. M. Molloy and M. Salavatipour, *The resolution complexity of random constraint satisfaction problems*. Submitted.
24. B. Pittel, J. Spencer and N. Wormald, *Sudden emergence of a giant k-core in a random graph*, Journal of Combinatorial Theory (B) **67** (1996) 111–151.
25. M. Talagrand, Concentration of mesure and isoperimetric inequalities, *Inst. Hautes Études Sci. Publ. Math.* 81 (1995) 73-205.
26. D. Waltz, *Understanding line drawings of scenes with shadows*, The Psychology of Computer Vision, McGraw-Hill, New York, (1975) 19-91.

Continuous-Time Quantum Walks on the Symmetric Group

Heath Gerhardt and John Watrous

Department of Computer Science
University of Calgary
Calgary, Alberta, Canada
{gerhardt, jwatrous}@cpsc.ucalgary.ca

Abstract. In this paper we study continuous-time quantum walks on Cayley graphs of the symmetric group, and prove various facts concerning such walks that demonstrate significant differences from their classical analogues. In particular, we show that for several natural choices for generating sets, these quantum walks do not have uniform limiting distributions, and are effectively blind to large areas of the graphs due to destructive interference.

1 Introduction

According to our current understanding of physics, quantum mechanics provides sources of true randomness, and mathematically speaking much of the underlying framework of quantum information and computation may be viewed as an extension of the study of random processes. The focus in quantum information and computation is often placed on finding information processing tasks that can be performed with the help of quantum information (such as factoring integers in polynomial time [20] or implementing unconditionally secure key distribution [7,21]) or on studying the distinctively non-classical aspects of quantum information (such as entanglement; see, for instance, [12]). However, it seems quite plausible that the study of quantum information and computation will also lead to new methods in the study of classical computation and random processes. Along these lines, Kerenidis and de Wolf [17] recently used quantum arguments to prove new results on (classical) locally decodable codes.

As a step toward understanding the possible implications of quantum methods for the study of random processes, it is natural to consider the differences between classical and quantum processes. One of the topics that has recently received attention in the quantum computing community that highlights these differences is the the study of quantum computational variants of random walks, or *quantum walks* [1,3,5,6,8,9,11,15,18,19,23]. (A recent survey on quantum walks by Kempe [16] is an ideal starting point for background on quantum walks.) In this paper we consider quantum walks on Cayley graphs of the symmetric group—a topic that has been suggested in at least two previous papers on quantum walks [16,3].

S. Arora et al. (Eds.): APPROX 2003+RANDOM 2003, LNCS 2764, pp. 290–301, 2003.

Two main variants of quantum walks have been considered: continuous-time quantum walks and discrete-time quantum walks. We restrict our attention to continuous-time quantum walks in this paper. Keeping in line with previous results on quantum walks, we find some significant differences between quantum and classical random walks on Cayley graphs of the symmetric group. In particular, we find that quantum walks on Cayley graphs of the symmetric group do not have uniform limiting distributions for several natural choices for the generators. This answers a question recently suggested by Ahmadi, Belk, Tamon, and Wendler [3] concerning non-uniform mixing of quantum walks.

One of the principle motivations for studying quantum walks has been that quantum walks may potentially be useful as algorithmic tools. This potential was recently demonstrated by Childs, Cleve, Deotto, Farhi, Gutmann and Spielman [8], who prove that there exists a black-box problem for which a quantum algorithm based on quantum walks gives an exponential speed-up over any classical randomized algorithm. The key to this algorithm is that a quantum walk is able to permeate a particular graph while any classical random walk (or any classical randomized algorithm, for that matter) cannot. One of the first problems that comes to mind as an obvious challenge for the quantum algorithms community is the graph isomorphism problem, and it is natural to ask whether quantum walks, and in particular quantum walks on Cayley graphs of the symmetric group, can be of any use for an algorithm for this problem. (While this was our primary motivation for studying quantum walks on the symmetric group, we have not found any way to apply our results to this problem.)

2   Definitions

2.1   Continuous-Time Quantum Walks on Graphs

A continuous-time quantum walk on an undirected graph $\Gamma = (V, E)$ can be defined in the following way. First, we let A be the $|V| \times |V|$ adjacency matrix of Γ, let D be the $|V| \times |V|$ diagonal matrix for which the diagonal entry corresponding to vertex v is $\deg(v)$, and let $L = D - A$. The matrix L is positive semidefinite and, under the assumption that Γ is connected, 0 is an eigenvalue with multiplicity 1; the uniform vector is a corresponding eigenvector. The quantum walk on Γ is then given by the unitary matrix $U(t) = e^{-itL}$ for $t \in \mathbb{R}$. If the quantum walk on Γ is run for time t starting at vertex u, then the amplitude associated with each vertex v is $U(t)[v, u]$, and thus measuring at this point (with respect to the standard basis) results in each vertex v with probability $|U(t)[v, u]|^2$. If instead of starting at a particular vertex u we have some quantum state described by $\psi : V \to \mathbb{C}$, and we run the quantum walk for time t, the new quantum state is described by $U(t)\psi$, and measuring results in each vertex v with probability $|(U(t)\psi)[v]|^2$. Other types of measurements can be considered, but we will focus just on this sort of measurement where the outcome is a vertex of the graph. To our knowledge, continuous-time quantum walks were first considered by Farhi and Gutmann [11].

Continuous-time quantum walks are analogous to continuous-time random walks on Γ, where the evolution is described by $M(t) = e^{-tL}$ rather than $U(t)$ as above. Specifically, if the continuous-time random walk is started at vertex u and run for time t, the probability of being at vertex v is given by $M(t)[v, u]$. Continuous-time random walks share many properties with their discrete-time variants [4].

This paper is concerned with quantum walks on Cayley graphs, which are regular graphs. In the case of regular graphs there is no difference between using the matrix L and the adjacency matrix for the definition of quantum walks, and we find it is more convenient to use the adjacency matrix for the graphs we are considering. (Of course one cannot replace L with the adjacency matrix when discussing the classical case, since this would not give rise to a stochastic process—the equivalence only holds for the quantum case.) The reasoning behind this equivalence is as follows. Because D and A commute for regular graphs, we see that $U(t) = e^{-itdI}e^{itA} = e^{-itd}e^{itA}$; the difference is a global phase factor, which has no significance when calculating the probabilities. So, from here after in this paper we will consider the unitary process given by $U(t) = e^{itA}$ rather than e^{-itL}.

In the case of classical random walks, there are various properties of random walks that are of interest. One of the most basic properties of a classical random walk is the limiting distribution (or stationary distribution). This distribution is the uniform distribution for random walks on connected, regular graphs, and in fact as a result of the way we have defined continuous-time random walks this distribution is uniform for any connected, undirected graph; this is apparent by considering the spectral decomposition of the matrix e^{-tL}.

As quantum walks are unitary (and therefore invertible) processes, they do not converge to any state, so one must be precise about what is meant by the limiting distribution. Suppose we have a quantum walk on some graph Γ and some vertex u has been designated as the starting vertex. The probability of measuring the walk at some vertex v after time t is, as described above, given by $P_t[v] = |U(t)[v, u]|^2$. If t is chosen uniformly from some range $[0, T]$ then the resulting distribution is

$$\bar{P}_T[v] = \frac{1}{T} \int_0^T P_t[v] \mathrm{d}t.$$

In the limit for large T these distributions converge to some distribution $\bar{P}$, which is the limiting distribution of the quantum walk. This notion of the limiting distribution for a quantum walk is discussed in [1].

2.2 Cayley Graphs and Representation Theory of the Symmetric Group

In this section we briefly discuss necessary background information on Cayley graphs of the symmetric group and on representation theory of the symmetric group, which is the main tool used in this paper to analyze quantum walks on Cayley graphs.

Let G be a finite group and let $R \subseteq G$ be a set of generators for G satisfying $g \in R \Leftrightarrow g^{-1} \in R$ for all $g \in G$. Then the Cayley graph of G with respect to R, which we denote by $\Gamma(G, R)$ in this paper, is an undirected graph defined as follows. The set of vertices of $\Gamma(G, R)$ coincides with G, and for any $g, h \in G$, $\{g, h\}$ is an edge in $\Gamma(G, R)$ if and only if $gh^{-1} \in R$. Equivalently, if $R = \{h_1, \ldots, h_d\}$ then each vertex g is adjacent to vertices $h_1 g, \ldots, h_d g$. Thus, $\Gamma(G, R)$ is a regular graph of degree $d = |R|$. We will restrict our attention to generating sets that form conjugacy classes. (The method we use for analyzing quantum walks on Cayley graphs is limited to such generating sets.) Recall that for some group G, elements g and h are conjugate if there exists some $a \in G$ such that $a^{-1} g a = h$. This is an equivalence relation that partitions G into conjugacy classes. A function $f : G \to \mathbb{C}$ is a *class function* if it is constant on conjugacy classes of G.

The conjugacy classes in S_n are determined by the cycle structures of elements when they are expressed in the usual cycle notation. Recall that a partition λ of n is a sequence $(\lambda_1, \ldots, \lambda_k)$ where $\lambda_1 \geq \cdots \geq \lambda_k \geq 1$ and $\lambda_1 + \cdots + \lambda_k = n$. The notation $\lambda \vdash n$ indicates that λ is a partition of n. There is one conjugacy class for each partition $\lambda \vdash n$ in S_n, which consists of those permutations having cycle structure described by λ. We denote by C_λ the conjugacy class of S_n consisting of all permutations having cycle structure described by λ.

A *representation* of a group G is a homomorphism from G to $\mathrm{GL}(d, \mathbb{C})$ for some positive integer d, where $\mathrm{GL}(d, \mathbb{C})$ denotes the general linear group of invertible $d \times d$ complex matrices. The *dimension* of such a representation is d, and we write $\dim(\rho)$ to denote the dimension of a given representation ρ. Two representations $\rho_1 : G \to \mathrm{GL}(d_1, \mathbb{C})$ and $\rho_2 : G \to \mathrm{GL}(d_2, \mathbb{C})$ are *equivalent* if there exists an invertible linear mapping $A : \mathbb{C}^{d_1} \to \mathbb{C}^{d_2}$ such that $A\rho_1(g) = \rho_2(g)A$ for all $g \in G$, otherwise they are *inequivalent*. A representation ρ of dimension d is *irreducible* if there are no non-trivial invariant subspaces of $\mathbb{C}^d$ under ρ. That is, if $W \subseteq \mathbb{C}^d$ is a subspace of $\mathbb{C}^d$ such that $\rho(g)W \subseteq W$ for all $g \in G$, then $W = \mathbb{C}^d$ or $W = \{0\}$. A collection of inequivalent, irreducible representations is said to be *complete* if every irreducible representation is equivalent to one of the representations in this set. It holds that any complete set of irreducible representations can be put into one-to-one correspondence with the conjugacy classes of the group in question.

The character corresponding to a representation ρ is a mapping $\chi_\rho : G \to \mathbb{C}$ obtained by taking the trace of the representation: $\chi_\rho(g) = \mathrm{tr}(\rho(g))$. Using the cyclic property of the trace it follows that the characters are constant on the conjugacy classes of a group. If we have a complete set of inequivalent, irreducible representations of a group, then the corresponding characters form an orthogonal basis for the space of all class functions.

The Fourier transform $\hat{f}$ of a complex-valued function f on G at a representation ρ is $\hat{f}(\rho) = \sum_{g \in G} f(g)\rho(g)$.

Fact 1 *Let f be a class function on a group G and ρ be an irreducible representation of G, then $\hat{f}(\rho) = \frac{1}{\dim(\rho)} \left(\sum_{g \in G} f(g)\chi_\rho(g) \right) I$.*

For the symmetric group on n elements there is a particular way of associating the partitions of n (which are in one-to-one correspondence with the conjugacy classes of S_n) with a complete set of inequivalent, irreducible representations of S_n. These particular representations are said to be in *Young normal form*. (Several text books describe the specific method for constructing these representations—see, for instance, James and Kerber [14]. It will not be important for this paper to discuss the actual construction of these representations.) These representations have the special property that all matrix entries in these representations are integers. Once we have these irreducible representations, it is possible to associate with each one an equivalent irreducible representation that has the property that $\rho(g)$ is a unitary matrix for every $g \in S_n$. The irreducible, unitary representation associated with a given partition $\lambda \vdash n$ will be denoted ρ_λ, and the corresponding character will be denoted χ_λ. The following fact will be a useful fact regarding these representations.

Fact 2 *Let λ and μ be partitions of n and let ρ_λ and ρ_μ be the associated unitary representations as described above. Then for all $1 \le i, j \le \dim(\rho_\lambda)$ and $1 \le k, l \le \dim(\rho_\mu)$,*

$$\sum_{g \in S_n} \rho_\lambda(g)[i,j]\overline{\rho_\mu(g)[k,l]} = \begin{cases} \frac{n!}{\dim(\rho_\lambda)} & \text{if } \lambda = \mu, \ i = k, \text{ and } j = l \\ 0 & \text{otherwise} \end{cases}$$

When $\lambda, \nu \vdash n$, we write $\chi_\lambda(\nu)$ to denote the character χ_λ evaluated at an arbitrary $g \in C_\nu$, and more generally if f is a class function we write $f(\nu)$ to mean $f(g)$ for any $g \in C_\nu$.

Fact 3 *The sum of the squares of the characters of a conjugacy class over any complete, irreducible set of representations of a group G multiplied by the order of the class is the order of G. Thus, we have $|C_\lambda| \sum_{\nu \vdash n} \chi_\nu(\lambda)^2 = n!$ for every $\lambda \vdash n$.*

It will be necessary for us to be able to evaluate the characters associated with the irreducible representations of the symmetric group in certain instances. The Murnaghan-Nakayama rule provides a tool for doing this—information on the Murnaghan-Nakayama rule can be found in [22].

3 Continuous-Time Quantum Walks on $\Gamma(S_n, C_\lambda)$

In this section we analyze the quantum walk on $\Gamma(S_n, C_\lambda)$ for $\lambda \vdash n$. Our analysis implies that for some natural choices for λ the quantum walk on $\Gamma(S_n, C_\lambda)$ does not have a uniform limiting distribution with respect to the definition discussed in the previous section. In essence, the quantum walk has a significant "blind spot" consisting of all n-cycles (i.e., permutations having cycle-structure consisting of a single n-cycle).

This section is divided into three subsections. First we prove a general result concerning the spectral decomposition of quantum walks on S_n. We then consider the case where the generating set consists of the set of all transposition, and finally the case where the generating set consists of all p-cycles for any choice of $p \in \{2, \ldots, n\}$.

3.1   Spectral Decomposition and Periodicity

Define $c_\lambda : S_n \to \mathbb{C}$ to be the unit vector that is uniform on the conjugacy class C_λ and zero everywhere else: $c_\lambda[g] = |C_\lambda|^{-1/2}$ if $g \in C_\lambda$, and $c_\lambda[g] = 0$ otherwise.

The analysis of quantum walks on $\Gamma(S_n, C_\lambda)$ is greatly simplified by the fact that these walks are constant on conjugacy classes, in the following sense.

Proposition 4 *Let $\alpha_t(g)$ denote the amplitude associated with vertex g after evolving the quantum walk on $\Gamma(S_n, C_\lambda)$ for time t, assuming the walk starts on a conjugacy class, i.e., $\alpha_t(g) = (U(t)c_\lambda)[g]$. Then for all t, α_t is a class function.*

The following theorem will be one of the main tools used in our analysis.

Theorem 5. *Assume $H[g, h] = f(g^{-1}h)$ for all $g, h \in S_n$, where f a class function on S_n, and let $U(t) = e^{itH}$ for all $t \in \mathbb{R}$. Then for any partitions $\lambda, \mu \vdash n$ we have*

$$c_\lambda^* U(t) c_\mu = \frac{\sqrt{|C_\lambda|}\sqrt{|C_\mu|}}{n!} \sum_{\nu \vdash n} \exp\left(\frac{it}{\dim(\rho_\nu)} \sum_{\gamma \vdash n} |C_\gamma| f(\gamma) \chi_\nu(\gamma) \right) \chi_\nu(\lambda) \chi_\nu(\mu).$$

In order to prove this theorem we will use the following lemma, by which a complete orthogonal set of eigenvectors and eigenvalues of $U(t)$ can be obtained.

Lemma 1. *Assume $H[g, h] = f(g^{-1}h)$ for all $g, h \in S_n$, where f is a class function on S_n. Define vectors $\psi_{\nu,i,j} : S_n \to \mathbb{C}$ for each $\nu \vdash n$, $1 \leq i, j \leq \dim(\rho_\nu)$ by $\psi_{\nu,i,j}[g] = \rho_\nu(g)[i, j]$ for all $g \in S_n$. Then each $\psi_{\nu,i,j}$ is an eigenvector of H with associated eigenvalue $\frac{1}{\dim(\rho_\nu)} \sum_{\gamma \vdash n} |C_\gamma| f(\gamma) \chi_\nu(\gamma)$. Moreover, these eigenvectors are pairwise orthogonal and span the space $\mathbb{C}^{S_n}$.*

Remark. The fact described in Lemma 1 is not new—for instance, it is discussed in Section 3E of [10] for general finite groups. A short proof of the lemma follows.

Proof of Lemma 1. For each $g \in S_n$ we have

$$(H\psi_{\nu,i,j})[g] = \sum_{h \in G} f(g^{-1}h)\rho_\nu(h)[i, j] = \sum_{h \in G} f(h)\rho_\nu(gh)[i, j].$$

Now, since ρ_ν is a homomorphism, we have $\rho_\nu(gh) = \rho_\nu(g)\rho_\nu(h)$, which implies

$$(H\psi_{\nu,i,j})[g] = \sum_{k=1}^{\dim(\rho_\nu)} \rho_\nu(g)[i, k] \left(\sum_{h \in S_n} f(h)\rho_\nu(h) \right) [k, j]$$

$$= \sum_{k=1}^{\dim(\rho_\nu)} \rho_\nu(g)[i, k] \hat{f}(\rho_\nu)[k, j].$$

By Fact 1 we see that

$$(H\psi_{\nu,i,j})[g] = \frac{1}{\dim(\rho_\nu)} \sum_{h \in S_n} f(h)\chi_\nu(h)\rho_\nu(g)[i,j]$$

$$= \left(\frac{1}{\dim(\rho_\nu)} \sum_{\gamma \vdash n} |C_\gamma| f(\gamma)\chi_\nu(\gamma) \right) \psi_{\nu,i,j}[g].$$

This establishes that the vectors $\psi_{\nu,i,j}$ are eigenvectors with associated eigenvalues as claimed. The fact that these eigenvectors are pairwise orthogonal follows from Fact 2 and the fact that they span the entire space $\mathbb{C}^{S_n}$ follows from this orthogonality along with Fact 3. ∎

Proof of Theorem 5. By Lemma 1 we may write

$$H = \sum_{\nu,j,k} \left(\frac{1}{\dim(\rho_\nu)} \sum_\gamma |C_\gamma| f(\gamma)\chi_\nu(\gamma) \right) \frac{\psi_{\nu,j,k}\psi^*_{\nu,j,k}}{\|\psi_{\nu,j,k}\|^2}$$

and therefore

$$U(t) = \sum_{\nu,j,k} \exp \left(\frac{it}{\dim(\rho_\nu)} \sum_\gamma |C_\gamma| f(\gamma)\chi_\nu(\gamma) \right) \frac{\psi_{\nu,j,k}\psi^*_{\nu,j,k}}{\|\psi_{\nu,j,k}\|^2}$$

Let $X_\lambda : S_n \to \mathbb{C}$ denote the characteristic function of C_λ for $\lambda \vdash n$. Then we have that

$$c^*_\lambda \psi_{\nu,j,k} = \frac{1}{\sqrt{|C_\lambda|}} \hat{X}_\lambda(\rho_\nu)[j,k] = \begin{cases} \frac{\sqrt{|C_\lambda|}\chi_\nu(\lambda)}{\dim(\rho_\nu)} & \text{if } j = k \\ 0 & \text{otherwise} \end{cases}$$

by Fact 1. By Fact 2 we have $\|\psi_{\nu,j,k}\|^2 = \frac{n!}{\dim(\rho_\nu)}$. So,

$$c^*_\lambda U(t)c_\mu$$

$$= \frac{1}{n!} \sum_\nu \exp \left(\frac{it}{\dim(\rho_\nu)} \sum_{\gamma \vdash n} |C_\gamma| f(\gamma)\chi_\nu(\gamma) \right) \dim(\rho_\nu) \sum_{j,k=1}^{\dim(\rho_\nu)} c^*_\lambda \psi_{\nu,j,k} \, \psi^*_{\nu,j,k} c_\mu$$

$$= \frac{\sqrt{|C_\lambda|}\sqrt{|C_\mu|}}{n!} \sum_{\nu \vdash n} \exp \left(\frac{it}{\dim(\rho_\nu)} \sum_{\gamma \vdash n} |C_\gamma| f(\gamma)\chi_\nu(\gamma) \right) \chi_\nu(\lambda)\chi_\nu(\mu),$$

which is what we wanted to show. ∎

Theorem 5 implies the following interesting fact.

Proposition 6 *Any continuous-time quantum walk on the Cayley graph of the symmetric group for which the generators form conjugacy classes is periodic, with period $2\pi/k$ for some $k \in \{1, 2, 3, \ldots\}$.*

Proof. Using Fact 1 we see that the quantity $|C_\gamma|\chi_\nu(\gamma)/\dim(\rho_\nu)$ is a sum of matrix elements of irreducible representations. This quantity is independent of the particular choice of the basis for the irreducible representations, so we may choose that basis that corresponds to Young's natural form, in which all of the matrix entries are integer valued, implying that the quantity itself is integer valued. Using Fact 3 and Theorem 5 therefore have that $U(2\pi) = U(0) = I$. Thus the period of the walk must divide 2π. ∎

We have not discussed mixing times in this paper, but the previous proposition implies that quantum walks on Cayley graphs of S_n reach their limiting distribution quickly, and when calculating the limiting distribution it is only necessary to average over times in the range $[0, 2\pi]$. Note that in terms of implementation, this does not mean that the walk mixes in constant time; some number of operations that is polynomial in the degree of the graph and in some accuracy parameter is required to implement such a walk, assuming the ability to compute the neighbors of each vertex. See [2,8] for further details.

3.2 Cayley Graphs of S_n Generated by Transpositions

For the Cayley graph of S_n generated by the transpositions, Theorem 5 has various implications that we discuss in this section. We will require explicit values for various characters of the symmetric group, which we now mention. Using the Murnaghan-Nakayama rule it can be shown that

$$\chi_\nu((n)) = \begin{cases} (-1)^{n-k} & \text{for } \nu = (k, 1, \ldots, 1), \ k \in \{1, \ldots, n\} \\ 0 & \text{otherwise} \end{cases}$$

and $\chi_{(k,1,\ldots,1)}(\mathrm{id}) = \dim(\rho_{(k,1,\ldots,1)}) = \binom{n-1}{k-1}$. For the characters at the transpositions, it is known [13] that

$$\chi_\nu(\tau) = \frac{\dim(\rho_\nu)}{\binom{n}{2}} \sum_j \left(\binom{\nu_j}{2} - \binom{\nu'_j}{2} \right).$$

Here, τ is any transposition, ν' is the partition generated by transposing the Young diagram of ν, while ν_j and ν'_j are the jth components of the partitions ν and ν'. Substituting these values into Theorem 5 gives

$$c_\lambda^* U(t) c_\mu = \frac{\sqrt{|C_\lambda|}\sqrt{|C_\mu|}}{n!} \sum_{\nu \vdash n} \exp\left(it \sum_j \left(\binom{\nu_j}{2} - \binom{\nu'_j}{2} \right) \right) \chi_\nu(\lambda)\chi_\nu(\mu)$$

for the quantum walk on $\Gamma(S_n, C_{(2,1,\ldots,1)})$, and specifically for the case where $\mu = (1, \ldots, 1)$ and $\lambda = (n)$ it follows that

$$c_{(n)}^* U(t) c_{(1,\ldots,1)}$$

$$= \frac{1}{\sqrt{n \cdot n!}} \sum_{k=1}^n \exp\left(it \left(\binom{k}{2} - \binom{n-k+1}{2} \right) \right) (-1)^{n-k} \binom{n-1}{k-1}$$

$$= \frac{(2i \sin(tn/2))^{n-1}}{\sqrt{n \cdot n!}}.$$

In particular,

$$\max_t \left| c^*_{(n)} U(t) c_{(1,\ldots,1)} \right|^2 = \frac{2^{2n-2}}{n \cdot n!}, \tag{1}$$

where the maximum occurs for $t = (2k+1)\pi/n$, $k \in \mathbb{Z}$.

Eq. 1 has the following interpretation. If we start a quantum walk on $\Gamma(S_n, C_{(2,1,\ldots,1)})$ at the identity element and evolve for any amount of time and measure, the probability to measure some n-cycle is at most $\frac{2^{2n-2}}{n \cdot n!}$ as opposed to probability approaching $\frac{1}{n}$ for the classical case. The probability to measure any particular n-cycle is therefore at most $\frac{2^{2n-2}}{(n!)^2}$, as opposed to some number approaching $\frac{1}{n!}$ classically. The probabilities in the quantum case are smaller by a factor that is exponential in n.

As discussed in Section 2.1, we will denote by P_t the distribution on S_n obtained by performing the quantum walk on $\Gamma(S_n, C_{(2,1,\ldots,1)})$ for time t starting at the identity then measuring. The above analysis gives a lower bound for the total variation distance of P_t from the uniform distribution:

$$\|P_t - \text{uniform}\| \geq \frac{1}{n} - \frac{2^{2n-2}}{n \cdot n!}$$

for all values of t. This bound follows from considering only the n-cycles, and we believe the true bound to be much larger. Numerical simulations support this claim, but thus far we only have exact expressions for the n-cycles.

Given that we have an exact expression for the probability $P_t[g]$ for any n-cycle g, it is easy to determine the probability associated with any n-cycle in the limiting distribution. By the periodicity of our walks, we have

$$\bar{P}[g] = \frac{1}{2\pi} \int_0^{2\pi} P_t[g] \, dt$$

for each $g \in S_n$, and thus for any $g \in C_{(n)}$ we have $\bar{P}[g] = \frac{\binom{2n-2}{n-1}}{(n!)^2}$. Somewhat surprisingly, this average probability associated with reaching a given n-cycle is not unique to the particular choice of $C_{(2,1,\ldots,1)}$ as a generating set, as shown in the next subsection.

3.3   Other Generating Sets

We have not been able to obtain tractable expressions for the amplitudes associated with quantum walks for other generating sets besides $C_{(2,1,\ldots,1)}$. However, we can prove some facts concerning the limiting distributions for such walks in the case that the generating set consists of all p-cycles for any choice of p. (In case p is odd, we must keep in mind that only the alternating group is being generated.) Again we will focus on the probability of reaching n-cycles starting from the identity.

Consider the quantum walk on $\Gamma(S_n, C_\gamma)$, where γ is any partition. According to Theorem 5, the probability associated with a given conjugacy class C_λ when starting from a uniform superposition on another class C_μ after time t is given by $|c_\lambda^* U(t) c_\mu|^2$, which may be written as

$$\frac{|C_\lambda||C_\mu|}{(n!)^2} \sum_{\nu,\eta} \exp\left(it\,|C_\gamma|\left(\frac{\chi_\nu(\gamma)}{\dim(\rho_\nu)} - \frac{\chi_\eta(\gamma)}{\dim(\rho_\eta)}\right)\right) \chi_\nu(\lambda)\chi_\nu(\mu)\chi_\eta(\lambda)\chi_\eta(\mu).$$

As before, we let $\bar{P}$ denote the limiting distribution of the walk when starting from the identity. Since our walks are periodic with period 2π, we therefore have

$$\bar{P}[g] = \frac{1}{(n!)^2} \sum_{\nu,\eta}^{*} \chi_\nu(g)\dim(\rho_\nu)\chi_\eta(g)\dim(\rho_\eta).$$

Here the asterisk denotes that the sum is over all partitions ν, η subject to the condition

$$\frac{\chi_\nu(\gamma)}{\dim(\rho_\nu)} = \frac{\chi_\eta(\gamma)}{\dim(\rho_\eta)}. \tag{2}$$

Observe that the choice of generators only affects the average distribution by determining what values other than $\nu = \eta$ are included in the sum. More generally, the average probability associated with obtaining some element in C_λ when starting the walk on the uniform superposition over C_μ is given by

$$\frac{|C_\lambda||C_\mu|}{(n!)^2} \sum_{\nu,\eta}^{*} \chi_\nu(\lambda)\chi_\nu(\mu)\chi_\eta(\lambda)\chi_\eta(\mu).$$

In the case that g is an n-cycle and $\gamma = (p, 1, \ldots, 1)$, the condition of Eq. 2 is relatively easy to characterize for those partitions ν and η for which $\chi_\nu(g)\chi_\eta(g) \neq 0$. Figure 1 summarizes the probability associated with each n-cycle g in the limiting distribution for the quantum walk on $\Gamma(S_n, C_\gamma)$. Due to space constraints, the derivation of these probabilities has been omitted. (See http://arxiv.org/abs/quant-ph/0305182 for a longer version of this paper containing these details.)

We have the following lower bounds on the total variation distance of the limiting distribution from the uniform distribution. As for the case of the quantum walk generated by the transpositions, this bound follows just from considering the n-cycles, and we believe the true distance from uniform to be much larger.

- Let $p \in \{2, \ldots, n\}$ be even, let $\gamma = (p, 1, \ldots, 1) \vdash n$ and let $\bar{P}$ denote the limiting distribution of the quantum walk on $\Gamma(S_n, C_\gamma)$. Then

$$\|\bar{P} - \mathrm{uniform}(S_n)\| \geq \frac{1}{n} - \frac{1}{n \cdot n!}\binom{2n-2}{n-1}.$$

- Let n be odd, let $p \in \{2, \ldots, n\}$ be odd, let $\gamma = (p, 1, \ldots, 1) \vdash n$ and let $\bar{P}$ denote the limiting distribution of the quantum walk on $\Gamma(S_n, C_\gamma)$. Then

$$\|\bar{P} - \mathrm{uniform}(A_n)\| \geq \frac{2}{n} - \frac{2}{n \cdot n!}\binom{2n-2}{n-1} + \frac{1}{n \cdot n!}\binom{n-1}{\frac{n-1}{2}}^2.$$

Parity of n	Parity of p	Range of p	Probability at each n-cycle
even or odd	even	$2 \leq p \leq \lceil \frac{n}{2} \rceil$	$\frac{1}{(n!)^2} \binom{2n-2}{n-1}$
even	even	$\frac{n}{2} + 1 \leq p \leq n-1$	$\frac{2}{(n!)^2} \sum_{k=1}^{n-p} \binom{n-1}{k-1}^2$
odd	even	$\frac{n+1}{2} + 1 \leq p \leq n-1$	$\frac{2}{(n!)^2} \sum_{k=1}^{n-p} \binom{n-1}{k-1}^2 + \frac{4}{(n!)^2} \binom{n-2}{p-1}^2$
even	even	$p = n$	$\frac{1}{(n!)^2} \binom{2n-2}{n-1}$
even	odd	—	0
odd	odd	$2 \leq p \leq \frac{n+1}{2}$	$\frac{2}{(n!)^2} \binom{2n-2}{n-1} - \frac{1}{(n!)^2} \binom{n-1}{\frac{n-1}{2}}^2$
odd	odd	$\frac{n+1}{2} + 1 \leq p \leq n-1$	$\frac{4}{(n!)^2} \sum_{k=1}^{n-p} \binom{n-1}{k-1}^2 + \frac{4}{(n!)^2} \binom{n-2}{p-1}^2$
odd	odd	$p = n$	$\frac{2}{(n!)^2} \binom{2n-2}{n-1} - \frac{1}{(n!)^2} \binom{n-1}{\frac{n-1}{2}}^2$

Fig. 1. Probabilities associated with each n-cycle in the limiting distribution for $\Gamma(S_n, C_{(p,1,\ldots,1)})$.

4   Conclusion

In this paper we have studied some of the properties of continuous-time quantum walks on Cayley graphs of the symmetric group. Many questions concerning these walks remain unanswered. One obvious question that we have not attempted to address in this paper is whether quantum walks on the symmetric group can be applied in the context of quantum algorithms. In terms of specific properties of these walks, we have focused on the limiting distribution—is the limiting distribution bounded away from uniform by a constant? Many other properties of these walks may be of interest as well. For instance, the effect of decoherence on these walks is an interesting topic to consider.

References

1. D. Aharonov, A. Ambainis, J. Kempe, and U. Vazirani. Quantum walks on graphs. In *Proceedings of the Thirty-Third Annual ACM Symposium on Theory of Computing*, pages 50–59, 2001.
2. D. Aharonov and A. Ta Shma. Adiabatic quantum state generation and statistical zero knowledge. In *Proceedings of the Thirty-Fifth Annual ACM Symposium on Theory of Computing*, 2003.
3. A. Ahmadi, R. Belk, C. Tamon, and C. Wendler. Mixing in continuous quantum walks on graphs. arXiv.org e-Print quant-ph/0209106, 2002.
4. D. Aldous and J. Fill. Reversible markov chains and random walks on graphs, 2002. See http://stat-www.berkeley.edu/users/aldous/RWG/book.html.
5. A. Ambainis, E. Bach, A. Nayak, A. Vishwanath, and J. Watrous. One-dimensional quantum walks. In *Proceedings of the Thirty-Third Annual ACM Symposium on Theory of Computing*, pages 60–69, 2001.

6. E. Bach, S. Coppersmith, M. Goldschen, R. Joynt, and J. Watrous. One-dimensional quantum walks with absorbing boundaries. arXiv.org e-Print quant-ph/0207008, 2002.

7. C. H. Bennett and G. Brassard. Quantum cryptography: Public key distribution and coin tossing. In *Proceedings of the IEEE International Conference on Computers, Systems, and Signal Processing*, pages 175–179, 1984.

8. A. Childs, R. Cleve, E. Deotto, E. Farhi, S. Gutmann, and D. Spielman. Exponential algorithmic speedup by quantum walk. In *Proceedings of the Thirty-Fifth Annual ACM Symposium on Theory of Computing*, 2003.

9. A. Childs, E. Farhi, and S. Gutmann. An example of the difference between quantum and classical random walks. *Quantum Information Processing*, 1(35), 2002.

10. P. Diaconis. *Group Representations in Probability and Statistics*. Lecture Notes-Monograph Series. Institute of Mathematical Statistics, 1988.

11. E. Farhi and S. Gutmann. Quantum computation and decision trees. *Physical Review A*, 58:915–928, 1998.

12. M. Horodecki, P. Horodecki, and R. Horodecki. Mixed-state entanglement and quantum communication. In *Quantum Information: An Introduction to Basic Theoretical Concepts and Experiments*, volume 173 of *Springer Tracts in Modern Physics*. Springer-Verlag, 2001.

13. R. Ingram. Some characters of the symmetric group. In *Proc. Amer. Math. Soc.*, volume 1, pages 358–369, 1950.

14. G. James and A. Kerber. *The Representation Theory of the Symmetric Group*. Addison-Wesley, 1981.

15. J. Kempe. Quantum random walks hit exponentially faster. arXiv.org e-Print quant-ph/0205083, 2002.

16. J. Kempe. Quantum random walks – an introductory overview. *Contemporary Physics*, 2003. To appear. See also arXiv.org e-Print quant-ph/0303081.

17. I. Kerenidis and R. de Wolf. Exponential lower bound for 2-query locally decodable codes via a quantum argument. In *Proceedings of the Thirty-Fifth Annual ACM Symposium on Theory of Computing*, 2003.

18. T. Mackay, S. Bartlett, L. Stephenson, and B. Sanders. Quantum walks in higher dimensions. arXiv.org e-Print quant-ph/0108004, 2001.

19. C. Moore and A. Russell. Quantum walks on the hypercube. In *Proceedings of the Sixth International Workshop on Randomization and Approximation Techniques in Computer Science (RANDOM)*, 2002.

20. P. Shor. Polynomial-time algorithms for prime factorization and discrete logarithms on a quantum computer. *SIAM Journal on Computing*, 26(5):1484–1509, 1997.

21. P. Shor and J. Preskill. Simple proof of security of the BB84 quantum key distribution protocol. arXiv.org e-Print quant-ph/0003004, 2000.

22. S. Sternberg. *Group Theory and Physics*. Cambridge University Press, 1994.

23. T. Yamasaki, H. Kobayashi, and H. Imai. Analysis of absorbing times of quantum walks. arXiv.org e-Print quant-ph/0205045, 2002.

Distribution-Free Property Testing

Shirley Halevy[1] and Eyal Kushilevitz[2]

[1] Department of Computer Science, Technion, Haifa 3200, Israel.
`shirleyh@cs.technion.ac.il`.
[2] Department of Computer Science, Technion, Haifa 3200, Israel.
`eyalk@cs.technion.ac.il`.

Abstract. We consider the problem of distribution-free property testing of functions. In this setting of property testing, the distance between functions is measured with respect to a *fixed but unknown* distribution D on the domain, and the testing algorithms have an oracle access to random sampling from the domain according to this distribution D. This notion of distribution-free testing was previously defined, but no distribution-free property testing algorithm was known for any (nontrivial) property. By extending known results (from "standard", uniform distribution property testing), we present the first such distribution-free algorithms for two of the central problems in this field:

  - A distribution-free testing algorithm for low-degree multivariate polynomials with query complexity $O(d^2 + d \cdot \epsilon^{-1})$, where d is the total degree of the polynomial.
  - A distribution-free monotonicity testing algorithm for functions $f : [n]^d \to A$ for low-dimensions (e.g., when d is a constant) with query complexity $O(\frac{\log^d n \cdot 2^d}{\epsilon})$.

The same approach that is taken for the distribution-free testing of low-degree polynomials is shown to apply also to several other problems.

1   Introduction

The classical notion of *decision problems* requires an algorithm to distinguish objects having some property $\mathcal{P}$ from those objects which do not have the property. *Property testing* is a recently-introduced relaxation of decision problems, where algorithms are only required to distinguish objects having the property $\mathcal{P}$ from those which are at least "ϵ-far" from every such object. The notion of property testing was introduced by Rubinfeld and Sudan [35] and since then attracted a considerable amount of attention. Property testing algorithms (or *property testers*) were introduced for problems in graph theory (e.g. [2, 23, 24, 30]), monotonicity testing (e.g. [9, 13, 14, 18, 19, 22]) and other properties (e.g. [1, 3, 4, 5, 7, 10, 12, 15, 17, 20, 27, 28, 29, 31, 32, 34]; the reader is referred to excellent surveys by Ron [33], Goldreich [21], and Fischer [16] for a presentation of some of this work, including some connections between property testing and other topics). The main goal of property testers is to avoid "reading" the whole object (which requires complexity at least linear in the size of its representation); i.e., to make the decision by reading a small (possibly, selected at

S. Arora et al. (Eds.): APPROX 2003+RANDOM 2003, LNCS 2764, pp. 302–317, 2003.

random) fraction of the input (e.g., a fraction of size polynomial in $1/\epsilon$ and poly-logarithmic in the size of the representation) and still having a good (say, at least 2/3) probability of success.

A crucial component in the definition of property testing is that of the *distance* between two objects. For the purpose of this definition, it is common to think of objects as being *functions* over some domain $\mathcal{X}$. For example, a graph G may be thought of as a function $f_G : V \times V \rightarrow \{0, 1\}$ indicating for each edge e whether it exists in the graph. The distance between functions f and g is then measured by considering the set $\mathcal{X}_{f \neq g}$ of all points x where $f(x) \neq g(x)$ and comparing the size of this set $\mathcal{X}_{f \neq g}$ to that of $\mathcal{X}$; equivalently, one may introduce a uniform distribution over $\mathcal{X}$ and measure the probability of picking $x \in \mathcal{X}_{f \neq g}$. Note that property testers access the input function (object) via *membership queries* (i.e., the algorithm gives a value x and gets $f(x)$).

It is natural to generalize the above definition of distance between two functions, to deal with arbitrary probability distributions D over $\mathcal{X}$, by measuring the probability of $\mathcal{X}_{f \neq g}$ according to D. Ideally, one would hope to get *distribution-free* property testers. A distribution-free tester for a given property $\mathcal{P}$ accesses the function using membership queries, as above, and by randomly sampling the **fixed but unknown** distribution D (this mimics similar definitions from *learning theory* and is implemented via an oracle access to D; see, e.g., [26] [3]). As before, the tester is required to accept the given function f with probability at least $\frac{2}{3}$ if f satisfies the property $\mathcal{P}$, and to reject it with probability at least $\frac{2}{3}$ if f is at least ϵ-far from $\mathcal{P}$ with respect to the distribution D.

Indeed, these definitions of distance with respect to an arbitrary distribution D and of distribution-free testing were already considered in the context of property testing [23]. However, to the best of our knowledge, no distribution-free property tester was known for any (non-trivial) property (besides testing algorithms that follow from the existence of proper learning algorithms in learning-theory [23]). Moreover, discouraging impossibility results, due to [23], show that for many graph-theoretic properties (for which testers that work with respect to the uniform distribution are known) no such (efficient) distribution-free tester exists. As a result, most previous work focused on testers for the uniform distribution; some of these algorithms can be generalized to deal with certain (quite limited) classes of distributions (e.g., product distributions [23]), and very few can be modified to be testers with respect to any **known** distribution (as was observed by [16] regarding the tester presented in [28]), but none is shown to be a distribution-free tester. Let us review some of the central problems, studied in the context of property testing, which are relevant to the current work.

Low-degree tests for polynomials. The first problem studied in the field of property testing was that of low-degree testing for multivariate polynomials over a

---

[3]   More precisely, distribution-free property testing is the analogue of the PAC+MQ model of learning (that was studied by the learning-theory community mainly via the EQ+MQ model); standard property testing is the analogue of the uniform+MQ model.

finite field, where one wishes to test whether a given function can be represented by a multivariate polynomial of total degree d, or is it ϵ-far from any such polynomial. Later, the problem of low-degree testing played a central role in the development of probabilistic checkable proofs (PCP), where the goal is to probabilistically verify the validity of a given proof. For the problem of low-degree testing, Rubinfeld and Sudan [35] presented a tester with query complexity of $O(d^2 + d \cdot \epsilon^{-1})$. This test was further analyzed in [8]. The reader is also referred to [10], where a linearity test (which tests whether a given function acts as an homomorphism between groups) is presented, and to [3, 6, 7, 20] for other related work.

Monotonicity testing. Monotonicity has also been a subject of a significant amount of work in the property testing literature (e.g. [9, 13, 14, 15, 18, 19, 22]). In monotonicity testing, the domain $\mathcal{X}$ is usually the d-dimensional cube $[n]^d$. A partial order is defined on this domain in the natural way (for $\boldsymbol{y}, \boldsymbol{z} \in [n]^d$, we say that $\boldsymbol{y} \leq \boldsymbol{z}$ if each coordinate of $\boldsymbol{y}$ is bounded by the corresponding coordinate of $\boldsymbol{z}$).[4] A function f over the domain $[n]^d$ is *monotone* if whenever $\boldsymbol{z} \geq \boldsymbol{y}$ then $f(\boldsymbol{z}) \geq f(\boldsymbol{y})$. Testers were developed to deal with both the low-dimensional and the high-dimensional cases (with respect to the uniform distribution over the domain). In what follows, we survey some of the known results on this problem. In the low-dimensional case, d is considered to be small compared to n (and, in fact, it is typically a constant); a successful algorithm for this case is typically one that is polynomial in $1/\epsilon$ and in $\log n$. The first paper to deal with this case is by Ergün et al. [14] which presented an $O(\frac{\log n}{\epsilon})$ algorithm for the line (i.e., the case $d = 1$), and showed that this query complexity cannot be achieved without using membership queries. This algorithm was generalized for any fixed d in [9]. For the case $d = 1$, there is a lower bound showing that testing monotonicity (for some constant ϵ) indeed requires $\Omega(\log n)$ queries [15]. In the high dimensional case, d is considered as the main parameter (and n might be as low as 2); a successful algorithm is typically one that is polynomial in $1/\epsilon$ and d. This case was first considered by Goldreich et al. [22] that showed an algorithm for testing monotonicity of functions over the boolean ($n = 2$) d-dimensional hyper-cube to a boolean range using $O(\frac{d}{\epsilon})$ queries. This result was generalized in [13] to arbitrary values of n, showing that $O(\frac{d \cdot \log^2 n}{\epsilon})$ queries suffice for testing monotonicity of general functions over $[n]^d$, which is the best known result so far.

1.1   Our Contributions

Our contributions are distribution-free testers for the two properties mentioned above: low-degree multivariate polynomials and low-dimensional monotone functions. We observe that the approach that stands behind the low-degree test can also be applied to the testing of other properties such as dictatorship and juntas functions [17, 32]. These algorithms are the first known distribution-free testers

---

[4] In the case $d = 1$ this yields a linear order.

for non-trivial properties. By this, we answer a natural question that has already been raised explicitly by Fischer [16, Subsection 9.3] and is implicit in [23]. We emphasize that our algorithms work for any distribution D without having any information about D.

Distribution-free low-degree testing for polynomials (and more). We show how to generalize the tester presented in [35] to a distribution-free tester with the same (up to a multiplicative constant factor of 2) query complexity ($O(d^2 + d \cdot \epsilon^{-1})$). The algorithm and its analysis are presented in Section 3.

The generalization of the uniform tester to a distribution-free one is done, in this case, by adding another stage to the uniform tester. In this new stage, after verifying that the input function f is close to some low-degree polynomial g with respect to the uniform distribution, we check that f is also close to this specific polynomial g with respect to the given distribution D. For this purpose, our approach requires that we will be able to calculate the values of g efficiently based on the values on f. This is a generalization of the notion of self-correctors for single functions (see [10]) to classes of functions (which was previously introduced in [35]). We observe that the same approach can be used for distribution-free testing of every property that is testable in the uniform distribution and has a self-corrector in the above sense. The full details of this generalization appear in Section 4.

Distribution-free monotonicity testing. We present a distribution-free monotonicity tester in the low-dimensional hyper-cube case. Specifically, we present an algorithm whose complexity is $O(\frac{\log^d n \cdot 2^d}{\epsilon})$ queries. This is done by first considering the one-dimensional case (the "line"). In this case, we prove that an algorithm of [14] can be slightly modified to deal with the distribution-free case with the same query complexity of $O(\frac{\log n}{\epsilon})$. Though it is possible to modify the original analysis for the distribution-free case, we choose to present a whole different analysis. We then show how to appropriately generalize this algorithm to deal with higher (yet, low) dimensions (a similar generalization approach was used in [9] for the uniform distribution case). The tester for the one-dimensional case and its generalization for higher dimensions appear in Section 5. Finally, we remark that it can be shown that distribution-free testing of monotonicity in the high-dimensional case cannot be done efficiently [11].

It is typical for known property testers to be quite simple and the analysis of why these algorithms work is where the property $\mathcal{P}$ in question requires understanding; indeed, Goldreich and Trevisan [25] proved that in certain settings this is an inherent phenomena: they essentially showed (with respect to the uniform distribution) that any graph-theoretic property that can be tested can also be tested (with a small penalty in the complexity) by a "generic" algorithm that samples a random subgraph and decides whether it has some property. Our work is no different in this aspect: our algorithms are similar to previously known algorithms and the main contribution is their analysis; in particular, that for the distribution-free case. Moreover, it is somewhat surprising that our distribution-free testers require no dramatically-different techniques than those used in the construction and the analysis of previous algorithms (that work for the uniform distribution case). We remark, however, that although all the distribution-free testers presented in this work can be viewed as variations of testers for the uniform distribution, the modifications of the uniform-distribution testers in the various problems are different.[5]

2   Definitions

In this section, we formally define the notion of being ϵ-far from a property $\mathcal{P}$ with respect to a given distribution D defined over $\mathcal{X}$, and of distribution-free testing. Assume that the range of the functions in question is $\mathcal{A}$.

Definition 1. *Let D and $\mathcal{X}$ be as above. The D-distance between functions $f, g : \mathcal{X} \to \mathcal{A}$ is defined by $dist_D(f, g) \overset{def}{=} \Pr_{x \sim D}\{f(x) \neq g(x)\}$.*
The D-distance of a function f from a property $\mathcal{P}$ (i.e., the class of functions satisfying the property $\mathcal{P}$) is $dist_D(f, \mathcal{P}) \overset{def}{=} \min_{g \in \mathcal{P}} dist_D(f, g)$.
We say that f is (ϵ, D)-far from a property $\mathcal{P}$ if $dist_D(f, \mathcal{P}) \geq \epsilon$.

When the distribution in question is the uniform distribution over $\mathcal{X}$, we either use U instead of D or (if clear from the context) we omit any reference to the distribution.

Next, we define the notion of distribution-free tester for a given property $\mathcal{P}$.

Definition 2. *A distribution-free tester for a property $\mathcal{P}$ is a probabilistic oracle machine M, which is given a distance parameter $\epsilon > 0$, and an oracle access to an arbitrary function $f : \mathcal{X} \to \mathcal{A}$ and to sampling of a fixed but unknown distribution D over $\mathcal{X}$, and satisfies the following two conditions:*
1. If f satisfies $\mathcal{P}$, then $\Pr\{M^{f,D} = Accept\} = 1$.
2. If f is (ϵ, D)-far from $\mathcal{P}$, then $\Pr\{M^{f,D} = Accept\} \leq \frac{1}{3}$.

We note that a more general definition of testers that allows two-sided errors (as discussed in the introduction) is not needed here; all our testers, like many previously known testers, have one-sided error and always accept any function that satisfies the property $\mathcal{P}$ in question.

The definition of a uniform distribution tester for a property $\mathcal{P}$ can be derived from the previous definition by omitting the sampling oracle (since the tester can sample in the uniform distribution by itself) and by measuring the distance with respect to the uniform distribution.

Notice that since the distribution D in question is arbitrary, it is possible that there are two different functions f and g such that $dist_D(f, g) = 0$. Specifically, it is possible that $f \notin \mathcal{P}$ and $g \in \mathcal{P}$. Since the notion of testing is meant to be a relaxation of the notion of decision problems, it is required that the algorithm accepts (with high probability) functions that satisfy $\mathcal{P}$, but may reject functions that have distance 0 from $\mathcal{P}$ (but do not satisfy $\mathcal{P}$). This definition

---

[5] Indeed, in light of [23], there can be no generic transformation of uniform-distribution testers into distribution-free ones.

of distribution-free testing was introduced in [23, Definition 2.1]. In addition, note that the algorithm is allowed to query the value of the input function also in points with probability 0 (which is also the case with membership queries in learning theory)[6].

3   Distribution-Free Low-Degree Testers for Polynomials

The first problem studied in the field of property testing was that of testing of multivariate polynomials (see [3, 6, 7, 10, 20, 35]). Let F be a finite field. In the problem of low-degree testing, with respect to the uniform distribution, the tester is given access to a function $f : F^m \rightarrow F$, a distance parameter ϵ, and a degree d, and has to decide whether f is a multivariate polynomial of total degree d, or is at least ϵ-far (with respect to the uniform distribution) from any degree d multivariate polynomial (i.e., one has to change the values of at least $\epsilon \times |F|^m$ points in order to transform f into a degree d multivariate polynomial; this implies that, for every degree d multivariate polynomial g, the probability that a *uniformly* drawn point x has a value $g(x)$ different than $f(x)$, is at least ϵ). Rubinfeld and Sudan ([35]) presented a tester for this problem with query complexity $O(d^2 + d \cdot \epsilon^{-1})$. We show how to modify this tester to a distribution-free tester with the same query complexity (up to a constant factor of 2).

3.1   Preliminaries

Fix some value for d and assume from now on that $|F| > 10d$. To describe the testers (both the one for the uniform distribution and our distribution-free one), we use the following terminology, from [35]:

A *line* in F^m is a set of $10d + 1$ points of the form $\{x, x + h, \ldots, x + 10dh\}$ for some $x, h \in F^m$. The line defined by x and h is denote $\ell_{x,h}$.

We say that a line $\ell_{x,h}$ *is an f-polynomial*, if there exists a univariate polynomial $P_{x,h}(i)$ of degree d, such that $f(x + ih) = P_{x,h}(i)$, for every $0 \le i \le 10d$.

Notice that if f is a multivariate polynomial of total degree at most d, then for every x and h, the line $\ell_{x,h}$ is an f-polynomial [7]. Given the values of f on a line $\ell_{x,h}$, testing whether this line is an f-polynomial can be done as follows:

- find, using interpolation, a univariate polynomial $P(i)$ of degree d, consistent with the values of f at the $d+1$ points $x, x+h, \ldots, x+dh$ (i.e., $P(i) = f(x+hi)$ for every $0 \le i \le d$).

---

[6] It is not known whether MQ are essential in general for testing even in the uniform case (see [33]); this is known only for specific problems such as monotonicity testing (see [14]).

[7] To see that, assume $f(x) = \sum_j a_j \prod_{l=1}^{d_j} x_{k_l^j}$, where a_j is the coefficient of the j'th term in f, d_j is the degree ($d_j \le d$), and k_l^j is the index of the l'th variable in that term (note that $k_{l_1}^j$ is not necessarily different than $k_{l_2}^j$ for $l_1 \neq l_2$). In this case, for every fixed $x = (x_1, \ldots, x_m)$ and $h = (h_1, \ldots, h_m)$ the value $f(x + ih) = \sum_j a_j \prod_{l=1}^{d_j} (x_{k_l^j} + ih_{k_l^j})$, which, of course, is a degree d univariate polynomial in i.

- check, for every $(d + 1) \leq i \leq 10d$, that $f(x + ih) = P(i)$. If so accept; otherwise reject.

We show how this basic test is used to build a uniform and a distribution-free low-degree test.

3.2   Low-Degree Test for the Uniform Distribution

The low-degree test for the uniform distribution is done by randomly sampling $O(d + \epsilon^{-1})$ lines (i.e., by uniformly choosing $x, h \in F^m$), and checking that each of these lines is an f-polynomial. The correctness of this algorithm follows immediately from the following theorem ([35, Theorem 9]).

Theorem 1. *There exists a constant c_U such that for $0 \leq \delta \leq \frac{1}{c_U \cdot d}$, if f is a function from F^m to F, such that all but at most δ fraction of the lines $\{\ell_{x,h} | x, h \in F^m\}$ are f-polynomials, then there exists a polynomial $g : F^m \to F$ of total degree at most d such that $dist_U(f, g) \leq (1 + o(1))\delta$ (provided that $|F| > 10d$).*

3.3   Distribution-Free Low-Degree Tester

Denote the class of multivariate polynomials of total degree d by $\mathcal{P}^d_{deg}$. In this section we show that the tester described in the previous subsection can be modified into a distribution-free tester for low-degree multivariate polynomials. That is, we present an algorithm with query complexity $O(d^2 + d \cdot \epsilon^{-1})$ that, given a distance parameter ϵ, a degree parameter d, and access to random sampling of F^m according to D and to membership queries of a function $f : F^m \to F$, distinguishes, with probability at least $\frac{2}{3}$, between the case that f is in $\mathcal{P}^d_{deg}$, and the case that f is (ϵ, D)-far from $\mathcal{P}^d_{deg}$.

The natural generalization of the uniform-distribution tester above for the distribution-free case would be to replace the sampling of the tested lines by sampling according to the distribution D; i.e. sample the $O(d + \epsilon^{-1})$ lines by choosing $x \sim D$ and $h \sim U$ and check that these lines are f-polynomials. However, we do not know whether this modification actually works. Instead, the algorithm we present consists of two stages – in the first stage we simply run the uniform distribution test as is, and check that the function f is ϵ-close to $\mathcal{P}^d_{deg}$ with respect to the uniform distribution; the second stage is the generalization suggested above. We prove that this combined strategy actually works.

Poly(ϵ, d)

Set $k \stackrel{\text{def}}{=} \max\{\epsilon^{-1}, c_U \cdot d\}$. Repeat $5k$ times:
- Choose $x, h \in_R F^m$. If the line $\ell_{x,h}$ is not an f-polynomial, **return FAIL**.
- Choose $x \in_D F^m, h \in_R F^m$. If the line $\ell_{x,h}$ is not an f-polynomial, **return FAIL**.

return PASS

Theorem 2. *Algorithm Poly(ϵ, d) is a distribution-free tester for $\mathcal{P}_{deg}^d$; its query complexity is $O(d^2 + d \cdot \epsilon^{-1})$.*

The correctness of the algorithm relies on the following lemma:

Lemma 1. *Let c_U be the constant as above. For every $0 \leq \delta \leq \frac{1}{c_U \cdot d}$, if f is a function from F^m to F such that*

- $\Pr_{x,h \sim U}\{\ell_{x,h}$ *is not an f polynomial*$\} \leq \delta$*, and*
- $\Pr_{x \sim D, h \sim U}\{\ell_{x,h}$ *is not an f polynomial*$\} \leq \delta$*,*

then there exists a polynomial $g : F^m \to F$ of total degree at most d such that $dist_D(f,g) \leq \frac{\delta}{1-40\delta} = (1+o(1))\delta$ (provided that $|F| > 10d$).

The proof of the above lemma is omitted for lack of space. The proof is similar to ones presented in [35] and will appear in the full version of the paper.

Proof. **of theorem 2.**
To prove that the algorithm is indeed a distribution-free tester for $\mathcal{P}_{deg}^d$, we prove the following two facts:

1. If f is in $\mathcal{P}_{deg}^d$, then the algorithm accepts f with probability 1.
2. If f is (ϵ, D)-far from $\mathcal{P}_{deg}^d$, then the algorithm $Poly(\epsilon, d)$ rejects f with probability at least $\frac{2}{3}$.

As explained before, if f is indeed a multivariate polynomial of total degree d, then every line is an f-polynomial. Hence, it follows that such f is accepted by the tester with probability 1. Assume from now on that f is (ϵ, D)-far from $\mathcal{P}_{deg}^d$: Notice that, by the definition of k, for $\epsilon' = \frac{1}{k}$, f is (ϵ', D)-far from $\mathcal{P}_{deg}^d$. Based on Lemma 1, either $\Pr_{x,h \sim U}\{\ell_{x,h}$ *is not an f polynomial*$\} > \frac{\epsilon'}{2+40\epsilon'}$, or $\Pr_{x \sim D, h \sim U}\{\ell_{x,h}$ *is not an f polynomial*$\} > \frac{\epsilon'}{2+40\epsilon'}$ (otherwise, it follows that there exists a degree d polynomial g such that $dist_D(f,g) \leq \frac{\epsilon'}{(2+40\epsilon') \cdot (1-40\frac{\epsilon'}{2+40\epsilon'})} = \frac{\epsilon'}{2} < \epsilon'$, contradicting the fact that the D-distance of f from any such polynomial is at least ϵ'). Assume that the first event occurs. Therefore, the probability that a randomly chosen line $\ell_{x,h}$ is an f-polynomial is at most $(1 - \frac{\epsilon'}{2+40\epsilon'})$. Hence, the probability that the algorithm accepts f is at most $(1 - \frac{\epsilon'}{2+40\epsilon'})^{5k} = (1 - \frac{1}{2k+40})^{5k} \leq \frac{1}{e}^2 \leq \frac{1}{3}$ (the first inequality follows since $c_U \geq 100$ [35] implying that $k \geq 100$). Similarly, if the second event occurs, the probability that a randomly chosen line $\ell_{x,h}$, where $x \sim D$ and $h \sim U$, is an f-polynomial is at most $(1 - \frac{\epsilon'}{2+40\epsilon'})$. Hence, as before, the probability that the algorithm accepts f is at most $(1 - \frac{\epsilon'}{2+40\epsilon'})^{5k} \leq \frac{1}{3}$. $\square$

4   Distribution-Free Testing of Properties with Self-corrector

A careful examination and manipulation of the distribution-free tester presented in the previous section shows that, in fact, the only two features of low-degree multivariate polynomials used in the construction are:

- the existence of a one-sided error uniform distribution tester for low-degree polynomials, and
- the ability to efficiently compute (with high probability), in every point x of the domain, the correct value of the polynomial g that is close to the input function f, if f is indeed close to a multivariate low-degree polynomial. We refer to this ability as "property self-correction".

We argue that it is possible to construct a distribution-free tester for every property $\mathcal{P}$ that satisfies these two conditions. We first define the notion of a "property self-correction" formally (it has already been defined implicitly and used in [35]), and then introduce a general scheme for obtaining distribution-free testers for a variety of properties that satisfy the conditions.

The notion of "property self-corrector" is a generalization of the notion of self-correctors for functions introduced by Blum, Luby and Rubinfeld in [10]. A self-corrector for a *specific function* f is a randomized algorithm that given oracle access to a function g which is ϵ-close to f, is able to compute the value of f in every point of the domain. This definition can be generalized to classes of functions, specifically demanding that all the functions in the class are self-correctable using the same algorithm.

Definition 3. *An ϵ self-corrector for a property $\mathcal{P}$ is a probabilistic oracle machine M, which is given an oracle access to an arbitrary function $f : \mathcal{X} \to \mathcal{A}$ and satisfies the following condition:*
If there exists a function $g \in \mathcal{P}$ such that $dist_U(f, g) \leq \epsilon$ (i.e., f is ϵ-close to $\mathcal{P}$), then $\Pr\{M^f(x) = g(x)\} \geq \frac{2}{3}$, for every $x \in \mathcal{X}$. If $f \in \mathcal{P}$, then $\Pr\{M^f(x) = f(x)\} = 1$ for every $x \in \mathcal{X}$.

Note that the definition of "property self-corrector" refers to distance measured only with respect to the uniform distribution, however, we still use these correctors for the construction of distribution-free testers. Observe that a necessary condition for the existence of an ϵ-self-corrector for a property $\mathcal{P}$ is that for every function f such that $dist_U(f, \mathcal{P}) \leq \epsilon$ (i.e., f is ϵ-close to $\mathcal{P}$ with respect to the uniform distribution), there exists a *unique* function $g \in \mathcal{P}$ that is ϵ-close to $\mathcal{P}$ (implying that ϵ cannot be too large). Notice that the property of monotonicity does not fulfill this requirement[8]. Hence, the distribution-free monotonicity tester that is presented in the next section requires a different approach.

---

[8] Consider for example the following function $f : [n] \to \{0, 1\}$: for every $1 \leq i \leq \frac{n}{2}$ set $f(i) = 1$, and for every $\frac{n}{2} + 1 \leq i \leq n$ set $f(i) = 0$. f is $\frac{1}{2}$-far from monotone, and it is $\frac{1}{2}$-close to both constant functions: 0 and 1.

Next, we describe the generalized distribution-free testing scheme. Let $\mathcal{P}$ be a property, let $T_{\mathcal{P}}$ be a uniform distribution tester for $\mathcal{P}$ with query complexity Q_T that has one-sided error, and let $C_{\mathcal{P}}$ be an ϵ' property self-corrector for $\mathcal{P}$ with query complexity Q_C. Let $\epsilon \le \epsilon'$, and $f : \mathcal{X} \to A$.

Tester$_D(\epsilon)$

Run $T_{\mathcal{P}}^f(\epsilon)$. If $T_{\mathcal{P}}^f(\epsilon) = $ FAIL, then **return FAIL**

Repeat $\frac{2}{\epsilon}$ times:

    Choose $x \in_D \mathcal{X}$.

    Repeat twice: Run $C_{\mathcal{P}}^f(x)$; If $f(x) \ne C_{\mathcal{P}}^f(x)$, then **return FAIL** .

return PASS

Theorem 3. *Algorithm Tester$_D(\epsilon)$ is a distribution-free tester for $\mathcal{P}$ with query complexity $Q_T(\epsilon) + \frac{2}{\epsilon} \cdot Q_C$.*

Proof. It is obvious that the query complexity of the algorithm Tester$_D(\epsilon)$ is indeed as required. Hence, we only have to prove the correctness of the algorithm. To do so, we prove the following two facts:

- if $f \in \mathcal{P}$ then f is accepted by the algorithm with probability 1.
- if f is (ϵ, D)-far from $\mathcal{P}$, then f is rejected by Tester$_D(\epsilon)$ with probability at least $\frac{2}{3}$.

If f is indeed in $\mathcal{P}$, then it passes the uniform test with probability 1, and the value returned by the self-corrector is always identical to the value of f. Hence, it is clear that in this case f is accepted by the algorithm. Assume from now on that f is (ϵ, D)-far from $\mathcal{P}$. In this case we distinguish between two possibilities:

If f is (ϵ, U)-far from $\mathcal{P}$, then the probability that it passes the uniform test is at most $\frac{1}{3}$.

If f is (ϵ, U)-close to $\mathcal{P}$, then there exists a function $g \in \mathcal{P}$ such that $dist(f, g) \le \epsilon$. However, since $dist_D(f, \mathcal{P}) \ge \epsilon$, we deduce that $dist_D(f, g) \ge \epsilon$ (in other words, $\Pr_{x \sim D}\{f(x) \ne g(x)\} \ge \epsilon$). If f is accepted by the algorithm then one of the two following events happened: either we failed to sample a point in which f and g differ, or we succeeded to sample such a point, but both runs of the self-corrector failed to compute the value of g in this point. The probability of the first event is at most $(1-\epsilon)^{\frac{2}{\epsilon}} \le \frac{1}{e}^2 \le \frac{1}{6}$, and by the definition of a property self-corrector the probability of the second event is at most $\frac{1}{3}^2 < \frac{1}{6}$. Therefore, the total probability that f is accepted by the algorithm is at most $\frac{1}{3}$.

Hence, in both cases the probability that f is accepted by the algorithm is at most $\frac{1}{3}$. $\qquad\square$

Remark 1. We used the assumption that there exists a uniform distribution test for the property $\mathcal{P}$ that has one-sided error. However, the same transformation can be applied also when the uniform distribution tester has two-sided error, only that the resulting distribution-free tester as well has two-sided error.

As was previously stated, the algorithm that was explicitly presented in Section 3 can actually be described as an application of this scheme for the class of low-degree multivariate polynomials. Hence, instead of fully describing the distribution-free tester and proving its correctness, it was enough to show that this property can be tested in the uniform distribution and that it can be self-corrected. This scheme, however, also implies the existence of distribution-free testers for other properties. Among these properties are low-degree multivariate polynomials over $GF(2)$, juntas and dictatorships functions. A function $f : \{0,1\}^n \to \{0,1\}$ is said to be a k-junta if there exists a subset of $\{x_1, \ldots, x_n\}$ of size k that determines the value of f (i.e., f is independent of the other variable). A special case of juntas are dictatorship functions, where a single variable determines the value of the function. These properties (and other related properties) have uniform distribution testers, as was shown in [3, 17, 32]. In addition, they are all subsets of the class of low-degree polynomials over $GF(2)$ which is self correctable (for example, k juntas are a special case of degree k multivariate polynomials), and thus are self correctable (see [3] and [10]). Therefore, we can apply the scheme described in this section to obtain distribution-free testers for these properties.

Remark 2. Notice that given two properties $\mathcal{P}$ and $\mathcal{P}'$ such that $\mathcal{P}' \subseteq \mathcal{P}$, the fact that $\mathcal{P}$ is testable in the uniform distribution does not imply that $\mathcal{P}'$ is thus testable (to see this observe, for example, that every property is a subset of the class of all functions that is clearly testable). However, the fact that $\mathcal{P}$ is self-correctable implies that $\mathcal{P}'$ is self-correctable (using the same correction algorithm).

5   Distribution-Free Monotonicity Testing on the d-Dimensional Cube

In this section, we present testers for monotonicity over the d-dimensional hypercube with respect to an arbitrary distribution D. As before, we assume D to be fixed but unknown, and beside the ability to sample according to D we assume no knowledge of D. For simplicity, we begin our discussion with the case $d = 1$, and show that given access to random sampling according to D and to membership queries, there is a distribution-free tester for monotonicity over $[n]$, whose query complexity is $O(\frac{\log n}{\epsilon})$. This algorithm can be generalized to a distribution-free tester for monotonicity over the d-dimensional hyper-cube whose query complexity is $O(\frac{\log^d n \cdot 2^d}{\epsilon})$.

We begin with a few notations and definitions. Denote by $[n]$ the set $\{1, \ldots, n\}$, and by $[n]^d$ the set of d-tuples over $[n]$. For every two points i and j in $[n]^d$ we say that $i \leq j$ if for every $1 \leq k \leq d$, $i_k \leq j_k$. Let $(A, <_A)$ be some linear order.

Definition 4. *We say that a function $f : [n]^d \to A$ is* monotone *if for every i and j if $i \leq j$ then $f(i) \leq_A f(j)$.*

Definition 5. *Let* $f : [n]^d \to A$ *be a function. A pair* (i, j) *is said to be an* f*-violation if* $i < j$ *and* $f(i) >_A f(j)$.

Let D be any distribution on $[n]^d$, and let S be a subset of $[n]^d$. Define $\Pr_D\{i\} \overset{\text{def}}{=} \Pr_{X \sim D}\{X = i\}$, and $\Pr_D\{S\} \overset{\text{def}}{=} \sum_{i \in S} \Pr_D\{i\}$.

5.1   Testing Monotonicity for the Line ($d = 1$)

In this section we consider the case $d = 1$. Our algorithm is a variant of the algorithm presented in [14] for testing monotonicity, with respect to the uniform distribution. However, the analysis presented here for this algorithm is quite different. The algorithm works in phases, in each phase a "center point" is selected according to the distribution D (in the original algorithm, the center point is selected uniformly), and the algorithm looks for a violation of the monotonicity with this center point. The search for a violation is done by randomly sampling in growing neighborhoods of the center point. In other words, in the case $d = 1$, the only change made in the original algorithm in order to adjust it to be distribution-free is that the choice of center points is made according to D. However, the search for violations remains unchanged. It is important to observe that, when dealing with an arbitrary distribution, there is no connection between the distance of the function from monotone (or the probability of the violation) and the number of pairs that form a violation of monotonicity[9]. Hence, the correctness of the algorithm for the uniform distribution (i.e., the fact that in a function that is far from monotone we find a violation of monotonicity with high probability) does not imply its correctness for the general case.

> Algorithm-monotone-1-dim$_D(f, \epsilon)$:
> **repeat** $\frac{2}{\epsilon}$ **times**
>   **choose** $i \in_D [n]$
>   **for** $k \leftarrow 0 \dots \lceil \log i \rceil$ **do**
>   **repeat** 8 **times**
>   **choose** $a \in_R [2^k]$
>   **if** $f(i - a) >_A f(i)$ **then return FAIL**
>   **for** $k \leftarrow 0 \dots \lceil \log(n - i) \rceil$ **do**
>   **repeat** 8 **times**
>   **choose** $a \in_R [2^k]$
>   **if** $f(i) >_A f(i + a)$ **then return FAIL**
> **return PASS**

Theorem 4. *Algorithm monotone-1-dim$_D$ is a distribution-free monotonicity tester over the line with query complexity* $O(\frac{\log n}{\epsilon})$.

To prove this theorem, we need the following definitions and lemmas.

---

[9] Observe, for example, the function $f : [n] \to [n]$ such that for every $0 \le i \le n - 2$ we set $f(i) = i$, $f(n - 1) = n$ and $f(n) = n - 1$. Set the distribution D to be $D(n - 1) = D(n) = \frac{1}{2}$.

Lemma 2. *Let $f : [n] \to A$ be a function, and let $S \subseteq [n]$ be a set. If for every f-violation (i, j) either $i \in S$ or $j \in S$, then there exists a monotone function f' that differs from f only on points in S.*

A similar claim was proved in [13]; proof omitted. An immediate conclusion of the above lemma is the following:

Lemma 3. *Let $f : [n] \to A$ be a function (ϵ, D)-far from monotone. Given $S \subseteq [n]$, if for every f-violation (i, j) either $i \in S$ or $j \in S$, then $\mathrm{Pr}_D\{S\} \geq \epsilon$.*

Definition 6. *For an f-violation (i, j), we say that i is* active *in this violation if*

$$|\{k \; : \; i < k < j \, , \; f(i) >_A f(k)\}| \; \geq \; \frac{j - i - 1}{2},$$

similarly, j is active *in this violation if* $|\{k \; : \; i < k < j \, , \; f(j) <_A f(k)\}| \geq \frac{j-i-1}{2}$.

That is, i is active in an f-violation (i, j), if for at least half of the points $i < k < j$, (i, k) is also an f-violation (i.e., $f(i) >_A f(k)$).

Observation 1: For every f-violation (i, j), at least one of i and j is active in (i, j). (Proof omitted)

Define the active set of f (denoted A_f) as the set of all points that are active in some f-violation. Following this observation and applying Lemma 3 to the set A_f, if f is (ϵ, D)-far from monotone then $\mathrm{Pr}_D\{A_f\} \geq \epsilon$. We turn now to prove Theorem 4.

Proof. It is easy to see that the query complexity of the algorithm is as required. Hence, we are left to prove that it is indeed a distribution-free tester. The fact that every monotone function f is accepted by the algorithm follows immediately from its definition. From now on, assume that f is (ϵ, D)-far from monotone; we prove that f is rejected with probability at least $\frac{2}{3}$. Our algorithm may fail to detect that f is not monotone if either one of the following two events occurs:

1. None of the points sampled by the algorithm according to D is in A_f.

2. The algorithm picked at least one point $i \in A_f$, but failed to detect that i belongs to some f-violation.

It is easily verified that the probability of the first event is at most $(1-\epsilon)^{\frac{2}{\epsilon}} \leq \frac{1}{e^2} \leq 1/6$. We now turn to bound the probability of the second event. By the definition of A_f, for every $i \in A_f$ there is a j such that either (i, j) or (j, i) is an f-violation and i is active in this violation. Assume w.l.o.g. that (i, j) is an f-violation. For $k = \min\{l \; : \; 2^l \geq j - i\}$ (i.e., k is the smallest integer s.t. $j \leq i + 2^k$), we can claim that $|\{l \mid i < l \leq i + 2^k, f(i) >_A f(l)\}|$ is more than $\frac{1}{4} \cdot 2^k$. This is due to the fact that $j - i > 2^{k-1}$, and since i is active in the f-violation (i, j), for at least half the points l between i and j (i.e., at least $\frac{2^{k-1}}{2}$ points) the pair (i, l) is an f-violation. The probability that the algorithm fails to find an f-violation for this k is at most $(\frac{3}{4})^8 \leq \frac{1}{6}$, and hence the probability of the second event is at most $\frac{1}{6}$, implying that the total probability that the algorithm will wrongly accept f is at most $\frac{1}{3}$. $\qquad\square$

Remark 3. In the journal version of [14], an additional tester for monotonicity on the line, called "Sort-Check-II", is presented. This algorithm can also be transformed to be a distribution-free monotonicity tester over the line. However, we do not know if it can be generalized to higher dimensions.

We saw how to test monotonicity over the one-dimensional hyper-cube (the line) when the distance is measured with respect to an arbitrary distribution. It is possible to generalize this algorithm to the d-dimensional case. The full details of the generalized algorithm and its analysis are omitted from this version and will appear in the full version of this paper.

References

[1] N. Alon, S. Dar, M. Parnas, and D. Ron, *Testing of clustering.* In *Proceedings of the 41st Annual Symposium on Foundations of Computer Science,* pages 240–251, 2000.

[2] N. Alon, E. Fischer, M. Krivelevich, and M. szegedy, *Efficient testing of large graphs.* In *Proceedings of the 40th Annual Symposium on Foundations of Computer Science,* pages 656–666, 1999.

[3] N. Alon, T. Kaufman, M. Krivelevich, S. Litsyn, and D. Ron, *Testing low-degree polynomials over GF(2).* To appear in *Proceedings of Random 2003.*

[4] N. Alon, M. Krivelevich, I. Newman, and M. Szegedy, *Regular languages are testable with a constant number of queries,* SIAM Journal on Computing 30:1842–1862, 2001 (also appeared in Proceedings of the 40th Annual Symposium on Foundations of Computer Science, pages 645–655, 1999).

[5] N. Alon and A. Shapira, *Testing satisfiablity.* In *Proccedings of 13th SODA,* 2001.

[6] S. Arora, C. Lund, R. Motwani, M. Sudan, and M. Szegedy, *Proof verification and the hardness of approximation problems,* JACM, 45(1):501–555, 1998, (preliminary version appeared in Proc. 33th FOCS, 1992).

[7] S. Arora and S. Safra, *Probabilistic checkable proofs: A new characterization of NP.* JACM, 45(1):70–122, 1998 (a preliminary version appeared in Proc. 33rd FOCS, 1992).

[8] S. Arora and M. Sudan, *Improved low-degree testing and its applications.* Proceedings of the 29th ACM STOC 485–495,1997.

[9] T. Batu, R. Rubinfeld, and P. White, *Fast approximation PCPs for multidimensional bin-packing problems,* Proceedings of the 3rd International Workshop on Randomization and Approximation Techniques in Computer Science 246–256,1999.

[10] M. Blum, M. Luby, and R. Rubinfeld, *Self testing/correcting with applications to numerical problems,* Journal of Computer and System Sceince 47:549–595, 1993.

[11] N. Bshouty, *Private communication.*

[12] A. Czumaj and C. Sohler, *Testing hypergraph coloring,* ICALP 2001, 493–505.

[13] Y. Dodis, O. Goldreich, E. Lehman, S. Raskhodnikova, D. Ron, and A. Samorodnitsky, *Improved testing algorithms for monotonicity,* Proceedings of the 3rd International Workshop on Randomized and Approximation Techniques in Computer Science, pages 97–108, 1999.

[14] E. Ergün, S. Kannan, R. Kumar, R. Rubinfeld, and M. Viswanathan, *Spot-checkers, Journal of Computing and System Science,* 60:717–751, 2000 (a preliminary version appeared in Proc. 30th STOC ,1998).

[15] E. Fischer, *On the strength of comparisons in property testing*, manuscript (available at ECCC TR00-083).

[16] E. Fischer, *The art of uninformed decisions: A primer to property testing*, The Computational Complexity Column of The bulletin of the European Association for Theoretical Computer Science, 75:97–126, 2001.

[17] E. Fischer, G. Kindler, D. Ron, S. Safra, and A. Samorodnitsky, *Testing Juntas*, Proceedings of the 43rd FOCS 103–112,2002.

[18] E. Fischer, E. Lehman, I. Newman, S. Raskhodnikova, R. Rubinfeld and, A. Samorodnitsky, *Monotonicity testing over general poset domains*, Proceedings of the 34th ACM STOC 474–483,2002.

[19] E. Fischer and I. Newman, *Testing of matrix properties*, Proceedings of the 33rd ACM STOC, pages 286–295, 2001.

[20] P. Gemmell, R. Lipton, R. Rubinfeld, M. Sudan, and A. Wigderson, *Self testing/correcting for polynomials and for approximate functions*. In Proceedings of the 23rd Annual ACM Symposium on Theory of Computing, pages 32–42, 1991.

[21] O. Goldreich, *Combinatorical property testing – a survey*, In: Randomized Methods in Algorithms Design (P. Pardalos, S. Rajasekaran and J. Rolim eds.), AMS-DIMACS pages 45–61, 1998 .

[22] O. Goldreich, S. Goldwasser, E. Lehman, D. Ron, and A. Samorodnitsky, *Testing Monotonicity*, Combinatorica, 20(3):301–337, 2000 (a preliminary version appeared in Proc. 39th FOCS, 1998).

[23] O. Goldreich, S. Goldwasser, and D. Ron, *Propert testing and its connection to learning and approximation*, Journal of the ACM, 45(4):653–750, 1998 (a preliminary version appeared in Proc. 37th FOCS, 1996).

[24] O. Goldreich and D. Ron, *Property testing in bounded degree graphs*. In Proccedings of the 31st Annual ACM Symposium on the Theory of Computing, pages 406–415, 1997.

[25] O. Goldreich and L. Trevisan, *Three theorems regarding testing graph properties*. In Proceedings of FOCS 2001, pages 460–469.

[26] M. J. Kearns and U. V. Vzirani, *An introduction to Computational Learning Theory*, MIT Press, 1994.

[27] Y. Kohayakawa, B. Nagle, and V. Rodl, *Efficient testing of hypergraphs*. In Proceedings of ICALP 2002.

[28] I. Newman, *Testing of functions that have small width branching programs*. In Proceedings of the 41st Annual Symposium of Foundations of Computer Science, pages 251–258, 2000.

[29] M. Parnas, and D. Ron, *Testing metric properties*. In Proceedings of the 33rd ACM STOC (2001), pages 276–285.

[30] M. Parnas, and D. Ron, *Testing the diameter of graphs*, RANDOM APPROX (1999), 85–96.

[31] M. Parnas, D. ron, and R. Rubinfeld, *Testing parenthesis languages*. In Proceedings of the 5th International Workshop on Randomization and Approximation Techniques in Computer Science (2001), pages 261–272.

[32] M. Parnas, D. Ron, and A. Samorodnitsky, *Proclaiming dictators and juntas ot testing boolean formulae*, RANDOM APPROX (2001), 273–284.

[33] D. Ron, *Property testing (a tutorial)*, In: Handbook of Randomized Computing (S.Rajasekaran, P. M. Pardalos, J. H. Reif and J. D. P. Rolin eds), Kluwer Press (2001).

[34] R. Rubinfeld, *Robust functional equations and their applications to program testing*. In: SIAM Journal on Computing, 28(6):1972–1997, 1999 (appeared in

Proceedings of the 35th Annual Symposium of Foundations of Computer Science, 1994).

[35] R. Rubinfeld and M. Sudan, *Robust characterization of polynomials with applocations to program testing*, *SIAM Journal of Computing*, 25(2):252–271, 1996. (first appeared as a technical report, Cornell University, 1993).

On the Graph-Density of Random 0/1-Polytopes

Volker Kaibel[1] and Anja Remshagen[2]

[1] DFG Research Center "Mathematics for key technologies", MA 6–2, TU Berlin,
Straße des 17. Juni 136, 10623 Berlin, Germany,
kaibel@math.tu-berlin.de
[2] Department of Computer Science, State University of West Georgia,
1600 Maple Street, Carrollton, GA 30118, USA,
anja@westga.edu

Abstract. Let $X_{d,n}$ be an n-element subset of $\{0,1\}^d$ chosen uniformly at random, and denote by $P_{d,n} := \operatorname{conv} X_{d,n}$ its convex hull. Let $\Delta_{d,n}$ be the density of the graph of $P_{d,n}$ (i.e., the number of one-dimensional faces of $P_{d,n}$ divided by $\binom{n}{2}$). Our main result is that, for any function $n(d)$, the expected value of $\Delta_{d,n(d)}$ converges (with $d \to \infty$) to one if, for some arbitrary $\varepsilon > 0$, $n(d) \leq (\sqrt{2} - \varepsilon)^d$ holds for all large d, while it converges to zero if $n(d) \geq (\sqrt{2} + \varepsilon)^d$ holds for all large d.

1 Introduction

Polytopes whose vertices have coordinates in $\{0,1\}$ (*0/1-polytopes*) are the objects of study in large parts of polyhedral combinatorics (see [10]). Since that theory has started to grow, people have been interested in the *graphs* (defined by the vertices and the one-dimensional faces) of the polytopes under investigation. The main reason for this interest was, of course, the role played by polytope graphs with respect to linear programming and, in particular, the simplex algorithm.

Later it was recognized that the graphs of the 0/1-polytopes associated with certain combinatorial objects (such as matchings in a graph or bases of a matroid) might also yield good candidates for neighborhood structures with respect to the construction of random walks for random generation of the respective objects. A quite important (yet unsolved) problem arising in this context is the question whether the graphs of 0/1-polytopes have good expansion properties (see [3,5,7]).

We are short of knowledge on the graphs of *general* 0/1-polytopes [13]. Among the few exceptions are results about their diameters [8] and their cycle structures [9]. Particularly striking is the fact that several *special* 0/1-polytopes associated with combinatorial problems have quite dense graphs. The most prominent example for this is probably the *cut polytope* CUT_k, i.e., the convex hull of the characteristic vectors of those subsets of edges of the complete graph K_k that form cuts in K_k. Barahona and Mahjoub [1] proved that the graph of CUT_k is complete, i.e., its density equals one (where the *density* of a graph $G = (V,E)$ is $|E|/\binom{|V|}{2}$). Since the dimension of CUT_k is $d = \binom{k}{2}$ and

S. Arora et al. (Eds.): APPROX 2003+RANDOM 2003, LNCS 2764, pp. 318–328, 2003.

there are $n = 2^{k-1}$ cuts in K_k, the cut polytopes yield an infinite series of d-dimensional 0/1-polytopes with (roughly) $c^{\sqrt{d}}$ vertices (for some constant c) and graph-density one.

In this paper, we investigate the question for the graph-density of a typical (i.e., random) 0/1-polytope. The (perhaps surprising) result is that in fact the high density of the graphs of several 0/1-polytopes important in polyhedral combinatorics (such as the cut polytopes) is not atypical at all. Our main result is the following theorem, where $\mathrm{Exp}[\,\cdot\,]$ denotes the expected value.

Theorem 1. *Let $n : \mathbb{N} \longrightarrow \mathbb{N}$ be a function, and let $P_{d,n(d)} := \mathrm{conv}\, X_{d,n(d)}$ with an $n(d)$-element subset $X_{d,n(d)}$ of $\{0,1\}^d$ that is chosen uniformly at random. Denote by $\Delta_{d,n(d)}$ the density of the graph of $P_{d,n(d)}$.*

(i) If there is some $\varepsilon > 0$ such that $n(d) \leq (\sqrt{2} - \varepsilon)^d$ for all sufficiently large d, then $\lim\limits_{d \to \infty} \mathrm{Exp}[\,\Delta_{d,n(d)}\,] = 1$.

(ii) If there is some $\varepsilon > 0$ such that $n(d) \geq (\sqrt{2} + \varepsilon)^d$ for all sufficiently large d, then $\lim\limits_{d \to \infty} \mathrm{Exp}[\,\Delta_{d,n(d)}\,] = 0$.

There is a similar threshold phenomenon for the volumes of random 0/1-polytopes. Let $\tilde{P}_{d,n(d)}$ be the convex hull of $n(d)$ points in $\{0,1\}^d$ that are chosen independently uniformly at random (possibly with repetitions). Dyer, Füredi, and McDiarmid [2] proved that the limit (for $d \to \infty$) of the expected value of the d-dimensional volume of $\tilde{P}_{d,n(d)}$ is zero if, for some $\varepsilon > 0$, $n(d) \leq (\frac{2}{\sqrt{e}} - \varepsilon)^d$ holds for all sufficiently large d, and it is one if, for some $\varepsilon > 0$, $n(d) \geq (\frac{2}{\sqrt{e}} + \varepsilon)^d$ holds for all sufficiently large d. Due to $\frac{2}{\sqrt{e}} < 1.214$ and $\sqrt{2} > 1.414$, one can deduce (we omit the details) the following result from this and Theorem 1. It may be a bit surprising due to the fact that the only d-dimensional 0/1-polytope with d-dimensional volume equal to one is the 0/1-cube $\mathrm{conv}\{0,1\}^d$, which has only graph-density $\frac{d}{2^d - 1}$.

Corollary 1. *For every $\delta > 0$ there are (infinitely many) 0/1-polytopes with both graph density and volume at least $(1 - \delta)$.*

Another threshold result that is related to our work is due to Füredi [4]. He showed that, in the setting of Theorem 1, the limit (for $d \to \infty$) of the probability that $P_{d,n(d)}$ contains the center of the 0/1-cube is zero if, for some $\varepsilon > 0$, $n(d) \leq (2 - \varepsilon) \cdot d$ holds for all sufficiently large d, and it is one if, for some $\varepsilon > 0$, $n(d) \geq (2 + \varepsilon) \cdot d$ holds for all sufficiently large d. The material in Sections 2.2, 2.3, and 2.4 of our paper is very much inspired by Füredi's work.

The aim of Sections 2 and 3 is to prove Theorem 1. Since it is a bit more convenient, we switch from 0/1-polytopes to polytopes whose vertices have coordinates in $\{-1, +1\}$ (± 1-*polytopes*). Recalling that the density of a graph equals the probability of a randomly chosen pair of its nodes to be adjacent, Propositions 4 and 5 (Section 3), together with Proposition 3, imply Theorem 1 (with the ε's in Propositions 4 and 5 replaced by $\log \frac{\sqrt{2}}{\sqrt{2}-\varepsilon}$ and $\log \frac{\sqrt{2}+\varepsilon}{\sqrt{2}}$, respectively). We close with a few remarks in Section 4.

2   The Long-Edge Probability $\tau(k, m)$

We define $Q_d := \{-1, +1\}^d$ and $Q_d^\star := Q_d \setminus \{-1, 1\}$ (where 1 is the all-one vector). For $v, w \in Q_d$, denote by $Q(v, w)$ the subset of all points in Q_d that agree with v and w in all components, where v and w agree. Thus, $Q(v, w)$ is the vertex set of the smallest face of conv Q_d containing v and w. The dimension of this face is

$$\text{dist}(v, w) := \#\{\, i \in \{1, \ldots, d\} \; : \; v_i \neq w_i \,\}$$

(the *Hamming distance* of v and w). Let $Q^\star(v, w) := Q(v, w) \setminus \{v, w\}$.

We refer to [12] for all notions and results from polytope theory that we rely on. For a polytope P, we denote by $V(P)$ and $E(P)$ the sets of vertices and edges of P, respectively. Recall that, for $X \subseteq Q_d$, we have $V(\text{conv } X) = X$.

The following fact is essential for our treatment. It can easily be deduced from elementary properties of convex polytopes.

Lemma 1. *For two vertices v and w of a ± 1-polytope $P \subset \mathbb{R}^d$ we have*

$$\{v, w\} \in E(P) \iff \text{conv}\{v, w\} \cap \text{conv}(P \cap Q^\star(v, w)) = \varnothing .$$

Throughout this section, let $Y_{k,m} \in \binom{Q_k^\star}{m}$ (the m-element subsets of $Q_k^\star$) be drawn uniformly at random and define

$$\tau(k, m) := \text{Prob}[\, \text{conv}(Y_{k,m}) \cap \text{conv}\{-1, 1\} = \varnothing\,] .$$

Thus, $\tau(k, m)$ is the probability that the "long edge" conv$\{-1, 1\}$ is an edge of the polytope conv$(Y_{k,m} \cup \{-1, 1\})$. The next lemma follows from Lemma 1.

Lemma 2. *Let $X_{d,n} \in \binom{Q_d}{n}$ be chosen uniformly at random, defining the polytope $P_{d,n} := \text{conv } X_{d,n}$. Choose a two-element subset $\{v, w\}$ of $X_{d,n}$ uniformly at random. Then, for every $k \in \{1, \ldots, d\}$ and $m \in \{0, \ldots, \min\{2^k - 2, n - 2\}\}$, we have the equation*

$$\text{Prob}[\, \{v, w\} \in E(P_{d,n}) \mid \text{dist}(v, w) = k, \#(X_{d,n} \cap Q^\star(v, w)) = m\,] \;=\; \tau(k, m) .$$

Via Lemma 2, asymptotic bounds on $\tau(k, m)$ will turn out to be important for the proofs in Section 3. In fact, we will basically compute (or estimate) the probability $\pi(d, n)$ (see Section 3) that two randomly chosen vertices of a d-dimensional random ± 1-polytope with n vertices are adjacent by partitioning the probability space into the events "dist$(v, w) = k$ and $\#(X_{d,n} \cap Q^\star(v, w)) = m$" for all $k \in \{1, \ldots, d\}$ and $m \in \{0, \ldots, \min\{2^k - 2, n - 2\}\}$.

For the study of $\tau(k, m)$, it is convenient to consider the conditional probability

$$\alpha(k, m) := \text{Prob}[\, \text{conv}(Y_{k,m}) \cap \text{conv}\{-1, 1\} = \varnothing \mid Y_{k,m} \cap (-Y_{k,m}) = \varnothing\,] ,$$

which is related to $\tau(k, m)$ in the following way.

Lemma 3. *For* $0 \leq m \leq 2^k - 2$ *we have*

$$\tau(k, m) \;=\; \frac{\binom{2^{k-1}-1}{m} \cdot 2^m}{\binom{2^k-2}{m}} \cdot \alpha(k, m) \;.$$

Proof. Clearly, $\mathrm{conv}(Y_{k,m}) \cap \mathrm{conv}\{-1, 1\} = \varnothing$ implies $Y_{k,m} \cap (-Y_{k,m}) = \varnothing$. Thus, the statement in the lemma is due to the fact that the number of sets $Y' \in \binom{Q_k^\star}{m}$ with $Y' \cap (-Y') = \varnothing$ is $\binom{2^{k-1}-1}{m} \cdot 2^m$.

We will first show that $\alpha(k, m)$ can be interpreted as a conditional probability that a random m-element subset of a certain vector configuration in $\mathbb{R}^{k-1}$ does not contain the origin in its convex hull (Section 2.1). The latter probability is then related to the expected number of chambers in a certain random hyperplane arrangement. This number of chambers is finally estimated via a well-known bound due to Harding (Section 2.2).

As a point of reference for the proofs in Section 3, let us state the following monotonicity result here, whose (straightforward) proof we omit.

Lemma 4. *For* $0 \leq m \leq 2^k - 3$, *we have* $\tau(k, m) \geq \tau(k, m + 1)$.

2.1 The Vector Configuration $\mathcal{V}_r$

Let $\varphi : \mathbb{R}^{r+1} \longrightarrow H_1 \longrightarrow \mathbb{R}^r$ denote the orthogonal projection of $\mathbb{R}^{r+1}$ onto the hyperplane $H_1 := \{x \in \mathbb{R}^{r+1} : \mathbf{1}^T x = 0\}$, followed by the orthogonal projection to the first r coordinates. We denote by $\mathcal{V}_r := \varphi(Q_{r+1}^\star)$ the image of $Q_{r+1}^\star$ under the projection φ. We omit the simple proof of the following result.

Lemma 5. *The projection* φ *is one-to-one on* $Q_{r+1}^\star$.

Lemma 6. *For* $Z_{r,m} \in \binom{\mathcal{V}_r}{m}$ *chosen uniformly at random, we have*

$$\alpha(r + 1, m) \;=\; \mathrm{Prob}[\, \mathbf{0} \notin \mathrm{conv}(Z_{r,m}) \mid Z_{r,m} \cap (-Z_{r,m}) = \varnothing \,] \;.$$

Proof. Since $\mathrm{conv}\, Y_{k,m} \cap \mathrm{conv}\{-1, 1\} = \varnothing$ holds if and only if $\mathbf{0} \notin \mathrm{conv}\, \varphi(Y_{k,m})$ holds, the claim follows from Lemma 5 (because $Y_{k,m} \cap (-Y_{k,m}) = \varnothing$ is equivalent to $\varphi(Y_{k,m}) \cap (-\varphi(Y_{k,m})) = \varnothing$).

With $\mathcal{V}_r^+ := \varphi\{v \in Q_{r+1}^\star : v_{r+1} = +1\}$, we have $\mathcal{V}_r = \mathcal{V}_r^+ \cup (-\mathcal{V}_r^+)$ and $\mathcal{V}_r^+ \cap (-\mathcal{V}_r^+) = \varnothing$. For any fixed finite subset $S \subset \mathbb{R}^r$, and a uniformly at random chosen $\varepsilon \in \{-1, +1\}^S$, denote $\alpha(S) := \mathrm{Prob}[\, \mathbf{0} \notin \mathrm{conv}\{\varepsilon_s s : s \in S\}\,]$.

Lemma 7. *Let* $Z_{r,m}^+ \in \binom{\mathcal{V}_r^+}{m}$ *be chosen uniformly at random. Then we have*

$$\alpha(r + 1, m) \;=\; \mathrm{Exp}[\, \alpha(Z_{r,m}^+)\,] \;.$$

Proof. This follows from Lemma 6.

2.2   Hyperplane Arrangements

For $s \in \mathbb{R}^r \setminus \{0\}$ let $H(s) := \{x \in \mathbb{R}^r : s^T x = 0\}$. The two connected components of $\mathbb{R}^r \setminus H(s)$ are denoted by $H^+(s)$ and $H^-(s)$, where $s \in H^+(s)$. For a finite subset $S \subset \mathbb{R}^r \setminus \{0\}$ denote by $\mathcal{H}(S) := \{H(s) : s \in S\}$ the *hyperplane arrangement* defined by S. The connected components of $\overline{\mathcal{H}(S)} := \mathbb{R}^r \setminus \bigcup_{s \in S} H(s)$ are the *chambers* of $\mathcal{H}(S)$. We denote the number of chambers of $\mathcal{H}(S)$ by $\chi(S)$.

Observation 1 *Let C be a chamber of $\mathcal{H}(S)$ for some finite subset $S \subset \mathbb{R}^r \setminus \{0\}$. For each $s \in S$, we have either $C \subseteq H^+(s)$ or $C \subseteq H^-(s)$. Defining $\varepsilon(C)_s := +1$ in the first, and $\varepsilon(C)_s := -1$ in the second case, we may assign a sign vector $\varepsilon(C) \in \{-1, +1\}^S$ to each chamber C of $\mathcal{H}(S)$. This assignment is injective.*

Lemma 8. *For each finite subset $S \subset \mathbb{R}^r \setminus \{0\}$, the following equation holds:*

$$\#\{ \varepsilon \in \{-1, +1\}^S : 0 \notin \mathrm{conv}\{ \varepsilon_s s : s \in S \} \} = \chi(S)$$

Proof. Let $S \subset \mathbb{R}^r \setminus \{0\}$ be finite. By the Farkas-Lemma (linear programming duality), for each $\varepsilon \in \{-1, +1\}^S$, we have $0 \notin \mathrm{conv}\{\varepsilon_s s : s \in S\}$ if and only if there is some $h \in \mathbb{R}^r$ such that $h^T(\varepsilon_s s) > 0$ holds for all $s \in S$, which in turn is equivalent to

$$h^T s \begin{cases} > 0 & \text{if } \varepsilon_s = +1 \\ < 0 & \text{if } \varepsilon_s = -1 \end{cases}$$

for all $s \in S$. Since the latter condition is equivalent to ε being the sign vector of some chamber of $\mathcal{H}(S)$, the statement of the lemma follows.

Lemma 7 and Lemma 8 immediately yield the following result.

Lemma 9. *For $Z_{r,m}^+ \in \binom{\mathcal{V}_r^+}{m}$ chosen uniformly at random, we have*

$$\alpha(r + 1, m) = \frac{1}{2^m} \cdot \mathrm{Exp}[\chi(Z_{r,m}^+)] .$$

The following upper bound on $\chi(\cdot)$ will (via Lemma 9) yield upper bounds on $\alpha(\cdot, \cdot)$ that are sufficient for our needs. We denote $b(p, q) := \sum_{i=0}^p \binom{q}{i}$.

Theorem 2 (Harding, see Winder [11, p. 816]). *For $S \in \binom{\mathbb{R}^r \setminus \{0\}}{m}$, we have*

$$\chi(S) \leq 2b(r - 1, m - 1) .$$

2.3   Bounds on $\tau(k, m)$

Proposition 1. *For $0 \leq m \leq 2^k - 2$ the following inequality holds:*

$$\tau(k, m) \leq \frac{b(k - 2, m - 1)}{2^{m-1}}$$

Proof. With $r = k - 1$, Lemma 3, Lemma 9, and Theorem 2 yield this.

In fact, one can prove that, if m is not too large relative to k, then the bound of Proposition 1 is asymptotically sharp as k tends to infinity. Since we do not need the result here, we omit the proof which (next to the theorem of Winder's cited in Theorem 2) relies on the fact that the probability of an $l \times l$ matrix with entries from $\{-1, +1\}$ (chosen uniformly at random) being singular converges to zero for l tending to infinity (see [6]).

Proposition 2. *For* $m(k) \in o(2^{\frac{k}{2}})$, *we have*

$$\lim_{k \to \infty} \left(\tau(k, m(k)) - \frac{b(k-2, m(k)-1)}{2^{m(k)-1}} \right) = 0 .$$

2.4 A Threshold for $\tau(k, m)$

For $x \in \mathbb{R}$, let

$$\Phi(x) := \frac{1}{\sqrt{2\pi}} \int_{-\infty}^{x} e^{-\frac{t^2}{2}} \, dt ,$$

i.e., Φ is the density function of the normal distribution.

Lemma 10 (de Moivre-Laplace theorem). *For each* $\mu \in \mathbb{R}$, *the following holds:*

$$\lim_{q \to \infty} \frac{b\left(\lfloor \frac{q}{2} + \mu\sqrt{q} \rfloor, q \right)}{2^q} = \Phi(2\mu)$$

Theorem 3. *For each* $\varepsilon > 0$, *we have*

$$\lim_{k \to \infty} \tau\big(k, \lceil (2+\varepsilon)k \rceil\big) = 0 .$$

Proof. Let $\varepsilon > 0$ be fixed, and define, for each k, $m_\varepsilon^+(k) := \lceil (2+\varepsilon)k \rceil$.

Let $\delta > 0$ be arbitrarily small, and choose $\mu < 0$ such that

$$\Phi(2\mu) < \frac{\delta}{2} . \tag{1}$$

Due to $\lim_{k \to \infty} \dfrac{m_\varepsilon^+(k)}{k} = 2 + \varepsilon$, we have, for large enough k,

$$k - 2 \leq \frac{m_\varepsilon^+(k) - 1}{2} + \mu\sqrt{m_\varepsilon^+(k) - 1} . \tag{2}$$

Due to Proposition 1, we have

$$\tau(k, m_\varepsilon^+(k)) \leq \frac{b(k-2, m_\varepsilon^+(k)-1)}{2^{m_\varepsilon^+(k)-1}} . \tag{3}$$

Since $b(\cdot, \cdot)$ is monotonically increasing in the first component, (2) yields that the right-hand side of (3) is bounded from above by

$$\frac{b\left(\frac{m_\varepsilon^+(k)-1}{2} + \mu\sqrt{m_\varepsilon^+(k) - 1} \, , \, m_\varepsilon^+(k) - 1 \right)}{2^{m_\varepsilon^+(k)-1}} . \tag{4}$$

By Lemma 10 (with q substituted by $m_\varepsilon^+(k) - 1$), (4) may be bounded from above by $\Phi(2\mu) + \frac{\delta}{2}$ for all large enough k (because of $\lim_{k\to\infty} m_\varepsilon^+(k) = \infty$). Thus, from (1) we obtain

$$\tau(k, m_\varepsilon^+(k)) < \delta$$

for all large enough k.

Exploiting Proposition 2, one can also prove the following result. It complements Theorem 3, but since we will not need it in our treatment, we do not give a proof here.

Theorem 4. *For each $\varepsilon > 0$ we have*

$$\lim_{k\to\infty} \tau(k, \lfloor(2-\varepsilon)k\rfloor) = 1$$

3 The Edge Probability $\pi(d, n)$

Throughout this section, let the set $X_{d,n} \in \binom{Q_d}{n}$ be drawn uniformly at random, $P_{d,n} := \operatorname{conv} X_{d,n}$, and let $\{v, w\} \in \binom{X_{d,n}}{2}$ be chosen uniformly at random as well. Our aim is to determine the probability

$$\pi(d, n) := \operatorname{Prob}[\{v, w\} \in E(P_{d,n})] .$$

Let us further denote

$$\pi_k(d, n) := \operatorname{Prob}[\{v, w\} \in E(P_{d,n}) \mid \operatorname{dist}(v, w) = k] .$$

Since $\{v, w\}$ is uniformly distributed over $\binom{Q_d}{2}$, the distance $\operatorname{dist}(v, w)$ has the same distribution as the number of positive components of a point chosen uniformly at random from $Q_d \setminus \{-\mathbf{1}\}$. Therefore, the following equation holds.

Lemma 11.

$$\pi(d, n) = \frac{1}{2^d - 1} \sum_{k=1}^{d} \binom{d}{k} \pi_k(d, n)$$

The following result, stating that $\pi(d, \cdot)$ is monotonically increasing, is quite plausible. Its straightforward proof is omitted here.

Proposition 3. *The function $\pi(d, \cdot)$ is monotonically decreasing, i.e., for $3 \le n \le 2^d - 1$, we have $\pi(d, n) > \pi(d, n+1)$.*

The next result implies part (i) of Theorem 1 (see the remarks at the end of Section 1).

Proposition 4. *For each $\varepsilon > 0$, we have*

$$\lim_{d\to\infty} \pi\left(d, \left\lfloor 2^{(\frac{1}{2}-\varepsilon)d}\right\rfloor\right) = 1 .$$

Proof. Let $\varepsilon > 0$, and define $n_\varepsilon^-(d) := \lfloor 2^{(\frac{1}{2}-\varepsilon)d} \rfloor$. For each $\mu > 0$, denote

$$K_\mu^{\leq}(d) := \{k \in \mathbb{Z} : 1 \leq k \leq \frac{d}{2} + \mu\sqrt{d}\}$$

and

$$\pi_\mu^-(d) := \min\{\pi_k(d, n_\varepsilon^-(d)) : k \in K_\mu^{\leq}(d)\}.$$

Then, due to Lemma 11, we have

$$\pi(d, n_\varepsilon^-(d)) \geq \sum_{k \in K_\mu^{\leq}(d)} \frac{\binom{d}{k}}{2^d} \cdot \pi_\mu^-(d).$$

For every $\nu > 0$, this implies (by Lemma 10) that

$$\pi(d, n_\varepsilon^-(d)) \geq (\Phi(2\mu) - \nu) \cdot \pi_\mu^-(d) \tag{5}$$

holds for all large enough d. Therefore, it remains to prove, for all $\mu > 0$,

$$\lim_{d \to \infty} \pi_\mu^-(d) = 1. \tag{6}$$

With

$$\xi_k := \mathrm{Prob}[X_{d,n_\varepsilon^-(d)} \cap Q^\star(v,w) = \varnothing \mid \mathrm{dist}(v,w) = k],$$

we have, for each $k \in K_\mu^{\leq}(d)$,

$$\pi_k(d, n_\varepsilon^-(d)) \geq \xi_k \geq \xi_{\lfloor \frac{d}{2} + \mu\sqrt{d} \rfloor} \tag{7}$$

(see Lemma 1). Clearly,

$$\mathrm{Exp}[\#(X_{d,n_\varepsilon^-(d)} \cap Q^\star(v,w)) \mid \mathrm{dist}(v,w) = k] = \frac{2^k - 2}{2^d - 2} \cdot (n_\varepsilon^-(d) - 2),$$

and thus, the estimation

$$\mathrm{Exp}[\#(X_{d,n_\varepsilon^-(d)} \cap Q^\star(v,w)) \mid \mathrm{dist}(v,w) = k] \leq 2^{k-(\frac{1}{2}+\varepsilon)d},$$

hold for each k. By Markov's inequality, this implies

$$\mathrm{Prob}[\#(X_{d,n_\varepsilon^-(d)} \cap Q^\star(v,w)) \geq d \cdot 2^{k-(\frac{1}{2}+\varepsilon)d} \mid \mathrm{dist}(v,w) = k] \leq \frac{1}{d} \tag{8}$$

for each d and k. For $k = \lfloor \frac{d}{2} + \mu\sqrt{d} \rfloor$, (8) yields

$$\mathrm{Prob}[\#(X_{d,n_\varepsilon^-(d)} \cap Q^\star(v,w)) \geq d \cdot 2^{\mu\sqrt{d}-\varepsilon d} \mid \mathrm{dist}(v,w) = \lfloor \frac{d}{2} + \mu\sqrt{d} \rfloor]$$
$$\leq \frac{1}{d} \tag{9}$$

for all d. Since $d \cdot 2^{\mu\sqrt{d}-\varepsilon d} < 1$ holds for large enough d, (9) implies $\xi_{\lfloor \frac{d}{2}+\mu\sqrt{d} \rfloor} \geq 1 - \frac{1}{d}$ for large enough d. Therefore,

$$\lim_{d \to \infty} \xi_{\lfloor \frac{d}{2}+\mu\sqrt{d} \rfloor} = 1$$

holds, which, by (7), finally implies (6).

The next result yields part (ii) of Theorem 1 (see the remarks at the end of Section 1).

Proposition 5. *For each $\varepsilon > 0$, we have*

$$\lim_{d \to \infty} \pi\left(d, \left\lceil 2^{(\frac{1}{2}+\varepsilon)d} \right\rceil\right) = 0 .$$

Proof. Let $\varepsilon > 0$, and define $n_\varepsilon^+(d) := \left\lceil 2^{(\frac{1}{2}+\varepsilon)d} \right\rceil$. For each $\mu > 0$, denote

$$K_\mu^{\geq}(d) := \{k \in \mathbb{Z} : \frac{d}{2} - \mu\sqrt{d} \leq k \leq d\} ,$$

and define

$$\pi_\mu^+(d) := \max\{ \pi_k(d, n_\varepsilon^+(d)) : k \in K_\mu^{\geq}(d) \} . \tag{10}$$

Then, due to Lemma 11, we have

$$\pi(d, n_\varepsilon^+(d)) \leq 2 \cdot \sum_{k=1}^{\lfloor \frac{d}{2} - \mu\sqrt{d} \rfloor} \frac{\binom{d}{k}}{2^d} + \pi_\mu^+(d) .$$

Thus, for every $\nu > 0$, by Lemma 10,

$$\pi(d, n_\varepsilon^+(d)) \leq \Phi(-2\mu) + \nu + \pi_\mu^+(d)$$

holds for all large enough d. Therefore, it remains to prove, for all $\mu > 0$,

$$\lim_{d \to \infty} \pi_\mu^+(d) = 0 . \tag{11}$$

For $k \in \{1, \ldots, d\}$ and $m \in \{0, \ldots, 2^k - 2\}$, we define

$$\xi_k(m) := \mathrm{Prob}[\#(X_{d, n_\varepsilon^+(d)} \cap Q^\star(v, w)) = m \mid \mathrm{dist}(v, w) = k]$$

(i.e., $\xi_k(0) = \xi_k$ in the proof of Proposition 4). Then we have (see Lemma 2)

$$\pi_k(d, n_\varepsilon^+(d)) = \sum_{m=0}^{2^k - 2} \xi_k(m)\tau(k, m) . \tag{12}$$

Since $\tau(k, \cdot)$ is monotonically non-increasing by Lemma 4, we thus can estimate

$$\pi_k(d, n_\varepsilon^+(d)) \leq \sum_{m=0}^{3k-1} \xi_k(m) + \tau(k, 3k) ,$$

for each $k \in K_\mu^{\geq}(d)$. This yields, again for for each $k \in K_\mu^{\geq}(d)$,

$$\pi_k(d, n_\varepsilon^+(d)) \leq 3d \cdot \max\{ \xi_k(m) : 0 \leq m \leq 3d - 1 \}$$
$$+ \max\{ \tau(k', 3k') : k' \in K_\mu^{\geq}(d) \} . \tag{13}$$

According to Theorem 3,

$$\lim_{d \to \infty} \max\{\tau(k', 3k') : k' \in K_\mu^\geq(d)\} = 0$$

holds. Hence, by (13) and (10), equation (11) can be proved by showing

$$\lim_{d \to \infty} (3d \cdot \max\{\xi_k(m) : 0 \leq m \leq 3d - 1, k \in K_\mu^\geq(d)\}) = 0. \tag{14}$$

Let us first calculate (using the notation $(a)_b := a(a-1)\cdots(a-b+1)$)

$$\xi_k(m) = \frac{\binom{2^k - 2}{m}\binom{2^d - 2^k}{n_\varepsilon^+(d)-m-2}}{\binom{2^d - 2}{n_\varepsilon^+(d)-2}}$$

$$= \binom{2^k - 2}{m} \cdot \frac{(2^d - 2^k)_{n_\varepsilon^+(d)-m-2}}{(2^d - 2)_{n_\varepsilon^+(d)-2}} \cdot \frac{(n_\varepsilon^+(d) - 2)!}{(n_\varepsilon^+(d) - m - 2)!}, \tag{15}$$

where the left, the middle, and the right factor of (15) may be bounded from above by $(2^d)^m$, $(2^d)^2 \cdot \left(\frac{2^d - 2^k}{2^d}\right)^{n_\varepsilon^+(d)}$, and $(2^d)^m$, respectively. Thus, we obtain, for $0 \leq m \leq 3d - 1$,

$$\xi_k(m) \leq 2^{\text{const} \cdot d^2} \cdot \left(1 - \frac{1}{2^{d-k}}\right)^{n_\varepsilon^+(d)}. \tag{16}$$

For $k \in K_\mu^\geq(d)$, we have

$$\left(1 - \frac{1}{2^{d-k}}\right)^{n_\varepsilon^+(d)} \leq \left(1 - \frac{1}{2^{\frac{d}{2}+\mu\sqrt{d}}}\right)^{2^{\left(\frac{1}{2}+\varepsilon\right)d}}$$

$$= \left[\left(1 - \frac{1}{2^{\frac{d}{2}+\mu\sqrt{d}}}\right)^{2^{\frac{d}{2}+\mu\sqrt{d}}}\right]^{2^{\varepsilon d - \mu\sqrt{d}}}. \tag{17}$$

For d tending to infinity, the expression in the square brackets of (17) converges to $\frac{1}{e} < \frac{1}{2}$ (where $e = 2.7182\cdots$ is Euler's constant). Therefore, (17) and (16) imply $\xi_k(m) \leq 2^{\text{const} \cdot d^2} \cdot (1/2)^{2^{\varepsilon d - \mu\sqrt{d}}}$ (for $k \in K_\mu^\geq(d)$, $0 \leq m \leq 3d - 1$, and for large enough d). This finally yields (14), and therefore completes the proof.

4 Remarks

The threshold for the function $\tau(\cdot, \cdot)$ described in Theorems 3 and 4 is much sharper than we needed for our purposes (proof of Proposition 5). The sharper result may, however, be useful in investigations of more structural properties of the graphs of random 0/1-polytopes. A particularly interesting such question is whether these graphs have good expansion properties with high probability.

Acknowledgements

We thank one of the referees for several suggestions that helped to improve the presentation.

References

1. F. Barahona and A. R. Mahjoub. On the cut polytope. *Math. Program.*, 36:157–173, 1986.
2. M. E. Dyer, Z. Füredi, and C. McDiarmid. Volumes spanned by random points in the hypercube. *Random Structures Algorithms*, 3(1):91–106, 1992.
3. T. Feder and M. Mihail. Balanced matroids. In *Proceedings of the 24th Annual ACM "Symposium on the theory of Computing" (STOC)*, pages 26–38, Victoria, British Columbia, 1992. ACM Press, New York.
4. Z. Füredi. Random polytopes in the d-dimensional cube. *Discrete Comput. Geom.*, 1(4):315–319, 1986.
5. M. Jerrum and A. Sinclair. The Markov chain Monte Carlo method. In D. Hochbaum, editor, *Approximation Algorithms*, pages 482–520. PWS, 1997.
6. J. Kahn, J. Komlós, and E. Szemerédi. On the probability that a random ±1-matrix is singular. *J. Amer. Math. Soc.*, 8(1):223–240, 1995.
7. V. Kaibel. On the expansion of graphs of 0/1-polytopes. Technical report, TU Berlin, 2001. To appear in: *The Sharpest Cut*. M. Grötschel (ed.), SIAM, 2003.
8. D. J. Naddef. The Hirsch conjecture is true for (0,1)-polytopes. *Math. Program., Ser. B*, 45(1):109–110, 1989.
9. D. J. Naddef and W. R. Pulleyblank. Hamiltonicity in (0-1)-polyhedra. *J. Comb. Theory, Ser. B*, 37:41–52, 1984.
10. A. Schrijver. *Combinatorial Optimization. Polyhedra and Efficiency. Vol. A–C*, volume 24 of *Algorithms and Combinatorics*. Springer-Verlag, Berlin, 2003.
11. R. O. Winder. Partitions of N-space by hyperplanes. *SIAM J. Appl. Math.*, 14:811–818, 1966.
12. G. M. Ziegler. *Lectures on Polytopes*, volume 152 of *Graduate Texts in Mathematics*. Springer-Verlag, New York, 1995. Revised edition: 1998.
13. G. M. Ziegler. Lectures on 0/1-polytopes. In *Polytopes—Combinatorics and Computation*, volume 29 of *DMV Sem.*, pages 1–41. Birkhäuser, Basel, 2000.

A Gambling Game Arising in the Analysis of Adaptive Randomized Rounding

Richard M. Karp[1] and Claire Kenyon[2]

[1] UC Berkeley and ICSI
[2] Ecole Polytechnique and IUF

Abstract. Let y be a positive real number and let $\{X_i\}$ be an infinite sequence of Bernoulli random variables with the following property: in every realization of the random variables, $\sum_{i=1}^{\infty} E[X_i|X_1, X_2, \cdots, X_{i-1}] \le y$. We specify a function $F(x, y)$ such that, for every positive integer x and every positive real y, $P(\sum_{i=1}^{\infty} X_i \ge x) \le F(x, y)$; moreover, for every x and y, $F(x, y)$ is the best possible upper bound. We give an interpretation of this stochastic process as a gambling game, characterize optimal play in this game, and explain how our results can be applied to the analysis of multi-stage randomized rounding algorithms, giving stronger results than can be obtained using the traditional Hoeffding bounds and martingale tail inequalities.

1 Introduction

Consider the following gambling game. A player starts with a *fortune* of y and a *goal* of x. At each step the player chooses a *bet* $p \in (0, 1]$ and tosses a coin with probability of heads p. His fortune is reduced by p, and he scores a *success* if the coin comes up heads. He wins the game if he achieves x successes while maintaining a nonnegative fortune. A function $G(x, y)$ can serve as the function F mentioned in the Abstract if and only if, for all (x, y), $G(x, y)$ is an upper bound on the success probability of all strategies with fortune y and goal x. Our main result is a uniformly optimum choice of this function.

Theorem 1. *Let x be any positive integer and y, any positive real number. Let $F(x, y)$ denote the supremum, over all strategies, of the probability of achieving x successes with fortune y. Then $F(x, y)$ is specified recursively as follows.*

$$
\begin{cases}
\text{if } x \le y & \text{then } F(x, y) = 1 \\
\text{if } y < x < y + 1 & \text{then } F(x, y) = (x - y) + (1 - x + y)F(x, x - 1) \\
\text{if } y + 1 \le x & \text{then } F(x, y) = \int_{z=0}^{y} exp(-z)F(x - 1, y - z)dz
\end{cases}
$$

Although we do not have a closed form for $F(x, y)$, we can easily compute an upper bound which is good enough for our purposes.

Corollary 1. *Let $s = (x - \lceil y \rceil)/y$. If $s > 1$, then $F(x, y) \le (e^{s-1}/s^s)^y$.*

S. Arora et al. (Eds.): APPROX 2003+RANDOM 2003, LNCS 2764, pp. 329–340, 2003.

Proof. $F(x, y)$ is less than or equal to the probability that a Poisson random variable with mean y is greater than or equal to $x - \lceil y \rceil$. The result follows using a Chernoff bound for the tail of the Poisson distribution.   □

Assume that infinitesimal bets are allowed – the precise meaning of an infinitesimal bet will be specified in Section 2. (If all bets must be positive reals then a success probability arbitrarily close to $F(x, y)$ can be achieved by placing suitably small positive bets instead of the infinitesimal bets in the strategy below.)

Theorem 2 (Best Strategy). *The following strategy achieves the success probability $F(x, y)$:*

Strategy $\mathcal{G}$ on (x, y):
if $y \geq x$ then bet 1;
if $x - 1 < y < x$ then bet $y - x + 1$;
if $y \leq x - 1$ then continue placing infinitesimal bets until a success occurs

---

In order to explain the link between the gambling game and multistage randomized rounding algorithms we first present an abstract setting for the traditional single-stage randomized rounding algorithms [5]. Consider a mixed integer program of the following form:

$$\text{Minimize } z \text{ subject to:}$$

Integrality Constraints: $x_i \in \{0, 1\}, i = 1, 2, \cdots, n$;
Covering Constraints: $\sum_{i \in S_i} x_i = 1, i = 1, 2, \cdots, t$;
Resource Constraints: $\sum_{j \in T_j} x_j \leq z c_j, j = 1, 2, \cdots, m$.

Each set S_i or T_j is a subset of $\{1, 2, \cdots, n\}$, and the sets S_i are disjoint.

Each x_i represents an activity, such as the selection of a path in a graph. Each covering constraint requires that one activity be selected from a specified set; for example, in an integer multicommodity flow problem we might require that a given source-sink pair be joined by a path. Each resource constraint represents a bound on some resource; in a multicommodity flow problem the resource might be a vertex or edge, with c_j representing its nominal capacity and T_j, the set of paths that consume a unit of that capacity. The variable z represents the maximum amount by which the capacity of any resource is exceeded.

Randomized rounding begins by solving a linear programming relaxation in which the integrality constraint on each variable x_i is replaced by the constraint $0 \leq x_i \leq 1$. Let $(y_1, y_2, \cdots, y_n)$ be the optimal solution to this linear program and let z^* be the optimal value. Randomization is then used to select exactly one variable in each set S_i to be set equal to 1. Variable x_r is selected with probability y_r. This rounding process gives a feasible solution to the integer program.

Let us consider the effect of this rounding process on the resource constraints. For the jth resource constraint let $p_{ij} = \sum_{r \in S_i \cap T_j} y_r$. Then p_{ij} is the probability that a unit of resource j is used to satisfy covering constraint i. Thus the total consumption of resource j is distributed as $\sum_i X_{ij}$, where the X_{ij} are independent Bernoulli random variables and $P(X_{ij} = 1) = p_{ij}$. The expected value of the sum of these random variables is at most $z^* c_j$. The Hoeffding bound on sums of independent Bernoulli random variables is used to bound the probability that the total usage of resource j exceeds a target value zc_j and a union bound is used to obtain an upper bound on the probability that some resource exceeds its target value.

In a *multistage randomized rounding algorithm*, a sequence of mixed integer programs of the above form is solved. The resources and their capacities are the same in all these programs, but in all other respects the structure of the kth program may depend on the solutions constructed for the $k - 1$ integer programs preceding it. We wish to bound the total usage of each resource over all the stages. For each resource j this total usage is a sum of Bernoulli random variables; for each set $S_i \cap T_j$ in the kth integer program there is a Bernoulli random variable with mean p_{ij}^k. Because of the adaptiveness in the choice of integer programs the Hoeffding bound, which requires that the Bernoulli random variables be independent, can only be used separately for each integer program, but not for the entire multistage process. However our gambling game, with the fortune defined as c_j times the sum of the optimal values of the linear programs, and the bets defined as the p_{ij}^k occurring in all the stages (for fixed resource j), is applicable to the multistage process because it allows the parameters of successive Bernoulli random variables to be dependent. In effect, the gambling game assumes that an adversary chooses the successive p_{ij} adaptively, with the goal of maximizing total resource usage, subject to a constraint on the sum of the p_{ij}^k over all stages. Note that martingale tail inequalities are not useful in this setting because they are not sensitive to this global constraint.

As a specific illustration we refine a bicriterion optimization result due to Ravi [6]. Motivated by the Telephone broadcast problem, Ravi gave a polynomial-time algorithm for constructing a spanning tree of small diameter and small maximum degree in a graph G. He showed that, if G has a spanning tree of diameter at most Δ and maximum degree at most D^*, then his algorithm produces a spanning tree of diameter $O(\Delta \log n)$ and maximum degree $O(D^* \log n + log^2 n)$ with high probability. Our analysis of the same algorithm using the gambling game shows that the algorithm produces a spanning tree of diameter $O(\Delta \log n)$ and maximum degree $O(D^* \log n)$ with high probability.

Subsequent to Ravi's paper, Bar-Noy, Guha, Naor and Schieber had also addressed the problem of constructing a short tree of small degree. In [1], they presented an algorithm which constructs a tree of diameter $O(\Delta \log n)$ and maximum degree $O((D^* + \Delta) \log n)$. The algorithm relies on a version of the randomized rounding theorem from [3], which exploits the fact the the sum of the absolute values of the entries of any column of the constraint matrix are small; we note that using the version from [7,4], which exploits matrices such that columns have few non-zero entries, would still give a bound on the degree that would depend on Δ. (The Telephone broadcast problem now has a much better approximation algorithm [2], but that algorithm is purely combinatorial and no longer relies either on linear programming or on trees of small height and degree; however we consider the bicriteria problem of constructing trees of small height and degree as interesting in its own right).

2   Definitions

Definition 1. *Given a non-negative integer x called the goal and a non-negative real number y called the fortune, a game is defined recursively as follows: if $x = 0$, the game is a win, if $x > 0$ and $y = 0$, the game is lost. If $x, y > 0$, then the game consists of a (finite or infinite) sequence of bets (p_i) such that $\sum_i p_i \leq y$, along with, for each i, a game for goal $x - 1$ and fortune $y - \sum_{j \leq i} p_i$. The success probability of a game is the probability that the game eventually ends up in a winning state.*

A game can be represented by a (possibly infinite) complete binary tree with labelled edges, where the two edges from the root are labelled $1 - p_1$ and p_1, the left child of the root is a game for $(x, y - p_1)$, and the right child of the root is a game for $(x - 1, y - p_1)$.

The success probability can be computed as follows.

Fact 1 *The success probability of a game has the following properties.*
If $x = 0$ then the probability equals 1.
If $x \geq 1$ and $y = 0$ then the probability equals 0.
Otherwise, the success probability of a game T is given by:

$$\Pr(T \text{ succeeds}) = \sum_i \prod_{j < i} (1 - p_j) p_i \Pr(T_i \text{ succeeds}), \tag{1}$$

where p_1 is the first bet of the game, p_i is the ith bet of the game if all previous bets were unsuccessful, and T_i is the remaining game played when the ith bet is the first successful bet.

Note that T_i is a game with goal $x - 1$ and fortune $y - \sum_{j \leq i} p_i$). The (possibly infinite) number of terms in the sum in Equation 1 is the maximum number of bets performed by the game while the goal is x.

Note that as defined, for a given intermediate state (x', y'), the game may decide to bet different amounts, depending on the past history of the game from its starting point. Indeed, if we label vertices of the game tree by the current goal and fortune, there may be several vertices with the same label (x', y'), and each of them is root of a game for (x', y'); these games may all be different from one another.

Definition 2. *A memoryless game is a game such that at every step, the bet placed depends only on the current goal x' and on the current fortune y'. A strategy $\mathcal{H}$ is a function $(x, y) \mapsto p$, where x is a positive integer, y is a positive real number, and $p \in (0, 1]$ is such that $p \leq y$.*

A memoryless game can naturally be extended into a strategy $\mathcal{H}$ by defining $\mathcal{H}(x, y) = \min(1, y)$ for every (x, y) which does not appear as a label of a tree node. Conversely, to any strategy naturally corresponds a game for each (x, y), which proceeds as follows: Consider the current state (x, y). If $x, y > 0$, we place the bet $p = \mathcal{H}(x, y)$. With probability p, the bet is successful and the new goal is $x' = x - 1$; with probability $1 - p$, the bet is unsuccessful and the goal is still $x' = x$. We then continue playing the game associated to $\mathcal{H}$ on the new state $(x', y - p)$.

If $H(x, y)$ denotes the success probability of the game associated to strategy $\mathcal{H}$, Equation 1 then becomes:

$$H(x, y) = \sum_i \prod_{j<i} (1 - p_j) p_i H(x - 1, y - \sum_{j \leq i} p_i), \tag{2}$$

where $p_1 = \mathcal{H}(x, y)$, and in general $p_i = \mathcal{H}(x, y - \sum_{j<i} p_j)$ if $y - \sum_{j<i} p_j$ is positive. When we talk about the success probability of a strategy, we refer to the success probability of the associated game.

Given x and y, we are interested in computing the supremum, over all games for (x, y), of the success probability of the game. Some easy cases may serve as a warmup: evidently this success probability equals 1 whenever $y \geq x$ since it is then sufficient to place x bets each equal to 1. Another easy case is when $x = 1$ and $y < 1$: the supremum is then reached by the strategy which bets y as we now explain.

Lemma 2. $\sup_{T \text{ game for } (1,y)} \Pr(T \text{ succeeds}) = \min(1, y)$.

Proof. Given that the fortune is y, the expected number of successes is always at most y, regardless of the game. Thus y is an upper bound to the probability that the number of successes is at least 1. This is reached by the strategy which makes a single bet equal to y. □

Definition 3. *A continuous strategy is an extension of strategies which in addition is allowed to place infinitesimal bets, of the form: "repeat betting infinitesimal bets until there is a success or until the fortune spent equals p", for some $p \in (0, y]$.*

We use the notation exp_z to mean the step: "repeat betting infinitesimal bets until there is a success or until the fortune spent equals z". For consistency, a continuous strategy obviously has: If $\mathcal{H}(x, y) = exp_z$, then $\mathcal{H}(x, y - t) = exp_{z-t}$ for every $t \in (0, z]$. This can be seen as the limit, as N tends to infinity, of the process which bets $(1/N, \ldots, 1/N)$ up to zN times or until first success. Since the binomial distribution converges to a Poisson process in the limit, the time to first success is distributed exponentially: for any real number $t \leq z$, we have:

$$\Pr(\text{fortune spent at the end of this step is } \geq t) = e^{-t},$$

and the probability that a success occurs during this step is $1 - e^{-z}$. We will often use the term "discrete strategies" as a synonym for strategies, to contrast them from continuous strategies.

These definitions formalize the Best Strategy Theorem stated in the introduction. Theorem 1, which is our main result, follows from the Best Strategy Theorem as a simple corollary.

3   Proof of the Best Strategy Theorem

This section is devoted to the proof of Theorem 2 for $y < x$ (the Theorem is obvious for $y \geq x$). We will prove that $F(x, y) = G(x, y)$. In subsection 3.1 we will prove that $F(x, y) \leq G(x, y)$. In subsection 3.2, we will prove that $F(x, y) \geq G(x, y)$.

3.1   The Upper Bound

Definition 4. *A game for (x, y) is finite if its game tree is finite. A strategy is finite if for every (x, y), the associated game is finite.*

Lemma 3. *A game if finite if and only if each tree T_i in Fact 1 is finite, and the number of such trees is finite. A discrete strategy is finite if and only if, for every (x, y), the number of terms in the sum in Equation 2 is finite.*

Proof. The statement of the lemma is obvious for games. As for the statement for strategies, one direction is obvious and the other one can be proved by infuction on x.   □

The following lemma shows a reduction from games to finite games.

Lemma 4. *Given x, y, ϵ and a discrete game T for (x, y), there exists a game U for (x, y), which is finite, and such that $\Pr(T \text{ succeeds}) \leq \Pr(U \text{ succeeds}) + x\epsilon$.*

Proof.
Given T, consider the following game U which simulates T.

Game U to simulate the game tree T:
Let (p_j) be the sequence of bets which would be placed by T on (x, y), if every bet was unsuccessful.
If $\sum_j p_j \geq \epsilon$ then place a bet $q = p_1$, and
if the bet is successful,recursively simulate the game represented by the left subtree;
if not,recursively simulate the game represented by the right subtree.
Otherwise, play the game associated to the strategy which bets $q = \min(1, y)$.

The proof is by induction on x. Note that $\sum_j p_j$ is at most y, hence the series converges. Let i_0 be the number of terms of the sum if that is finite, or else the smallest index such that $\sum_{j > i_0} p_j < \epsilon$. Game U coincides with T for the first i_0 bets, and makes at most $i_0 + 1$ bets while the goal is x, hence U is a finite game by induction on x and by Lemma 3. A short calculation concludes the proof.
□

Fix $\epsilon > 0$. Consider a game T for (x, y). By Lemma 4, there exists a finite game U such that $\Pr(T \text{ succeeds}) \leq \Pr(U \text{ succeeds}) + x\epsilon$. Since this holds for every ϵ, we deduce that $\Pr(T \text{ succeeds}) \leq \sup_U$ finite game $\Pr(U \text{ succeeds})$. Since this holds for every T, we deduce that $F(x, y) = \sup_U$ finite game $\Pr(U \text{ succeeds})$.

The following lemma shows a reduction from finite games to finite memoryless games.

Lemma 5. *Given x, y and a finite game T, there exists a finite game U which is memoryless, and such that $\Pr(T \text{ succeeds}) \leq \Pr(U \text{ succeeds})$.*

From Lemma 5, we get that $F(x, y) = \sup_{\mathcal{H}}$ finite strategy $H(x, y)$. To finish the proof of the upper bound, all we need is to prove the following Proposition, to which we will devote the rest of this section.

Proposition 1. $\mathcal{G}$ *is better than any finite strategy.*

We start by observing that G is convex.

Lemma 6 (Convexity). *If $x \geq y + 1$ then $G(x+1, y) + G(x-1, y) \geq 2G(x, y)$.*

Proof. The proof uses induction on $x + \lceil y \rceil$. The base case $x = 1, y = 0$ is easy. Consider the general case. Run the three processes $G(x+1, y)$, $G(x, y)$ and $G(x-1, y)$ so as to couple the Poisson processes.

<u>Case 1</u>: If $y \leq x - 2$ then all three processes start with an exponential waiting time to first success. We use straightforward induction on G for $x - 1$ and the remaining fortune at the time of first success.

<u>Case 2</u>: If $y \in (x - 2, x - 1]$ then we let $y' = x - 2$ and observe the three processes until the remaining fortune is y'. For a shorthand, let $G_z = G(z, y')$ for any z. Let $a = y - y'$. After some calculations, we get:

$$G(x-1, y) + G(x+1, y) - 2G(x, y) = e^{-a}(G_{x-1} + G_{x+1} - 2G_x) + ae^{-a}(1 + G_x - 2G_{x-1}).$$

By induction hypothesis for $y' = x - 2$ (noting that $\lceil y' \rceil < \lceil y \rceil$), both quantities within brackets are non negative, hence the lemma.

$\square$

The following is a technical Lemma which will be used in the sequel. It uses the notion of continuous games, similar to the notion of continuous strategies.

Definition 5. *Given (x, y), a continuous game is defined recursively as follows: if $x = 0$, the game is a win, if $x > 0$ and $y = 0$, the game is lost. If $x, y > 0$, then the game consists of a (finite or infinite) sequence of steps, where step i consists either of bet $p_i > 0$ or of the repetition of infinitesimal bets until there is a success or until the fortune spent equals p_i; we must have $\sum_i p_i \leq y$. For each i, if step i was a positive bet p_i, then we also have a game for goal $x - 1$ and fortune $y - \sum_{j \leq i} p_i$; if step i was a sequence of infinitesimal bets up to p_i, then we also have, for each t such that $p_1 + \cdots + p_{i-1} < t \leq p_1 + \cdots + p_i$, a game for $(x - 1, y - t)$.*

Lemma 7. *Let H and K be two continuous games for (x,y) which both go through a state where the remaining fortune is $y' < x - 2$, after having had 0, 1 or 2 successes, and then continue with strategy $\mathcal{G}$ after that point. Then $K(x,y) \geq H(x,y)$ if and only if the probability of having had 0 successes before y' is greater for K than for H.*

Proof. Uses the convexity lemma. □

The following lemma is the core of the proof of Theorem 2.

Lemma 8. *Consider a continuous game whose first bet is arbitrary positive and which then continues by using strategy $\mathcal{G}$. Then its success probability is less than or equal to $G(x,y)$.*

Proof. The proof is by induction on x. If $x = 0$, then there is nothing to prove. Consider $x \geq 1$. Let p be the first bet placed by the game on (x,y). Let $K(x,y)$ denote the success probability of the game. There are several cases.

Case 1: $y \leq x - 1$. Then $\mathcal{G}$ starts by making infinitesimal bets.

Subcase 1.1: Assume $y - p < x - 2$. Let $y' = y - p$. We compare K to the following game L. L makes infinitesimal bets until first success or y'; in the former case, let t be the remaining fortune at that time of first success: L then places a bet of $t - y'$ to get to fortune y'. Once the fortune is y', L continues by following strategy $\mathcal{G}$.

We appeal to Lemma 7 to compare K and L. The probability that K has had 0 successes by the time the fortune is y' is $1 - p$. The probability that L has had 0 successes is $e^{-p} > 1 - p$, hence $L(x,y) > K(x,y)$.

It is now easy to compare L to $\mathcal{G}$: $L(x,y) = \int_0^p e^{-z} L'(x - 1, y - z) dz + e^{-p} G(x, y - p)$. Game L' places a first bet of $(y - z) - y'$ and then continues using strategy $\mathcal{G}$. By induction applied to x and L', we have $L'(x - 1, y - z) \leq G(x - 1, y - z)$. Thus $L(x,y) \leq \int_0^p e^{-z} G(x - 1, y - z) dz + e^{-p} G(x, y - p) = G(x,y)$.

Together, these inequalities imply $K(x,y) \leq L(x,y) \leq G(x,y)$.

Subcase 1.2: Assume $y - p \geq x - 2$. Let $y' = x - 2$. Our game K first bets p, bringing its fortune down to $y - p$, then applies $\mathcal{G}$: if the first bet was successful, it bets $r = (y - p) - y'$, bringing its fortune down to y'. If it was unsuccessful, it makes infinitesimal bets until a first success (when the remaining fortune is t) or y', and in the first case, bets $t - y'$, bringing the fortune down to y'.

We compare K to the following game L: makes infinitesimal bets until first success (when the remaining fortune is t') or y', and in the former case, bets $t' - y'$, bringing the fortune down to y'. Once the fortune is y', L continues by following strategy $\mathcal{G}$.

We appeal to Lemma 7 to compare K and L. The probability that K has had 0 successes by the time the fortune is y' is $(1 - p)e^{-r}$. The probability that L has had 0 successes is $e^{-(p+r)} = e^{-p}e^{-r} > (1 - p)e^{-r}$, hence $L(x,y) > K(x,y)$.

The comparison of L to G is similar to Subcase 1.1. **Case 2:** $y \geq x - 1$. (Of course, we still have $y < x$.) Then $\mathcal{G}$ starts by betting $y - (x - 1)$.

Subcase 2.1: $p < y - (x - 1)$. Let $z = x - 1$. Game K first bets p, bringing its fortune down to $y - p$, then applies $\mathcal{G}$ by betting $r = (y - p) - z$ to bring the fortune down to z, then continues applying $\mathcal{G}$. We compare K to the game associated to strategy $\mathcal{G}$: make a single bet of $y - z$ to bring the fortune down to z, then continue applying $\mathcal{G}$.

The winning probability of K is $K(x, y) = p + (1-p)r + (1-p)(1-r)G(x, z)$. The winning probability of $\mathcal{G}$ is $(p + r) + (1-p-r)G(x, z)$, and one easily checks that $\mathcal{G}$ is better than K.

Subcase 2.2: $p > y - (x - 1)$. Let $y' = x - 2$. Game K bets p, bringing the fortune down to $y - p$, then, in case of success, bets $(y-p) - y'$; in case of failure, it makes infinitesimal bets until first success (when the remaining fortune is t) or y', and in the former case, bets $t - y'$. Strategy $\mathcal{G}$ first bets $u = y - (x - 1)$, then, in case of success, bets 1 to bring the fortune down to y'; in case of failure, it makes infinitesimal bets until first success (when the remaining fortune is t) or y', and in the former case, it bets $t - y'$ ro bring the fortune down to y'.

We appeal to Lemma 7 to compare K and $\mathcal{G}$. The probability that K has had 0 successes is $p_K = (1 - u - v)e^{-(1-v)}$, where $u = y - (x - 1)$ and $u + v = p$. The probability that $\mathcal{G}$ has had 0 successes is $p_{\mathcal{G}} = (1 - u)e^{-1}$. The ratio is

$$\frac{p_K}{p_{\mathcal{G}}} = (1 - \frac{v}{1-u})e^v < (1-v)e^v < e^{-v}e^v = 1,$$

hence $\mathcal{G}$ is better than K.   □

Proposition 1 then follows by induction on N, the maximum number of steps of the finite strategy applied to (x, y), and by appealing to Lemma 8.

3.2   The Lower Bound

Lemma 9. *Let T be a continuous game for (x, y). For each positive ϵ, there exists a discrete game U for (x, y) such that $\Pr(U \text{ succeeds}) \geq \Pr(T \text{ succeeds}) - x\epsilon$.*

Proof. We will compare T to the following randomized game U.

Game U to simulate T on (x, y):
If T places a positive bet $p > 0$, then bet p;
   if successful, recursively simulate the right subtree of T;
   if not, recursively simulate the left subtree of T.
Otherwise (T places an infinitesimal bet spending up to p), bet $\alpha = \min(p, \epsilon)$;
   if unsuccessful, recursively simulate the game for $(x, y - \alpha)$;
   otherwise, with probability $[e^{-\alpha} - (1 - \alpha)]/(1 - \alpha)$, recursively simulate the game for $(x, y - \alpha)$; and with the remaining probability, recursively simulate the game for $(x - 1, y - T)$, where the random variable $T \in [0, \alpha]$ has density function $e^{-t}/(1 - e^{-\alpha})$.

It is easy to see that U simulates T exactly (except for "giving away" a success with probability $(1 - \alpha) \times [e^{-\alpha} - (1 - \alpha)]/(1 - \alpha)$) while spending only a little bit more fortune: every time T has a success, game U spends up to ϵ more fortune than T. But T needs only x successes to reach the goal: so, if T on (x, y) still has a remaining fortune of at least $x\epsilon$ when it reaches its goal, then an initial fortune of y will be sufficient for game U. The probability that T reaches its goal while spending the last $x\epsilon$ part of its fortune, is bounded by the probability that it has one or more success while spending that last $x\epsilon$ fortune; that is less than or equal to the expected number of successes during that time, i.e. less than or equal to $x\epsilon$. Hence $\Pr(U \text{ succeeds}) \geq \Pr(T \text{ succeeds}) - x\epsilon$.

Finally, it is easy to de-randomize U: just pick some T' such that the success probability for the game for $(x-1, y-T')$ is greater than or equal to the expected value, over T, of the game for $(x-1, y-T)$. We thus obtain a strategy satisfying the Lemma.   □

From Lemma 9, we get that $F(x, y) = \sup_T$ continuous game $\Pr(T \text{ succeeds})$, which is obviously greater than or equal to $G(x, y)$, and the proof of Theorem 2 is complete.

4   A Randomized Rounding Application

In [6], Ravi presented an algorithm to build a spanning tree of small diameter and small maximum degree in a given graph. Here, using the framework of gambling games, we present a finer analysis of Ravi's algorithm, thus improving on his approximation bounds. Here is the algorithm.

Input: a graph G with vertex set $V(G)$ and a bound Δ on the desired diameter.

Output: a spanning tree G

Dynamic variables in the algorithm are a subgraph K of G and a set $C \subseteq V$ of cluster centers.

1. Initialize K to a graph with vertex set $V(G)$ and no edges; Initialize C to $V(G)$.
2. While there is more than one cluster center do:
   (a) Set up an integer program of the type described in the Introduction, where:
      i. For every path P of length at most Δ directed from one cluster center to another there is a 0-1 variable $x(P)$;
      ii. For every cluster center $c \in C$ there is a covering constraint of the form $\sum x(P) = 1$, where the sum is over all paths P directed out of c;
      iii. For every vertex v there is a resource constraint of the form $\sum x(P) \leq z$, where the sum is over paths P incident with vertex v.
   (b) Solve the linear programming relaxation of this integer program (Ravi shows that this can be done in polynomial time);
   (c) Use randomized rounding to obtain a feasible solution to the integer program, giving, for each cluster center, a path of length at most Δ directed to some other cluster center;
   (d) Consider the graph H with one vertex for each cluster center in C and one directed edge (c, c') for each path in the solution to the integer program. Each vertex in H has out-degree exactly 1. By elementary graph theory,

find a subgraph H' of H containing at least $\frac{|V(H)|}{3}$ edges and consisting of a disjoint union of "stars," where each star consists of a root vertex c' and one or more vertices c such that (c, c') is an edge of H;

(e) For each edge (c, c') in H', add the corresponding path to K, and delete c from C, the set of cluster centers.

3. Let c be the unique remaining cluster center. By breadth-first search from c in K, construct a tree T spanning $V(G)$.

Theorem 3 (Ravi). *Assume that G has a spanning tree T^* of diameter at most Δ and maximum degree at most D^*. Then with high probability the above algorithm will produce a spanning tree of height $O(\Delta \log n)$ and maximum degree $O(D^* \log n + \log^2 n)$.*

Proof. The height of T equals the height of K. Since the "while" loop is executed $O(\log n)$ times, each vertex $v \in G$ is linked to c in K by a sequence of at most $O(\log n)$ flow paths, each of length at most Δ. Hence T has height $O(\Delta \log n)$. Using T^*, it is easy to construct a multicommodity flow of length at most Δ and value at most $2D^*$. Hence the solution of the LP in step 2a has value at most $2D^*$. By the randomized rounding Theorem (which is based on a Hoeffding bound), the integral multicommodity flow in step 2b has value at most $2D^* + O(\log n)$ with high probability, and so the union of the flow paths taken in step 2e also has maximum degree at most $2D^* + O(\log n)$. Since the "while" loop is executed $O(\log n)$ times, the resulting graph K has maximum degree $O(D^* \log n + \log^2 n)$ (with high probability), and hence the output T also has maximum degree $O(D^* \log n + \log^2 n)$ (with high probability). □

We will use our gambling game to provide a more refined analysis of Ravi's algorithm.

Theorem 4. *Assume that G has a spanning tree T^* of diameter at most Δ and maximum degree at most D^*. Then the above algorithm will produce a spanning tree of height $O(\Delta \log n)$ and maximum degree $O(D^* \log n)$ (with high probability).*

Proof. Fix a vertex ℓ of G. We play the gambling game: the initial fortune is $2D^* t$ where t is the number of integer programs solved in the algorithm, and the goal is $6\alpha D^* \log n$, where α is chosen to guaranteed that $t \leq \alpha \log n$. The bets are done in phases corresponding to the successive integer programs in the algorithm. In each phase there is a bet for each cluster center, equal to the probability that vertex ℓ will lie in the path from that cluster center selected by the integer program. The sum of these bets is just the sum of the fractional variables in the linear program corresponding to directed paths passing through ℓ. Since the value of the linear program is at most $2D^*$ the sum of the bets in each phase is at most $2D^*$ and the sum of all bets does not exceed the fortune $2D^* t$. The degree of ℓ in the tree T is at most twice the number of selected paths through ℓ in the course of the algorithm, and the number of selected paths is equal to the number of successes in the gambling game. Hence the probability that the degree of ℓ is T is greater than or equal to $6\alpha D^* \log n$ is at most $F(3\alpha D^* \log n,$ $2\alpha D^* \log n)$, which can be shown to be exponentially small in n.   □

References

1. Amotz Bar-Noy, Sudipto Guha, Joseph (Seffi) Naor, and Baruch Schieber, *Multicasting in Heterogeneous Networks*, SIAM J. Comput. 30(2): 347-358 (2000).
2. Michael Elkin and Guy Kortsarz, *Sublogarithmic Approximation for Telephone Multicast: Path out of Jungle*, SODA 2003, to appear.
3. R. M. Karp, F. T. Leighton, C. D. Thompson, U. V. Vazirani, and V. V. Vazirani, *Global wire routing in two-dimensional arrays*, Algorithmica, 2, 113-129, 1987.
4. New Algorithmic Aspects of the Local Lemma with Applications to Routing and Partitioning, F. T. Leighton, C.-J. Lu S. B. Rao and Aravind Srinivasan. SIAM Journal on Computing, Vol. 31, 626-641, 2001.
5. P. Raghavan and C.D. Thompson, *Randomized rounding: a technique for provably good algorithms and algorithmic proofs*, Combinatorica 7 (1987), 365-374.
6. R. Ravi, *Rapid Rumor Ramification: Approximation the minimum broadcast time*, 35th IEEE Symposium on Foundations of Computer Science (FOCS), 1994.
7. Aravind Srinivasan, *An Extension of the Lovasz Local Lemma and its Applications to Integer Programming*, Proc. ACM-SIAM Symposium on Discrete Algorithms (SODA), pages 6-15, 1996.

Tight Bounds for Testing Bipartiteness in General Graphs

Tali Kaufman[1]*, Michael Krivelevich[2]**, and Dana Ron[3]***

[1] School of Computer Science, Tel Aviv University,Tel Aviv 69978 Israel.
kaufmant@post.tau.ac.il
[2] Department of Mathematics, Tel Aviv University, Tel Aviv 69978, Israel.
krivelev@post.tau.ac.il
[3] Department of Electrical Engineering-Systems, Tel Aviv University, Tel Aviv 69978, Israel.
danar@eng.tau.ac.il

Abstract. In this paper we consider the problem of testing bipartiteness of general graphs. The problem has previously been studied in two models, one most suitable for dense graphs, and one most suitable for bounded-degree graphs. Roughly speaking, dense graphs can be tested for bipartiteness with constant complexity, while the complexity of testing bounded-degree graphs is $\tilde{\Theta}(\sqrt{n})$, where n is the number of vertices in the graph. Thus there is a large gap between the complexity of testing in the two cases.

In this work we bridge the gap described above. In particular, we study the problem of testing bipartiteness in a model that is suitable for all densities. We present an algorithm whose complexity is $\tilde{O}(\min(\sqrt{n}, n^2/m))$ where m is the number of edges in the graph, and match it with an almost tight lower bound.

1 Introduction

Property testing algorithms [16, 8] are algorithms that perform *approximate decisions*. Namely, for a predetermined property P they should decide whether a given object O has property P or is *far* from having property P. In order to perform this approximate decision they are given *query access* to the object O. Property testing problems are hence defined by the type of objects in question, the property tested, the type of queries allowed, and the notion of distance to having a property. Much of the focus of property testing has been on testing properties of graphs. In this context several models have been considered. In all models, for a fixed graph property P, the algorithm is required to accept graphs that have P and to reject graphs that are ϵ-far from having P, for a given distance parameter ϵ. In all cases the algorithm is allowed a constant probability of failure. The models differ in the type of queries they allow and in the notion of distance

---

* This work is part of the author's Ph.D. thesis prepared at Tel Aviv University under the supervision of Prof. Noga Alon, and Prof. Michael Krivelevich.
** Research supported in part by a USA Israeli BSF grant and by a grant from the Israel Science Foundation.
*** Research supported by the Israel Science Foundation (grant number 32/00-1).

S. Arora et al. (Eds.): APPROX 2003+RANDOM 2003, LNCS 2764, pp. 341–353, 2003.

they use (which underlies the definition of being ϵ-far from having the property). The complexity of the algorithm is measured by the number of queries to the object Q it performs.

1.1 Models for Testing Graph Properties

The first model, introduced in [8], is the adjacency-matrix model. In this model the algorithm may perform queries of the form: "Is there an edge between vertices u and v in the graph?" That is, the algorithm may probe the adjacency matrix representing the graph. We refer to such queries as *vertex-pair* queries. The notion of distance is also linked to this representation: a graph is said to be ϵ-far from having property P if more than ϵn^2 edge modifications should be performed on the graph so that it obtains the property, where n is the number of vertices in the graph. In other words, ϵ measures the fraction of entries in the adjacency matrix of the graph that should be modified. This model is most suitable for *dense* graphs in which the number of edges m is $\Theta(n^2)$. This model was studied in [8, 3, 2, 1, 4, 11, 7].

The second model, introduced in [9], is the (bounded-degree) incidence-lists model. In this model, the algorithm may perform queries of the form: "Who is the i'th neighbor of vertex v in the graph?" That is, the algorithm may probe the incidence lists of the vertices in the graph, where it is assumed that all vertices have degree at most d for some fixed degree-bound d. We refer to these queries as *neighbor* queries. Here too the notion of distance is linked to the representation: A graph is said to be ϵ-far from having property P if more than ϵdn edge modifications should be performed on the graph so that it obtains the property. In this case ϵ measures the fraction of entries in the incidence lists representation (among all dn entries), that should be modified. This model is most suitable for graphs with $m = \Theta(dn)$ edges; that is, whose maximum degree is of the same order as the average degree. In particular, this is true for *sparse* graphs that have *constant degree*. This model was studied in [10, 9, 6].

In [15] it was suggested to decouple the questions of representation and type of queries allowed from the definition of distance to having a property. Specifically, it was suggested to measure the distance simply with respect to the number of edges, denoted m, in the graph. Namely, a graph is said to be ϵ-far from having a property, if more than ϵm edge modifications should be performed so that it obtains the property. In [15] the algorithm was allowed the same type of queries as in the bounded-degree incidence-lists model, but no fixed upper-bound was assumed on the degrees and the algorithm could query the degree of any vertex. The main advantage of this model over the bounded-degree incidence-lists model is that it is suitable for graphs whose degrees may vary significantly.

The Model Studied in this Paper. In this work we are interested in a model that may be useful for testing all types of graphs: dense, sparse, and graphs that lie in-between the two extremes. As is discussed in more detail in the next subsection, the two extremes sometimes exhibit very different behavior in terms of the complexity of testing the same property. We are interested in understanding the transformation from testing sparse (and in particular bounded-degree) graphs to testing dense graphs.

Recall that a model for testing graph properties is defined by the distance measure used and by the queries allowed. The model of [15] is indeed suitable for all graphs in terms of the distance measure used, since distance is measured with respect to the actual number of edges m in the graph.[4] Thus this notion of distance adapts itself to the density of the graph, and we shall use it in our work.

The focus in [15] was on testing properties that are of interest in sparse (but not necessarily bounded-degree) graphs, and hence they allowed only neighbor queries. However, consider the case in which the graph is not sparse (but not necessarily dense). In particular suppose that the graph has $\omega(n^{1.5})$ edges, and that we are seeking an algorithm that performs $o(\sqrt{n})$ queries. While in the case of sparse graphs, there is no use in asking vertex-pair queries (i.e., is there an edge between a particular pair of vertices), such queries may become helpful when the number of edges is sufficiently large. Hence, we allow our algorithms to perform both neighbor queries and vertex-pair queries.

1.2 Testing Bipartiteness

One of the properties that has received quite a bit of attention in the context of property testing, is *bipartiteness*. Recall that a graph is bipartite if it is possible to partition its vertices into two parts such that there are no edges with both endpoints in the same part. This property was first studied in [8] where it was shown that bipartiteness can be testing by a simple algorithm using $\tilde{O}(1/\epsilon^3)$ queries. This was improved in [3] to $\tilde{O}(1/\epsilon^2)$ queries. The best lower bound known in this model is $\tilde{\Omega}(1/\epsilon^{1.5})$, due to [7]. Thus the complexity of this problem is independent of the number of vertices n and polynomial in $1/\epsilon$.

The complexity of testing bipartiteness changes significantly when considering the bounded-degree incidence-lists model. In [10] a lower bound of $\Omega(\sqrt{n})$ is established in this model, for constant ϵ and d (the degree bound). An almost matching upper bound of $\tilde{O}(\sqrt{n} \cdot \text{poly}(1/\epsilon))$ is shown in [9]. Thus, in the case of bipartiteness there is a large gap between the results that can be obtained for dense graphs and for constant-degree graphs. Here we venture into the land of graphs that are neither necessarily sparse, nor necessarily dense, and study the complexity of testing bipartiteness. Other graph properties exhibit similar (and sometimes even larger) gaps, and hence we believe that understanding the transformation from sparse to dense graphs is of general interest.

1.3 Our Results

In this work we present two complementary results for n-vertex graphs having m edges:

- We describe and analyze an algorithm for testing bipartiteness in general graphs whose query complexity (and running time) is $O(\min(\sqrt{n}, n^2/m) \cdot \text{poly}(\log n/\epsilon))$.

---

[4] We assume for simplicity that the number of vertices, n, and the number of edges, m, are both given to the testing algorithm. If they are not known exactly, the algorithm can work using upper bounds on these values. The tightness of these bounds will naturally affect the performance of the algorithm.

The algorithm has a one-sided error (i.e., it always accepts bipartite graphs). Furthermore, whenever it rejects a graph it provides *evidence* that the graph is not bipartite in the form of an odd-length cycle of length poly$(\log n/\epsilon)$.

- We present an almost matching lower bound of $\Omega(\min(\sqrt{n}, n^2/m))$ (for a constant ϵ). This bound holds for all testing algorithms (that is, for those which are allowed a two-sided error and are adaptive). Furthermore, the bound holds for regular graphs.

As seen from the above expressions, as long as $m = O(n^{1.5})$, that is, the average degree is $O(\sqrt{n})$, the complexity of testing is $\tilde{\Theta}(\sqrt{n})$. Once the number of edges goes above $n^{1.5}$, we start seeing a decrease in the query complexity which in this case is at most $O((n^2/m) \cdot \text{poly}(\log n/\epsilon))$. In terms of our algorithm, this is exactly the point where our algorithm starts exploiting its access to vertex-pair queries. Our lower bound shows that this behavior of the query complexity is not only an artifact of our algorithm but is inherent in the problem.

Note that even if the graph is sparse then we obtain a new result that does not follow from [9]. Namely, we have an algorithm with complexity $\tilde{O}(\sqrt{n} \cdot \text{poly}(1/\epsilon))$ for sparse graphs with varying degrees.

1.4   Our Techniques

We present our algorithm in two stages. First we describe an algorithm that works for almost-regular graphs, that is, graphs in which the maximum degree is of the same order as the average degree. The algorithm and its analysis closely follow the algorithm and analysis in [9]. Indeed, as long as the degree d of the graph is at most $\sqrt{n}$, we execute the [9] algorithm. The place where we depart from [9] is in the usage of vertex-pair queries once $d > \sqrt{n}$. We refer to our first algorithm as Test-Bipartite-Reg.

In the second stage we show how to reduce the problem of testing bipartiteness of general graphs to bipartiteness of almost-regular graphs. Namely, we show how, for every given graph G, it is possible to define a graph G' such that: (1) G' has roughly the same number of vertices and edges as G, and its maximum degree is of the same order as its average degree (which is roughly the same as the average degree in G); (2) If G is bipartite then so is G', and if G is far from bipartite then so is G'. We then show how to emulate the execution of the algorithm Test-Bipartite-Reg on G' given query access to G, so that we may accept G if it accepts G', and reject G if it rejects G'.

In the course of this emulation we are confronted with the following interesting problem: We would like to sample vertices in G according to their degrees (which aids us in sampling vertices uniformly in G', a basic operation that is required by Test-Bipartite-Reg). The former is equivalent to sampling *edges* uniformly in G. In order not to harm the performance of our testing algorithm, we are required to perform this task in $\tilde{O}(\min(\sqrt{n}, n^2/m))$ queries. If m is sufficiently large (once again, if $m \geq n^{1.5}$), this can be performed simply by sampling sufficiently many pairs of vertices in G. However, we do not know how to perform this task exactly (in an efficient manner) when the number of edges is significantly smaller than $n^{1.5}$. Nonetheless, we provide a sampling procedure that selects edges according to a distribution that approximates the desired uniform distribution on edges, and is sufficient for our purposes. The approximation is such that for all but a small fraction of the m edges, the probability of selecting an edge is $\Omega(1/m)$. This procedure may be of independent interest.

We also conjecture that variants of our construction of G' (and in particular a prob-abilistic construction we suggest in the long version of this paper [12]), may be useful in transforming other results that hold for graphs whose maximum degree is similar to their average degree, to results that hold for graphs with varying degrees.

We establish our lower bound by describing, for every pair n, d (n even, $d \geq 64$), two distributions over d-regular graphs. In one distribution all graphs are bipartite by construction. For the other distribution we prove that almost all graphs are far from bipartite. We then show that every testing algorithm that can distinguish between a graph chosen randomly from the first distribution (which it should accept with proba-bility at least $2/3$), and a graph chosen randomly from the second distribution (which it should reject with probability at least $2/3$), must perform $\Omega(\min(\sqrt{n}, n/d)) = \Omega(\min(\sqrt{n}, n^2/m))$ queries. In the lower bound proof we show the necessity of both *neigbhor queries* and *vertex-pair queries*. Specifically by using only one type of queries the lower bound increases.

1.5   Further Research

As noted previously, there are other problems that exhibit a significant gap between the query complexity of testing dense graphs (in the adjacency-matrix model) and the complexity of testing sparse, bounded-degree graphs (in the bounded-degree incidence-lists model). In particular this is true for testing k-colorability. It is possible to test dense graphs for k-colorability using $\text{poly}(k/\epsilon)$ queries [8, 3], while testing sparse graphs requires $\Omega(n)$ queries [6]. We stress that these bounds are for query complexity, where we put time complexity aside. We would like to understand this transformation from essentially constant complexity (for constant k and ϵ) to linear complexity, and we would like to know whether any intermediate results can be obtained for graphs that are neither sparse nor dense. Other problems of interest are testing whether a graph has a relatively large clique [8], testing acyclicity of directed graphs [5], and testing that a graph does not contain a certain subgraph [1].

2   Preliminaries

Let $G = (V, E)$ be an undirected graph with n vertices labeled $1, ..., n$, and let $m = m(G) = |E(G)|$ be the total number of edges in G. Unless stated otherwise, we assume that G contains no multiple edges. For each vertex $v \in V$ let $\Gamma(v)$ denote its set of neighbors, and let $\deg(v) = |\Gamma(v)|$ denote its degree. The edges incident to v (and their end-points, the neighbors of v), are labelled from 1 to $\deg(v)$. Note that each edge has two, possibly different, labels, one with respect to each of its end-points. We hence view edges as quadruples. That is, if there is an edge between v and u, and it is the i-th edge incident to v and the j-th edge incident to u, then this edge is denoted by (u, v, i, j). When we want to distinguish between the quadruple (u, v, i, j) and the pair (u, v) then we refer to the latter as an *edge-pair*. We let $d_{\max} = d_{\max}(G)$ denote the maximum degree in the graph G and $d_{\text{avg}} = d_{\text{avg}}(G)$ denote the average degree in the graph (that is, $d_{\text{avg}}(G) = 2m(G)/n$).

Distance to having a property. Consider a fixed graph property $\mathcal{P}$. For a given graph G, let $e_{\mathcal{P}}(G)$ be the minimum number of edges that should be added to G or removed from G so that it obtain property $\mathcal{P}$. The distance of G to having property $\mathcal{P}$ is defined as $e_{\mathcal{P}}(G)/m(G)$. In particular, we say that graph G is ϵ-*far* from having the property $\mathcal{P}$ for a given distance parameter $0 \le \epsilon < 1$, if $e_{\mathcal{P}}(G) > \epsilon \cdot m(G)$. Otherwise, it is ϵ-*close* to having property $\mathcal{P}$. In some cases we may define the distance to having a property with respect to an upper bound $m_{\max} \ge m(G)$ on the number of edges in the graph (that is, the distance to having property $\mathcal{P}$ is defined as $e_{\mathcal{P}}(G)/m_{\max}$). For example, if the graph is dense, so that $m(G) = \Omega(n^2)$ then we set $m_{\max} = n^2$, and alternatively, if the graph has some bounded degree d, then we set $m_{\max} = d \cdot n$. (In the latter case we could set $m_{\max} = (d \cdot n)/2$, but for simplicity we set the slightly higher upper bound.) If $e_{\mathcal{P}}(G)/m_{\max} > \epsilon$ then we shall say that the graph is ϵ-far from property $\mathcal{P}$ *with respect to* $m_{\max}$.

Testing algorithms. A testing algorithm for a graph property $\mathcal{P}$ is required to accept with probability at least $2/3$ every graph that has property $\mathcal{P}$ and to reject with probability at least $2/3$ every graph that is ϵ-far from having property $\mathcal{P}$, where ϵ is a given distance parameter. If the algorithm always accepts graphs that have the property then it is a *one-sided error* algorithm. The testing algorithm is given the number of vertices in the graph, the number of edges in the graph, or an upper bound on this number, and it is provided with *query access* to the graph. Specifically we allow the algorithm the following types of queries.

- The first type of queries are *degree* queries. That is, for any vertex u of its choice, the algorithm can obtain $\deg(u)$. We assume that a degree query has cost one. In fact it can be easily implemented using neighbor queries with cost $O(\log d_{\max}) = O(\log n)$.

- The second type of queries are *neighbor* queries. Namely, for every vertex u and index $1 \le i \le \deg(u)$, the algorithm may obtain the i-th neighbor of vertex u.

- The third type of queries are *vertex-pair* queries. Namely, for any pair of vertices (u, v), the algorithm can query whether there is an edge between u and v in G.

Bipartiteness. In this work we focus on the property of *bipartiteness*. Let (V_1, V_2) be a partition of V. We say that an edge $(u, v) \in E$ is a *violating* edge with respect to (V_1, V_2), if u and v belong to the same subset V_b, (for some $b \in \{1, 2\}$). A graph is *bipartite* if there exists a partition of its vertices with respect to which there are no violating edges. By definition, a graph is ϵ-far from bipartite if for every partition of its vertices, the number of violating edges with respect to the partition is greater than $\epsilon \cdot m$. Recall that a graph is bipartite if and only if it contains no odd-length cycles.

3   The Algorithm for the Almost-Regular Case

In this section we describe an algorithm that accepts every bipartite graph and that rejects with probability at least $2/3$ every graph that is ϵ-far from bipartite with respect to an upper bound $m_{\max} = d_{\max}n$ on the number of edges. Namely, this algorithm rejects (with probability at least $2/3$) graphs for which the number of edges that need to be removed so that they become bipartite is greater than $\epsilon \cdot m_{\max} = \epsilon \cdot d_{\max} n$. The query complexity (and running time) of this algorithm is $O(\min(\sqrt{n}, n/d_{\max}) \cdot \text{poly}(\log n/\epsilon))$.

In the case where the graph is almost-regular, that is, the maximum degree of the graph $d_{\max}$ is of the same order as the average degree, d_{avg}, then we essentially obtain a tester as desired (since in such a case $\epsilon d_{\max} n = O(\epsilon m)$). However, in general, $d_{\max}$ may be much larger d_{avg} (for example, it is possible that $d_{\max} = \Theta(n)$ while $d_{\text{avg}} = \Theta(1)$). To deal with the general case we show in the next section (Section 4) how to reduce the problem in the general case to the special case of $d_{\max} = O(d_{\text{avg}})$.

A High Level Description of the Algorithm. Throughout this section let $d = d_{\max}$. Our algorithm builds on the testing algorithm for bipartiteness described in [9] whose query complexity is $O(\sqrt{n} \cdot \text{poly}(\log n/\epsilon))$ (and which works with respect to $m_{\max} = dn$ as well). In fact, as long as $d \leq \sqrt{n}$ our algorithm is equivalent to the algorithm in [9]. In particular, as in [9], our algorithm selects $\Theta(1/\epsilon)$ *starting vertices* and from each it performs several random walks (using neighbor queries), each walk of length $\text{poly}(\log n/\epsilon)$. If $d \leq \sqrt{n}$ then the number of these walks is $O(\sqrt{n} \cdot \text{poly}(\log n/\epsilon))$, and the algorithm simply checks whether an odd-length cycle was detected in the course of these random walks (possibly relying on information from more than one random walk to find an odd cycle).

If $d > \sqrt{n}$ then there are two important modifications: (1) The number of random walks performed from each vertex is reduced to $O(\sqrt{n/d} \cdot \text{poly}(\log n/\epsilon))$; (2) For each pair of end vertices reached in these walks with the same parity, the algorithm performs a vertex-pair query. Similarly to the $d \leq \sqrt{n}$ case, the graph is rejected if an odd-length cycle is found in the subgraph induced by all queries performed. Pseudo-code for the algorithm is shown in Figure 1.

Random Walks and Paths in the Graph. The random walks performed are defined as follows: At each step, if the degree of the current vertex v is $d' \leq d$, then the walk *remains* at v with probability $1 - \frac{d'}{2d} \geq \frac{1}{2}$, and for each $u \in \Gamma(v)$, the walk *traverses* to u with probability $\frac{1}{2d}$. The important property of the random walk is that the stationary distribution it induces over the vertices is uniform.

For every walk (or, more generally, for any sequence of steps), there corresponds a *path* in the graph. The path is determined by those steps in which an edge is traversed (while ignoring all steps in which the walk stays at the same vertex). Such a path is not necessarily simple, but does not contain self loops. Note that when referring to the length of a walk, we mean the total number of steps taken, including steps in which the walk remains at the current vertex, while the length of the corresponding path does not include these steps.

Theorem 1 *The algorithm Test-Bipartite-Reg accepts every graph that is bipartite, and rejects with probability at least $2/3$ every graph that is ϵ-far from bipartite with respect to $m_{\max} = d_{\max} n$. Furthermore, whenever the algorithm rejects a graph it outputs a* certificate *to the non-bipartiteness of the graph in form of an odd-length cycle of length $\text{poly}(\log n/\epsilon)$. The query complexity and running time of the algorithm are $O(\min(\sqrt{n}, n/d_{\max}) \cdot \text{poly}(\log n/\epsilon))$.*

Test-Bipartite-Reg$(n, d_{\max}, \epsilon)$

- Repeat $T = \Theta(\frac{1}{\epsilon})$ times:
  1. Uniformly select s in V.
  2. If Odd-Cycle(s) returns found then output reject.
- In case no call to Odd-Cycle returned found then output accept.

Odd-Cycle(s)

1. If $d = d_{\max} \leq \sqrt{n}$ then let $K \overset{\text{def}}{=} \Theta(\frac{\log^{1/2}(n/\epsilon)}{\epsilon^3})$ and $L \overset{\text{def}}{=} \Theta(\frac{\log(n/\epsilon)^3}{\epsilon^5})$. Otherwise $(d > \sqrt{n})$, let $K \overset{\text{def}}{=} \Theta\left(\frac{\log^{1/2}(n/\epsilon)\cdot\sqrt{n/d}}{\epsilon^8}\right)$, and $L \overset{\text{def}}{=} \Theta\left(\frac{\log^6(n/\epsilon)}{\epsilon^8}\right)$.
2. Perform K random walks starting from s, each of length L.
3. Let A_0 (A_1) be the set of vertices that appear on the ends of the K walks whose paths are of even (odd) length.
4. If $d \leq \sqrt{n}$ then check whether $A_0 \cap A_1 \neq \emptyset$. If the intersection is non-empty then return found, otherwise return not-found.
5. Else $(d > \sqrt{n})$, perform vertex-pair queries between every pair of vertices $u, v \in A_0$ $(u, v \in A_1)$. If an edge is detected then return found, otherwise return not-found.

Fig. 1. Algorithm Test-Bipartite-Reg for testing bipartiteness with respect to the upper bound $m_{\max} = d_{\max} \cdot n$ on the number of edges, and the procedure Odd-Cycle for detecting odd-length cycles in the graph G.

Note that the algorithm can work when G contains self-loops and multiple-edges. The latter will be of importance in the next section. The corollary below will become useful in the next section as well.

Corollary 2 *If G is ϵ-far from bipartite with respect to $m_{\max} = d_{\max}n$, then $\Omega(\epsilon)$-fraction of its vertices s are such that* Odd-Cycle(s) *returns* found *with probability at least $\frac{2}{3}$.*

Since the proof of Theorem 1 has similar structure to the proof given in [9], we omit it from this extended abstract. All details of this proof, as well as other proofs, can be found in the full version of this paper [12].

4   The Algorithm for the General Case

In this section we build on the testing algorithm presented in the previous section and show a one-sided error bipartite testing algorithm that works with respect to the actual number of edges $m = m(G)$. Hence this algorithm is suitable for general graphs (for which $d_{\max}$ may vary significantly from d_{avg}). The query complexity and running time of the algorithm are of the same order of magnitude as for Test-Bipartite-Reg, that is, $O(\min(\sqrt{n}, n^2/m) \cdot \text{poly}(\log n/\epsilon))$. We note that once the graph becomes very dense, that is $m = \Omega(n^2/\log^c n)$ (where c is approximately 4), it is preferable to use the adjacency-matrix model algorithm [8, 3] with distance parameter $\epsilon/(n^2/m)$.

A High Level Description of the Algorithm. The basic idea is to reduce the problem of testing with respect to the actual number of edges m to the problem of testing with respect to the upper bound $m_{\max} = d_{\max} \cdot n$. Specifically, for any graph G we show how to define a graph G' over $\Theta(n)$ vertices that has the following useful properties. First, the maximum degree in G' is roughly the same as the average degree, and furthermore, this degree is roughly the same as the average degree in G. In particular this implies that the two graphs have roughly the same number of edges. Second, G' approximately preserves the distance of G to bipartiteness. More precisely, if G is bipartite then so is G', but if G is far from bipartite with respect to $m(G)$, then G' is far from bipartite with respect to $m_{\max} = d_{\max}(G')n'$. Thus G' can be viewed as a kind of "regularized-degree version" of G.

If we had direct access to G', then by the above we would be done: by running the algorithm Test-Bipartite-Reg on G' we could decide whether G is bipartite or far from bipartite. However, we only have access to G. Nonetheless, given query access to G we can efficiently "emulate" queries in G'. This would almost suffice for running Test-Bipartite-Reg on G'. One more issue is the uniform selection of starting vertices in G', required by Test-Bipartite-Reg. As we shall see, selecting a vertex uniformly from G' is (roughly) equivalent to uniformly selecting an edge in G. We shall approximate the latter process.

In what follows we assume that $m \geq n'$ and that there are no multiple edges (where we can actually deal with the case in which there are multiple edges but they do not constitute more than a constant fraction of the total number of edges).
The main theorem of this subsection follows.

Theorem 3 *For every graph G having n vertices and $m \geq n$ edges, we can define a graph G' having n' vertices and m' edges for which the following holds:*

1. *$n \leq n' \leq 3n$, $m \leq m' \leq 6m$, and $d_{\max}(G') \leq 2d_{\mathrm{avg}}(G)$.*
2. *If G is bipartite then G' is bipartite, and if G is ϵ-far from bipartite with respect to m, then G' is ϵ'-far from bipartite with respect to $m_{\max}(G') = d_{\max}(G')n'$ for $\epsilon' = \Theta(\epsilon)$.*
3. *Given a starting vertices s in G', it is possible to emulate random walks in G' starting from s, by performing queries to G. The amortized cost of each random walk step is $O(\log^2 n)$ (degree and neighbor) queries in G. By emulating these random walks it is possible to execute a slight variant of Odd-Cycle(s) in G' which we denote Odd-Cycle'(s). This variant is such that $\Pr[\text{Odd-Cycle'}(s)=\text{found}] \geq \Pr[\text{Odd-Cycle}(s)=\text{found}]$, where if Odd-Cycle'($s$) returns found, then we can obtain an odd-length cycle of length $\mathrm{poly}(\log n/\epsilon)$ in the original graph G.*
4. *There exists a procedure Sample-Vertices-Almost-Uniformly-in-G' that for any given parameter $0 < \delta \leq 1$, performs $\tilde{O}(\min(\sqrt{n/\delta}, n^2/m))$ queries in G and returns a vertex in G' such that the following holds: For all but at most $\delta n'$ of the vertices x in G', the probability that x is selected by the procedure is $\Omega(1/n')$.*

We note that for every graph G there is actually a *family* of graphs G' with the above properties (all defined over the same set of vertices). When we run algorithm Test-Bipartite-Gen, we construct one such (arbitrary) graph G' in the family as we go along. As a corollary to Theorem 3 and Corollary 2 we obtain:

Corollary 4 *Algorithm Test-Bipartite-Gen (see Figure 2) accepts every graph G that is bipartite, and rejects with probability at least 2/3 every graph G that is ϵ-far from bipartite (with respect to $m(G)$). Furthermore, whenever the algorithm rejects a graph it outputs a* certificate *to the non-bipartiteness of the graph G in form of an odd-length cycle of length* poly$(\log n/\epsilon)$.
The query complexity and running time of the algorithm are $O\left(\min(\sqrt{n}, n^2/m) \cdot \text{poly}(\log n/\epsilon)\right)$.

Test-Bipartite-Gen$(n, d_{\text{avg}}, \epsilon)$

- Repeat $T = \Theta(\frac{1}{\epsilon})$ times:
  1. Set $\epsilon' = \epsilon/108$.
  2. Select a vertex s in G' by calling the procedure Sample-Vertices-Almost-Uniformly-in-G' with $\delta = \epsilon'/c$ (where c is a sufficiently large constant).
  3. Apply Odd-Cycle'(s).
  4. If Odd-Cycle'(s) returns **found** then output **reject**.
- In case no call to Odd-Cycle' returned **found** then output **accept**.

Fig. 2. Algorithm Test-Bipartite-Gen for testing bipartiteness with respect to the actual number of edges $m = m(G)$ in the graph G.

4.1 Defining G' and Proving the First Item in Theorem 3

In all that follows, let $d = d_{\text{avg}}(G)$, and let $d' = d_{\max}(G')$. We shall assume that d is a sufficiently large constant. If $d_{\text{avg}}(G)$ is not sufficiently large then we still set d in the construction below to be sufficiently large, and run the algorithm with ϵ set to $\epsilon/(d/d_{\text{avg}}(G))$.

The Construction of G'. For each vertex v in G such that $\deg(v) \leq d$, we have a single vertex in G'. For each vertex v in G such that $\deg(v) > d$ we have in G' a subgraph, denoted $H(v)$. It is a bipartite graph over two subsets of vertices, one denoted $X(v)$, the *external* part, and one denoted $I(v)$, the *internal* part. Both parts consist of $\lceil \deg(v)/d \rceil$ vertices. Every vertex in $X(v)$ represents up to d specific neighbors of v according to some fixed, *but arbitrary* partition of the neighbors of v. We refer to the vertices in the two subsets by $\{X_i(v)\}_{i=1}^{\lceil \deg(v)/d \rceil}$ and $\{I_i(v)\}_{i=1}^{\lceil \deg(v)/d \rceil}$, respectively. The edges in $H(v)$ are determined as follows. In case $\deg(v)/d < d$ then we have $\lceil d^2/\deg(v) \rceil$-multiple edges between every internal vertex and every external vertex in $H(v)$. In case $\deg(v)/d \geq d$, denote $s = \lceil \deg(v)/d \rceil$ and let $H(v)$ be a bipartite expander where each of its sides has s vertices ($s \geq d$). Each vertex in $H(v)$ has degree d. All eigenvalues of the adjacency matrix of H, but the largest one and the smallest one (which are equal to d and $-d$, respectively), are at most $d/4$ in their absolute values. Explicit constructions of such expanders can be found, e.g., in [14, 13]. Furthermore, these constructions allow the determination of the i-th neighbor of any given vertex in constant time.

We have described how vertices of G are transformed into vertices of G'. It remains to describe the relevant transformation to the edges of G. Consider an edge $(u, v) \in E(G)$ where v is the i-th neighbor of u and u is the j-th neighbor of v. Let $X_k(u)$ and $X_\ell(v)$ be the external vertices representing the i-th neighbor of u, and the j-th neighbor of v, respectively. Then, there is an edge $(X_k(u), X_\ell(v))$ in G'. It directly follows that every vertex in G' has degree at most $2d$ and that $n' = |V(G')| \leq \sum_{v \in G} 2\lceil \deg(v)/d \rceil \leq 3n$, and $m' = m(G') \leq 3dn = 6m$.

In the long version of this paper [12] we suggest the following alternative probabilistic construction of G' that establishes Theorem 3. Every vertex of G is transformed into $\lceil \deg(v)/d \rceil$ vertices. Denote by $X(v)$ the vertices in G' related to a vertex $v \in V(G)$. The vertices in $X(v)$ are denoted by $X_i(v), 1 \leq i \leq \lceil \deg(v)/d \rceil$. Thus, $n' = |V(G')| \leq \sum_{v \in G} \lceil \frac{\deg(v)}{d} \rceil \leq 2n$. The edges of G' are determined as follows: an edge $(u, v) \in E(G)$ chooses independently uniformly at random a vertex from $X(v)$ and a vertex from $X(u)$. In G' there will be an edge between these two randomly chosen vertices. Clearly, $m' = |E(G')| = |E(G)| = (nd)/2$.

The probabilistic construction is simpler and more robust than the deterministic one, and it may be applicable to other problems as well. However in this construction we need that $d = \Omega(1/\epsilon)$.

4.2   Establishing Items 2 and 3 in Theorem 3

The proofs of these two items are ommitted from this extended abstract, and can be found in [12]. We note that Item 2 builds on the expander graphs defined in the construction of G'.

4.3   Establishing Item 4 in Theorem 3

In this subsection we provide a sketch for the proof of the last item in Theorem 3. Consider the construction of G'. Sampling a vertex uniformly at random from G' is equivalent to sampling a vertex v from G with probability proportional to its degree (and then taking randomly and uniformly one of the vertices belong to $H(v)$). The latter is equivalent to sampling randomly uniformly an edge from G, and taking one of its end-points at random. Thus, the proof of this item is based on a presentation of a procedure for sampling edges almost uniformly from G.

We consider two cases: $d > \sqrt{\delta n}$ and $d \leq \sqrt{\delta n}$ (recall that $d = d_{\mathrm{avg}}(G)$ is the average degree in G and that our goal is to use $\tilde{O}(\min(\sqrt{n/\delta}, n/d))$ queries to G). The first case is easy since if G contains sufficiently many edges then we simply sample $\Theta(n/d) = \Theta(n^2/m)$ pairs of vertices in order to obtain an edge.

In the second case, where G contains fewer edges ($d \leq \sqrt{\delta n}$), we do not have an algorithm that selects an edge uniformly from G (using relatively few queries). However, we can show the following lemma, from which Item 4 in Theorem 3 can be derived. The proof of this lemma can be found in [12].

Lemma 1 *There exists a procedure* Sample-Edges-almost-Uniformly-in-G *that uses* $\tilde{O}(\sqrt{n/\delta})$ *degree and neighbor queries in G and for which the following holds: For all but $(\delta/4)m$ of the edges e in G, the probability that the procedure outputs e is at least*

$1/(64m)$. *Furthermore, there exists a subset $U_0 \subset V(G)$, $|U_0| \leq (\delta n/2)$, such that for all edges $e = (u, v)$ that are output with probability less than $1/(64m)$, we have $u, v \in U_0$.*

5   A Lower Bound

In this section we present a lower bound on the number of queries necessary for testing bipartiteness. Similarly to the lower bound presented in [9], this lower bound holds for testing algorithms that are allowed a two-sided error, and the graphs used for the lower bound construction are regular graphs. However, the lower bound of $\Omega(\sqrt{n})$ (for constant ϵ) established in [9], holds for graphs having constant degree (e.g., degree 3), and when the algorithm is allowed only neighbor queries. Our lower bound is more general in that it allows the algorithm to perform both neighbor queries and vertex-pair queries, and it is applicable to all graphs.

Theorem 5 *Every algorithm for testing bipartiteness with distance parameter $\epsilon \leq 2^{-4}$ must perform $\Omega(\min(\sqrt{n}, n^2/m))$ queries.*

The high-level structure of our proof is similar to other lower-bound proofs for testing, which can be traced back to [17]. We present two distributions over graphs, where all graphs generated by one distribution are bipratite (and hence should be accepted), while with very high probability a graph generated according to the other distribution is far from bipartite. We then show that any algorithm with query complexity below the lower bound, cannot distinguish between the two distributions (and hence must have a large failure probability).

Specifically, both distributions, denoted $\mathcal{G}(n, d)$, and $\mathcal{G}(n/2, n/2, d)$, are over d-regular graphs having n vertices, where we assume for simplicity that n is even. A graph generated according to $\mathcal{G}(n, d)$ is obtained by selecting, uniformly and independently, d perfect matchings between the n vertices. A graph generated according to $\mathcal{G}(n/2, n/2, d)$ is obtained by first randomly partitioning the n vertices into two equal parts, and then selecting, uniformly and independently, d perfect matchings between the two parts. By definition, all graphs in the support of $\mathcal{G}(n/2, n/2, d)$ are bipartite, and we prove that graphs generated according to $\mathcal{G}(n, d)$ are ϵ-far from bipartite with high probability, for $\epsilon \leq 1/16$ and $d \geq 64$

We then show that the following two claims hold when a graph is generated either according to $\mathcal{G}(n, d)$ or according to $\mathcal{G}(n/2, n/2, d)$: (1) Any algorithm that asks $o(n^2/m) = o(n/d)$ queries, will not detect an edge by any vertex-pair query with very high probability. (2) Any algorithm that asks $o(\sqrt{n})$ queries will not receive as an answer to any neighbor query, a vertex it has already observed in a previous query (with very high probability as well). From this we can conclude that any algorithm that asks $o(\min(\sqrt{n}, n^2/m))$ queries cannot distinguish between the two distributions, as desired. In the lower bound proof we show the necessity of both *neigbhor queries* and *vertex-pair queries*. Specifically by using only one type of queries the lower bound increases.

References

[1] N. Alon. Testing subgraphs of large graphs. *Random Structures and Algorithms*, 21:359–370, 2002.

[2] N. Alon, E. Fischer, M. Krivelevich, and M Szegedy. Efficient testing of large graphs. *Combinatorica*, 20:451–476, 2000.

[3] N. Alon and M. Krivelevich. Testing *k*-colorability. *SIAM Journal on Discrete Math*, 15(2):211–227, 2002.

[4] N. Alon and A. Shapira. Testing subgraph in directed graphs. Submitted, 2002.

[5] M. Bender and D. Ron. Testing properties of directed graphs: Acyclicity and connectivity. *Random Structures and Algorithms*, pages 184–205, 2002.

[6] A. Bogdanov, Kenji Obata, and L. Trevisan. A lower bound for testing 3-colorability in bounded-degree graphs. In *Proceedings of the Forty-Third Annual Symposium on Foundations of Computer Science*, pages 93–102, 2002.

[7] A. Bogdanov and L. Trevisan. Lower bounds for testing bipartiteness in dense graphs. Submitted, 2002.

[8] O. Goldreich, S. Goldwasser, and D. Ron. Property testing and its connection to learning and approximation. In *Proceedings of the Thirty-Seventh Annual Symposium on Foundations of Computer Science*, pages 339–348, 1996.

[9] O. Goldreich and D. Ron. A sublinear bipartite tester for bounded degree graphs. *Combinatorica*, 19(3):335–373, 1999.

[10] O. Goldreich and D. Ron. Property testing in bounded degree graphs. *Algorithmica*, pages 302–343, 2002.

[11] O. Goldreich and L. Trevisan. Three theorems regarding testing graph properties. In *Proceedings of the Forty-Second Annual Symposium on Foundations of Computer Science*, pages 460–469, 2001.

[12] T. Kaufman, M. Krivelevich, and D. Ron. Tight bounds for testing bipartiteness in general graphs. http://www.eng.tau.ac.il/~danar/papers.html.

[13] A. Lubotzky, R. Phillips, and P. Sarnak. Explicit expanders and the ramanujan conjectures. In *Proceedings of the Eighteenth Annual ACM Symposium on Theory of Computing*, pages 240 –246, 1986.

[14] Gregory A. Margulis. Explicit constructions of expanders. *Problemy Peredachi Informatsii*, 9(4):71–80, 1973. expanders construction.

[15] M. Parnas and D. Ron. Testing the diameter of graphs. *Random Structures and Algorithms*, 20(2):165–183, 2002.

[16] R. Rubinfeld and M. Sudan. Robust characterization of polynomials with applications to program testing. *SIAM Journal on Computing*, 25(2):252–271, 1996.

[17] A.C. Yao. Probabilistic computation, towards a unified measure of complexity. In *Proceedings of the Eighteenth Annual Symposium on Foundations of Computer Science*, pages 222–227, 1977.

Discrete Quantum Walks Hit Exponentially Faster

Julia Kempe

CNRS-LRI, UMR 8623
Université de Paris-Sud, 91405 Orsay, France and
Computer Science Division and Dept. of Chemistry
University of California, Berkeley, CA 94709

Abstract. This paper addresses the question: what processes take polynomial time on a quantum computer that require exponential time classically? We show that the hitting time of the discrete time quantum random walk on the n-bit hypercube from one corner to its opposite is polynomial in n. This gives the first exponential quantum-classical gap in the hitting time of discrete quantum walks. We provide the basic framework for quantum hitting time and give two alternative definitions to set the ground for its study on general graphs. We outline a possible application to sequential packet routing.

1 Introduction

Random walks form one of the cornerstones of theoretical computer science as well as the basis of a broad variety of applications in mathematics, physics and the natural sciences. In computer science they are frequently used in the design and analysis of randomized algorithms. Markov chain simulations provide a paradigm for exploring an exponentially large set of combinatorial structures (such as assignments to a Boolean formula or matchings in a graph) by a sequence of simple, local transitions. As algorithmic tools they have been applied to a variety of central problems, such as approximating the permanent [JS89], finding satisfying assignments for Boolean formulas [Sch99] and the estimation of the volume of a convex body [DFK91]. Other well-known examples of algorithms based on random walks include 2-SAT, Graph Connectivity and probability amplification [MR95, Pap94].

Recently the study of quantum walks has been initiated, with the hope of bringing new powerful algorithmic tools into the setting of quantum computing. To this day nearly all efficient quantum algorithms are based on the Quantum Fourier Transform (QFT), like Simon's period-finding algorithm [Sim97] or Shor's celebrated algorithms for Factoring and Discrete Log [Sho97]. However, it seems that the power of the QFT might be limited to solve similar problems on non-Abelian groups, like for the symmetric group for Graph Isomorphism [HRT00, GS+01]. It seems crucial to develop new algorithmic tools.

Several striking differences between classical and quantum discrete walks have already been observed for walks on the cycle [AA+01], the line [AB+01] and the hypercube [MR02]. The reason for this is quantum interference. Whereas there cannot be destructive interference in a classical random walk, in a quantum walk two separate

S. Arora et al. (Eds.): APPROX 2003+RANDOM 2003, LNCS 2764, pp. 354–369, 2003.
© Springer-Verlag Berlin Heidelberg 2003

paths leading to the same point may be out of phase and cancel out. The focus of previous work has been primarily on the mixing time of a discrete quantum walk. It has been shown that quantum walks on a large class of graphs can mix nearly quadratically faster than their classical counterparts. Since mixing times are an important quantity for many classical algorithms, this has raised the question of whether quantum walks can mix exponentially faster. However in [AA$^+$01] a lower bound on the mixing time of any local quantum walk has been obtained, which implies in essence that quantum walks can mix at most quadratically faster than classical walks (this is exactly true for bounded degree graphs; for graphs of maximal degree d this speed-up may be enhanced by a factor of $1/d$). This result showed that in all likelihood quantum walks cannot drastically enhance mixing times of classical walks.

In this paper we set the stage to exactly analyze another crucial quantity of discrete time walks: the hitting time. The hitting time is important in many algorithmic applications of classical random walks, like k-SAT or Graph Connectivity. For instance the most efficient known solution to 3-SAT is based on the hitting time of a random walk [Sch99]. In the algorithmic context, the question whether a quantum process can achieve an exponentially faster penetration of graphs has first been raised by Farhi and Gutmann [FG98]. For the continuous time quantum walk, a different model from the one we analyze, Farhi et al. gave a mixture of analytical and numerical evidence of an exponential gap in hitting behavior [FG98, CFG02]. After our work has been completed very recently Childs et al. succeeded to give an oracle-based algorithmic exponential speed-up between classical and quantum query complexity based on the quantum continuous-time walk [CC$^+$02]. They are able to construct a family of random graphs with two special nodes such that on average any classical algorithm that needs to find the sink node starting form the the source node requires an exponential number of queries, whereas the quantum algorithm succeeds in polynomial time. The continuous-time quantum walk at the base of that example is different from the discrete time model we analyze and it is a priori not clear how both models are related. Even though their beautiful result proves a rigorous separation between the classical and the quantum setting, the wider applicability of their example is questionable at the moment. It is important to rigorously establish the notions and methods for hitting behaviour of quantum walks, in particular in the discrete case, and to analyze it for other graphs and structures. Our work provides a step in this direction.

The hitting time h_{uv} of node v starting from node u measures the expected time it takes until the walk hits v for the first time. In the quantum case we face a dilemma: as is well known, observations of the quantum system (like "Has the walk hit node v?") influence the state of the quantum system. In particular if one were to observe the position of the quantum walk at each time it would lose its quantum coherence and reduce ("collapse") to the standard classical random walk, in which case we cannot expect any non-classical behavior or speed-ups. We give two alternatives out of this dilemma and establish two different notions of "quantum hitting time". In the first case the walk is not observed at all. Started at node u the position of the walk is measured at a (previously determined) time T. If the probability p to be at node v at time T is sufficiently large (an inverse polynomial in the graph size) we call T a "one-shot p hitting time". In the second case ("concurrent measurement") we do not require any previous knowledge of when to measure the position of the walk. Starting from node u at every step of the walk a partial measurement is performed (only the question "Is the position v or not v?" is asked). If the walk is found to have hit node v, it is stopped, otherwise the next step follows. This measurement perturbs the walk slightly but does not kill all the quantum coherence at once. If after a time T the probability p to halt is bounded below by an inverse polynomial in the size of the graph, we call T a "concurrent p hitting time".

After having made these notions rigorous we are able to show that on the hypercube both definitions of quantum hitting time lead to polynomial quantities for the walk from one corner to the opposite corner. This is in stark contrast to the classical case, where the corner-to-corner hitting time is exponential. Our result provides the first fully analytical classical-quantum exponential gap for a discrete quantum walk on a graph. It opened the possibility that quantum algorithms based on random walks may significantly improve upon classical algorithms. We will state similar results for the continuous-time quantum walk and also outline a possible application of rapid hitting on the hypercube: "quantum-random" sequential routing in a network.

It is interesting to know how much the exponential speed-up of the quantum walk depends on the choice of initial and final position. We establish two bounds: a lower bound on the size of the neighborhood of one corner from which we still achieve polynomial hitting behavior to the opposite corner and an upper bound on this neighborhood. This latter derives from a lower bound on quantum unstructured search algorithms [BB+97].

While quantum walks are very easy to describe, they appear to be quite difficult to analyze. Standard techniques for analyzing classical random walks are apparently of little use. Whereas in the classical case most quantities depend only on the gap between the first and second largest eigenvalue of the underlying chain, in the quantum case all eigenvalues seem to play an equally important role and new methods are needed. We hope that establishing the rigorous notions and necessary techniques will help to analyze quantum walks on a variety of graphs.

Related Work: Various quantum walk variants have previously been studied by several authors. The general framework for discrete quantum walks is introduced in [Mey96, Wat01, AA+01, AB+01] . The mixing time of the quantum random walk on the hypercube has been analysed in [MR02], both in the discrete and continuous time setting. We use the spectral decomposition of [MR02] in our analysis. However, the results in [MR02] regard only the mixing time of the walk and do not deal with hitting times. In [AB+01] a notion of "halting" and intermediate partial measurement similar to our concurrent measurement is used, but the results regard the total halting probability of the quantum walk, and not the expected hitting time. Numerical studies of the hitting time on the hypercube have been communicated to us by Tomohiro Yamasaki [Yam01] (published in [YKI02] after our work has been completed). A quantum search algorithm based on the discrete walk on the hypercube has recently been found [SKW03].

A different model of quantum random walks, so called continuous time walks, has been introduced by Farhi and Gutmann [FG98]. They are defined via a Hamiltonian that stems from the generating matrix of the classical continuous random walk. Until now it is not clear how their model is related to the discrete case we analyze. For their random walk model Farhi and Gutmann first exhibited an infinite tree and a walk that hits a set of leaves with inverse polynomial probability in polynomial time (similar to our notion of "one-shot hitting time"), where the classical analog has exponential hitting time. Later in [CFG02] another finite graph with a similar property is presented; both proofs are partly analytic and partly numeric, however. After the completion of the present work Childs et al. [CC+02] where able to construct a family of graphs based on the one in [CFG02] and to show that the continuous-time random walk gives rise to an exponential algorithmic speed-up between average case classical query complexity and its quantum version for the problem to find a very specific node in this graph.

Structure of the paper: We begin by reviewing in Sec. 2 the necessary background on classical random walks, quantum computation and quantum discrete time walks on graphs and in particular on the hypercube. In Sec. 3 we introduce the relevant definitions of quantum hitting times, and state and prove the upper bounds on quantum hitting times on the hypercube. In Sec. 4 we provide upper and lower bounds on the size of the neighborhood of a node from which the quantum random walk has polynomial hitting behavior to the opposite corner. In Sec. 5 we outline a quantum routing application. In Appendix A we compare continuous-time random walks to discrete walks and establish analogous results for their hitting time.

2   Background

2.1   Random Walks

Here we will state a few specific definitions and theorems as they are relevant to the present work to compare the behavior of classical and quantum walks (for a more complete treatment see e.g. [MR95, AF01]).

Simple Random Walk: A simple random walk on an undirected graph $G(V, E)$, is described by repeated applications of a stochastic matrix P, where $P_{u,v} = \frac{1}{d_u}$ if (u, v) is an edge in G and d_u the degree of u. If G is connected and non-bipartite, then the distribution of the random walk, $D^t = P^t D^0$ converges to a stationary distribution π which is independent of the initial distribution D^0. If a simple random walk on a bipartite graph has some periodicity (there is a state i and an initial distribution D^0 such that $D_i^t > 0$ iff t belongs to the arithmetic progression $\{a + ms | m \geq 0\}$ for some integer a) the introduction of a resting probability will make the walk aperiodic and convergent to π. For $d-$regular graphs G (all nodes of same degree d), the limiting probability distribution is uniform over the nodes of the graph.

Hitting Time: Given an initial state i, the probability that the first transition *into* a state j occurs at time t is denoted by r_{ij}^t. The hitting time h_{ij} is the expected number of steps to reach state j starting from state i and is given by $h_{ij} = \sum_{t>0} t r_{ij}^t$. For *aperiodic* simple random walks the Fundamental Theorem of Markov Chains implies that the number of times a state i is visited in the stationary state is $1/\pi_i$ and $h_{ii} = 1/\pi_i$.

Hypercube: The stationary distribution of the simple aperiodic random walk on the n-bit hypercube is given by $\pi_i = 1/2^n$. The hitting time from one node i to the opposite corner of the cube j is exponential in n, $h_{ij} = 2^n (1 + \frac{1}{n} + \frac{1}{O(n^2)}))$.

Continuous time walk: The theory of continuous time Markov chain closely parallels discrete time chains. A continuous chain is specified by non-negative transition rates

q_{ij}. Given that the state of the system at time t is $X_t = i$, the probability that $X_{t+dt} = j$ is $q_{ij}dt$. One can define $q_{ii} = -\sum_{j \neq i} q_{ij}$ to obtain a matrix Q. The state of the system with initial state D^0 is then given by $D^t = exp(Qt)D^0$. All the results on convergence and hitting essentially carry over to the continuous case with only slight modifications. To transition form discrete to continuous one can "discretize" a continuous chain by setting $P = exp(Q)$ or make a discrete chain continuous by setting $q_{ij} = p_{ij}$ for $i \neq j$. Stationary distribution and mean hitting times remain unchanged.

2.2 Quantum Computation

The model. Consider a finite Hilbert space $\mathcal{H}$ with an orthonormal set of basis states $|s\rangle$ for $s \in \Omega$. The states $s \in \Omega$ may be interpreted as the possible classical states of the system described by $\mathcal{H}$. In general, the state of the system, $|\alpha\rangle$, is a unit vector in the Hilbert space $\mathcal{H}$, and can be written as $|\alpha\rangle = \sum_{s \in \Omega} a_s |s\rangle$, where $\sum_{s \in \Omega} |a_s|^2 = 1$. $|\alpha^*\rangle$ denotes the conjugate and $\langle \alpha |$ denotes the conjugate transpose of $|\alpha\rangle$. $\langle \beta | \alpha \rangle$ denotes the inner product of $|\alpha\rangle$ and $|\beta\rangle$. For more details on quantum computing see e.g. [NC00]. A quantum system can undergo two basic operations: unitary evolution and measurement. **Unitary evolution**: Quantum physics requires that the evolution of quantum states is unitary, that is the state $|\alpha\rangle$ is mapped to $U|\alpha\rangle$, where U satisfies $U \cdot U^\dagger = I$, and $U^\dagger$ denotes the transpose complex conjugate of U. Unitary transformations preserve norms, can be diagonalized with an orthonormal set of eigenvectors, and the corresponding eigenvalues are all of absolute value 1.
Measurement: We will describe here only projective (von Neuman) measurements, defined by a set of orthogonal projectors $\{\Pi_i : i \in I\}$ ($\Pi_i^\dagger = \Pi_i$, $\Pi_i^2 = \Pi_i$ and $\Pi_i \Pi_j = \delta_{ij}\Pi_i$) such that $\sum_{i \in I} \Pi_i = \mathbf{1}$. The output of the measurement of the state $|\alpha\rangle$ is an element $i \in I$ with probability $||\Pi_i|\alpha\rangle||^2$, we then say that Π_i was measured. Moreover, the new state of the system after the measurement with outcome i is the (normalized) state $(||\Pi_i|\alpha\rangle||)^{-1}\Pi_i|\alpha\rangle$. We denote the projectors on one basis state $|s\rangle$ by $|s\rangle\langle s|$.
Combining two quantum systems: If $\mathcal{H}_A$ and $\mathcal{H}_B$ are the Hilbert spaces of two systems, A and B, then the joint system is described by the tensor product of the Hilbert spaces, $\mathcal{H}_A \otimes \mathcal{H}_B$. If the basis states for $\mathcal{H}_A$, $\mathcal{H}_B$ are $\{|a\rangle\}, \{|v\rangle\}$, respectively, then the basis states of $\mathcal{H}_A \otimes \mathcal{H}_B$ are $\{|a\rangle \otimes |v\rangle\}$. We use the abbreviated notation $|a, v\rangle$ for the state $|a\rangle \otimes |v\rangle$. This coincides with the interpretation by which the set of basis states of the combined system A, B is spanned by all possible classical configurations of the two classical systems A and B.

2.3 Discrete-Time Quantum Random Walk

It is not possible to define the quantum random walk naïvely in analogy to the classical walk as a move in all directions "in superposition". It is easy to verify [Mey96] that a translationally invariant walk which preserves unitarity is necessarily proportional to a translation in one direction. If the particle has an extra degree of freedom that assists in its motion, however, then it is possible to define more interesting homogeneous local unitary processes. Following [AA+01] we call the extra space the "coin-space" alluding to the classical coin that decides upon the walk direction.

More specifically let $G(V,E)$ be a graph, and let $\mathcal{H}_V$ be the Hilbert space spanned by states $|v\rangle$ where $v \in V$. We denote by N, or $|V|$ the number of vertices in G. We will only consider d-regular graphs G here, but slightly modified definitions can be made in the general case. Let $\mathcal{H}_C$ be the "coin"-Hilbert space of dimension d spanned by the states $|1\rangle$ through $|d\rangle$. Let $\mathbf{C}$ be a unitary transformation on $\mathcal{H}_C$ (the "coin-tossing operator" which we will define later). Label each directed edge with a number between 1 and d, such that for each a, the directed edges labeled a form a permutation. For Cayley graphs the labeling of a directed edge is simply the generator associated with the edge. Now we can define a shift operator $\mathbf{S}$ on $\mathcal{H}_C \otimes \mathcal{H}_V$ such that $\mathbf{S}|a,v\rangle = |a,u\rangle$ where u is the a-th neighbor of v. Note that since the edge labeling is a permutation, S is unitary. One step of the quantum walk is given by a local transformation acting on the coin-space only, followed by a conditional shift which leaves the coin-space unchanged [AA⁺01]:
$\mathbf{U} = \mathbf{S} \cdot (\mathbf{C} \otimes \mathbf{I_N})$.

Random Walk on the Hypercube: The hypercube of dimension n is a Cayley graph with $N = 2^n$ vertices. The position states are bit-strings $|x\rangle$ of length n. The directions can be labeled by the n basis-vectors $\{|1\rangle, \dots, |n\rangle\}$, corresponding to the n vectors of Hamming weight 1 $\{|e_1\rangle, \dots, |e_n\rangle\}$, where e_i has a 1 in the ith position.

To mimic the permutation symmetry of the classical simple random walk we need to define the $n \times n$ coin operator $\mathbf{C}$ such that $\mathbf{U}$ is invariant to permutations of bits. As pointed out in [MR02] the symmetry of the hypercube defines the coin operator $\mathbf{C}$ to be of the form $C_{ij} = a$ if $i = j$ and $C_{ij} = b$ if $i \neq j$ with two parameters $a, b \in \mathbb{C}$. Unitarity of $\mathbf{C}$ further imposes two quadratic constraints on a and b, so that finally up to an overall phase all symmetric coins are characterized by one real parameter $1 - 2/n \leq |a| \leq 1$. Among all these coins the one farthest away from the identity operator $\mathbf{1}_n$ is given by $a = 2/n - 1$ and $b = 2/n$ [MR02]. We will call this latter coin $\mathbf{G}$ and use it as our coin in the rest of this paper. It is not hard to see that using another coin (with constant a, b) from the set of permutation invariant coins (except $\mathbf{1}_n$ of course) only slows down the walk by a constant factor and does not change the order of magnitude of the hitting behavior. To respect symmetry we will also impose permutation invariance for the initial state of the walk.

Definition 1 (Discrete time walk on the hypercube). *The symmetric discrete time walk $\mathbf{U}$ on the n - dimensional hypercube is acting on a $n \cdot 2^n$ dimensional space $\mathcal{H}_n \otimes \mathcal{H}_2^{\otimes n}$ as $\mathbf{U} = \mathbf{S} \cdot (\mathbf{G} \otimes \mathbf{1_N})$ where the shift operator $\mathbf{S}$ is defined as $\mathbf{S} : |i,x\rangle \Rightarrow |i, x \oplus e_i\rangle$, i.e. $\mathbf{S} = \sum_{i=1}^n |i\rangle\langle i| \otimes S_i$ with $S_i|x\rangle = |x \oplus e_i\rangle$. The initial state of the walk is chosen to be symmetric with respect to bit-permutations. For a walk starting in $|x\rangle$ the initial state is $\frac{1}{\sqrt{n}} \sum_{i=1}^n |i\rangle \otimes |x\rangle$.*

Note that this discrete-time quantum walk reduces to the classical symmetric walk if we perform a measurement in the coin-space in the direction-basis after every step of the walk. The resulting classical walk with last step in direction i will uniformly change to one of the $n - 1$ directions $j \neq i$ with probability $|b|^2 = 4/n^2$ and will return back to the node it came from (direction i) with probability $|a|^2 = 1 - 4/n + 4/n^2$. This type of classical random walk has a "direction-memory" one step back in time, but can be modeled by a (memoryless) Markov chain if we add a directional space to the position space. In other words each node v is blown up into n nodes v_i where i is the direction the walk came from. This resulting walk has a preference to oscillate back and forth between two adjacent nodes and has obviously still an exponential hitting time from one corner to its opposite.

The walk as defined is *periodic*: nodes with even Hamming weight are visited at even times only, nodes with odd Hamming weight at odd times. The inclusion of a "resting" coin-state $|0\rangle$ and a $n+1 \times n+1$ coin allowing for a self-loop transition amplitude of $a = 2/(n+1) - 1$ make this walk aperiodic. To simplify the analysis we will only show the results for the periodic case, though; they hold with very slight modification in the aperiodic case as well.

3   Hitting Times on the Hypercube

For classical random walks the *hitting time* of a node v of a walk starting at an initial node i is defined as the expected time it takes the walk to reach v for the first time starting from i. Alternatively one can let the classical walk stop upon reaching the node v and define the *stopping-time* of the walk as the expected time for this walk to stop. In the classical case both notions are clearly the same. Care has to be applied to define an analogous notion for a quantum walk. To define "reaching" v we have to circumvent the measurement problem. Namely if we were to measure the position of the walk after each step we will kill the quantum coherences and collapse the walk onto the corresponding classical walk. There are two alternatives: either to let the walk evolve and measure the position of the walk after T iterations ("one-shot measurements"), or to perform a partial measurement, described by the two projectors $\Pi_0 = |v\rangle\langle v|$ and $\Pi_1 = \mathbf{1} - \Pi_0$ (where $|v\rangle$ is some specific position we wish to "hit") after every step of the iteration ("concurrent measurement"). A priori these two notions can be very different in the quantum case.

Definition 2 (One-shot hitting time). *A quantum random walk U has a (T,p) one-shot $(|\phi_0\rangle, |x\rangle)$ hitting time if the probability to measure state $|x\rangle$ at time T starting in $|\phi_0\rangle$ is larger than p, i.e. $\|\langle x|U^T|\phi_0\rangle\|^2 \geq p$.*

Definition 3 ($|x\rangle$-stopped walk). *A $|x\rangle$-stopped walk from U starting in state $|\phi_0\rangle$ is the process defined as the iteration of a measurement with the two projectors $\Pi_0 = \Pi_x = |x\rangle\langle x|$ and $\Pi_1 = \mathbf{1} - \Pi_0$ and, if Π_1 is measured, an application of U. If Π_0 is measured the process is stopped.*

Definition 4 (Concurrent hitting time). *A quantum random walk U has a (T,p) concurrent $(|\phi_0\rangle, |x\rangle)$ hitting-time if the $|x\rangle$-stopped walk from U and initial state $|\phi_0\rangle$ has a probability $\geq p$ of stopping at a time $t \leq T$.*

These two notions presuppose very different behavior of an algorithm exploiting them. In the one-shot case we have to know exactly *when* to measure the walk, which usually means that we have to know the dimension of the hypercube or, in more general applications, the shape of the graph. The advantage of the concurrent case is that we do not need any knowledge of when the walk will "hit" the target state. We simply continuously query the walk at the target state until we measure a "hit"; we do not need to have a priori information about the graph; probably ultimately more useful for algorithmic applications.

Note also that in the concurrent case if (T, p) is a hitting-time then for $T' \geq T$ (T', p) is also a hitting-time, i.e. hitting with probability at least p is a monotone property in time. In the one-shot case this is not at all true; we will see that for the hypercube there are certain windows in time where the probability to measure a certain node is high, followed by times where this probability is very low - yet another difference to the classical case.

3.1 One-Shot Hitting Time

We will now state and prove our first main result for the symmetric discrete-time quantum walk on the hypercube $\mathbf{U}$ of dimension n. Times T and t are always understood to be the closest integers of the same parity as n. We denote with $\bar{x}$ the opposite corner to x (obtained by conjugating all bits).

Theorem 1. $\mathbf{U}$ has a (T, p) one-shot $(|x\rangle, |\bar{x}\rangle)$ hitting time with (1) $T = \frac{\pi}{2}n$ and $p = 1 - O(\frac{\log^3 n}{n})$, (2) $T = \frac{\pi}{2}n \pm O(n^\beta)$ and $p = 1 - O(\frac{\log n}{n^{1-2\beta}})$ with $0 < \beta < 1/2$, (3) $T \in [\frac{\pi}{2}n - O(\frac{\sqrt{n}}{\log n}), \frac{\pi}{2}n + O(\frac{\sqrt{n}}{\log n})]$ and $p = 1 - O(\frac{\log \log n}{\log n})$).

Remark: The "$\sqrt{n}$"-window around the exact one-shot measurement time of $\pi n/2$ makes the algorithm more robust to slight perturbations in the exact time of the measurement.

Proof of Theorem 1: Note that by the symmetry of the hypercube and the walk $\mathbf{U}$ the hitting time is the same for all $(|x\rangle, |\bar{x}\rangle)$ with $x \in \{0, 1\}^n$. So w.l.o.g. we set $|x\rangle = |00 \ldots 0\rangle$. We will use the following facts from [MR02]: The $n \cdot 2^n$ eigenstates of $\mathbf{U}$ are of the form $|v_k^i\rangle \otimes |\tilde{k}\rangle$ where $|\tilde{k}\rangle = \frac{1}{\sqrt{2^n}} \sum_{x \in \{0,1\}^n} (-1)^{k \cdot x} |x\rangle$ is the $\mathbb{Z}_2^n$-Fourier transform for $k \in \mathbb{Z}_2^n$ and the n vectors $\{|v_k^i\rangle : i = 1 \ldots n\}$ for each k are the eigenvectors of the matrix $\mathbf{S_k} \cdot \mathbf{G}$, where $\mathbf{S_k}$ is the diagonal $n \times n$ matrix with $(\mathbf{S_k})_{lm} = \delta_{lm}(-1)^{k_l}$.

The symmetric initial state is $|\Phi_0\rangle := |\Psi_{in}\rangle \otimes |00 \ldots 0\rangle := \frac{1}{\sqrt{n}} \sum_{i=1}^{n} |i\rangle \otimes |00 \ldots 0\rangle$ (see Def. 1). For all k, only two of the n eigenvectors $|v_k^i\rangle$ have non-zero inner product with $|\Psi_{in}\rangle$ [MR02]. These two eigenvectors are complex conjugates, call them $|w_k\rangle$ and $|w_k^*\rangle$, and their corresponding eigenvalues are λ_k and λ_k^* with $\lambda_k = 1 - \frac{2|k|}{n} + i\frac{2}{n}\sqrt{|k|(n - |k|)}$ where $|k|$ is the Hamming weight of k. Let $\lambda_k = e^{i\omega_{|k|}} = \cos \omega_{|k|} + i \sin \omega_{|k|}$ where $\cos \omega_m = 1 - 2m/n$. The entries of $|w_k\rangle$ are $(w_k)_l = \frac{-i}{\sqrt{2}\sqrt{n-|k|}}$ if $k_l = 0$ and $(w_k)_l = \frac{1}{\sqrt{2}\sqrt{|k|}}$ if $k_l = 1$. (If $k = 0$ and $k = n$ there is only one eigenvector, the uniform superposition over all directions, with eigenvalue $\lambda_0 = 1$ and $\lambda_n = -1$. When we write out the general eigenvectors this special case will be self-understood.) The initial state is a superposition over $2^{n+1} - 2$ eigenvectors $|\Phi_0\rangle = \sum_{k \in \{0,1\}^n} (a_k |w_k\rangle + a_k^* |w_k^*\rangle) \otimes |\tilde{k}\rangle$ with $a_k = \frac{1}{\sqrt{n \cdot 2^{n+1}}}(\sqrt{|k|} - i\sqrt{n - |k|})$. Let us denote by $|\Phi_t\rangle = \mathbf{U}^t(|\Psi_{in}\rangle \otimes |00 \ldots 0\rangle)$ the state of the system after t iterations. Note that because both the walk $\mathbf{U}$ and its initial state preserve the bit-permutation symmetry of the hypercube, the only consistent coin-state for position $|11 \ldots 1\rangle$ is the completely symmetric state over all directions: $\frac{1}{\sqrt{n}} \sum_{i=1}^{n} |i\rangle = |\Psi_{in}\rangle$. Let us call $|f\rangle = |\Psi_{in}\rangle \otimes |11 \ldots 1\rangle$ the "target" state. With these quantities in place, α_t, the amplitude at time t of the particle being in $|11\ldots1\rangle$, the opposite corner, is

$$\alpha_t := \alpha_t^{11\ldots1} = \langle f|\Phi_t\rangle = \sum_{k\in\{0,1\}^n} (a_k\lambda_k^t\langle\Psi_{in}|w_k\rangle + a_k^*\lambda_k^{*t}\langle\Psi_{in}|w_k^*\rangle)\cdot\langle11\ldots1|\tilde{k}\rangle$$

$$= \sum_{k\in\{0,1\}^n} \frac{1}{\sqrt{n\cdot2^{n+1}}} \frac{2n\cos(\omega_k t)}{\sqrt{2}\sqrt{n}} \frac{(-1)^{|k|}}{\sqrt{2^n}} = \frac{1}{2^n}\sum_{m=0}^{n}\binom{n}{m}(-1)^m\cos(\omega_m t). \quad (1)$$

Claim 1. For $t \in [\frac{\pi}{2}n - O(n^\beta), \frac{\pi}{2}n + O(n^\beta)]$ s.t. $t - n$ is even, $|\alpha_t|$ is lower bounded by $1 - O(\frac{\log n}{n^{1-2\beta}})$ for $0 < \beta < 1/2$.

Proof of Claim 1: Let us split the sum (1) into two parts, one where the index $m \in M := [(1-\delta)n/2, (1+\delta)n/2]$ and one where $m \notin M$, with $\delta < 1$ specified later. By standard Chernoff bounds on the tail probabilities of the binomial distribution we can upper-bound the absolute value of all the contributions from $m \notin M$ as

$$\left|\frac{1}{2^n}\sum_{m\notin M}\binom{n}{m}(-1)^m\cos(\omega_m t)\right| \leq \frac{1}{2^n}\sum_{m\notin M}\binom{n}{m} \leq 2e^{-\frac{\delta^2 n}{2}}. \quad (2)$$

Let us set $\delta = \sqrt{\frac{g(n)}{n}}$ with $g(n) = \Omega(\log n)$, in which case (2) is upper bounded by $2e^{-\Omega(\log n)/2}$. Let us write $t = \frac{\pi}{2}n \pm \varepsilon$ (i.e. $\varepsilon = O(n^\beta)$). The second term in the sum will come from contributions $m \in M$, so the terms $\cos\omega_m = 1 - 2m/n \in [-\delta, \delta]$ will be small. Call $v_m = \frac{\pi}{2} - \omega_m$, so $\cos\omega_m = \cos(\frac{\pi}{2} - v_m) = v_m - O(v_m^3)$ which means $v_m = 1 - 2m/n \pm O(\delta^3)$. Then

$$\cos(\omega_m t) = \cos[(\frac{\pi}{2} - v_m)(\frac{\pi}{2}n \pm \varepsilon)] = \cos[(\frac{t-n}{2} + m)\pi \mp \varepsilon(1 - \frac{2m}{n}) \pm tO(\delta^3)]$$
$$= (-1)^{\frac{t-n}{2}}(-1)^m\cos[\mp\varepsilon(1 - \frac{2m}{n}) \pm O(n\delta^3)] = (-1)^{\frac{t-n}{2}}(-1)^m[1 - O(\varepsilon^2\delta^2) - O(n^2\delta^6)] \quad (3)$$

and the second sum

$$\frac{1}{2^n}\sum_{m\in M}\binom{n}{m}(-1)^m\cos(\omega_m t) = (-1)^{\frac{t-n}{2}}[1 - O(\varepsilon^2\delta^2) - O(n^2\delta^6)]\frac{1}{2^n}\sum_{m\in M}\binom{n}{m}.$$

Since $\frac{1}{2^n}\sum_{m\in M}\binom{n}{m} \geq 1 - 2e^{-g(n)/2}$ we have

$$|\alpha_t| \geq \left|\frac{1}{2^n}\sum_{m\in M}\binom{n}{m}(-1)^m\cos(\omega_m t)\right| - 2e^{\frac{-g(n)}{2}} \geq 1 - O(\frac{g(n)}{n^{1-2\beta}}) - O(\frac{g^3(n)}{n}) - 4e^{\frac{-g(n)}{2}} \quad (4)$$

Set $g(n) = 2\log n$ to prove the claim for $0 < \beta < 1/2$. ∎

To prove Theorem 1 note that the probability of measuring the system in $|11\ldots1\rangle$ is $p = |\alpha_t|^2$. Set $\beta = \frac{1}{2}(1 - \frac{\log\log n}{\log n})$ and use Eq. (4) with $g(n) = 2\log\log n$ to get $p \geq 1 - O(\frac{\log\log n}{\log n})$. For $\beta = 0$ set $g(n) = 2\log n$ to get a lower bound of $1 - O(\log^3 n/n)$. ∎

Remark: Note that if we set $T = (2m+1)n\pi/2$ we obtain a similar result to the $m = 0$ case as long as T is sufficiently small so that $O(T^2\delta^6)$ terms do not matter, i.e.

$m = O(n)$. We can think of the walk returning to $|11\ldots\rangle$ every πn steps, which is in stark contrast to the classical case where the expected number of times a walk returns to some node i is $1/\pi_i = 2^n$ (see Sec. 2.1). Observe that hitting with probability at least p is not a monotone property.

3.2   Concurrent Hitting Time

Our second result relates to the concurrent version of hitting time for the symmetric walk **U** on the hypercube of dimension n. It implies that even without information on when to measure we retain a polynomial hitting behavior:

Theorem 2. **U** has a $(\frac{\pi}{2}n, \Omega(\frac{1}{n\log^2 n}))$ concurrent $(|x\rangle, |\bar{x}\rangle)$ hitting time.

Remarks: (1) Amplification: If the probability p in Defs. 2 and 4 is an inverse polynomial $p(n)$ in the size of the instance, we can use standard classical amplification to boost this probability to be exponentially close to 1. We just restart the random walk from scratch and repeat it $O(1/p(n))$ times. So the amplified coined symmetric discrete-time quantum walk on the hypercube of dimension n has a $(O(n^2\log^2 n), 1 - 2^{-O(n)})$ concurrent $(|x\rangle, |\bar{x}\rangle)$ hitting time.

(2) To be fair we should compare our results to *tail-bounds* for the hitting time in the classical case. It is very easy to show, however, that for the simple random walk on the hypercube starting in a node i the probability to hit the opposite corner j in a polynomial number of steps is exponentially small since each of the probabilities r_{ij}^t to be at j at time t (see Sec. 2.1) is exponentially small.

Proof of Theorem 2: The strategy of the proof is to compare the hitting probabilities at time t of the $|11\ldots1\rangle$-stopped walk to the unmeasured walk and to show that the perturbation caused by the measurement of the walk only gives a polynomial "loss" in hitting amplitude.

For the $|11\ldots1\rangle$-stopped walk (see Def. 3) the same symmetry arguments as before apply, since the measurement projectors Π_0 and $\Pi_1 = I - \Pi_0$ are also symmetric with respect to bit permutations. So the only possible "target" state is again $|f\rangle = \frac{1}{\sqrt{n}}\sum_{i=0}^{n}|i\rangle \otimes |11\ldots1\rangle$ and we may assume that we measure with $\{\Pi_0 = |f\rangle\langle f|, \Pi_1 = 1 - \Pi_0\}$. Let $|\Phi_t\rangle$, $|\Phi_0\rangle$ and $\alpha_t = \langle f|\Phi_t\rangle$ be the same quantities as before for the *unmeasured* walk **U**. Since the walk has non-zero transition amplitude only between nearest neighbors, the first time $\alpha_t \neq 0$ is for $t = n$ and since the walk is 2-periodic $\alpha_t = 0$ whenever t and n have different parity.

Let us define $|\tilde{\Phi}_t\rangle = (U\Pi_1)^t(|\Psi_{in}\rangle \otimes |00\ldots0\rangle)$ as the *non-normalised* state we get at time t given the walk has not stopped before t and $\beta_t := \langle f|\tilde{\Phi}_t\rangle$. Note that for $t \leq n$ we have $|\Phi_t\rangle = |\tilde{\Phi}_t\rangle$ and $\alpha_t = \beta_t$.

Claim 2. The probability to stop at some time $t \leq T$ is given by $p_T = \sum_{t=0}^{T}|\langle f|\tilde{\Phi}_t\rangle|^2 = \sum_{t=0}^{T}|\beta_t|^2$.

Proof of Claim 2: As in previous work [AB+01] it is easy to see that calculating with the renormalized state gives the *unconditional* probability to stop. If we do not renormalize our states we get exactly the conditional probability to stop at time t given we have not stopped before. ∎

We now want to relate the α_t from the *unmeasured* walk to the actual β_t of the *measured* walk.

Claim 3. $|\tilde{\Phi}_{n+k}\rangle = |\Phi_{n+k}\rangle - \sum_{i=0}^{k-1} \beta_{n+i} U^{k-i}|f\rangle$ *and* $\beta_{n+k} = \alpha_{n+k} - \sum_{i=1}^{k} \beta_{n+k-i} \cdot \gamma_i$ *with* $\gamma_t = \langle f|U^t|f\rangle$.

Proof of Claim 3: By induction on k. We have $|\Phi_t\rangle = |\tilde{\Phi}_t\rangle$ and $\alpha_t = \beta_t$ for $t \leq n$. Further $|\tilde{\Phi}_{n+1}\rangle = U|\Phi_n\rangle - U\alpha_n|f\rangle = |\Phi_{n+1}\rangle - \beta_n U|f\rangle$ so $\beta_{n+1} = \langle f|\Phi_{n+1}\rangle - \alpha_n\langle f|U|f\rangle = \alpha_{n+1} - \beta_n\langle f|U|f\rangle$. Write $|\tilde{\Phi}_{n+k+1}\rangle = U|\tilde{\Phi}_{n+k}\rangle - \beta_{n+k}U|f\rangle$ and apply the induction hypothesis to $|\tilde{\Phi}_{n+k}\rangle$. The claim on β_{n+k} follows immediately. ∎

Claim 4. Let $T = \lceil \frac{\pi}{2}n \rceil$ or $\lfloor \frac{\pi}{2}n \rfloor$ s.t. $T - n$ is even, let $0 \leq 2t \leq T - n$ and define $\tilde{\gamma}_{2t} = (-1)^t\gamma_{2t}$.
  1. $\gamma_t = \frac{1}{2^n}\sum_{m=0}^{n}\binom{n}{m}\cos(\omega_m t)$ *and* $\gamma_{2t+1} = 0$,
  2. $|\tilde{\gamma}_{2t} - \tilde{\gamma}_{2(t+1)}| = O(\frac{\log n}{\sqrt{n}})$,
  3. $\exists c$ s.t. for $t_c = \lfloor c\sqrt{n} \rfloor$ we have $|\alpha_{T-2t_c}| \leq \frac{1}{2}$.

Proof of Claim 4: Omitted. A complete proof will be given in another version.

We now can give a lower bound on $|\beta_t|$ in terms of quantities of the unmeasured walk:

Claim 5. Let t_c be as in Claim 4.3. If $\sum_{i=0}^{\frac{T-n}{2}-t_c}|\beta_{n+2i}| = o(\frac{1}{\log n})$ then $|\beta_{n+2t}| \geq |\alpha_{n+2t}| - |\alpha_{n+2t-2}| - o(\frac{1}{\sqrt{n}})$ for $T - n - 2t_c \leq 2t \leq (T - n)$.

Proof of Claim 5: Omitted. Will appear in another version.

If the assumption of Claim 5 is not true, then $\Omega(\frac{1}{\log n}) = \sum_{i=0}^{\frac{T-n}{2}-t_c}|\beta_{n+2i}| \leq \sqrt{\frac{T-n}{2}\sum_{i=0}^{\frac{T-n}{2}}|\beta_{n+2i}|^2} \leq \sqrt{np_T}$ which means $p_T = \Omega(\frac{1}{n\log^2 n})$.

The rest of Th. 2 follows from Claim 2 and Claim 5, $p_T = \sum_{t=n}^{T}|\beta_t|^2 \geq \sum_{t=T-\lfloor c\sqrt{n}\rfloor}^{T}|\beta_t|^2$

$$p_T \geq \frac{1}{c\sqrt{n}}\left(\sum_{t=T-\lfloor c\sqrt{n}\rfloor}^{T}|\beta_t|\right)^2 \geq \frac{1}{c\sqrt{n}}\left(\sum_{t=T-\lfloor c\sqrt{n}\rfloor}^{T}|\alpha_t| - |\alpha_{t-1}| - o(\frac{1}{\sqrt{n}})\right)^2$$

$$= \frac{\left(|\alpha_T| - \left|\alpha_{T-\lfloor c\sqrt{n}\rfloor - 1}\right| - o(1)\right)^2}{c\sqrt{n}} \geq \frac{(|\alpha_T| - 1/2 - o(1))^2}{c\sqrt{n}}$$

From Theorem 1 we know $|\alpha_T| = 1 - O(\frac{\log^3 n}{n})$ which establishes $p_T \geq \frac{1/4}{c\sqrt{n}} - o(\frac{1}{\sqrt{n}}) = \Omega(\frac{1}{\sqrt{n}})$ if the assumption of Claim 5 is true or $p_T = \Omega(\frac{1}{n\log^2 n})$ if it is not, in both cases proving the theorem. ∎

4   Dependence on the Initial State

One might wonder how much this polynomial hitting time depends on the fact that the walk is from one vertex to exactly the opposite corner of the hypercube. What if the two states where not exactly opposite? It is easy to see that if we start the walk in a neighbor of $|x\rangle$ we still obtain a polynomial hitting time to $|\bar{x}\rangle$, since after one step the walk spreads evenly to all its neighbors and a polynomial ($O(1/n)$) fraction of the amplitude will be on $|x\rangle$. This in turn implies that after $T = \pi/2n$ steps a polynomial fraction of the amplitude will be on $|\bar{x}\rangle$. This type of argument shows that for polynomially sized neighborhoods of $|x\rangle$ we get polynomial hitting times. But how large can the "polynomially $|\bar{x}\rangle$ hitting" region around $|x\rangle$ be? It turns out that a polynomial hitting time can not be true in general. We give a limit that comes from the lower bound on quantum unstructured search ([BB+97]).

Theorem 3. *Let S_x be a neighborhood of x (defined e.g. by a cut-off Hamming distance from x) s.t. for $y \in S_x$ the quantum walk has a $(O(poly(n)), \Omega(1/poly(n))$ concurrent $(|y\rangle, |\bar{x}\rangle)$ hitting time. Then $|S_x| = O(poly(n) \cdot \sqrt{2^n})$.*

Proof of Theorem 3: We will think of S_x as a ball around x, but the neighborhood of a node can be defined in any arbitrary way, the arguments go through for all of them. So $S_x = \{y : d_H(x,y) \leq d_c\}$ where d_H is the Hamming distance and d_c is a cut-off such that all $y \in S_x$ have $(O(p(n)), \Omega(1/q(n))$ concurrent $(|y\rangle, |\bar{x}\rangle)$ hitting time. By symmetry, $|S_x| =: S$ is independent of x. Let us cover the hypercube with K balls of size $|S|$, where each of the balls is centered around a node $x_1, x_2, \ldots, x_K$. A simple probabilistic argument shows that we can achieve this with $K = O(n \cdot 2^n/|S|)$ balls. Define a quantum search algorithm as follows: starting in $|x_1\rangle$ launch an $|x\rangle$-stopped quantum random walk as in Def. 3, where $|x\rangle$ is the marked state we are searching for. That means at every step we query the oracle with the current state of the walk and the question "Is this the marked state or not?". (We can adapt the standard oracle in Grover's algorithm [Gro96] to behave this way by measuring the auxiliary output qubit of the oracle.) We iterate this quantum walk for $p(n)$ steps and use classical amplification (repeat $q(n)$ many times). We repeat the amplified walk for each initial state $|x_i\rangle : i = 1 \ldots K$. With probability close to 1 one of the walks will find the marked state. The whole algorithm takes $O(p(n) \cdot q(n) \cdot K)$ queries. From the query lower bound of $\Omega(\sqrt{2^n})$ for any unstructured quantum search algorithm [BB+97] it follows that $K = \Omega(\sqrt{2^n}/poly(n))$ which yields the upper bound on $|S|$. ∎

5   Quantum Routing

Let us apply rapid hitting of the quantum random walk to sequential routing of a packet in a noisy network with a possible adversary trying to prevent the arrival of the packet. Assume the time when the packet is launched from node x is given only approximatively to the other nodes. We focus on both robustness of the algorithm against random noise (edge deletion, faulty nodes) as well as malicious attacks (adversary choses the most vulnerable edges/nodes to delete). The nodes of the network are bit-strings of length n and each node is connected to all nodes that differ by exactly one bit, so that the network has the topology of the hypercube. Consider quantum routing from node x to node y as follows:

(1) Let $d = d_H(x,y)$. We route on the sub-cube of dimension d spanned by the support of $x \oplus y$ (i.e. all strings z s.t. $z_i = x_i$ whenever $x_i = y_i$). The coin-space of the quantum random walk is d-dimensional; call the coin operator C_d. We assume that each

node v is capable to locally apply $C_d \otimes |v\rangle\langle v|$ (e.g. the bitpositions $x \oplus y$ of the sub-cube are broadcast). Nodes can locally implement the conditional shift (which requires only interactions between nearest neighbors). Both operations are local in the topology of the hypercube and can be implemented in a quantum network.

(2) The quantum random walk is applied $T = d\frac{\pi}{2}$ times (rounded appropriately).

(3) After T steps the state of the system is measured. With probability $1 - O(\frac{\log^3 d}{d})$ the packet is at y.

(3') At each time step node y performs the partial measurement to see if it has received the packet or not. After T steps the probability that the packet is at y is $\Omega(1/n\log^2 n)$. In case of failure the packet can be resent ($O(n\log^2 n)$ times) to boost the success probability close to 1.

Let us state the quantum advantages of this algorithm when x and y differ in $\Omega(n)$ bits (almost surely for random x and y).

Classically we could route the packet deterministically (by fixing the path in advance). We need to broadcast either the path or x (or y) so the nodes know which bit to flip when they receive the packet; the non-exact start makes it otherwise impossible to deduce this. This strategy is fast ($T = O(d)$) but neither secure against failure of one of the routing nodes/transversed edges nor against adversarial attacks. It suffices to affect one node/edge on the fixed path and the routing will fail. A fast randomized algorithm can flip the necessary bits in some random order. This strategy is robust against deletion of a subexponential number of random edges or nodes but requires again common knowledge of y. This in turn makes it vulnerable to adversarial attacks (it suffices to delete all the edges incident to y). A fully randomized classical routing algorithm, corresponding to a simple random walk on the cube, is robust against adversarial attacks but takes exponential time. It is here that quantum routing has an advantage. The nodes do not have to know the origin x and destination y of the packet, only $x \oplus y$. In the one-shot case (3') even the node at y does not have to know that it is the target - only at the measurement stage will it receive the packet. (This might enforce a more cooperative behavior of each of the routing nodes since they all could be the target). Knowledge of $x \oplus y$ alone is not sufficent to identify the most vulnerable edges (those incident or close to x and y) which reduces the adversary to random noise.

If a subexponential number of edges is deleted at random or a subexponential number of random nodes does not cooperate in the process, the success probability of the quantum routing algorithm changes only by an exponentially small amount.

To account for edge deletion we can assume that the deleted edge is replaced by a self-loop at each of its incident nodes. A faulty node v could apply any local operation $O_v \otimes |v\rangle\langle v|$ (including measurements) instead of $C_d \otimes |v\rangle\langle v|$. Almost surely the deleted edges or faulty nodes will be in a region of the hypercube of Hamming weight $\frac{d}{2} \pm O(\sqrt{d})$. In this region there is an exponential number of nodes for each Hamming weight. Since the walk spreads symmetrically over all states of same Hamming weight, the amplitude of each single state is exponentially small and perturbing a subexponential number of them in each step can induce only an exponentially small perturbation to the state of the walk. The walk is only $O(d)$ steps long so these exponential perturbations cannot add up to anything significant. ∎

Note that the fact that all the adversary can do is essentially random allows us to use this type of argument. If even an exponentially small change at each step happens outside the region around Hamming weight $d/2$ the resulting perturbation can be large - this is precisely the difficulty in proving Theorem 2 from Theorem 1.

It is important to see the quantum routing algorithm not only in terms of its advantages over classical routing. It is conceivable that quantum nets will be available in the near future and new routing strategies might have to be applied e.g. to distribute qubits to establish secret keys between certain nodes in the network. Our algorithm is a first step in this spirit.

Acknowledgments: I wish to thank Ronald de Wolf for many helpful discussions and improvements to Theorem 3, Tomohiro Yamasaki for mentioning his numerical results [Yam01], Cris Moore for extended conversations and Wim van Dam for generously sharing his ideas on continuous time random walks. Thanks to Dorit Aharonov, Andris Ambainis, Daniel Gottesman, Neil Shenvi and Birgitta Whaley for inspirational discussions. Partial support by DARPA and USAF under agreements number F030602-01-2-0524 and FDN00014-01-1-0826 is acknowledged.

References

[AA⁺01] D. Aharonov, A. Ambainis, J. Kempe, and U. Vazirani. Quantum walks on graphs. In *Proc. 33th STOC*, pages 50–59, New York, NY, 2001. ACM.

[AB⁺01] A. Ambainis, E. Bach, A. Nayak, A. Vishwanath, and J. Watrous. One-dimensional quantum walks. In *Proc. 33th STOC*, pages 60–69, New York, NY, 2001. ACM.

[AF01] D. Aldous and J. Fill. Reversible markov chains and random walks on graphs. Unpublished, preprint available at http://stat-www.berkeley.edu/users/aldous/RWG/book.html, 2001.

[BB⁺97] C.H. Bennett, E. Bernstein, G. Brassard, and U. Vazirani. Strengths and weaknesses of quantum computing. *Siam Journal on Computing*, 26:1510, 1997.

[CC⁺02] A.M. Childs, R. Cleve, E. Deotto, E. Farhi, S. Gutmann, and D.A. Spielman. Exponential algorithmic speedup by quantum walk. In *Proc. 35th STOC*, to appear 2003.

[CFG02] A. Childs, E. Farhi, and S. Gutmann. An example of the difference between quantum and classical random walks. *Quantum Information Processing*, 1:35, 2002.

[DFK91] M. Dyer, A. Frieze, and R. Kannan. A random polynomial-time algorithm for approximating the volume of convex bodies. *Journal of the ACM*, 38(1):1–17, January 1991.

[FG98] E. Farhi and S. Gutmann. Quantum computation and decision trees. *Phys. Rev. A*, 58:915–928, 1998.

[Gro96] L. Grover. A fast quantum mechanical algorithm for database search. In *Proc. 28th STOC*, pages 212–219, Philadelphia, Pennsylvania, 1996. ACM.

[GS⁺01] M. Grigni, L. Schulman, M. Vazirani, and U. Vazirani. Quantum mechanical algorithms for the nonabelian hidden subgroup problem. In *Proc. 33th STOC*, pages 68–74, New York, NY, 2001. ACM.

[HRT00] S. Hallgren, A. Russell, and A. Ta-Shma. Normal subgroup reconstruction and quantum computation using group representations. In *Proc. 32nd STOC*, pages 627–635, 2000.

[JS89]   M. Jerrum and A. Sinclair. Approximate counting, uniform generation and rapidly mixing Markov chains. *Information and Computation*, 82(1):93–133, 1989.

[Mey96]   D. Meyer. From quantum cellular automata to quantum lattice gases. *J. Stat. Phys.*, 85:551–574, 1996.

[MR95]   R. Motwani and P. Raghavan. Randomized Algorithms. Cambridge University Press, 1995.

[MR02]   C. Moore and A. Russell. Quantum walks on the hypercube. In *Proc. RANDOM 2002*, pages 164–178, Cambridge, MA, 2002. Springer.

[NC00]   M.A. Nielsen and I.L. Chuang. *Quantum Computation and Quantum Information*. Cambridge University Press, Cambridge, UK, 2000.

[Pap94]   C. Papadimitriou. *Computational Complexity*. Addison Wesley, Reading, Massachusetts, 1994.

[Sch99]   U. Schöning. A probabilistic algorithm for *k*-SAT and constraint satisfaction problems. In *Proc. 40th FOCS*, pages 410–414. IEEE, 1999.

[Sho97]   P.W. Shor. Polynomial-time algorithms for prime factorization and discrete logarithms on a quantum computer. *SIAM J. Comp.*, 26(5):1484–1509, 1997.

[Sim97]   D. Simon. On the power of quantum computation. *SIAM J. Comp.*, 26(5):1474–1483, 1997.

[SKW03]   N. Shenvi, J. Kempe, and K.B. Whaley. A quantum random walk search algorithm. *Phys. Rev. A*, 67(5):052307, 2003.

[Wat01]   J. Watrous. Quantum simulations of classical random walks and undirected graph connectivity. *Journal of Computer and System Sciences*, 62(2):376–391, 2001.

[Yam01]   T. Yamasaki. personal communication, 2001.

[YKI02]   T. Yamasaki, H. Kobayashi, and H. Imai. An analysis of absorbing times of quantum walks. In *Proc. 3rd UMC*, volume 2509 of *Lecture Notes in Computer Science*, pages 315–330. Springer, 2002.

A   Continuous-Time Quantum Random Walk

The continuous-time walk has been defined in [FG98] as a quantum version of the classical continuous-time walk (see Sec. 2.1). To make the classical continuous walk with generator Q quantum one simply sets $U(t) = exp(iQt)$, which is unitary as long as $Q = Q^\dagger$ (which is the case for simple random walks on undirected graphs). This walk works directly with the space formed by the nodes of the graph and does not require auxiliary coin spaces. To date it is not clear how the continuous and the discrete time walk are related.

For the hypercube the continuous time quantum walk is described by the following transformation on the space spanned by n-bit strings [MR02]:

$$\mathbf{U}_{cont}(t) = e^{i\frac{t}{n}(X_1+X_2+\cdots+X_n)} = e^{i\frac{t}{n}X_1} \cdot e^{i\frac{t}{n}X_2} \cdot \ldots \cdot e^{i\frac{t}{n}X_n}$$

where X_i acts only on the ith bit as $X|0\rangle = |1\rangle$ and $X|1\rangle = |0\rangle$. The expression in the exponential corresponds to the adjacency matrix of the hypercube. $\mathbf{U}_{cont}(t)$ can be simulated uniformly by a quantum circuit with $O(n)$ local gates. It is now straightforward to prove the following theorem:

Theorem 4 (One - shot hitting time). $\mathbf{U}_{cont}$ has a $(T = \frac{\pi n}{2}, 1)$ and a $(T = \frac{\pi n}{2} \pm n^\beta, 1 - O(1/n^{1-2\beta}))$ one shot hitting time for $\beta = const < 1/2$.

Proof: Omitted.

Definition 5 ($|x\rangle$**-stopped walk, concurrent hitting time:**). *The* $|x\rangle$*-stopped walk is the iterative process where first a measurement with* $\{\Pi_0 = |x\rangle\langle x|, \Pi_1 = 1 - \Pi_0\}$ *is performed. If* $|x\rangle$ *is measured the walk is stopped, otherwise* $\mathbf{U}_{cont}(1)$ *is applied and the procedure repeated. The walk has a* (T, p) *concurrent hitting time if the probability to stop before time* T *is* $> p$.

Theorem 5 (Concurrent hitting time). *The continuous time walk has a* $(T = \frac{\pi n}{2}, \Omega(\frac{1}{\sqrt{n}}))$ *concurrent hitting time.*

Proof: Omitted.

Approximate Testing of Visual Properties

Sofya Raskhodnikova

MIT Laboratory for Computer Science, Cambridge MA 02139, USA,
sofya@theory.lcs.mit.edu,
http://theory.lcs.mit.edu/~sofya/

Abstract. We initiate a study of property testing as applied to visual properties of images. Property testing is a rapidly developing area investigating algorithms that, with a small number of local checks, distinguish objects satisfying a given property from objects which need to be modified significantly to satisfy the property. We study visual properties of discretized images represented by $n \times n$ matrices of binary pixel values. We obtain algorithms with query complexity independent of n for several basic properties: being a half-plane, connectedness and convexity.

1 Introduction

> We chose to investigate *connectedness*
> because of a belief that this predicate is
> nonlocal in some very deep sense; therefore
> it should present a serious challenge to any
> basically local, parallel type of computation.
>
> PERCEPTRONS
> *Marvin Minsky and Seymour Papert*

Images are typically so large that it is impractical to read every single bit of them. It is natural to ask what properties of an image can be detected by *sublinear* algorithms that read only a small portion of the image. In general, most problems are not solvable exactly with that restriction. Property testing [16,11] (see [15,9] for surveys) is a notion of approximation tailored for decision problems and widely used for studying sublinear algorithms. Property tests distinguish inputs with a given property from those that are *far* from satisfying the property. *Far* means that many characters of the input must be changed before the property arises in it. The query complexity of a property test is the number of characters it reads. The goal is to design tests with sublinear complexity.

Image analysis is one area potentially well suited to the property testing paradigm. Some salient features of an image may be tested by examining only a small part thereof. Indeed, one motivation for this study is the observation that the eye focuses on relatively few places within an image during its analysis. The analogy is not perfect due to the eye's peripheral vision, but it suggests that property testing may give some insight into the visual system.

S. Arora et al. (Eds.): APPROX 2003+RANDOM 2003, LNCS 2764, pp. 370–381, 2003.
© Springer-Verlag Berlin Heidelberg 2003

In this paper, we present tests for a few properties of images. All our tests have complexity independent of the image size, and therefore work well even for huge images. We use image representation popular in learning theory (see, e.g., [14,13]). Each image is represented by an $n \times n$ matrix M of pixel values. We focus on black and white images given by binary matrices with black denoted by 1 and white denoted by 0. To keep the correspondence with the plane, we index the matrix by $\{0, 1, \ldots, n-1\}^2$, with the lower left corner being $(0, 0)$ and the upper left corner being $(0, n-1)$. The object is a subset of $\{0, 1, \ldots, n-1\}^2$ corresponding to black pixels; namely, $\{(i, j)|M_{i,j} = 1\}$.

1.1 Property Testing in the Pixel Model

The *distance* between two images of the same size is defined as the number of pixels (matrix entries) on which they differ. (Two matrices of different size are considered to have infinite distance.) The *relative distance* is the ratio of the distance and the number of pixels in the image. A *property* is defined as a collection of pixel matrices. The distance of an image (matrix) M to a property $\mathcal{P}$ is $\min_{M' \in \mathcal{P}} dist(M, M')$. Its *relative distance* to $\mathcal{P}$ is its distance to $\mathcal{P}$ divided by the size of the image matrix. An image is ε-far from $\mathcal{P}$ if its relative distance to $\mathcal{P}$ is at least ε. If the image is not ε-far from $\mathcal{P}$, it is ε-close to it.

A property is (ε, q)-*testable* if there is a randomized algorithm that for every input matrix M queries at most q entries of M and with probability at least $\frac{2}{3}$ distinguishes between matrices with the property and matrices which are ε-far from having it. The algorithm is referred to as an (ε, q)-*test*. This definition allows tests to have *2-sided error*. An algorithm has *1-sided error* if it always accepts an input that has the property.

1.2 Our Results

We present tests for three visual properties: being a half-plane, convexity and connectedness. The number of queries in all tests is independent of the size of the input. The algorithm for testing if the input is a half-plane is a 1-sided error test with $\frac{2 \ln 3}{\varepsilon} + o(\frac{1}{\varepsilon})$ queries. The convexity test has 2-sided error and makes $O(1/\varepsilon^2)$ queries. Finally, the connectedness test has 1-sided error and makes $O\left(\frac{1}{\varepsilon^2} \log^2 \frac{1}{\varepsilon}\right)$ queries.

1.3 Related Results in Property Testing

Previous papers on property testing in computational geometry [7,6] consider a model different from ours, where the input is the set of object points and a query i produces coordinates of the ith point. Their results, in general, are incomparable to ours. In their model, the problems we consider would have query complexity dependent on the number of points in the object. But they are able to study properties which are trivially testable in our model because all instances are either close to having the property or close to not having it. An example is the property that a given graph is a Euclidean minimum spanning tree of a given point set in the plane [7].

Another related work is [10] which studies properties of d-dimensional matrices. It gives a class of properties which are testable with a number of queries polynomial in $1/\varepsilon$. It does not seem applicable to our geometric properties.

Goldreich and Ron [12] study property testing in bounded degree graphs represented by adjacency lists. Note that an image in the pixel model can be viewed as a graph of degree 4 where vertices correspond to black pixels and they are connected by an edge if the corresponding entries in the image matrix are adjacent. (See the definition of the *image graph* in the beginning of section 4.) Goldreich and Ron measure distance between graphs as the ratio of the number of edges that need to be changed to transform one graph into the other over the maximum possible number of edges in the graphs with the given number of vertices and degree. In our case, the distance between two image graphs corresponds to the fraction of points (vertices) on which they differ, i.e. the edge structure of the graphs is fixed, and only vertices can be added or removed to transform one graph into another. Our connectedness test is exactly the same as the connectivity test in [12], with one minor variation due to different input representation and the fact that the pixel model allows graphs with a small number of vertices. (In the bounded degree graph model, the number of vertices is a part of the input.) However, since our distance measures are different, their proof of correctness of the algorithm does not apply to the pixel model.

One more paper that studies fast algorithms for connectedness in graphs is [5]. It shows how to approximate the number of connected components in an arbitrary graph in a sublinear time.

1.4   Related Results in Learning

In property testing terminology, a PAC (probably approximately correct) learning algorithm [17] is given oracle access (or access via random samples) to an unknown *target* object with the property $\mathcal{P}$ and has to output a *hypothesis* which is within relative distance ε to the target with high probability. If the hypothesis is required to have the property $\mathcal{P}$, the learning algorithm is *proper*. As proved in [11], a proper PAC learning algorithm for $\mathcal{P}$ with sampling complexity $q(\varepsilon)$ implies a (2-sided error) $(\varepsilon, q(\varepsilon/2) + O(1/\varepsilon))$-test for $\mathcal{P}$.

Learning half-planes exactly is considered in [14]. This work gives matching upper and lower bound of $\Theta(\log n)$ for the problem. In the PAC model, a proper learning algorithm with $O(1/\varepsilon \log(1/\varepsilon))$ sampling complexity follows from [3]. Together with the [11] result above, it implies a (2-sided error) $(\varepsilon, O(1/\varepsilon \log(1/\varepsilon)))$-test for the half-plane property. Our result for testing half-planes is a modest improvement of shaving off the log factor and making the error 1-sided.

The generic approach of [11] for transforming PAC proper learners into property tests does not seem to work well for convexity and connectedness. The complexity of PAC learning algorithms is at least proportional to Vapnik Chervonenkis (VC) dimension[1][8]. Since VC dimension of convexity is $\Theta(n)$ and VC dimension of connectedness is $\Theta(n^2)$, the corresponding tests obtained by the generic approach have query complexity guarantee $O(n)$ and $O(n^2)$, respectively. Our tests for these properties have query complexity independent of n.

2   Testing if an Image Is a Half-Plane

First we present an algorithm for testing whether the image is a half-plane. An image is a *half-plane* if there is a vector $w \in \mathbb{R}^2$ and a number $a \in \mathbb{R}$ such that a pixel x is black if and only if $w^T x \geq a$. The algorithm first finds a small region within which the dividing line falls. Then it checks if pixels on one side of the region are white and on the other side are black.

Call pixels $(0,0), (0, n-1), (n-1, 0), (n-1, n-1)$ *corners*. Call the first and the last row and the first and the last column of the matrix *sides*. For a pair of pixels p_1, p_2, let $\ell(p_1, p_2)$ denote the line[2] through p_1, p_2. Let $R_1(p_1, p_2)$ and $R_2(p_1, p_2)$ denote the regions into which $\ell(p_1, p_2)$ partitions the image pixels not on the line.

HALF-PLANE TEST $T_1(\varepsilon)$

Given access to an $n \times n$ pixel matrix,

1. Query the four corners. Let s be the number of sides with differently colored corners.
   (a) If $s = 0$ (all corners are of the same color c), query $\frac{\ln 3}{\varepsilon}$ pixels independently at random. Accept if all of them have color c. Reject otherwise.
   (b) If $s = 2$,
      i. For both sides with differently colored corners, do binary search of pixels on the side to find two differently colored pixels within distance less than $\varepsilon n/2$. For one side, call the white pixel w_1 and the black pixel b_1. Similarly, define w_2 and b_2 for the second side.
      ii. Let $W_i = R_i(w_1, w_2)$ and $B_i = R_i(b_1, b_2)$ for $i = 1, 2$. W.l.o.g., suppose W_2 and B_1 intersect while W_1 and B_2 do not. Query $\frac{2 \ln 3}{\varepsilon}$ pixels from $W_1 \cup B_2$ independently at random. Accept if all pixels from W_1 are white, all pixels from B_2 are black. Otherwise, reject.
   (c) If s is not 0 or 2, reject.

---

[1]   The *VC dimension* is the cardinality of the largest set $X \subseteq \{0, \ldots, n-1\}^2$ shattered by $\mathcal{P}$. A set $X \subseteq \{0, \ldots, n-1\}^2$ is *shattered* by $\mathcal{P}$ if for every partition (X_0, X_1) of X, $\mathcal{P}$ contains a matrix M with $M_x = 1$ for all $x \in X_1$ and $M_x = 0$ for all $x \in X_0$.

[2]   Whenever a geometric notion (e.g., line, angle, convex hull) is used without a definition, it refers to the standard continuous notion. All discretized notions are defined.

Theorem 1. *Algorithm T_1 is a 1-sided error $(\varepsilon, \frac{2\ln 3}{\varepsilon} + o(\frac{1}{\varepsilon}))$-test for the half-plane property.*

Proof. The algorithm queries at most $\frac{2\ln 3}{\varepsilon} + O(\log(1/\varepsilon))$ pixels. To prove correctness, we need to show that all half-planes are always accepted, and all images that are ε-far from being half-planes are rejected with probability at least $2/3$.

Case (a) [0 differently colored sides]: The image is a half-plane if and only if it is unicolored. If it is unicolored, the test always accepts since it never finds pixels of different colors. If the image is ε-far from being a half-plane, it has at least εn^2 pixels of a wrong color. Otherwise, it can be made unicolored, and hence a half-plane, by changing less than an ε-fraction of pixels. The test fails to find an incorrectly colored pixel and accepts with probability at most $(1-\varepsilon)^{\ln 3/\varepsilon} < 1/3$.

Case (b) [2 differently colored sides]: The test always accepts all half-planes because it rejects only if it finds two white pixels and two black pixels such that the line through the white pixels intersects the line through the black pixels.

It remains to show that if an image is ε-far from being a half-plane, it is rejected with probability $\geq 2/3$. We prove the contrapositive, namely, that if an image is rejected with probability $< 2/3$, modifying an ε fraction of pixels can change it into a half-plane.

Suppose that an image is accepted with probability $\geq 1/3 = e^{-\ln 3} > (1 - \varepsilon/2)^{2\ln 3/\varepsilon}$. That means that $< \varepsilon/2$ fraction of pixels from which we sample in step 1(b)ii differ from the color of their region (white for W_1 and black for B_2). Note also that there are at most $\varepsilon n/2$ pixels outside of $W_1 \cup B_2$. Changing the color of all black pixels in W_1 and all white pixels in B_2 and making all pixels outside of those regions white, creates a half-plane by changing $< \varepsilon$ fraction of the pixels, as required.

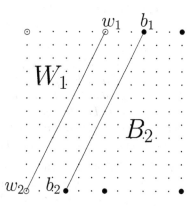

Fig. 1. Half-plane test

Case (c) [everything else]: The number of image sides with differently colored corners is even (0, 2, or 4). That holds because the cycle $((0,0), (n-1,0), (n-1, n-1), (0, n-1), (0,0))$ visits a vertex of a different color every time it moves along such a side. So, the only remaining case is 4 differently colored sides. In this case, the image cannot be a half-plane. The test always rejects. $\square$

3 Convexity Testing

The image is *convex* if the convex hull of black pixels contains only black pixels. The test for convexity first roughly determines the object by querying pixels on the $n/u \times n/u$ grid with a side of size $u = \Theta(\varepsilon n)$. Then it checks if the object corresponds to the rough picture it obtained.

For all indices i, j divisible by u, call the set $\{(i', j')| \ i' \in [i, i+u], j' \in [j, j+u]\}$ a *u-square*. We refer to pixels $(i, j), (i + u, j)(i + u, j + u)$, and $(i, j + u)$ as its corners.

CONVEXITY TEST $T_2(\varepsilon)$

Given access to an $n \times n$ pixel matrix,

1. Query all pixels with both coordinates divisible by $u = \lfloor \varepsilon n/120 \rfloor$.
2. Let B be the convex hull of discovered black pixels. Query $\frac{5}{\varepsilon}$ pixels from B independently at random. Reject if a white pixel in B is found in steps 1 or 2.
3. Let W be the union of all u-squares which contain no pixels from B. Query $\frac{5}{\varepsilon}$ pixels from W independently at random. Reject if a black pixel is found. Otherwise accept.

Lemma 1, used in the analysis of the convexity test, asserts that the number of pixels outside $B \cup W$ is small.

Lemma 1. *In an $n \times n$ image, let B be the convex hull of black pixels with coordinates divisible by u. Let W be the union of u-squares which contain no pixels from B. Let the "fence" F be the set of pixels not contained in B or W. Then F contains at most $4un$ pixels.*

Proof. Intuitively, F is the largest when it contains all u-squares along the sides of the image. We call u-squares that are not fully contained in B or W *fence u-squares*. Note that F is covered by *fence u-squares*. Therefore, to prove the lemma it is enough to show that there are at most $4n/u$ *fence u-squares*.

To count the *fence u-squares*, we define a cyclic ordering on them. To do that, we describe a walk that connects centers of all *fence u-squares*. The walk goes from one center to the next by traveling left, right, up or down. It visits the centers of *fence u-squares* by traveling clockwise and keeping the boundary between F and W on the left-hand side. Each *fence u-square* is visited because it intersects with some u-square in W in at least one pixel.

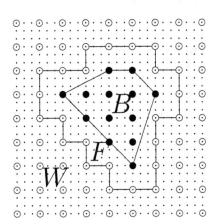

Fig. 2. Convexity test

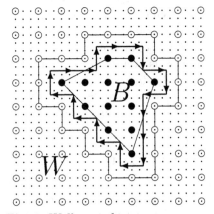

Fig. 3. Walk over fence u-squares

There are n/u rows of u-squares. We claim that from each of these rows the walk can travel up at most once. Suppose for contradiction that it goes up twice, from ℓ_1 to ℓ_2 and from r_1 to r_2, where ℓ_1 and r_1 are *fence u-squares* with centers in row $(k+0.5)u$, and ℓ_2 and r_2 are *fence u-squares* with centers in row $(k+1.5)u$ for some integer k.

W.l.o.g. suppose that the centers of l_1, l_2 are in a column with a lower index than the centers of r_1, r_2. Since the walk keeps the boundary between W and F on the left-hand side, the left corners of ℓ_1, ℓ_2, r_1, r_2 are in W. By definition of *fence u-squares*, ℓ_1, ℓ_2, r_1, r_2 each contain a pixel from B. The common left corner of r_1 and r_2 is also in B, since B is convex. But this is a contradiction because W and B are disjoint.

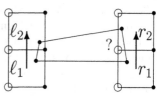

Thus, the walk can travel up only once per row. Similarly, it can travel down only once per row, and travel left (right) only once per column. Since there are n/u rows (columns) of u-squares, the walk can have at most $4n/u$ steps. As it visits all *fence u-squares*, there are at most $4n/u$ of them. Since each u-square contributes u^2 pixels, the number of pixels in F is at most $4nu$. □

The analysis of the convexity test uses the fact that if an image is convex, W contains only a small number of black pixels. Proposition 1 proves this fact for a special case of an image which is "invisible" on the big grid. Later, we use the proposition to handle the general case in lemma 2.

Proposition 1. *In an $n \times n$ convex image, if all pixels with both coordinates divisible by u are white, then the image contains less than $2un$ black pixels.*

Proof. Let $black(r)$ denote the number of black pixels in a row r. If each row contains fewer than $u - 1$ pixels, the total number of black pixels is at most un. Otherwise, consider a row r with $black(r) \geq u$. Let integers k and t be such that $r = ku + t$ and $0 \leq t < u$. Since the image is convex, black pixels of every fixed row must have consecutive column indices. Since every pixel with both coordinates divisible by u is white, $black(ku) < u$ and $black((k+1)u) < u$.

Because of the convexity of the object, if $black(r_1) < black(r)$ for a row $r_1 > r$ then $black(r_2) \leq black(r_1)$ for all rows $r_2 > r_1$. Similarly, if $black(r_1) < black(r)$ for a row $r_1 < r$ then $black(r_2) \leq black(r_1)$ for all rows $r_2 < r_1$. Thus, all rows r_2 excluding $ku + 1, ku + 2, \ldots, (k+1)u - 1$ have $black(r_2) < u$. Together, they contain $< (n - u)u$ black pixels. Cumulatively, the remaining $u - 1$ rows contain $< (u - 1)n$ pixels. Therefore, the image contains less than $2un$ black pixels. □

Lemma 2. *In an $n \times n$ convex image, let W be the union of all u-squares which contain no pixels from B. Then W contains less than $8un$ black pixels.*

Proof. As before, let F be the set of all pixels not contained in B or W. We call pixels on the boundary between F and W with both coordinates divisible by u *fence posts*. Since all *fence posts* are white, any portion of the object protruding into W has to squeeze between the *fence posts*. We show that there are at most three large protruding pieces, each of which, by proposition 1, contains less than $2un$ pixels. All other sticking out portions fall close to the fence and are covered by the area containing less than $2un$ pixels.

Let O be the boundary of our convex object. O can be viewed as a piecewise linear trajectory on the plane that turns $360°$. Whenever O leaves region F to go into W, it has to travel between two *fence posts*. Whenever O comes back into F, it has to return between the same fence posts because the object is convex and *fence posts* do not belong to it. The figure depicts an excursion of O into W with accumulated turn α.

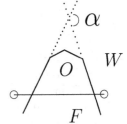

Notice that since O turns $360°$ total, at most 3 excursion into W have accumulated turn $> 90°$. Each of them can be viewed as delineating a part of our convex object, cut off by the line between the *fence posts*. This part of the object is convex, and therefore, by proposition 1, has $< 2un$ pixels. This gives us a total of $< 6un$ pixels for the protruding parts where O turns more than $90°$.

Consider any excursion into W where O leaves F between *fence posts* p_1 and p_2 and turns $\leq 90°$ before coming back. Any such trajectory part lies inside the circle of diameter u containing p_1 and p_2. The half of the circle protruding into W is covered by a half of a u-square. By an argument identical to counting *fence squares* in lemma 1, there are at most $4n/u$ segments of the F/W boundary between adjacent fence posts. Therefore, the total number of pixels that might be touched by the parts of the object, described by O's excursions into W that turn $\leq 90°$ is at most $4n/u \cdot u^2/2 = 2un$.

Thus, the total number of black pixels in W is at less than $8un$. $\square$

Theorem 2. *Algorithm T_2 is a $(\varepsilon, O(1/\varepsilon^2))$-test for convexity.*

Proof. The test makes $(n/u)^2 + O(1/\varepsilon) = O(1/\varepsilon^2)$ queries. We bound failure probability, considering convex and far from convex images separately.

If the input image is convex, B contains only black pixels. The test never rejects in step 2. By lemma 2, the fraction of black pixels in W is $< 8u/n = \varepsilon/15$. By the union bound, the probability that the test rejects in step 3 is $< \frac{\varepsilon}{15}\frac{5}{\varepsilon} = \frac{1}{3}$.

If the input image is ε-far from convex, it has $\geq 2\varepsilon n^2/5$ white pixels in B or $\geq 2\varepsilon n^2/5$ black pixels in W. Otherwise, we could make the image convex by making all pixels in W white and all remaining pixels black. It would require $< 2\varepsilon n^2/5$ changes in B, $< 2\varepsilon n^2/5$ changes in W, and by lemma 1, $\leq 4un < \varepsilon n^2/5$ changes in F. Thus, the distance of the image to convex would be less than εn^2.

Suppose w.l.o.g. that there are $\geq 2\varepsilon/5$ black pixels in W. Step 3 will fail to find a black pixel with probability $\leq (1 - \frac{2\varepsilon}{5})^{5/\varepsilon} \leq e^{-2} < \frac{1}{3}$. $\square$

4   Connectedness Testing

Define the *image graph* $G_M = (V, E)$ of image matrix M by $V = \{(i,j)|M_{i,j} = 1\}$ and $E = \{((i_1, j), (i_2, j))| \ |i_1 - i_2| = 1\} \cup \{((i, j_1), (i, j_2))| \ |j_1 - j_2| = 1\}$. In other words, the image graph consists of black pixels connected by the grid lines. The image is *connected* if its image graph is connected. When we say that the image has k connected components, we are also referring to its image graph.

The test for connectedness looks for isolated components of size less than $d = 4/\varepsilon^2$. We prove that a significant fraction of pixels are in such components if the image is far from connected. When a small isolated component is discovered, the test rejects if it finds a black pixel outside of the component. Lemma 3 implies that if an image is far from connected, it has a large number of connected components. An averaging argument in lemma 4 demonstrates that many of them have to be small. This gives rise to a simple test T_3, which is later improved to test T_4 with more careful accounting in proposition 2.

Both tests for connectedness and proposition 2 are adopted from [12]. The only change in the tests, besides parameters, is that after finding a small component, we make sure there is some point outside of it before concluding that the image is far from connected.

CONNECTEDNESS TEST $T_3(\varepsilon)$

Let $\delta = \frac{\varepsilon^2}{4} - o(1)$ and $d = 4/\varepsilon^2$. Given access to an $n \times n$ pixel matrix,

1. Query $2/\delta$ pixels independently at random.
2. For every pixel (i, j) queried in step 1, perform a breadth first search (BFS) of the image graph starting from (i, j) until d black pixels are discovered or no more new black pixels can be reached; i.e., for each discovered black pixel query all its neighbors if they haven't been queried yet. If no more new black pixels can be reached, a small connected component has been found.
3. If a small connected component is discovered for some (i, j) in step 2, query $2/\varepsilon$ pixels outside of the square $[i - d, i + d] \times [j - d, j + d]$ independently at random. If a black pixel is discovered, reject. Otherwise (if no small connected component is found or if no black pixel is discovered outside of the small component), accept.

Lemma 3. *If an $n \times n$ image contains at most p connected components, they can be linked into one connected component by changing at most $n(\sqrt{2p} + O(1))$ pixel values from white to black.*

Proof. Let $s = n\sqrt{2/p}$. To turn the image into one connected component, we first add the comb-like set $S = \{(i, j) \mid j = n - 1 \text{ or } i = n - 1 \text{ or } s \text{ divides } i\}$. Now every connected component is linked to S by adding at most $s/2$ pixels leading to the nearest "tooth of the comb". That is, if a component contains a pixel $(ks + \ell, j)$ for an integer k and $0 \le \ell \le s/2$, add pixels $(ks + 1, j), (ks + 2, j), \ldots, (ks + \ell - 1, j)$. Otherwise (a component contains a pixel $(ks + \ell, j)$ for integer k and $s/2 < \ell < s$), add pixels $(ks + \ell + 1, j), (ks + \ell + 2, j), \ldots, (ks + s - 1, j)$. The first stage adds $|S| = n(n/s + O(1))$ pixels and the second, less than $s/2$ per connected component, adding the total of $n(n/s + O(1)) + ps/2 = n\sqrt{2p} + O(1)$ pixels. $\square$

Lemma 4. *If an image is ε-far from connected, at least an $\frac{\varepsilon^2}{4} - o(1)$ fraction of its pixels are in connected components of size less than $d = 4/\varepsilon^2 + o(1)$.*

Proof. Consider an $n \times n$ ε-far from connected image with p connected components. By lemma 3, changing $\leq n(\sqrt{2p} + O(1))$ pixels makes it connected. Then $n(\sqrt{2p} + O(1)) \geq \varepsilon n^2$, and $p \geq \varepsilon^2 n^2/2 - O(n)$. Let b be the number of black pixels. The average component size is $b/p \leq n^2/(\varepsilon^2 n^2/2 - O(n)) = 2/\varepsilon^2 + o(1)$. Thus, the fraction of components of size up to $d = \frac{4}{\varepsilon^2} + o(1)$ is $\geq 1/2$. That is, there are $\geq p/2 = \varepsilon^2 n^2/4 - O(n)$ such components. Since each connected component contains a pixel, $\geq \varepsilon^2/4 - o(1)$ fraction of pixels are in connected components of size d. $\square$

Theorem 3. *Algorithm T_3 is a 1-sided $(\varepsilon, O(\varepsilon^{-4}))$-test for connectedness.*

Proof. The algorithm accepts all connected images because it rejects only if an isolated component and some pixel outside of it are found.

It remains to show that an ε-far from connected image is rejected with probability at least $2/3$. By lemma 4, such an image has at least a δ fraction of its pixels in connected components of size less than d. The probability that step 1 fails to find a pixel from a small connected component is $(1 - \delta)^{2/\delta} \leq e^{-2}$. In step 2, $3d - 1$ queries are sufficient to discover that a component of size $d - 1$ is isolated because it has at most $2d$ neighboring white pixels. There are at least $\varepsilon n^2 - 4d^2$ black pixels outside of the $2d \times 2d$ square containing the small isolated component. Step 3 will fail to find a black pixel with probability $(1 - \varepsilon)^{2\varepsilon} \leq e^{-2}$. By the union bound, the failure probability is at most $2/e^2 < 1/3$.

The number of queries is at most $2/\delta \times (3d - 1) + 2/\varepsilon = O(\varepsilon^{-4})$. $\square$

The algorithm can be improved by employing the Goldreich-Ron trick [12] of considering small components of different sizes separately. The following proposition is adopted from [12].

Proposition 2. *If an image has at least C connected components of size less than d, there is $\ell \leq \log d$ such that at least $\frac{C \cdot 2^{\ell-1}}{\log d}$ points are in connected components of size between $2^{\ell-1}$ and $2^\ell - 1$.*

Proof. For some $\ell \leq \log d$, the image has at least $C/\log d$ connected components of size between $2^{\ell-1}$ and $2^\ell - 1$. Each of them contains at least $2^{\ell-1}$ points. $\square$

(IMPROVED) CONNECTEDNESS TEST $T_4(\varepsilon)$

Let $\delta = \frac{\varepsilon^2}{4} - o(1)$ and $d = 4/\varepsilon^2$. Given access to an $n \times n$ pixel matrix,

1. For $\ell = 1$ to $\log d$
   (a) Query $\frac{4 \log d}{\delta 2^\ell}$ pixels independently at random.
   (b) For every pixel (i, j) queried in step 1a, perform a BFS of the image graph starting from (i, j) until 2^ℓ black pixels are discovered or no more new black pixels can be reached (a small connected component has been found).
2. If a small connected component is discovered for some (i, j) in step 1, proceed as in step 3 of algorithm T_3.

Theorem 4. *Algorithm T_4 is a 1-sided $\left(\varepsilon, O\left(\frac{1}{\varepsilon^2}\log^2\frac{1}{\varepsilon}\right)\right)$-test for connectedness.*

Proof. The algorithm accepts all connected images because it rejects only if an isolated component and some pixel outside of it are found.

If an $n \times n$ image is ε-far from connected, by the proof of lemma 4, it has at least a δn^2 connected components of size less than d. Proposition 2 implies that for some $\ell < \log d$, at least an $\frac{\delta \cdot 2^{\ell-1}}{\log d}$ fraction of its points are in connected components of size between $2^{\ell-1}$ and $2^\ell - 1$. For this ℓ, the probability that step 1 fails to find a pixel from a component of size between $2^{\ell-1}$ and $2^\ell - 1$ is at most e^{-2}. The rest of the correctness analysis is the same as in theorem 3.

The number of queries is at most $\log d \cdot O\left(\frac{\log d}{\delta}\right) + 2/\varepsilon = O\left(\frac{1}{\varepsilon^2}\log^2\frac{1}{\varepsilon}\right)$. $\square$

5   Conclusion and Open Problems

Employing the Paradigm from the Half-Plane Test The strategy employed in the half-plane test of section 2 is very simple. First we approximately learn the position of the dividing line. Then, using the fact that all half-planes consistent with our knowledge of the dividing line differ only on a fixed $\varepsilon/2$ fraction of the pixels, we randomly check if the matrix corresponds to these half-planes on the remaining pixels.

This suggests a general paradigm for transforming PAC learning algorithms into property tests with *1-sided error*. Namely, consider a property $\mathcal{P}$ where all objects with $\mathcal{P}$ which are $\varepsilon/2$-close to a given object are the same on all but $\varepsilon/2$ fraction of the points. In addition, assume there is a proper PAC learning algorithm with sampling complexity $q(n, \varepsilon)$. Then the following test for $\mathcal{P}$ has 1-sided error and query complexity $q(n, \varepsilon/2) + O(1/\varepsilon)$: learn the property within relative error of $\varepsilon/2$ and then randomly test the object on points where all objects $\varepsilon/2$-close to the hypothesis coincide. The proof of this fact is very similar to the case 2 of the analysis of the half-plane test.

Extensions and Lower Bounds We restricted our attention to images representable by binary matrices. However, in real life images have many colors (or intensity values). Property tests for images represented by integer-valued matrices would be a natural generalization. For example, one can generalize convexity in the following way. Call an image represented by an $n \times n$ matrix with values in R *convex* if the corresponding function $\{0, 1, \ldots, n-1\}^2 \rightarrow R$ is convex.

A straightforward extension of our tests to d dimensions seems to give tests with dependence on d, and thus dependent on the size of the image. It would be interesting to investigate if this dependence is necessary.

It is known that testing some properties requires a number of queries linear in the size of the input [4,2]. However, known hard properties do not seem to have a natural geometric interpretation. It would be nice to find natural 2-dimensional visual properties which are hard to test. One such result follows directly from [1], which shows that testing whether a string of length n is a shift of another string requires $\Omega(n^{1/2})$ queries. This implies that testing whether the lower half of an $n \times n$ image is a shift of the upper half requires $\Omega(n^{1/2})$ queries. It would be interesting to find even harder visual properties.

Acknowledgements The author would like to thank Michael Sipser for proposing the problem and many useful discussions. She is also very grateful to Piotr Indyk for help and moral support.

References

1. T. Batu, F. Ergun, J. Kilian, A. Magen, S. Raskhodnikova, R. Rubinfeld, and R. Sami, A Sublinear Algorithm for Weakly Approximating Edit Distance, *Proceedings of the* 35th *ACM STOC* (2003)
2. E. Ben-Sasson, P. Harsha, and S. Raskhodnikova, 3CNF Properties are Hard to Test, *Proceedings of the* 35th *ACM STOC* (2003)
3. A. Blumer, A. Ehrenfeucht, D. Haussler, M. Warmuth, Learnability and the Vapnik-Chervonenkis dimension, *Journal of the Association for computing machinery* **36(4)** (1989) 929–965
4. A. Bogdanov, K. Obata, L. Trevisan, A linear lower bound on the query complexity of property testing algorithms for 3-coloring in bounded-degree graphs, *Proceedings of the* 42nd *IEEE FOCS* (2002)
5. B. Chazelle, R. Rubinfeld, and L. Trevisan, Approximating the minimum spanning tree weight in sublinear time, *Proceedings of ICALP* (2001)
6. A. Czumaj and C. Sohler, Property testing with geometric queries, *Proceedings of the 9th European Symposium on Algorithms* (2001) 266–277
7. A. Czumaj, C. Sohler and M. Ziegler, Property testing in computational geometry, *Proceedings of the 8th European Symposium on Algorithms* (2000) 155–166
8. A. Ehrenfeucht, D. Haussler, M. Kearns and L. Valiant, A General Lower Bound on the Number of Examples Needed for Learning, *Information and Computation* **82(3)** (1989) 247–261
9. E. Fischer, The art of uninformed decisions: A primer to property testing, *The Computational Complexity Column of The Bulletin of the European Association for Theoretical Computer Science* **75** (2001) 97–126
10. E. Fischer and I. Newman, Testing of matrix properties, *Proceedings of the* 33rd *ACM STOC* (2001) 286–295
11. O. Goldreich, S. Goldwasser and D. Ron, Property testing and its connection to learning and approximation, *Journal of the ACM* **45** (1998) 653–750
12. O. Goldreich and D. Ron, Property Testing in Bounded Degree Graphs, *Proceedings of the* 28th *ACM STOC* (1997)
13. E. Kushilevitz and D. Roth, On Learning Visual Concepts and DNF Formulae, *Machine Learning* (1996)
14. W. Maass and G. Turan, On the complexity of learning from counterexamples, *Proceedings of the* 30th *IEEE FOCS* (1989) 262–267
15. D. Ron, Property testing (a tutorial), In *Handbook of Randomized Computing* (S. Rajasekaran, P. M. Pardalos, J. H. Reif and J. D. P. Rolimeds), Kluwer Press (2001)
16. R. Rubinfeld and M. Sudan, Robust characterization of polynomials with applications to program testing, *SIAM Journal of Computing* **25** (1996) 252–271
17. L. Valiant, A theory of the learnable, *Communications of the ACM* **27** (1984) 1134–1142

Faster Algorithms for MAX CUT and MAX CSP, with Polynomial Expected Time for Sparse Instances

Alexander D. Scott[1] and Gregory B. Sorkin[2]

[1] Department of Mathematics, University College London,
London WC1E 6BT, UK.
scott@math.ucl.ac.uk
[2] IBM T.J. Watson Research Center, Department of Mathematical Sciences,
Yorktown Heights NY 10598, USA.
sorkin@watson.ibm.com

Abstract. We show that a random instance of a weighted maximum constraint satisfaction problem (or MAX 2-CSP), whose clauses are over pairs of binary variables, is solvable by a deterministic algorithm in polynomial expected time, in the "sparse" regime where the expected number of clauses is half the number of variables. In particular, a maximum cut in a random graph with edge density $1/n$ or less can be found in polynomial expected time.

Our method is to show, first, that if a MAX 2-CSP has a connected underlying graph with n vertices and m edges, the solution time can be deterministically bounded by $2^{(m-n)/2}$. Then, analyzing the tails of the distribution of this quantity for a component of a random graph yields our result. An alternative deterministic bound on the solution time, as $2^{m/5}$, improves upon a series of recent results.

1   Introduction

In this paper we prove that a maximum cut of a sparse random graph can be found in polynomial expected time.

Theorem 1. *For any $c \leq 1$, a maximum cut of a random graph $G(n, c/n)$ can be found in time whose expectation is* $\mathrm{poly}(n)$, *and using space $O(m+n)$, where m is the size of the graph.*

Our approach is to give a deterministic algorithm and bound its running time on any graph in terms of size and cyclomatic number. We then bound the expected running time for random instances by bounding the distribution of cyclomatic number in components of a sparse random graph.

Theorem 2. *Let G be a connected graph with m edges and n vertices. There is an algorithm that finds a maximum cut of G in time $O(m+n) \min\{2^{m/5}, 2^{(m-n)/2}\}$, and in space $O(m+n)$.*

S. Arora et al. (Eds.): APPROX 2003+RANDOM 2003, LNCS 2764, pp. 382–395, 2003.
© Springer-Verlag Berlin Heidelberg 2003

We remark that the bound in Theorem 2 is of independent interest, and improves on previous algorithms giving bounds of $2^{m/4}\operatorname{poly}(m+n)$ [KF02] and $2^{m/3}\operatorname{poly}(m+n)$ [GHNR].

In fact, the algorithm employs several local reductions that take us outside the class of MAX CUT problems. We therefore work with the larger class MAX 2-CSP of weighted maximum constraint satisfaction problems consisting of constraints on pairs (and singletons) of variables, where each variable may take two values. Theorems 1 and 2 are then special cases of the more general Theorems 3 and 5 below.

1.1   Context

Our results are particularly interesting in the context of phase transitions for various maximum constraint-satisfaction problems. Since the technicalities are not relevant to our result, but only help to put it into context, we will be informal. It is well known that a random 2-SAT formula with density $c < 1$ (where the number of clauses is c times the number of variables) is satisfiable with probability tending to 1, as the number n of variables tends to infinity, while for $c > 1$, the probability of satisfiability tends to 0 as $n \to \infty$ [CR92, Goe96, FdlV92]; for more detailed results, see [BBC+01]. More recently, MAX 2-SAT has been shown to exhibit similar behavior, so for $c < 1$, only an expected $\Theta(1/n)$ clauses go unsatisfied, while for $c > 1$, $\Theta(n)$ clauses are unsatisfied [CGHS03, CGHS].

For a random graph $G(n, c/n)$, with $c < 1$ the graph almost surely consists solely of small trees and unicyclic components, while for $c > 1$, it almost surely contains a "giant", complex component, of order $\Theta(n)$ [Bol01]. Again, [CGHS] proves the related facts that in a maximum cut of such a graph, for $c < 1$ only an expected $\Theta(1)$ edges fail to be cut, while for $c > 1$ it is $\Theta(n)$.

Theorem 3 is concerned with algorithms that run in polynomial expected time. Results on coloring random graphs in polynomial expected time can be found in [KV02, COMS, TCO03]. For both MAX CUT and MAX 2-SAT, it seems likely that the mostly-satisfiable (or mostly-cuttable) sparse instances are algorithmically easy, while the not-so-satisfiable dense instances are algorithmically hard. While, as far as we are aware, little is known about the hardness of dense instances, our results here confirm that not only are typical sparse MAX CUT instances easy, but even the atypical ones can be accommodated in polynomial expected time; see the Conclusions for further discussion.

1.2   Outline of Proof

Our proof of Theorem 3 has a few main parts. Since the maximum cut of a graph is the combination of maximum cuts of each of its connected components, it suffices to bound the expected time to partition the component containing a fixed vertex.

In Theorem 5 we show that Algorithm A's running time on a component is bounded by a function of the component's cyclomatic number, the number of edges less the number of vertices plus one. For brevity we will call this the "excess" (a slight abuse of the standard meaning, which is just edges minus vertices). Theorem 5 also gives a $2^{m/5} \operatorname{poly}(m+n)$ bound on the running time.

In the randomized setting, Lemma 8 provides a bound on the exponential moments of the excess of a component. It does so by "exploring" the component as a branching process, dominating it with a similar process, and analyzing the latter as a random walk. This gives stochastic bounds on the component order u and, conditioned upon u, the "width" w (to be defined later); the excess is easily stochastically bounded in terms of u and w.

Finally, we combine the running times, which are exponentially large in the excess, with the exponentially small large-deviation bounds on the excess, to show that Algorithm A runs in polynomial expected time.

2   Solving a Maximum Constraint-Satisfaction Instance

We begin by defining a class of weighted maximum constraint satisfaction problems, or MAX CSPs, generalizing MAX CUT, and (in Theorem 5) bounding their running time in terms of parameters of an instance.

2.1   Weighted Maximum Constraint-Satisfaction Problems

We may think of MAX CUT as a MAX CSP in which the constraints simply prefer opposite "colors" on the endpoints of each edge, and all constraints have the same "weight". We generalize this not only for the sake of a more general result but because we need to: intermediate steps of Algorithm A, applied to a MAX CUT instance, generate instances of more general type.

For our purposes, a general instance of a (weighted) MAX 2-CSP consists of a graph $G = (V, E)$, and a score function consisting of: a sum of "monadic constraint" scores of each vertex and its color, "dyadic" scores of each edge and the pair of colors at its endpoints, and (for notational convenience) a single "niladic" score (a constant). Specifically, there is a (niladic) score s_0; for each $x \in V$ (monad) there is a pair of scores s_R^x, s_B^x corresponding to the two ways that the vertex could be colored; and for each edge $e = \{x, y\} \in E$ (dyad) there is a 4-tuple of scores $s_{BB}^{xy}, s_{BR}^{xy}, s_{RB}^{xy}, s_{RR}^{xy}$ corresponding to the four ways that the edge could be colored, and the score of a coloring $\phi : V \to \{R, B\}$ is

$$S(\phi) := s_0 + \sum_{x \in V} s_{\phi(x)}^x + \sum_{\{x,y\} \in E} s_{\phi(x)\phi(y)}^{xy}.$$

(Note that for any $C, D \in \{R, B\}$, s_{CD}^{xy} and s_{DC}^{yx} refer to the same score, and thus must be equal.) Let S refer to the full collection of scores s_C^x and s_{CD}^{xy} as above. Then MAX(V, E, S) is the computational problem of finding a coloring ϕ achieving $\max_\phi S(\phi)$.

As one quick example, MAX 2-SAT is such a MAX CSP. Using colors T (true) and F (false), a SAT constraint $\bar{X} \vee Y$ is modelled as a dyadic constraint mapping (T, F) to score 0 (unsatisfied) and any other coloring to score 1 (satisfied).

Another example is MAX DICUT, the problem of partitioning a directed graph to maximize the number of edges passing from the left side to the right.

Our main result is that a weighted MAX 2-CSP on a random graph $G(n, c/n)$, $c < 1$, can be solved in polynomial expected time, per the following theorem.

Theorem 3. *For any $c \leq 1$ and any n, let $G(n, c/n)$ be a random graph, and let (G, S) be any weighted MAX 2-CSP instance over this graph. Then (G, S) can be solved exactly in expected time $\mathrm{poly}(n)$, and in space $O(m + n)$.*

2.2   Algorithm A

In this section we give an algorithm for solving instances of weighted MAX 2-CSP. The algorithm will use 3 types of reductions. We begin by defining these reductions. We then show how the algorithm fixes a sequence in which to apply the reductions by looking at the underlying graph of the CSP. This sequence defines a tree of CSPs, which can be solved bottom-up to solve the original CSP. Finally, we bound the algorithm's time and space requirements.

Reductions The first two reductions each produce equivalent problems with fewer vertices, while the third produces a pair of problems, both with fewer vertices, one of which is equivalent to the original problem.

Reduction I Let y be a vertex of degree 1, with neighbor x. Reducing (V, E, S) on y results in a new problem (V', E', S') with $V' = V \setminus y$ and $E' = E \setminus xy$. S' is the restriction of S to V' and E', except that for $C, D \in \{R, B\}$ we set

$$s'^{x}_{C} = s^{x}_{C} + \max_{D}\{s^{xy}_{CD} + s^{y}_{D}\},$$

i.e., we set

$$s'^{x}_{R} = s^{x}_{R} + \max\{s^{xy}_{RR} + s^{y}_{R}, s^{xy}_{RB} + s^{y}_{B}\}$$
$$s'^{x}_{B} = s^{x}_{B} + \max\{s^{xy}_{BB} + s^{y}_{B}, s^{xy}_{BR} + s^{y}_{R}\}.$$

Note that any coloring ϕ' of V' can be extended to a coloring of V in two ways, namely ϕ_R and ϕ_B (corresponding to the two colorings of x); and the defining property of the reduction is that $S'(\phi') = \max\{S(\phi_R), S(\phi_B)\}$. In particular, $\max_{\phi'} S'(\phi') = \max_{\phi} S(\phi)$, and an optimal coloring ϕ' for the problem MAX(V', E', S') can be extended to an optimal coloring ϕ for MAX(V, E, S), in constant time.

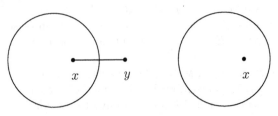

Reduction II Let y be a vertex of degree 2, with neighbors x and z. Reducing (V, E, S) on y results in a new problem (V', E', S') with $V' = V \setminus y$ and $E' = (E \setminus \{xy, yz\}) \cup \{xz\}$. S' is the restriction of S to V' and E', except that for $C, D, E \in \{R, B\}$ we set

$$s'^{xz}_{CD} = s^{xz}_{CD} + \max_E \{s^{xy}_{CE} + s^{yz}_{ED} + s^y_E\}$$

i.e., we set

$$s'^{xz}_{RR} = s^{xz}_{RR} + \max\{s^{xy}_{RR} + s^{yz}_{RR} + s^y_R, s^{xy}_{RB} + s^{yz}_{BR} + s^y_B\}$$
$$s'^{xz}_{RB} = s^{xz}_{RB} + \max\{s^{xy}_{RR} + s^{yz}_{RB} + s^y_R, s^{xy}_{RB} + s^{yz}_{BB} + s^y_B\}$$
$$s'^{xz}_{BR} = s^{xz}_{BR} + \max\{s^{xy}_{BR} + s^{yz}_{RR} + s^y_R, s^{xy}_{BB} + s^{yz}_{BR} + s^y_B\}$$
$$s'^{xz}_{BB} = s^{xz}_{BB} + \max\{s^{xy}_{BR} + s^{yz}_{RB} + s^y_R, s^{xy}_{BB} + s^{yz}_{BB} + s^y_B\},$$

where our notation presumes that if xz was not an edge in E, then $s^{xz}_{CD} = 0$ for all colors C and D. As in Reduction I, any coloring ϕ' of V' can be extended to V in two ways, ϕ_R and ϕ_B, and S' picks out the larger of the two scores. Also as in Reduction I, $\max_{\phi'} S'(\phi') = \max_\phi S(\phi)$, and an optimal coloring ϕ' for MAX(V', E', S') can be extended to an optimal coloring ϕ for MAX(V, E, S), in constant time. (Note that neither multiple edges nor loops are created by this reduction, nor the next one.)

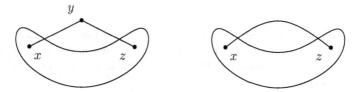

Reduction III Let y be a vertex of degree 3 or higher. Where reductions I and II each had a single reduction of (V, E, S) to (V', E', S'), here we define a pair of reductions of (V, E, R), to (V', E', S^R) and (V', E', S^B), corresponding to assigning the color R or B to y. We define $V' = V \setminus y$, and E' as the restriction of E to $V \setminus y$. For $C, D, E \in \{R, B\}$, S^C is the restriction of S to $V \setminus y$, except that we set

$$(s^C)_0 = s_0 + s^y_C,$$

and, for every neighbor x of y,

$$(s^C)^x_D = s^x_D + s^{xy}_{DE}.$$

In other words, S^R is the restriction of S to $V \setminus y$, except that we set $(s^C_0) = s_0 + s^y_C$ and, for every neighbor x of y,

$$(s^R)^x_R = s^x_R + s^{xy}_{RR} + s^y_R$$
$$(s^R)^x_B = s^x_B + s^{xy}_{BR} + s^y_R.$$

Similarly S^B is given by $(s^B)_0 = s_0 + s^y_B$ and, for every neighbor x of y,

$$(s^B)^x_R = s^x_R + s^{xy}_{RB} + s^y_B$$
$$(s^B)^x_B = s^x_B + s^{xy}_{BB} + s^y_B.$$

As in the previous reductions, any coloring ϕ' of $V \setminus y$ can be extended to V in two ways, ϕ_R and ϕ_B, corresponding to the color given to y, and now (this is different!) $S_R(\phi') = S(\phi_R)$ and $S_B(\phi') = S(\phi_B)$. Furthermore,

$$\max\{\max_{\phi'} S_R(\phi'), \max_{\phi'} S_B(\phi')\} = \max_{\phi} S(\phi),$$

and an optimal coloring on the left can be extended to an optimal coloring on the right in time $O(\deg(y))$.

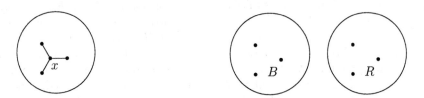

Defining Algorithm A in terms of these reductions is straightforward, and it should come as no surprise that the running time is polynomial in n and m, times 2 raised to the power of the number of times reduction III is employed. We now detail this.

Setup Phase: Choosing a Sequence of Reductions First, observe that the two problems generated by reduction III have different score sets, but the same underlying graph. Thus each of the three reductions, considering only the graphs and ignoring the scores, reduces a graph to a subgraph of smaller order.

Given an input graph G of order n, Algorithm A begins by constructing a sequence $G_1, G_2, \ldots, G_i$, of at most n graphs, where $G_1 = G$ is the input graph, each subsequent graph is a reduction of its predecessor graph (ignoring scores), and the final graph G_i has no edges.

Specifically, with an ordering on the vertices of G: if G has minimum degree 1, apply reduction I to the first vertex of degree 1; if G has minimum degree 2, apply reduction II to the first vertex of degree 2; and otherwise, apply reduction III to the first vertex of maximum degree.

The precise running time of this setup procedure clearly depends on the data structures employed, but it is clearly polynomial. Maintaining a list of vertices of each degree, and the neighbors of each vertex, and storing only the changes at each step rather than the new graph, the time can be limited to $O(n+m)$ in the RAM model (where the length of an integer's binary representation is ignored).

Solving the Tree of csps The sequence of graphs, along with another sequence specifying one binary value for each type-III reduction, determines a sequence of CSPs; the collection of all 2^r binary sequences (where r is the number of type-III reductions) naturally defines a tree of CSPs, having depth i (we generate a child even for type-I and -II reductions) and 2^r leaves (each type-III reduction producing 2 children for each CSP in the current generation). Given an optimal solution to a CSP's child/children, an optimal solution to the CSP can be found by trying both extensions to the vertex "y", in time $O(\deg(y))$.

Starting from the leaf problems, and propagating their solutions upwards, solves the original problem.

Analysis The foregoing procedure runs in time $O(m+n)2^r$. Moreover, the tree can be stored and traversed implicitly, as a path with nodes corresponding to the graph reductions, and at each type-III node a state corresponding to which of the two reductions is currently being explored, yielding a space bound of $O(m+n)$. Thus we have the following lemma.

Lemma 4. *Given a weighted* MAX *2-CSP whose underlying graph G is connected, and an order on the vertices of G, Algorithm A returns an optimal solution in time $O(m+n)2^r$ and space $O(m+n)$, where $r(G)$ is the (order-dependent) number of type-III reductions taken for G.*

3   Parametric Complexity

The following theorem bounds the running time of Algorithm A in terms of parameters of the graph underlying the CSP.

Theorem 5. *Given a weighted* MAX *2-CSP whose underlying graph G is connected, has order n, size m, and excess $\kappa = m - n$, Algorithm A returns an optimal solution in time $O(m+n)2^{\min\{m/5,\kappa/2\}}$.*

We remark that to prove our expected-time result (Theorem 3), we use only the $2^{\kappa/2}$ bound. However, the $2^{m/5}O(m+n)$ bound, for arbitrary MAX 2-CSPs, is of independent interest. For MAX CUT it improves on the $2^{m/4}\operatorname{poly}(m+n)$ of [KF02], and for MAX 2-SAT it matches the $2^{m/5}\operatorname{poly}(m+n)$ bound of [GHNR] (which also gave a $2^{m/3}\operatorname{poly}(m+n)$ bound for MAX CUT). These works also used algorithms based on reductions.

In light of Lemma 4, it suffices to prove that (for any order on the vertices of G), the number of type-III reduction steps $r(G)$ is bounded by both $m/5$ and $\kappa/2$. These two claims are proved in the following two subsections.

3.1   Bounding in Terms of Excess

Claim 6. *For a connected graph G with excess κ, the number of type-III reduction steps of Algorithm A is $r \leq \max\{0, \kappa/2\}$.*

Proof. The proof is by induction on the order of G. If G has excess 0 (it is unicyclic) or excess -1 (it is a tree), then type-I and -II reductions destroy all its edges, so $r = 0$.

Otherwise, the first type-III reduction reduces the number of edges by at least 3 and the number of vertices by exactly 1, thus reducing the excess to $\kappa' \leq \kappa - 2$. If G' has components $G'_1, \dots, G'_I$, then $r(G) = 1 + \sum_i r(G'_i)$. Given that we applied a type-III reduction, G had minimum degree ≥ 3, so G' has minimum degree ≥ 2. Thus each component G'_i has minimum degree ≥ 2, and so excess $\kappa'_i \geq 0$. Then, by induction, $r(G) = 1 + \sum_i r(G'_i) \leq 1 + \sum_i \kappa'_i/2 \leq 1 + \kappa'/2 \leq \kappa/2$. Note that the inductive step $r(G'_i) \leq \kappa'_i/2$ used the fact that $\kappa'_i \geq 0$. $\qquad\square$

3.2 Bounding in Terms of Size

Claim 7. *For a graph G with m edges, the number of type-III reduction steps of Algorithm A is at most $m/5$.*

Proof. Since type-I and type-II steps cannot increase the number of edges, it is enough to show that each type-III step, on average, reduces the number of edges by 5 or more. As long as the maximum degree is $d \geq 5$ this is clear, since each type-III reduction immediately destroys d edges. Thus it suffices to consider graphs of maximum degree $d \leq 4$; since the reductions never increase the degree of any vertex, the maximum degree will then remain at most 4.

Given a graph of maximum degree at most 4, suppose that Algorithm A performs r type-III reduction steps, consisting of r_3 reductions on vertices of degree 3, and r_4^k reductions on vertices of degree 4 having k neighbors of degree 3 and $r - k$ neighbors of degree 4. (If a neighbor had degree more than 4 we should have chosen it in preference to y; degree 2 or less and we should have applied a type-I or -II reduction instead.)

How many edges are destroyed by the $r = r_3 + \sum_{k=0}^{r} r_4^k$ type-III reductions? Each "r_3-reduction" deletes the 3 edges incident on y, each of which went to a vertex also of degree 3 (4 or more and we would have chosen it in preference to y, 2 or less and we would have applied a type-I or -II reduction), changing their degrees to 2 and subjecting each to a type-II reduction, and so destroying 3 more edges. (A type-II reduction destroys edges yx and yz, and if edge xz was not previously present it creates it, thus reducing the number of edges by at least 1, and possibly 2.) Similarly, each "r_4^k reduction", on a degree-4 vertex adjacent to k degree-3 vertices, along with the k type-II reductions it sets up, destroys $4 + k$ edges. Thus the average number of edges destroyed per step is at least

$$\frac{6r_3 + \sum_{k=0}^{4}(4+k)r_4^k}{r_3 + \sum_{k=0}^{4} r_4^k}. \tag{1}$$

Clearly this ratio is at least 5 unless the value of r_4^0 can be made large, but we now show that the r_4^k values must satisfy an additional condition which effectively prohibits this.

Note that each r_3-reduction decreases the number of degree-3 vertices by 4 (itself and its 3 neighbors), while each r_4^k-reduction decreases it by $2k - 4$ (destroying k degree-3 neighbors, but also turning $4 - k$ old degree-4 neighbors into new degree-3 vertices). Type-I and -II reductions do not affect the number of degree-3 vertices. Since the number of degree-3 vertices is initially non-negative, and finally 0, the decrease must be non-negative, i.e.,

$$\sum_k r_4^k(2k - 4) + 4r_3 \geq 0. \tag{2}$$

Subject to the constraint given by (2), how small can the ratio (1) be? To be (slightly) pessimistic, we may let the values r_3 and r_4^k range over the non-negative reals. Multiplying the set of values by any constant affects neither the constraint nor the ratio, so without loss of generality we may set the denominator of (1) to 1. That is, we add a constraint

$$r_3 + \sum r_4^k = 1, \tag{3}$$

and minimize

$$6r_3 + \sum_{k=0}^{4} (4 + k)r_4^k. \tag{4}$$

This is simply a linear program (LP) with objective function (4) and the two constraints (2) and (3). The LP's optimal objective value is 5, and the LP dual solution of $(\frac{1}{4}, 5)$ establishes 5 as a lower bound. That is, adding $\frac{1}{4}$ times constraint (2) to 5 times constraint (3) gives

$$\frac{1}{4}\left(\sum(2k - 4)r_4^k + 4r_3\right) + 5\left(r_3 + \sum r_4^k\right) = 6r_3 + \sum(4 + k/2)r_4^k \geq 5,$$

so (4), which is $6r_3 + \sum(4 + k)r_4^k$, must be at least this large.

This establishes that the number of edges destroyed by type-III reductions is at least 5 times the number of such reductions, concluding the proof. □

We note that the upper bound of $m/5$ is achievable; that is, $m/5$ type-III reductions are needed by some graphs. An easy example is K_5, with 10 edges, reduced by two type-III reductions to K_4 and K_3, the latter reduced to the empty graph by type-I and -II reductions.

4   Stochastic Size and Excess of a Random Graph

We stochastically bound the excess κ of a component of a random graph G through a standard "exposure" process. Given a graph G and a vertex x_1 in G, together with a linear order on the vertices of G, the exposure process finds a spanning tree of the component G_1 of G that contains x_1 and, in addition, counts the number of non-tree edges of G_1 (i.e., calculates the excess).

At each step of the process, vertices are classified as "living", "dead", and "unexplored", beginning with just x_1 living, and all other vertices unexplored. At the ith step, the process takes the earliest living vertex x_i. All edges from x_i to unexplored vertices are added to the spanning tree, and the number of non-tree edges is increased by 1 for each edge from x_i to a living vertex. Unexplored vertices adjacent to x_i are then reclassified as living, and x_i is made dead. The process terminates when there are no live vertices.

Now suppose G is a random graph in $\mathcal{G}(n, c/n)$, with the vertices ordered at random. Let $w(i)$ be the number of live vertices at the ith step and define the *width* $w = \max w(i)$. Let $u = |G_1|$, so that $w(0) = 1$ and $w(u) = 0$. The number of non-tree edges uncovered in the ith step is binomially distributed as $B(w(i) - 1, c/n)$, and so, conditioning on u and $w(1), \ldots, w(u)$, the number of excess edges is distributed as $B(\sum_{i=1}^{u}(w(i)-1), c/n)$. Since $\sum_{i=1}^{u}(w(i)-1) \leq uw$, the (conditioned, and therefore also the unconditioned) number of excess edges is dominated by the random variable $B(uw, c/n)$.

At the ith stage of the process, there are at most $n - i$ unexplored vertices, and so the number of new live vertices is dominated by $B(n - i, 1/n)$. Consider now a variant of the exposure process in which at each step we add enough special "red" vertices to bring the number of unexplored vertices to $n - i$. Let $h(i)$ be the number of living vertices at the ith stage. Then $h(0) = 1$, and $h(i)$ is distributed as $h(i - 1) + B(n - i, c/n) - 1$. Let $X = n \wedge \min\{t : h(t) = 0\}$ and $H = \max_{i \leq X} h(i)$.

By considering the second process as an extension of the first (and exploring the added vertices in the second process only when no other vertices remain), we obtain a coupling between the two processes such that $u \leq X$ and $w \leq H$. Thus the excess of G_1 is dominated by $B(XH, 1/n)$.

Since the running time of Algorithm A is at most $\mathbb{E}(O(m + n)2^{\kappa/2})$, it can be bounded by the quantity $O(n^2)\mathbb{E}(\sqrt{2}^{(B(XH,1/n)}))$. It is useful to note that

$$\mathbb{E}z^{B(n,p)} = \sum_{i=0}^{n}\binom{n}{i}z^i p^i (1-p)^{n-i} = (pz + (1-p))^n = (1 + p(z-1))^n \leq \exp(p(z-1)n).$$

In particular, $\mathbb{E}\sqrt{2}^{B(n,p)} \leq \exp((\sqrt{2} - 1)np)$. In the following, we therefore focus on bounding quantities of form $\Pr(X = x, H = h)\exp((\sqrt{2} - 1)xh/n)$.

Lemma 8. *With $h(t)$ the random process defined above, for all times $i = 1, 2, \ldots$ parametrized as $\alpha n = i$,*

$$\Pr(h(\alpha n) \geq 0) \leq \exp\left(-3\alpha^3 n/(24 - 8\alpha)\right). \tag{5}$$

Furthermore, for any height h parametrized as $h = \beta n$, with $\alpha^2/(8-4\alpha) \leq \beta \leq \alpha$,

$$\Pr(\max_{t \leq \alpha n} h(t) \geq \beta n \mid h(\alpha n) = 0) \leq O(n^{3/2})\exp\left(-\left(\beta - \frac{\alpha^2/4}{2 - \alpha}\right)^2 \frac{7n}{8\alpha}\right). \tag{6}$$

In order to prove the lemma, we shall make use of the following fairly standard bound.

Claim 9. *With $N = ni - \binom{i+1}{2}$, let $Z_1, Z_2, \ldots, Z_N$, be a random sequence of binomial random variables conditioned upon $\sum_{j=1}^{N} Z_j = i - 1$. Parametrize $i = \alpha n$. Suppose that β is in the range $\alpha^2/(8 - 4\alpha) \leq \beta \leq \alpha$, and $t \leq i$. Then, writing $N' = nt - \binom{t+1}{2}$,*

$$\Pr(\sum_{i=1}^{N'} Z_i \geq \beta n + (t - 1)) \leq O(\sqrt{n}) \exp\left(-\left(\beta - \frac{\alpha^2}{8 - 4\alpha}\right)^2 \frac{7n}{8\alpha}\right). \quad (7)$$

We omit the proof.

Proof (of Lemma 8). We first prove (5). Note that

$$h(i) = B\left((n - 1) + \cdots + (n - i), 1/n\right) - i + 1 = B\left(ni - \binom{i+1}{2}, 1/n\right) - i + 1$$

and so $h(i) \geq 0$ means that

$$B\left(ni - \binom{i+1}{2}, 1/n\right) \geq i + 1 = \alpha n + 1. \quad (8)$$

This binomial r.v. has expectation

$$\left(\alpha n^2 - \binom{\alpha n + 1}{2}\right)\frac{1}{n} \leq (\alpha - \alpha^2/2)n. \quad (9)$$

Thus if (8) holds, the r.v. differs from its expectation by at least $\alpha^2 n/2$.

We use the inequality that for a sum of independent 0-1 Bernoulli random variables with parameters $p_1, \ldots, p_n$ and expectation $\mu = \sum_{i=1}^{n} p_i$, $\mathbb{P}(X \geq \mu + t) \leq \exp\left(-t^2/(2\mu + 2t/3)\right)$. Together with (9) this implies that (8) has probability at most $\exp\left(-(\alpha^4 n^2/4)/(2\alpha n(1 - \alpha/2) + \alpha^2 n/3)\right) = \exp\left(-3\alpha^3 n/(24 - 8\alpha)\right)$.

To prove (6), we bound the conditional probability

$$\Pr(\max_{t \leq \alpha n} h(t) \geq \beta n \mid h(\alpha n) = 0). \quad (10)$$

In this part, rather than thinking of $h(i)$ as $B(ni - \binom{i+1}{2}, 1/n) - i + 1$, we think of it as a sum of $N = ni - \binom{i+1}{2}$ independent Bernoulli random variables Z_i each with distribution $B(1/n)$, plus $-i + 1$. Note that, conditional on the sum of the Z_is, any particular assignment of 0s and 1s is equally likely: the collection of Z_is is a random binomial sequence conditioned upon $h(\alpha n) = 0$, i.e., upon having sum $\alpha n - 1$. We apply Claim 9 to show that for any given t, the probability of each of the events comprising that in (10) is bounded by (7), namely $\Pr(h(t) \geq \beta n \mid h(\alpha n) = 0) \leq O(\sqrt{n}) \exp\left(-\left(\beta - \frac{\alpha^2}{8-4\alpha}\right)^2 \frac{7n}{8\alpha}\right)$.

Summing over $1 \leq t = \gamma n \leq \alpha n$, the required bound (6) follows. $\quad\square$

Recall the random process h defined before Lemma 8, with stopping time X and maximum height H.

Lemma 10.
$$\mathbb{E}\left[\exp\left((\sqrt{2}-1)XH/n\right)\right] \leq n^{9/2}.$$

Proof. We show that each possible pair $X \in \{1, \ldots, n-1\}$ and $H \in \{1, \ldots, \frac{1}{2}n^2 + O(1)\}$ contributes at most $O(n^{3/2})$ to the expectation. Specifically, we show that for all α and β, $\exp\left((\sqrt{2}-1)\alpha\beta n\right)\Pr(X = \alpha n)\Pr(Y = \beta n) = O(n^{3/2})$.

Case 1. If $\beta < \alpha^2/(8 - 4\alpha)$ then, from Lemma 8,

$$\Pr(X = \alpha n) \leq \Pr(h(\alpha n) = 0) \leq \Pr(h(\alpha n) \geq 0) \leq \exp\left(-3\alpha^3 n/(24 - 8\alpha)\right) \tag{11}$$

and so

$$\exp\left((\sqrt{2}-1)\alpha\beta n\right)\Pr(X = \alpha n) \leq \exp\left((\sqrt{2}-1)\frac{\alpha^3 n}{8 - 4\alpha} - \frac{3\alpha^3 n}{24 - 8\alpha}\right).$$

This is less than 1 provided that

$$\frac{\sqrt{2}-1}{8 - 4\alpha} \leq \frac{3}{24 - 8\alpha},$$

which is easily verified to hold for all $\alpha \in [0, 1]$.

Case 2. If $\beta \geq \alpha^2/(8 - 4\alpha)$ then, from Lemma 8, in addition to (11), we have that

$$\Pr(H = \beta n \mid X = \alpha n) \leq \Pr(H \geq \beta n \mid X = \alpha n)$$
$$\leq O(n^{3/2})\exp\left(-\left(\beta - \frac{\alpha^2/4}{2 - \alpha}\right)^2 \frac{7n}{8\alpha}\right).$$

So in this case it suffices to show that

$$\exp\left(\left[(\sqrt{2}-1)\alpha\beta n\right] - \left[3\alpha^3 n/(24 - 8\alpha)\right] - \left[\left(\beta - \frac{\alpha^2/4}{2 - \alpha}\right)^2 \frac{7n}{8\alpha}\right]\right) \leq 1, \tag{12}$$

i.e., that

$$\left[(\sqrt{2}-1)\alpha\beta\right] - \left[3\alpha^3/(24 - 8\alpha)\right] - \left[\left(\beta - \frac{\alpha^2/4}{2 - \alpha}\right)^2 \frac{7}{8\alpha}\right] \tag{13}$$

is at most 0.

For fixed $a \in (0, 1]$, (13) is maximized by

$$\beta = \frac{4}{7}(\sqrt{2}-1)\alpha^2 + \frac{\alpha^2}{8 - 4\alpha}.$$

Substituting this value of β into (13), and multiplying by the (positive) quantity $(\alpha - 2)(\alpha - 3)/\alpha^3$ gives a quadratic which is easily seen to be negative on $(0, 1]$.

Thus, in both Case 1 and Case 2, for any α and β, the contribution of the $X = \alpha n$, $H = \beta n$ term to the expectation of $(\sqrt{2} - 1)^{XH/n}$ is at most $O(n^{3/2})$, and the sum of all $O(n^3)$ such contributions (recalling that X and H may take on $O(n)$ and $O(n^2)$ possible values, respectively) is $O(n^{9/2})$. □

We can now prove Theorem 3.

Proof (of Theorem 3). By Theorem 5, and the remarks before Lemma 8, Algorithm A runs in expected time $\mathbb{E}(O(m + n)\sqrt{2}^\kappa \leq O(n^2)\mathbb{E}(\sqrt{2}^{B(XH)}) \leq O(n^2)\mathbb{E}(\exp((\sqrt{2}-1)XH/n))$. But it follows from Lemma 10 that this is $O(n^{13/2})$. □

5   Conclusions

In the present paper we focus on MAX CUT. Our result for "sparse" instances is strong in that it applies right up to $c = 1$, and we expect it could be extended through the scaling window, to $c = 1+\lambda n^{-1/3}$ (at the expense of a constant factor depending on λ in the run time, and additional complication in the analysis). We also believe that our methods can be extended to MAX 2-SAT, but the analysis is certainly more complicated. In fact our results already apply to any MAX CSP, and in particular to MAX 2-SAT, but only in the regime where there are about $n/2$ clauses on n variables; since it is likely that random instances with up to about n clauses can be solved efficiently on average (the 2-SAT phase transition occurs around n clauses), our present result for MAX 2-SAT is relatively weak.

Since MAX CUT is in general NP-hard (and even NP-hard to approximate to better than a 16/17 factor [TSSW00]), it would be interesting to resolve whether dense instances of MAX CUT as well as sparse ones can be solved in polynomial expected time (thus separating the average-case hardness from the worst-case hardness) or whether random dense instances are hard. Precisely the same questions can be asked about MAX 2-SAT, and in both cases we would guess that dense instances are hard, even on average.

References

[BBC+01] Béla Bollobás, Christian Borgs, Jennifer T. Chayes, Jeong Han Kim, and David B. Wilson, *The scaling window of the 2-SAT transition*, Random Structures Algorithms **18** (2001), no. 3, 201–256.

[Bol01] Béla Bollobás, *Random graphs*, Cambridge Studies in Advanced Mathematics, vol. 73, Cambridge University Press, Cambridge, 2001.

[CGHS] Don Coppersmith, David Gamarnik, Mohammad Hajiaghayi, and Gregory B. Sorkin, *Random MAX SAT, random MAX CUT, and their phase transitions*, Submitted for publication. 49 pages.

[CGHS03]   Don Coppersmith, David Gamarnik, Mohammad Hajiaghayi, and Gregory B. Sorkin, *Random MAX SAT, random MAX CUT, and their phase transitions*, Proceedings of the 14th Annual ACM–SIAM Symposium on Discrete Algorithms (Baltimore, MD, 2003), ACM, New York, 2003.

[COMS]   Amin Coja-Oghlan, C. Moore, and V. Sanwalani, *Max k-cut and approximating the chromatic number of random graphs*, To appear.

[CR92]   Vašek Chvátal and Bruce Reed, *Mick gets some (the odds are on his side)*, 33th Annual Symposium on Foundations of Computer Science (Pittsburgh, PA, 1992), IEEE Comput. Soc. Press, Los Alamitos, CA, 1992, pp. 620–627.

[FdlV92]   Wenceslas Fernandez de la Vega, *On random 2-SAT*, Manuscript, 1992.

[GHNR]   Jens Gramm, Edward A. Hirsch, Rolf Niedermeier, and Peter Rossmanith, *New worst-case upper bounds for MAX-2-SAT with an application to MAX-CUT*, Discrete Applied Mathematics, In Press.

[Goe96]   Andreas Goerdt, *A threshold for unsatisfiability*, J. Comput. System Sci. **53** (1996), no. 3, 469–486.

[KF02]   A. S. Kulikov and S. S. Fedin, *Solution of the maximum cut problem in time $2^{|E|/4}$*, Zap. Nauchn. Sem. S.-Peterburg. Otdel. Mat. Inst. Steklov. (POMI) **293** (2002), no. Teor. Slozhn. Vychisl. 7, 129–138, 183.

[KV02]   Michael Krivelevich and Van H. Vu, *Approximating the independence number and the chromatic number in expected polynomial time*, J. Comb. Optim. **6** (2002), no. 2, 143–155.

[TCO03]   Anusch Taraz and Amin Coja-Oghlan, *Colouring random graphs in expected polynomial time*, Proceedings of STACS 2003, LNCS 2607, 2003, pp. 487–498.

[TSSW00]   Luca Trevisan, Gregory B. Sorkin, Madhu Sudan, and David P. Williamson, *Gadgets, approximation, and linear programming*, SIAM J. Comput. **29** (2000), no. 6, 2074–2097.

A Nearly Linear Size 4-Min-Wise Independent Permutation Family by Finite Geometries

Jun Tarui[1], Toshiya Itoh[2], and Yoshinori Takei[3]

[1] Dept. of Information and Communication Eng., Univ. of Electro-Comm,
1-5-1 Chofu-gaoka, Chofu, Tokyo 182-8585, Japan
tarui@ice.uec.ac.jp
[2] Global Scientific Inform. and Comput. Center, Tokyo Institute of Technology,
2-12-1 O-okayama, Meguro-ku, Tokyo 152-8550, Japan
titoh@dac.gsic.titech.ac.jp
[3] Dept. of Electrical Engineering, Nagaoka Univ. of Technology,
1603-1 Kamitomioka-machi, Nagaoka 940-2188, Japan
takei@nagaokaut.ac.jp

Abstract. Informally, a family $\mathcal{F} \subseteq S_n$ of permutations is k-restricted min-wise independent if for any $X \subseteq [0, n-1]$ with $|X| \leq k$, each $x \in X$ is mapped to the minimum among $\pi(X)$ equally likely, and a family $\mathcal{F} \subseteq S_n$ of permutations is k-rankwise independent if for any $X \subseteq [0, n-1]$ with $|X| \leq k$, all elements in X are mapped in any possible order equally likely. It has been shown that if a family $\mathcal{F} \subseteq S_n$ of permutations is k-restricted min-wise (resp. k-rankwise) independent, then $|\mathcal{F}| = \Omega(n^{\lfloor (k-1)/2 \rfloor})$ (resp. $|\mathcal{F}| = \Omega(n^{\lfloor k/2 \rfloor})$). In this paper, we construct families $\mathcal{F} \subseteq S_n$ of permutations of which size are close to those lower bounds for $k = 3, 4$, i.e., we construct a family $\mathcal{F} \subseteq S_n$ of 3-restricted (resp. 4-restricted) min-wise independent permutations such that $|\mathcal{F}| = O(n \lg^2 n)$ (resp. $|\mathcal{F}| = O(n \lg^3 n)$) by applying the affine plane $AG(2, q)$, and a family $\mathcal{F} \subseteq S_n$ of 4-rankwise independent permutations such that $|\mathcal{F}| = O(n^3 \lg^6 n)$ by applying the projective plane $PG(2, q)$. Note that if a family $\mathcal{F} \subseteq S_n$ of permutations is 4-rankwise independent, then $|\mathcal{F}| = \Omega(n^2)$. Since a family $\mathcal{F} \subseteq S_n$ of 4-rankwise independent permutations is 4-restricted min-wise independent, our family $\mathcal{F} \subseteq S_n$ of 4-restricted min-wise independent permutations is the *witness* that properly separates the notion of 4-rankwise independence and that of 4-restricted min-wise independence.

1 Introduction

1.1 Definitions and Known Results

The notion of k-restricted min-wise independence was introduced by Broder, et al [3] to estimate the *resemblance* between two documents [2] for detecting almost identical documents on the Web (and a similar notion was implicitly used by Mulmuley to reduce the amount of randomness used by algorithms [9,4]). In fact, Broder, et al [3] showed that a family $\mathcal{F} \subseteq S_n$ of permutations precisely estimates the resemblance between two documents of size not greater than $k \geq 1$ iff it is

S. Arora et al. (Eds.): APPROX 2003+RANDOM 2003, LNCS 2764, pp. 396–408, 2003.

k-restricted min-wise independent. For any pair of integers $i \leq j$, let $[i,j] = \{i, i+1, \ldots, j\}$. We use S_n to denote the set of all permutations on $[0, n-1]$ and use $|A|$ to denote the cardinality of a finite set A. Broder, et al [3] defined a notion of k-restricted min-wise independent permutations as follows:

Definition 1 ([3]). *A family $\mathcal{F} \subseteq S_n$ of permutations is said to be k-restricted min-wise independent if for any subset $X \subseteq [0, n-1]$ with $|X| \leq k$ and any $x \in X$, $\Pr\{\min\{\pi(X)\} = \pi(x)\} = 1/|X|$, when $\pi \in \mathcal{F}$ is chosen uniformly at random.*

Itoh, Takei, and Tarui [6] showed how to construct a family $\mathcal{F} \subseteq S_n$ of k-restricted min-wise independent permutations such that $|\mathcal{F}_n| \leq (2n)^k \operatorname{lcm}(k, k-1, \ldots, 1)$. For the case that a biased (rather than the uniform) sampling of permutations is allowed, Broder, et al [3] showed that there exists a family $\mathcal{F} \subseteq S_n$ of k-restricted min-wise independent permutations such that $|\mathcal{F}_n| \leq \sum_{j=1}^{k} j\binom{n}{j}$ (and this is improved to $|\mathcal{F}_n| \leq 1 + \sum_{j=2}^{k}(j-1)\binom{n}{j}$ by Matoušek and Stojaković [10]).

For any $X \subseteq [0, n-1]$ and any $x \in X$, let $\mathrm{LT}(x, X) = \{y \in X : y < x\}$ and define the *rank* of x in X by $\mathrm{RANK}\{x, X\} = |\mathrm{LT}(x, X)|$. Itoh, Takei, and Tarui [6] defined the following notion stronger than k-restricted min-wise independence.

Definition 2 ([6]). *A family $\mathcal{F} \subseteq S_n$ of permutations is said to be k-rankwise independent if for any subset $X = \{x_1, x_2, \ldots, x_k\} \subseteq [0, n-1]$ and any k distinct values $r_1, r_2, \ldots, r_k \in [0, k-1]$, $\Pr[\bigwedge_{i=1}^{k} \mathrm{RANK}\{\pi(x_i), \pi(X)\} = r_i] = 1/k!$, when $\pi \in \mathcal{F}$ is chosen uniformly at random.*

Itoh, Takei, and Tarui [6] showed how to construct a family $\mathcal{F} \subseteq S_n$ of k-rankwise independent permutations such that $|\mathcal{F}_n| \leq n^{O(k^2/\ln k)}$ if $(k-1)! \leq n$.

For any pair of integers $n \geq d \geq 0$, define $m(n, d)$ to be $m(n, d) = \sum_{j=0}^{d/2} \binom{n}{j}$ if d is even; $m(n, d) = \sum_{j=0}^{(d-1)/2} \binom{n}{j} + \binom{n-1}{(d-1)/2}$ if d is odd. For the lower bounds of the family size of k-restricted min-wise and k-rankwise independent permutations, we have the following results (and for the related works, see [3,6,11,10]).

Theorem 1 ([7]). *For any pair of integers $n \geq k \geq 1$, if a family $\mathcal{F} \subseteq S_n$ of permutations is k-restricted min-wise independent, then $|\mathcal{F}| \geq m(n-1, k-1)$.*

Theorem 2 ([7]). *For any pair of integers $n \geq k \geq 1$, let $s = \lfloor k/2 \rfloor$. If a family $\mathcal{F} \subseteq S_n$ of permutations is k-rankwise independent, then $|\mathcal{F}| \geq m(n-1, k-1)$ if $k \leq 3$; $|\mathcal{F}| \geq \max\{m(n-1, k-1), \lfloor n/s \rfloor \binom{n-\lfloor n/s \rfloor}{s-1}\}$ if $k \geq 4$.*

1.2 Main Results

In this paper, we construct a family $\mathcal{F} \subseteq S_n$ of 3-restricted and 4-restricted min-wise independent permutations by applying the affine plane $\mathrm{AG}(2, q)$.

Theorem 3. *For any integer $n \geq 3$, there exists a 3-restricted min-wise independent permutation family $\mathcal{F}_n \subseteq S_n$ such that $|\mathcal{F}_n| \leq 12\sqrt{e}(1 + o(1)) \cdot n \lg^2 n$.*

Theorem 4. *For any integer $n \geq 4$, there exists a 4-restricted min-wise independent permutation family $\mathcal{F}_n \subseteq S_n$ such that $|\mathcal{F}_n| \leq 12\sqrt{e}(1 + o(1)) \cdot n \lg^3 n$.*

So from Theorem 1, it follows that if a family $\mathcal{F} \subseteq S_n$ of permutations is 3-restricted or 4-restricted min-wise independent, then $|\mathcal{F}| = \Omega(n)$. Thus the upper bounds of Theorems 3 and 4 are within poly($\lg n$) factor to the lower bound of Theorem 1. We also construct a family $\mathcal{F} \subseteq S_n$ of 4-rankwise independent permutations by applying the projective plane PG(2, q).

Theorem 5. *For any integer $n \geq 4$, there exists a 4-rankwise independent permutation family $\mathcal{F}_n \subseteq S_n$ such that $|\mathcal{F}_n| \leq 15e(1 + o(1)) \cdot n^3 \lg^6 n$.*

From Theorem 2, it follows that if a family $\mathcal{F} \subseteq S_n$ of permutations is 4-rankwise independent, then $|\mathcal{F}| = \Omega(n^2)$. Thus the result of Theorem 5 is close to that of Theorem 2. Note that for any family $\mathcal{F} \subseteq S_n$ of permutations, if it is k-rankwise independent, then it is k-restricted min-wise independent. So it follows from Theorem 2 that our family $\mathcal{F} \subseteq S_n$ of 4-restricted min-wise independent permutations given in Theorem 4 is the *witness* that properly separates the notion of 4-rankwise independence and that of 4-restricted min-wise independence.

2   Preliminaries

We use $\mathsf{RV}_n \in S_n$ to denote a *reverse* permutation, i.e., for each $x \in [0, n-1]$, $\mathsf{RV}_n(x) = n-1-x$. For each $\pi \in S_n$, let $\mathsf{RV}_n^0 \circ \pi = \pi$ and $\mathsf{RV}_n^1 \circ \pi = \mathsf{RV}_n \circ \pi$. Let $m \geq n \geq 1$ be integers and $\pi \in S_m$. Define $\widetilde{\pi} : [0, n-1] \to [0, n-1]$ such that for each $x \in [0, n-1]$, $\widetilde{\pi}(x) = \mathsf{RANK}\{\pi(x), [0, m-1] - \pi([n, m-1])\}$. Note that $\widetilde{\pi} \in S_n$ and we use $\mathsf{T}_{m,n}$ to denote the transform $S_m \ni \pi \mapsto \widetilde{\pi} \in S_n$.

Proposition 1 ([6]). *For any integer $m > 0$, let $\mathcal{F} \subseteq S_m$ be a family of k-restricted min-wise (resp. k-rankwise) independent permutations. For any integer $n \leq m$, let $\mathcal{G} = \{\widetilde{\pi} : \widetilde{\pi} = \mathsf{T}_{m,n}\pi, \pi \in \mathcal{F}\}$. Then the family $\mathcal{G} \subseteq S_n$ of permutations is k-restricted min-wise (resp. k-rankwise) independent and $|\mathcal{G}| = |\mathcal{F}|$.*

Our constructions of permutation families are *recursive*, and the rest of this section is concerned with technicalities necessary for the analysis of the recursion. So the discussion below is not needed to understand the constructions and the readers may skip this part. Let $g : \mathbf{Z} \to \mathbf{Z}$ be a (partial) function such that for any integer $q' = 2^t$, $g(q') = \sqrt{q'}$ if t is even; $g(q') = \sqrt{2q'}$ otherwise. For any integer $q = 2^{2t} \geq 4$, we define a sequence of integers $q_\ell = q, q_{\ell-1} = g(q_\ell), \ldots, q_1 = g(q_2) = 4 = 2^2$. Note that $\ell - 1 \leq 1 + \lg t = \lg \lg q$. For each $i \in [1, \ell-1]$, we have two cases: (i) $q_{i+1} = q_i^2$; (ii) $2q_{i+1} = q_i^2$. Notice that for any integer $q = 2^{2^h} \geq 4$, the case (ii) never occurs. In general, we have that for any integer $q = 2^{2t} \geq 4$, the case (ii) occurs at most $\lg t = (\lg \lg q) - 1$ times. Thus

$$\prod_{i=1}^{\ell-1} q_i = \frac{q_\ell}{q_1} \cdot \prod_{i=1}^{\ell-1} \frac{q_i^2}{q_{i+1}} \begin{cases} = \dfrac{q}{4} & \text{if } q = 2^{2^h} \geq 4; \\[2mm] \leq \dfrac{q}{4} \cdot 2^{\lg \lg q - 1} = \dfrac{q}{8} \cdot \lg q & \text{if } q = 2^{2t} \geq 4. \end{cases} \tag{1}$$

From the definition of the sequence $\{q_i\}_{i \in [1,\ell]}$, we have that $q_i \geq 2^{i+1}$. Thus

$$\prod_{i=1}^{\ell-1} \left(1 + \frac{1}{q_i}\right) \leq \prod_{i=1}^{\ell-1} \left(1 + \frac{1}{2^{i+1}}\right) \leq \left\{\sum_{i=1}^{\ell-1} \frac{1 + 2^{-(i+1)}}{\ell-1}\right\}^{\ell-1} \leq \sqrt{e}. \qquad (2)$$

3 3-Restricted Min-Wise Independent Permutations

For a family $\mathcal{F} \subseteq S_n$ of 3-restricted min-wise independent permutations, it follows from Theorem 1 that $|\mathcal{F}| \geq m(n-1, 3-1) = n$. To show Theorem 3, we construct a family $\mathcal{F}_n \subseteq S_n$ of 3-restricted min-wise independent permutations such that $|\mathcal{F}_n| = n^{1+o(1)}$, which is close to the result of Theorem 1.

3.1 Affine Planes

Let $\mathbf{F}_q$ be a field of q elements, where q is a prime power. An affine plane $\mathrm{AG}(2, q)$ is a 2-dimensional vector space over $\mathbf{F}_q$ consisting of q^2 points and $q^2 + q$ lines [8]. In $\mathrm{AG}(2, q)$, there exist $q+1$ parallel line classes $\{C_0, C_1, \ldots, C_q\} = \mathcal{C}_q$. Note that each $C_i \in \mathcal{C}_q$ contains q parallel lines, i.e., $C_i = \{L_0^i, L_1^i, \ldots, L_{q-1}^i\}$ and each $L_j^i \in C_i$ has q points, i.e., $L_j^i = \{p_{j,0}^i, p_{j,1}^i, \ldots, p_{j,q-1}^i\}$. For simplicity, we identify $\mathbf{F}_q$ with $[0, q-1]$. Arrange q^2 points in $\mathrm{AG}(2, q)$ in a natural manner. For each parallel line class $C_i \in \mathcal{C}_q$, let $f_L^i, f_p^i : \mathrm{AG}(2, q) \to [0, q-1]$ be functions such that for each $x \in \mathrm{AG}(2, q)$, $f_L^i(x) = u$ and $f_p^i(x) = v$ if $x = p_{u,v}^i$, i.e., there exists a line $L_u^i \in C_i$ in which $x = p_{u,v}^i \in L_u^i$. For each parallel line class $C_i \in \mathcal{C}_q$, we have that for any pair of $x, y \in \mathrm{AG}(2, q)$, $x \neq y$ iff $(f_L^i(x), f_p^i(x)) \neq (f_L^i(y), f_p^i(y))$.

3.2 Recursive Construction: 3-Restricted Min-Wise Independence

For any $t \geq 2$, let $q = 2^t$. Assume that there exists a family $\mathcal{G}_q \subseteq S_q$ of 3-restricted min-wise independent permutations. Our construction of 3-restricted min-wise independent permutations can be viewed as follows: For each parallel line class $C_i \in \mathcal{C}_q$, q lines $L_j^i \in C_i$ are permuted in 3-restricted min-wise independent manner; q points $p_{j,h}^i \in L_j^i$ are permuted in 3-restricted min-wise independent manner and those permuted q points in L_j^i are reversed. More formally,

(1) For each parallel line class $C_i \in \mathcal{C}_q$, each $\pi \in \mathcal{G}_q$, and each $X \in \{0, 1\}$, define $\sigma : [0, q^2 - 1] \to [0, q^2 - 1]$ such that for each $x \in [0, q^2 - 1]$, $\sigma(x) = \pi(f_L^i(x))q + \mathrm{RV}_q^X \circ \pi(f_p^i(x))$.

(2) Let $\mathcal{G}_{q^2} = \{\sigma : i \in [0, q], \pi \in \mathcal{G}_q, X \in \{0, 1\}\}$.

It is not difficult to see that $\mathcal{G}_{q^2}$ is a permutation family, i.e., $\mathcal{G}_{q^2} \subseteq S_{q^2}$. To show Theorem 3, the following lemma is applied recursively.

Lemma 1. *For any prime power q, if a family $\mathcal{G}_q \subseteq S_q$ of permutations is 3-restricted min-wise independent, then the family $\mathcal{G}_{q^2} \subseteq S_{q^2}$ of permutations is 3-restricted min-wise independent.*

3.3   Proof of Lemma 1

Consider the following cases: For any distinct points $x_1, x_2, x_3 \in [0, q^2 - 1]$, (i) there exists the unique parallel line class $C_a \in \mathcal{C}_q$ and the unique line $L_b^a \in C_a$ such that $x_1, x_2, x_3 \in L_b^a$, i.e., x_1, x_2, x_3 are *colinear*; (ii) for any parallel line class $C_\alpha \in \mathcal{C}_q$, there exists no line $L_\beta^\alpha \in C_\alpha$ such that $x_1, x_2, x_3 \in L_\beta^\alpha$, i.e., x_1, x_2, x_3 are in *general position*. For each $1 \leq h \leq 3$, let $A_h = \{\sigma \in \mathcal{G}_{q^2} : \min\{\sigma(x_1, x_2, x_3)\} = \sigma(x_h)\}$ be the event when $\sigma \in \mathcal{G}_{q^2}$ is chosen uniformly at random. For the proof of Lemma 1, it suffices to show the following claims.

Claim 31. *For the case* (i), $\Pr[A_1] = \Pr[A_2] = \Pr[A_3] = 1/3$.

Proof: We have the following events: For each $i \in [0, q - 1]$, (i-1) $i = a$; (i-2) $i \neq a$ (see Figure 1). For the event (i-1), there exists a unique line $L_b^a \in C_a$ such that $x_1, x_2, x_3 \in L_b^a$. Since a family $\mathcal{G}_q \subseteq S_q$ of k-restricted min-wise independent permutations is k-restricted *max-wise* independent [5, Theorem 2], any of the points $x_1, x_2, x_3 \in [0, q^2 - 1]$ can be the minimum with probability $1/3$ by $\pi \in \mathcal{G}_q$ and $\mathrm{RV}_q \circ \pi$. So it follows that any of the points $x_1, x_2, x_3 \in [0, q^2 - 1]$ can be the minimum with probability $1/3$ by $\sigma \in \mathcal{G}_{q^2}$. So for each $1 \leq h \leq 3$, $\Pr[A_h \wedge$ event (i-1)] $= \frac{1}{q+1} \cdot \frac{1}{3}$. For the event (i-2), we have that $x_1 \in L_{j_1}^i$, $x_2 \in L_{j_2}^i$, and $x_3 \in L_{j_3}^i$, where $j_1, j_2, j_3 \in [0, q - 1]$ are distinct. Since $f_L^i(x_1) = j_1$, $f_L^i(x_2) = j_2$, and $f_L^i(x_3) = j_3$, any of $j_1, j_2, j_3 \in [0, q-1]$ can be the minimum with probability $1/3$ by $\pi \in \mathcal{G}_q$. Then any of the points $x_1, x_2, x_3 \in [0, q^2 - 1]$ can be the minimum with probability $1/3$ by $\sigma \in \mathcal{G}_{q^2}$. So for each $1 \leq h \leq 3$, $\Pr[A_h \wedge$ event (i-2)] $= \frac{q}{q+1} \cdot \frac{1}{3}$. Thus we have that $\Pr[A_1] = \Pr[A_2] = \Pr[A_3] = 1/3$. ∎

Claim 32. *For the case* (ii), $\Pr[A_1] = \Pr[A_2] = \Pr[A_3] = 1/3$.

Proof: We have the following events: For each $i \in [0, q - 1]$, (ii-1) $x_1, x_2, x_3 \in [0, q^2 - 1]$ are on different three lines in C_i; (ii-2) only two of $x_1, x_2, x_3 \in [0, q^2 - 1]$ are on the same line (see Figure 2). For the event (ii-1), we can show that any of the points $x_1, x_2, x_3 \in [0, q^2 - 1]$ can be the minimum with probability $1/3$ by $\sigma \in \mathcal{G}_{q^2}$ in a way similar to the event (i-2) of Claim 3.1. So we have that for each $1 \leq$

(i-1)   (i-2)

Fig. 1. Events for the Case (i) — 3-Restricted Min-Wise Independence

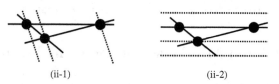

(ii-1)   (ii-2)

Fig. 2. Events for the Case (ii) — 3-Restricted Min-Wise Independence

$h \leq 3$, $\Pr[A_h \wedge \text{event (ii-1)}] = \frac{q+1-3}{q+1} \cdot \frac{1}{3}$. For the event (ii-2), consider the following subevents: For distinct $i_1, i_2, i_3 \in [0, q]$, (ii-2.1) $x_1 \notin L_{j_1}^{i_1}$ and $x_2, x_3 \in L_{j_1}^{i_1}$; (ii-2.2) $x_2 \notin L_{j_2}^{i_2}$ and $x_1, x_3 \in L_{j_2}^{i_2}$; (ii-2.3) $x_3 \notin L_{j_3}^{i_3}$ and $x_1, x_2 \in L_{j_3}^{i_3}$. For the subevent (ii-2.1), the point x_1 can be the minimum with probability $1/2$. For the subevents (ii-2.2) and (ii-2.3), the point x_1 can be the minimum with probability $(1/2)^2 = 1/4$. So we have that $\Pr[A_1 \wedge \text{subevent (ii-2)}] = \frac{1}{q+1} \cdot \frac{1}{2} + \frac{2}{q+1} \cdot \frac{1}{4} = \frac{1}{q+1}$. In a way similar to the argument above, we can show that $\Pr[A_2 \wedge \text{subevent (ii-2)}] = \Pr[A_3 \wedge \text{subevent (ii-2)}] = \frac{1}{q+1}$. Thus $\Pr[A_1] = \Pr[A_2] = \Pr[A_3] = 1/3$. ∎

3.4 Proof of Theorem 3

For any integer $q = 2^{2t} \geq 4$, define a sequence $\{q_i\}_{i \in [1, \ell]}$ of integers by the function g defined in Section 2. As in Subsection 3.2, construct a family $\mathcal{G}_{q_{i-1}^2} \subseteq S_{q_{i-1}^2}$ of 3-restricted min-wise independent permutations from the family $\mathcal{G}_{q_{i-1}} \subseteq S_{q_{i-1}}$ of 3-restricted min-wise independent permutations. Note that for each $i \in [2, \ell]$, $q_i \leq q_{i-1}^2$. By Proposition 1, transform the family $\mathcal{G}_{q_{i-1}^2} \subseteq S_{q_{i-1}^2}$ of 3-restricted min-wise independent permutations to a family $\mathcal{G}_{q_i} \subseteq S_{q_i}$ of 3-restricted min-wise independent permutations. We can start with any family $\mathcal{G}_{q_1} \subseteq S_{q_1} = S_4$ of 3-restricted min-wise independent permutations. Then we take $\mathcal{G}_{q_1} = S_4$, i.e., $|\mathcal{G}_{q_1}| = |S_4| = 4! = 24$. Recall that $\ell - 1 \leq \lg \lg q$. So we have that

$$|\mathcal{G}_{q_\ell}| = 2(q_{\ell-1} + 1) \cdot |\mathcal{G}_{q_{\ell-1}}| = 2^{\ell-1} \left\{ \prod_{i=1}^{\ell-1} (q_i + 1) \right\} \cdot |\mathcal{G}_{q_1}|$$

$$\leq 24 \cdot \lg q \cdot \left\{ \prod_{i=1}^{\ell-1} q_i \right\} \cdot \left\{ \prod_{i=1}^{\ell-1} \left(1 + \frac{1}{q_i}\right) \right\}. \tag{3}$$

For any integer $n \geq 4$, let $q = 2^{2t}$ be the minimum integer such that $n \leq q$. Note that $q \leq 4n$. So from Ineq.(3), Eqs.(1) and (2), and Proposition 1, it follows that for any integer $n \geq 4$, there exists a family $\mathcal{F}_n \subseteq S_n$ of 3-restricted min-wise independent permutations such that

$$|\mathcal{F}_n| = |\mathcal{G}_q| \leq 3\sqrt{e} \cdot q \lg^2 q \leq 12\sqrt{e} \cdot n(2 + \lg n)^2 = 12\sqrt{e}(1 + o(1)) \cdot n \lg^2 n.$$

In particular, we have the following corollary from Ineq.(3) and Eqs.(1) and (2).

Corollary 1. *For any integer $n = 2^{2^h} \geq 4$, there exists a family $\mathcal{F}_n \subseteq S_n$ of 3-restricted min-wise independent permutations such that $|\mathcal{F}_n| \leq 6\sqrt{e} \cdot n \lg n$.*

4 4-Restricted Min-Wise Independent Permutations

For a family $\mathcal{F} \subseteq S_n$ of 4-restricted min-wise independent permutations, it follows from Theorem 1 that $|\mathcal{F}| \geq m(n-1, 4-1) = 2n - 2$. To show Theorem 4, we construct a family $\mathcal{F}_n \subseteq S_n$ of 4-restricted min-wise independent permutations such that $|\mathcal{F}_n| = n^{1+o(1)}$, which is close to the result of Theorem 1. For the case that $k = 4$, Theorem 2 implies that if a family $\mathcal{F} \subseteq S_n$ of permutations is 4-rankwise independent, then $|\mathcal{F}| = \Omega(n^2)$. Thus our family $\mathcal{F}_n \subseteq S_n$ of permutations given in Theorem 4 shows that the notion of 4-rankwise independence is *strictly* stronger than that of 4-restricted min-wise independence.

4.1   Recursive Construction: 4-Restricted Min-Wise Independence

For any $t \geq 2$, let $q = 2^t$. Assume that there exists a family $\mathcal{G}_q \subseteq S_q$ of 4-restricted min-wise independent permutations. Our construction of 4-restricted min-wise independent permutations is similar to that of 3-restricted min-wise independent permutations in Subsection 3.2 and can be viewed as follows: For each parallel line class $C_i \in \mathcal{C}_q$, q lines $L^i_j \in C_i$ are permuted in 4-restricted min-wise independent manner and those permuted q lines in C_i are reversed; q points $p^i_{j,h} \in L^i_j$ are permuted in 4-restricted min-wise independent manner and those permuted q points in $L^i_{j,h}$ are reversed. More formally,

(1) For each parallel line class $C_i \in \mathcal{C}_q$, each $\pi \in \mathcal{G}_q$, and each $X, Y \in \{0, 1\}$, define $\sigma : [0, q^2 - 1] \to [0, q^2 - 1]$ such that for each $x \in [0, q^2 - 1]$, $\sigma(x) = \mathrm{RV}^X_q \circ \pi(f^i_L(x))q + \mathrm{RV}^Y_q \circ \pi(f^i_p(x))$.

(2) Let $\mathcal{G}_{q^2} = \{\sigma : i \in [0, q], \pi \in \mathcal{G}_q, X \in \{0, 1\}, Y \in \{0, 1\}\}$.

It is not difficult to see that $\mathcal{G}_{q^2}$ is a permutation family, i.e., $\mathcal{G}_{q^2} \subseteq S_{q^2}$. To show Theorem 4, the following lemma is applied recursively.

Lemma 2. *For any prime power q, if a family $\mathcal{G}_q \subseteq S_q$ of permutations is 4-restricted min-wise independent, then the family $\mathcal{G}_{q^2} \subseteq S_{q^2}$ of permutations is 4-restricted min-wise independent.*

4.2   Proof of Lemma 2

Consider the following cases: For any distinct points $x_1, x_2, x_3, x_4 \in [0, q^2 - 1]$, (i) there exists a unique line $L^a_b \in C_a$ such that $x_1, x_2, x_3, x_4 \in L^a_b$, i.e., x_1, x_2, x_3, x_4 are *colinear*; (ii) there exists a unique line $L^a_b \in C_a$ including only three points $x_{h_1}, x_{h_2}, x_{h_3} \in \{x_1, x_2, x_3, x_4\}$, i.e., $x_{h_1}, x_{h_2}, x_{h_3}$ are *colinear*; (iii) for any parallel line class $C_\alpha \in \mathcal{C}_q$, there exists no line $L^\alpha_\beta \in C_\alpha$ such that for some three points $x_{h_1}, x_{h_2}, x_{h_3} \in \{x_1, x_2, x_3, x_4\}$, $x_{h_1}, x_{h_2}, x_{h_3} \in L^\alpha_\beta$. For any $1 \leq h \leq 4$, let $B_h = \{\sigma \in \mathcal{G}_{q^2} : \min\{\sigma(x_1, x_2, x_3, x_4)\} = \sigma(x_h)\} \subseteq \mathcal{G}_{q^2}$ be the event when $\sigma \in \mathcal{G}_{q^2}$ is chosen uniformly at random. To show Lemma 2, the following claims suffice.

Claim 41. *For the case (i), $\Pr[B_1] = \Pr[B_2] = \Pr[B_3] = \Pr[B_4] = 1/4$.*

Proof: We have the following events: For each $i \in [0, q - 1]$, (i-1) $i = a$; (i-2) $i \neq a$ (see Figure 3). In a way similar to the proof of Claim 3.1, it is immediate that for each $1 \leq h \leq 4$, $\Pr[B_h \wedge \text{event (i-1)}] = \frac{1}{q+1} \cdot \frac{1}{4}$ and $\Pr[B_h \wedge \text{event (i-2)}] = \frac{q+1-1}{q+1} \cdot \frac{1}{4} = \frac{q}{q+1} \cdot \frac{1}{4}$. Thus $\Pr[B_1] = \Pr[B_2] = \Pr[B_3] = \Pr[B_4] = 1/4$. ∎

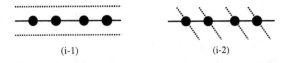

(i-1)   (i-2)

Fig. 3. Events for the Case (i) — 4-Restricted Min-Wise Independence

(ii-2.{1,2,3})   (ii-2.4)

Fig. 4. Subevents of (ii-2) — 4-Restricted Min-Wise Independence

Claim 42. *For the case* (ii), $\Pr[B_1] = \Pr[B_2] = \Pr[B_3] = \Pr[B_4] = 1/4$.

Proof: Without loss of generality, assume that there exists a unique line $L_b^a \in C_a$ such that $x_1 \notin L_b^a$ and $x_2, x_3, x_4 \in L_b^a$ (the other cases can be handled analogously). For each $C_i \in C_q$, we have the following events: (ii-1) $i = a$; (ii-2) $i \neq a$. For the event (ii-1), it is immediate that $\Pr[B_1 \wedge \text{event (ii-1)}] = \frac{1}{q+1} \cdot \frac{1}{2}$ and for each $2 \leq h \leq 4$, $\Pr[B_h \wedge \text{event (ii-1)}] = \frac{1}{q+1} \cdot \frac{1}{6}$. For the event (ii-2), we have three lines $L_{j_2}^{i_2} \in C_{i_2}, L_{j_3}^{i_3} \in C_{i_3}, L_{j_4}^{i_4} \in C_{i_4}$ such that $x_1, x_2 \in L_{j_2}^{i_2}$, $x_1, x_3 \in L_{j_3}^{i_3}$, $x_1, x_4 \in L_{j_4}^{i_4}$. For each $C_i \in C_q$, we have the following subevents: (ii-2.1) $i = i_2$; (ii-2.2) $i = i_3$; (ii-2.3) $i = i_4$; (ii-2.4) $i \notin \{i_2, i_3, i_4\}$ (see Figure 4), each of which occurs with the probability shown in Table 1. So $\Pr[B_1] = \Pr[B_2] = \Pr[B_3] = \Pr[B_4] = 1/4$. ∎

Claim 43. *For the case* (iii), $\Pr[B_1] = \Pr[B_2] = \Pr[B_3] = \Pr[B_4] = 1/4$.

Proof: Without loss of generality, consider the following subcases: (iii-1) there exist the two unique parallel line classes $C_a, C_b \in C_q$ for which there exist a pair of lines $L_{j_1}^a, L_{j_2}^a \in C_a$ such that $x_1, x_2 \in L_{j_1}^a$ and $x_3, x_4 \in L_{j_2}^a$ and a pair of lines $L_{h_1}^b, L_{h_2}^b \in C_b$ such that $x_1, x_4 \in L_{h_1}^b$ and $x_2, x_3 \in L_{h_2}^b$; (iii-2) there exists only a unique parallel line class $C_a \in C_q$ for which there exists a pair of lines $L_{j_1}^a, L_{j_2}^a \in C_a$ such that $x_1, x_2 \in L_{j_1}^a$ and $x_3, x_4 \in L_{j_2}^a$; (iii-3) there exists no parallel line class $C_a \in C_q$ such that there exists a pair of lines $L_{j_1}^a, L_{j_2}^a \in C_a$, each of which includes two points of x_1, x_2, x_3, x_4. For the subcase (iii-1), consider the following events: For each $C_i \in C_q$, (iii-1.1) $i \in \{a, b\}$; (iii-1.2) $i \in [0, q] - \{a, b\}$ such that there exists $L_j^i \in C_i$ including the points x_1, x_3; (iii-1.3) $i \in [0, q] - \{a, b\}$ such that there exists $L_j^i \in C_i$ including the points x_2, x_4; (iii-1.4) $i \in [0, q] - \{a, b\}$ such that there exists $L_j^i \in C_i$ including any of two points of x_1, x_2, x_3, x_4 (see Figure 5), where each of the subevents occurs with the probability as shown in Table 2. Thus for the subcase (iii-1), $\Pr[B_1] = \Pr[B_2] = \Pr[B_3] = \Pr[B_4] = 1/4$. For the subcases (iii-2) and (iii-3), we can show the claim analogously. ∎

4.3   Proof of Theorem 4

We can show Theorem 4 in a way similar to the proof of Theorem 3.

Table 1. Probability of Subevents — The Case (ii)

	B_1	B_2	B_3	B_4
subevent (ii-2.1)	$\frac{1}{q+1}\cdot\frac{1}{6}$	$\frac{1}{q+1}\cdot\frac{1}{6}$	$\frac{1}{q+1}\cdot\frac{1}{3}$	$\frac{1}{q+1}\cdot\frac{1}{3}$
subevent (ii-2.2)	$\frac{1}{q+1}\cdot\frac{1}{6}$	$\frac{1}{q+1}\cdot\frac{1}{3}$	$\frac{1}{q+1}\cdot\frac{1}{6}$	$\frac{1}{q+1}\cdot\frac{1}{3}$
subevent (ii-2.3)	$\frac{1}{q+1}\cdot\frac{1}{6}$	$\frac{1}{q+1}\cdot\frac{1}{3}$	$\frac{1}{q+1}\cdot\frac{1}{3}$	$\frac{1}{q+1}\cdot\frac{1}{6}$
subevent (ii-2.4)	$\frac{q-3}{q+1}\cdot\frac{1}{4}$	$\frac{q-3}{q+1}\cdot\frac{1}{4}$	$\frac{q-3}{q+1}\cdot\frac{1}{4}$	$\frac{q-3}{q+1}\cdot\frac{1}{4}$

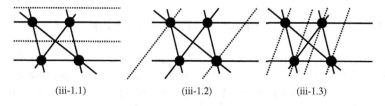

(iii-1.1)   (iii-1.2)   (iii-1.3)

Fig. 5. Events for the Case (iii-1) — 4-Restricted Min-Wise Independence

For any integer $q = 2^{2t} \geq 4$, we define a sequence $\{q_i\}_{i\in[1,\ell]}$ of integers by the function g in Section 2. We start with a family $\mathcal{G}_{q_1} = S_{q_1} = S_4$, i.e., $|\mathcal{G}_{q_1}| = |S_4| = 4! = 24$, and construct a family $\mathcal{G}_q \subseteq S_q$ of 4-restricted min-wise independent permutations as shown in Subsection 3.4. Recall that $\ell - 1 \leq \lg\lg q$. So

$$|\mathcal{G}_{q_\ell}| = 4\,(q_{\ell-1}+1)\cdot|\mathcal{G}_{q_{\ell-1}}| = 2^{2(\ell-1)}\left\{\prod_{i=1}^{\ell-1}(q_i+1)\right\}\cdot|\mathcal{G}_{q_1}|$$

$$\leq 24\cdot\lg^2 q\cdot\left\{\prod_{i=1}^{\ell-1}q_i\right\}\cdot\left\{\prod_{i=1}^{\ell-1}\left(1+\frac{1}{q_i}\right)\right\}. \tag{4}$$

For any integer $n \geq 4$, let $q = 2^{2t}$ be the minimum integer such that $n \leq q$. Note that $q \leq 4n$. So from Ineq.(4), Eqs.(1) and (2), and Proposition 1, it follows that for any integer $n \geq 4$, there exists a family $\mathcal{F}_n \subseteq S_n$ of 4-restricted min-wise independent permutations such that

$$|\mathcal{F}_n| = |\mathcal{G}_q| \leq 3\sqrt{e}\cdot q\lg^3 q \leq 12\sqrt{e}\cdot n(2+\lg n)^3 = 12\sqrt{e}(1+o(1))\cdot n\lg^3 n.$$

In particular, we have the following corollary from Ineq.(4) and Eqs.(1) and (2).

Corollary 2. *For any integer $n = 2^{2^h} \geq 4$, there exists a family $\mathcal{F}_n \subseteq S_n$ of 4-restricted min-wise independent permutations such that $|\mathcal{F}_n| \leq 6\sqrt{e}\cdot n\lg^2 n$.*

Table 2. Probability of Subevents — The Subcase (iii-1)

	B_1	B_2	B_3	B_4
subevent (iii-1.1)	$\frac{2}{q+1} \cdot \frac{1}{4}$	$\frac{2}{q+1} \cdot \frac{1}{4}$	$\frac{2}{q+1} \cdot \frac{1}{4}$	$\frac{2}{q+1} \cdot \frac{1}{4}$
subevent (iii-1.2)	$\frac{1}{q+1} \cdot \frac{1}{6}$	$\frac{1}{q+1} \cdot \frac{1}{3}$	$\frac{1}{q+1} \cdot \frac{1}{3}$	$\frac{1}{q+1} \cdot \frac{1}{6}$
subevent (iii-1.3)	$\frac{1}{q+1} \cdot \frac{1}{3}$	$\frac{1}{q+1} \cdot \frac{1}{6}$	$\frac{1}{q+1} \cdot \frac{1}{6}$	$\frac{1}{q+1} \cdot \frac{1}{3}$
subevent (iii-1.4)	$\frac{q-3}{q+1} \cdot \frac{1}{4}$	$\frac{q-3}{q+1} \cdot \frac{1}{4}$	$\frac{q-3}{q+1} \cdot \frac{1}{4}$	$\frac{q-3}{q+1} \cdot \frac{1}{4}$

5   4-Rankwise Independent Permutations

For a family $\mathcal{F} \subseteq S_n$ of 4-rankwise independent permutations, it follows from Theorem 2 that $|\mathcal{F}| \geq \lfloor n/2 \rfloor \binom{n-\lfloor n/2 \rfloor}{1} = \Omega(n^2)$. To show Theorem 5, we construct a 4-rankwise independent permutation family $\mathcal{F}_n \subseteq S_n$ such that $|\mathcal{F}_n| = n^{3+o(1)}$, which is close to the result of Theorem 2.

5.1   Projective Planes

Let q be a prime power. A projective plane $\mathrm{PG}(2, q)$ of order q consists of $q^2 + q + 1$ points and has the following properties [8]:

Property 1. A projective plane $\mathrm{PG}(2, q)$ of order q satisfies (P1) every line has $q + 1$ points; (P2) any two points lie on a unique line; (P3) any point lies on $q + 1$ lines; (P4) there are $q^2 + q + 1$ lines; (P5) any two lines meet in a unique point.

From P3 of Property 1, it follows that for each point $s \in \mathrm{PG}(2, q)$, there exists a set $L^s = \{\ell_0^s, \ell_1^s, \ldots, \ell_q^s\}$ of $q + 1$ lines, each of which intersects the point s. From P1 of Property 1, we have that each line $\ell_i^s \in L^s$ consists of $q + 1$ points, i.e., $\ell_i^s = \{p_{i,0}^s, p_{i,1}^s, \ldots, p_{i,q-1}^s, s\}$. Arrange $q^2 + q + 1$ points of $\mathrm{PG}(2, q)$ naturally. For each $s \in \mathrm{PG}(2, q)$, let $f_L^s : \mathrm{PG}(2, q) - \{s\} \to [0, q]$ and $f_p^s : \mathrm{PG}(2, q) - \{s\} \to [0, q-1]$ be functions such that for any point $x \in \mathrm{PG}(2, q) - \{s\}$, $f_L^s(x) = i$ and $f_p^s(x) = j$ if $x = p_{i,j}^s$, i.e., there exists a line $\ell_i^s \in L^s$ on which $x = p_{i,j}^s \in \ell_i^s$. For each point $s \in \mathrm{PG}(2, q)$, note that for any pair of points $x, y \in \mathrm{PG}(2, q) - \{s\}$, $x \neq y$ iff $(f_L^s(x), f_p^s(x)) \neq (f_L^s(y), f_p^s(y))$.

5.2   3-Wise Independent 0/1-Random Variables

Alon, Babai, and Itai [1] showed the following result on the construction of k-wise independent random variables with a small sample space.

Proposition 2 ([1]). *Let $m = 2^h - 1$ and $k = 2t + 1 \leq m$ be integers. Then there exist k-wise independent random variables $X_0, X_1, \ldots, X_{m-1} : \Omega \to \{0,1\}$ for which the distribution on the sample space Ω is uniform; $|\Omega| = 2(m+1)^t$; $\Pr[X_0 = 1] = \Pr[X_1 = 1] = \cdots = \Pr[X_{m-1} = 1] = 1/2$.*

To construct a small family of 4-rankwise independent permutations, we apply the following proposition as a special case of Proposition 2, i.e.,

Proposition 3. *For any integer $n \geq 3$, let $m = 2^h - 1 \geq n$. Then there exist 3-wise independent random variables $X_0, X_1, \ldots, X_{n-1} : \Omega \to \{0,1\}$ for which the distribution on the sample space Ω is uniform; $|\Omega| = 2(m+1)$; $\Pr[X_0 = 1] = \Pr[X_1 = 1] = \cdots = \Pr[X_{n-1} = 1] = 1/2$.*

5.3   Construction of 4-Rankwise Independent Permutations

For any $t \geq 2$, let $q = 2^t$ and $m = 2^{t+1} - 1$. For convenience, we identify $PG(2, q)$ with $[0, q^2 + q]$. Note that for any $t \geq 2$, $q + 2 \leq m$ and $m + 1 = 2q$. It follows from Proposition 3 that there exist 3-wise independent random variables $X_0, X_1, \ldots, X_{q+1} : \Omega \to \{0,1\}$ for which the distribution on the sample space Ω is uniform; $|\Omega| = 2(m+1) = 4q$; $\Pr[X_0 = 1] = \Pr[X_1 = 1] = \cdots = \Pr[X_{q+1} = 1] = 1/2$. Assume that there exists a (small) family $\mathcal{G}_{q+1} \subseteq S_{q+1}$ of 4-rankwise independent permutations. Informally, our construction of 4-rankwise independent permutations can be viewed as follows: Choose a point $s \in PG(2, q)$ uniformly at random; map the point $s \in PG(2, q)$ to the minimum or maximum among $PG(2, q)$ with probability $1/2$; $q + 1$ lines $\ell_i^s \in L^s$ are permuted in 4-rankwise independent manner; q points $p_{i,j}^s$ on each $\ell_i^s - \{s\}$ are permuted in 4-rankwise independent manner; the line permutation and the point permutation are reversed in 3-wise independent manner. More formally,

(1) For each $s \in PG(2, q)$, $X_0, X_1, \ldots, X_{q+1} : \Omega \to \{0,1\}$, each $X \in \{0,1\}$, and each $\pi \in \mathcal{G}_{q+1}$, define $\sigma : PG(2, q) \to PG(2, q)$ such that $\sigma(s) = (q^2 + q)X$ and for each $x \in PG(2, q) - \{s\}$,

$$\sigma(x) = \left\{ \mathsf{RV}_{q+1}^{X_{q+1}} \circ \pi(f_L^s(x)) \right\} q + \mathsf{RV}_q^{X_{f_L^s(x)}} \circ \mathsf{T}_{q+1,q} \pi(f_p^s(x)) + 1 - X.$$

(2) Let $\mathcal{G}_{q^2+q+1} = \{\sigma : s \in PG(2, q), X \in \{0,1\}, \pi \in \mathcal{G}_{q+1}, \bigwedge_{i=0}^{q+1} X_i \in \{0,1\}\}$.

It is not difficult to see that $\mathcal{G}_{q^2+q+1}$ is a family of permutations, i.e., $\mathcal{G}_{q^2+q+1} \subseteq S_{q^2+q+1}$. To show Theorem 5, the following lemma is applied recursively.

Lemma 3. *For any prime power q, if a family $\mathcal{G}_{q+1} \subseteq S_{q+1}$ of permutations is 4-rankwise independent, then the family $\mathcal{G}_{q^2+q+1} \subseteq S_{q^2+q+1}$ of permutations is 4-rankwise independent.*

Proof (Sketch): For any distinct $x_1, x_2, x_3, x_4 \in PG(2, q)$, we consider the following cases: (i) there exists a unique line $\ell(x_1, x_2, x_3, x_4)$ including x_1, x_2, x_3, x_4; (ii) there exists a unique line $\ell(x_i, x_j, x_k)$ including $x_i, x_j, x_k \subseteq \{x_1, x_2, x_3, x_4\}$; (iii) there exists no line $\ell(x_i, x_j, x_k)$ including $x_i, x_j, x_k \subseteq \{x_1, x_2, x_3, x_4\}$. Let $E = \{\sigma \in \mathcal{G}_{q^2+q+1} : \sigma(x_1) < \sigma(x_2) < \sigma(x_3) < \sigma(x_4)\} \subseteq \mathcal{G}_{q^2+q+1}$ be the event when $\sigma \in \mathcal{G}_{q^2+q+1}$ is chosen uniformly at random. By Property 1, we can show that for each of the cases (i), (ii), and (iii), $\Pr[E] = 1/4! = 1/24$. ∎

5.4   Proof of Theorem 5

For any integer $q = 2^{2t} \geq 4$, define a sequence $\{q_i\}_{i \in [1,\ell]}$ of integers by the function g given in Section 2. Assume that for any $i \in [2, \ell]$, there exists a family $\mathcal{G}_{q_{i-1}+1} \subseteq S_{q_{i-1}+1}$ of 4-rankwise independent permutations. As in Subsection 5.3, construct a family $\mathcal{G}_{q_{i-1}^2+q_{i-1}+1} \subseteq S_{q_{i-1}^2+q_{i-1}+1}$ of 4-rankwise independent permutations from the family $\mathcal{G}_{q_{i-1}+1} \subseteq S_{q_{i-1}+1}$ of 4-rankwise independent permutations. Note that for each $i \in [2, \ell]$, $q_i + 1 \leq q_{i-1}^2 + q_{i-1} + 1$. By Proposition 1, we transform $\mathcal{G}_{q_{i-1}^2+q_{i-1}+1} \subseteq S_{q_{i-1}^2+q_{i-1}+1}$ to a family $\mathcal{G}_{q_i+1} \subseteq S_{q_i+1}$ of 4-rankwise independent permutations. We can start with any family $\mathcal{G}_{q_1+1} \subseteq S_{q_1+1} = S_5$ of 4-rankwise independent permutations. Then we take $\mathcal{G}_{q_1+1} = S_5$, i.e., $|\mathcal{G}_{q_1+1}| = |S_5| = 5! = 120$. Recall that $\ell - 1 \leq \lg\lg q$. So we have that

$$|\mathcal{G}_{q_\ell+1}| = 8\left(q_{\ell-1}^2 + q_{\ell-1} + 1\right) q_{\ell-1} |\mathcal{G}_{q_{\ell-1}+1}| \leq 2^3 q_{\ell-1}^3 \left(1 + \frac{1}{q_{\ell-1}}\right)^2 |\mathcal{G}_{q_{\ell-1}+1}|$$

$$\leq 2^{3(\ell-1)} \cdot \left\{\prod_{i=1}^{\ell-1} q_i\right\}^3 \cdot \left\{\prod_{i=1}^{\ell-1}\left(1 + \frac{1}{q_i}\right)\right\}^2 \cdot |\mathcal{G}_{q_1+1}|$$

$$\leq 120\lg^3 q \cdot \left\{\prod_{i=1}^{\ell-1} q_i\right\}^3 \cdot \left\{\prod_{i=1}^{\ell-1}\left(1 + \frac{1}{q_i}\right)\right\}^2. \tag{5}$$

For any integer $n \geq 4$, let $q = 2^{2t}$ be the minimum integer such that $n \leq q + 1$. Note that $q \leq 4n$. So from Ineq.(5), Eqs.(1) and (2), and Proposition 1, it follows that for any integer $n \geq 4$, there exists a family $\mathcal{F}_n \subseteq S_n$ of 4-rankwise independent permutations such that

$$|\mathcal{F}_n| = |\mathcal{G}_{q+1}| \leq \frac{15e}{2^6} \cdot q^3 \lg^6 q \leq 15e \cdot n^3 (2 + \lg n)^6 = 15e(1 + o(1)) \cdot n^3 \lg^6 n.$$

In particular, we have the following corollary from Ineq.(5) and Eqs.(1) and (2).

Corollary 3. *For any integer* $n = 2^{2^h} \geq 4$, *there exists a family* $\mathcal{F}_n \subseteq S_n$ *of 4-rankwise independent permutations such that* $|\mathcal{F}_n| \leq (15e/8) \cdot n^3 \lg^3 n$.

References

1. Alon, N., Babai, L., and Itai, A., A Fast and Simple Randomized Parallel Algorithm for the Maximal Independent Set Problem, *J. of Algorithms*, 7:567–583, 1986.
2. Broder, A., On the Resemblance and Containment of Documents, in *Proc. of Compression and Complexity of Sequences*, 21–29, 1998.
3. Broder, A., Charikar, M., Frieze, A., and Mitzenmacher, M., Min-Wise Independent Permutations, in *Proc. of the 30th Annual ACM Symposium on Theory of Computing*, 327–336, 1998.
4. Broder, A., Charikar, M., and Mitzenmacher, M., A Derandomization Using Min-Wise Independent Permutations, in *Proc. of RANDOM'98*, Lecture Notes in Computer Science 1518, Springer, 15–24, 1998.

5. Broder, A. and Mitzenmacher, M., Completeness and Robustness Properties of Min-Wise Independent Permutations, *Random Structures and Algorithms*, 18:18–30, 2001.
6. Itoh, T., Takei, Y., and Tarui, J., On Permutations with Limited Independence, in *Proc. of the 11th Annual ACM-SIAM Symposium on Discrete Algorithms*, 137–146, 2000.
7. Itoh, T., Takei, Y., and Tarui, J., On the Sample Size k-Restricted Min-Wise Independent Permutations and Other k-Wise Distributions, in *Proc. of the 35th Annual ACM Symposium on Theory of Computing*, 710–719, 2003.
8. Graham, R., Grötschel, M., and Lovász, L., *Handbook of Combinatorics*, North-Holland, 1995.
9. Mulmuley, K., Randomized Geometric Algorithms and Pseudorandom Generators, *Algorithmica*, 16:450–463, 1996.
10. Matoušek, J. and Stojaković, M., On Restricted Min-Wise Independence of Permutations, *Preprint*, 2002. http://kam.mff.cuni.cz/~matousek/preprints.html/
11. Norin, S., A Polynomial Lower Bound for the Size of any k-Min-Wise Independent Set of Permutation, *Zapiski Nauchnyh Seminarov POMI*, 277:104–116, 2001 (in Russian). http://www.pdmi.ras.ru/znsl/

Author Index